Land Surveyor Reference Manual

Third Edition

Andrew L. Harbin

Professional Publications, Inc. • Belmont, CA

How to Get Online Updates for This Book

I wish I could claim that this book is 100% perfect, but 25 years of publishing have taught me that textbooks seldom are. Even if you only took one course in college, you are familiar with the issue of mistakes in textbooks.

I am inviting you to log on to Professional Publications' website at **www.ppi2pass.com** to obtain a current listing of known errata in this book. From the website home page, click on "Errata." Every significant known update to this book will be listed as fast as we can say "HTML." Suggestions from readers (such as yourself) will be added as they are received, so check in regularly.

PPI and I have gone to great lengths to ensure that we have brought you a high-quality book. Now, we want to provide you with high-quality after-publication support. Please visit us at **www.ppi2pass.com**.

Michael R. Lindeburg, PE
Publisher, Professional Publications, Inc.

LAND SURVEYOR REFERENCE MANUAL
Third Edition

Printed in the United States of America

Professional Publications, Inc.
1250 Fifth Avenue, Belmont, CA 94002
(650) 593-9119
www.ppi2pass.com

Current printing of this edition: 2

ISBN 1-888577-50-9

The CIP data is pending.

Table of Contents

Preface and Acknowledgments

As we hurtle full-tilt into the digital age, our knowledge and technological capability seem to increase at a faster pace every year. Advances in surveying technology such as satellites and computers, and concomitant progress in accompanying fields such as law, seem to follow one upon the other at an almost geometric rate. On the one hand, of course, this is a good thing. Recent technological advances have made surveying processes easier, and surveying results more accurate, than ever before. On the other hand, this amazing rate of change requires more time and effort on the part of both practicing surveyors and candidates for the professional licensure examinations to keep up with the current state of the profession.

As a result, we present the third edition of the *Land Surveyor Reference Manual*, which has been revised and expanded to reflect recent changes in surveying technology and practice and create a more comprehensive and useful reference volume. The most extensive changes to this edition are as follows.

- Full solutions for all practice problems have been added to all chapters.

- A section on geodetic azimuths was added to Chapter 19 (Astronomic Observations).

- Chapter 20 (Map Projections and State Plane Coordinate Systems) was changed to provide information regarding NAD83 and to cover high-accuracy reference networks.

- Chapter 21 (Property Law) was reorganized and sections were added on the legal doctrine of *stare decisis* and the gradient boundary methodology. In addition, the section on tidal boundaries was updated.

- New chapters were added for Numbering Systems (Chapter 25), Computer Hardware (Chapter 26), and Data Structures and Program Design (Chapter 27).

- In the Appendices, tables were added for conversion factors and constants, and for mensuration of two-dimensional and three-dimensional areas. In addition, a Manning equation nomograph was added for channel flow calculations.

Writing a book this large is, almost inevitably, a collaborative effort. I would like to thank David R. Doyle of the National Geodetic Society for information and advice concerning material used in Chapter 20; my son-in-law Seth S. Searcy III, Esq., friend and neighbor Susan N. Kelly, Esq., and friend and former student Mark A. Haines, registered Professional Land Surveyor, for information and advice concerning Chapter 21; and George Cole for his excellent work in technically reviewing the entire manuscript.

I would also like to extend my appreciation to the staff at Professional Publications, Inc., especially Publisher Michael R. Lindeburg, Acquisitions Editor Aline Magee, and Production Manager Cathy Schrott, who tirelessly led the entire long, complex publishing process that turned notebooks full of handwritten papers into a coherent hardbound book.

Finally, I would like to ask you to take a minute, when you have finished going through this book, to become part of this collaborative process yourself by filling out the comment card that came with your book and mailing it to PPI. Revisions to this edition were greatly influenced by readers' comments on the previous edition, and revisions to the fourth edition will be determined just as surely by your comments on this one. Please take the time to help those who will follow you into the surveying profession by letting us know what you liked and disliked about this book, and what we can do to make the fourth edition even better. Thank you.

Introduction

How to Use This Book

This book is intended to serve a dual purpose.

First, as the title suggests, it is intended to serve as a comprehensive manual for practicing land surveyors. If you are a practicing surveyor, you will find it to be an invaluable and frequently used addition to your library. I keep one copy on the bookshelf next to my desk in my office and another copy in my home library. That way, it is always available for easy reference.

Second, because of its comprehensive coverage, it is an ideal resource for people preparing for the Fundamentals of Land Surveying (FLS) examination. (It is also an ideal textbook for instructors teaching review courses for the FLS.) This book may also be used as a study aid for the Principles and Practices of Land Surveying (PLS) exam, but it should be recognized that since it is intended as a comprehensive manual of fundamental principles rather than an in-depth analysis of any particular area of surveying, it may not be suitable as a stand-alone study guide for all portions of the PLS.

About the Examinations

The FLS and PLS examinations are standardized tests prepared by the National Council of Examiners for Engineering and Surveying (NCEES) to ensure that only qualified candidates are allowed to legally practice as land surveyors. To ensure the reliability and validity of the tests, the FLS and PLS exams are based on input from committees of professional land surveyors throughout the United States.

The FLS tests general entry-level surveying principles the candidate is expected to have gained through academic study. In most states, passing the FLS is a requirement for registration as a Surveyor-in-Training or Surveyor Intern. The PLS tests the candidate's ability to apply those principles to the kinds of problems typically encountered in professional practice. Passing the PLS is required in most states for full licensure as a professional land surveyor. This is necessary if, among other reasons, your company requires licensure as a requisite for employment or advancement, your state requires registration before you may use the title "Land Surveyor," or you wish to be an independent consultant. The PLS is usually administered in conjunction with a test on surveying practices and regulations specific to the state administering the exam.

Although both the FLS and the PLS are prepared by the NCEES, they are administered by the licensing boards of the various states, which usually also require candidates to have attained a certain level of education and (for the PLS) experience before being allowed to take the exams. For information regarding these requirements in the state in which you plan to become licensed, and to apply for licensure, contact that state's board. Current addresses and phone numbers for each state board may be obtained at Professional Publications' website, www.ppi2pass.com.

Examination Formats

The NCEES Fundamentals of Land Surveying (FLS) examination is an eight-hour examination concentrating on the fundamentals and basics of surveying. The test consists of two four-hour sections, separated by a one-hour break. Each of the two portions of the examination contains 85 questions, for a total of 170. For each question, you will be asked to select the best answer from four choices. Sheets are provided for recording answers, which are scored by machine.

The NCEES Principles and Practice of Land Surveying (PLS) examination consists of a four-hour section containing 67 questions and, following a one-hour break, a two-hour section containing 33 questions. This examination tests the candidate's ability to apply basic principles of land surveying to typical problems from surveying practice. As in the FLS exam, machine-scored sheets are provided for recording answers.

Examination Content

The FLS examination involves questions from 20 subject areas, while the PLS examination includes questions based on seven functional areas. The areas of coverage as well as the approximate percentage of questions from each area for both exams are provided in the following list, compiled from recent NCEES publications. Typical surveying tasks using each subject are also provided.

Fundamentals of Land Surveying (FLS)

1. Algebra and Trigonometry (6%)

(units of measurement; formula development; formula manipulation; solving systems of equations; basic mensuration formulas for length, area, volume; quadratic equations; trigonometric functions; right triangle solutions; oblique triangle solutions; trigonometric identities; spherical coordinates; and trigonometry)

- Perform astronomic measurements.
- Perform trigonometric and differential leveling.
- Compute survey data.
- Compute areas and volumes.
- Determine and prepare lot and street patterns for land division.
- Design horizontal and vertical alignment for roads within a subdivision.

2. Higher Math (beyond trigonometry) (4%)

(analytical geometry; linear algebra; equations of a line, circle, parabola, ellipse; differentiation of functions; integration of elementary functions; infinite series; mathematical modeling)

- Perform geodetic surveys using conventional methods and geodetic and/or plane surveys using GPS methods.
- Perform astronomic measurements.
- Compute, analyze and adjust survey data.
- Design horizontal and vertical alignment for roads within a subdivision.

3. Probability and Statistics (4%)

(standard deviation; variance; standard deviation of unit weight; tests of significance; concepts of probability and confidence intervals; error ellipses; data distributions; and histograms)

- Determine levels of precision and order of accuracy.
- Perform geodetic and/or plane surveying using GPS methods.
- Compute, analyze, and adjust survey data.

4. Basic Sciences (3%)

(light and wave propagation; basic electricity; optics; gravity; refraction; mechanics; forces; kinematics; temperature and heat; biology; dendrology; geology; plant science)

- Calibrate instruments.

5. Geodesy and Survey Astronomy (4%)

(reference ellipsoids; gravity fields; geoid; geodetic datums; direction and distance on the ellipsoid; conversion from geodetic heights to elevation; orbit determination and tracking; determination of azimuth using common celestial bodies; time systems)

- Select appropriate vertical and/or horizontal datum and basis of bearing.
- Perform geodetic surveys using conventional methods and geodetic and/or plane surveys using GPS methods.
- Perform astronomic measurements.

- Perform hydrographic surveys.
- Perform differential leveling.
- Analyze and adjust survey data.

6. Computer Operations and Programming (5%)

(operating systems; graphical user interfaces (windows); data communication by serial or parallel interface; bits and bytes; Internet; computer architecture; keyboard programming of a hand-held calculator; programming a computer in a compiled language; order of arithmetic operations; programming concepts such as decision statements; flow charts; looping; arrays)

- Perform hydrographic surveys.
- Produce survey data using photogrammetric methods.
- Utilize survey data produced from photogrammetric methods.
- Compute survey data.
- Utilize computer-aided drafting systems.

7. Written and Verbal Communication (6%)

(written communication; grammar; sentence structure; punctuation; bibliographical referencing; verbal and nonverbal communication)

- Evaluate project elements to define scope of work.
- Prepare and negotiate proposals and/or contracts.
- Consult and coordinate with allied professionals and/or regulatory agencies.
- Facilitate regulatory review and approval of project documents and maps.
- Determine and secure entry rights.
- Gather parol evidence.
- Perform boundary surveys.
- Advise clients regarding boundary uncertainties.
- Testify as an expert witness.
- Review documents with clients and/or attorneys.
- Prepare sketches and/or preliminary plats, survey maps, plats, reports, and land descriptions.

8. Cadastral Law and Administration (6%)

(land descriptions; real property rights; concepts of land ownership; case law; statute law; conveyancing; official records; public/quasi-public/private land record sources; tax assessment; legal instruments of title; U.S. Public Land Survey System; Colonial/metes and bounds survey system; subdivision survey system; other cadastral systems)

- Facilitate regulatory review and approval of project documents and maps.
- Determine and secure entry rights.

- Research and evaluate court records and case law and evidence from private and public record sources.

- Gather and evaluate parol evidence.

- Perform boundary and condominium surveys.

- Reconcile survey and record data.

- Identify and evaluate field evidence for possession, boundary line discrepancies, and potential adverse possession claims.

- Identify riparian and/or littoral boundaries.

- Apply Public Land and other Survey System principles.

- Evaluate the priority of conflicting title elements.

- Determine the location of boundary lines and encumbrances.

- Advise clients regarding boundary uncertainties.

- Testify as an expert witness.

- Review documents with clients and/or attorneys.

- Determine subdivision development requirements and constraints.

- Determine and prepare lot and street patterns for land division.

- Perpetuate and/or establish monuments and their records.

- Document potential possession claims.

- Prepare and file record of survey.

9. Boundary Law (7%)

(rules of evidence relative to land boundaries and court appearance; boundary control and legal principles; order of importance of conflicting title elements; possession principles; conflict resolution; riparian/littoral/water boundaries; boundary evidence; simultaneous and sequential conveyance)

- Research and evaluate court records and case law and evidence from private and public record sources.

- Gather and evaluate parol evidence.

- Identify pertinent physical features, landmarks, and existing monumentation.

- Perform boundary, route and right-of-way surveys.

- Reconcile survey and record data.

- Identify and evaluate field evidence for possession, boundary line discrepancies, and potential adverse possession claims.

- Identify riparian and/or littoral boundaries.

- Apply public land and other survey system principles.

- Evaluate the priority of conflicting title elements.

- Determine the location of boundary lines and encumbrances.

- Advise clients regarding boundary uncertainties.

- Testify as an expert witness.

- Review documents with clients and/or attorneys.

- Determine subdivision development requirements and constraints.

- Perpetuate and/or establish monuments and their records.

- Document potential possession claims.

- Prepare survey maps, plats, reports, and land descriptions.

10. Business Law, Management, Economics, and Finance (4%)

(sole proprietorship, corporation, and partnership structures; contract law; tax structure; employment law; liability; operation analysis and optimization; land economics; appraisal science; critical path analysis; personnel management principles; cost/benefit analysis of a project or operation; econometric modeling; time value of money; budgeting)

- Evaluate project elements to define scope of work.

- Prepare and negotiate proposals and/or contracts.

- Consult and coordinate with allied professionals and/or regulatory agencies.

- Consult with and advise clients and/or their agents.

- Facilitate regulatory review and approval of project documents and maps.

11. Surveying and Mapping History (4%)

(surveying/mapping instruments and their development; prominent events and personalities; history of cartography; photogrammetric instruments and their development; history of the profession)

- Research and evaluate court records and case law and evidence from private and public record sources.

- Gather parol evidence.

- Recover horizontal/vertical control.

- Calibrate instruments.

- Apply Public Land and other survey system principles.

- Evaluate the priority of conflicting title elements.

- Determine locations of boundary lines and encumbrances.

12. Field Data Acquisition and Reduction (6%)

(field notes and electronic data collection; measurement of distances, angles and directions; modern instruments and their construction and use; precise levels; theodolites; total stations; EDMs; precision tapes; global positioning system; hydrographic data collection instruments; construction layout instruments and procedures for routes; structures)

- Determine levels of precision and order of accuracy.

- Recover horizontal/vertical control.

- Identify pertinent physical features, landmarks, and existing monumentation.

- Calibrate instruments.

- Perform geodetic and plane surveys using conventional methods and geodetic and/or plane surveys using GPS methods.

- Perform astronomic measurements.

- Perform record, as-built, ALTA/ACSM, hydrographic, photogrammetric control surveys.

- Perform trigonometric or differential leveling.

- Perform field verification of photogrammetric maps.

- Produce survey data using photogrammetric methods.

- Perform boundary, route and right-of-way, topographic, flood plain, construction, and condominium surveys.

- Perpetuate and/or establish monuments and their records.

13. Photo/Image Data Acquisition and Reduction (3%)

(cameras; image scanners; digitizers; stereo plotters; orientation; editing; ortho-image production; geo-rectification; airborne GPS; image processing; raster/vector data conversions)

- Determine levels of precision and orders of acuracy.

- Perform record, as-built, ALTA/ACSM, photogrammetric control, topographic, and flood plain surveys.

- Perform field verification of photogrammetric maps.

- Produce survey data using photogrammetric methods.

- Utilize survey data produced from photogrammetric methods.

- Prepare survey maps, plats, and reports.

14. Graphical Communication, Mapping (5%)

(principles of effective graphical display of spatial information; preparation of sketches; scaled drawings; survey plats and maps; interpretation of features on

three-dimensional drawings; principles of cartography and map projections; computer mapping; use of overlays)

- Perform record, as-built, and ALTA/ACSM surveys.

- Produce survey data using photogrammetric methods.

- Utilize survey data produced from photogrammetric methods.

- Prepare work sheets for analysis of surveys.

- Utilize computer-aided drafting systems.

- Determine and prepare lot and street patterns for land division.

- Design horizontal and vertical alignment for roads within a subdivision.

- Prepare sketches and/or preliminary plats.

- Prepare and file records of survey.

- Prepare survey maps, plats, and reports.

- Develop and/or provide data for LIS/GIS.

15. Plane Survey Calculation (7%)

(computation and adjustment of traverses; COGO computation of boundary; route alignments; construction; subdivision plats; calculation of route curves and volumes)

- Determine levels of precision and orders of accuracy.

- Calibrate instruments.

- Perform geodetic and plane surveys using conventional methods and geodetic and/or plane surveys using GPS methods.

- Perform astronomic measurements.

- Perform record, as-built, ALTA/ACSM, hydrographic, photogrammetric control, boundary, route, right-of-way, topographic, flood plain surveys, construction, and condominium surveys.

- Perform trigonometric and differential leveling.

- Compute, analyze and adjust survey data.

- Reconcile survey and record data.

- Convert survey data to appropriate datum.

- Prepare worksheets for analysis of surveys.

- Determine locations of boundary lines and encumbrances.

- Determine and prepare lot and street patterns for land division.

- Design horizontal and vertical alignment for roads within a subdivision.

- Develop and/or provide data for LIS/GIS.

16. Geodetic Survey Calculation (5%)

(calculation of position on a recognized coordinate system such as latitude/longitude; state plane coordinate systems and UTM; coordinate transformation; scale factors; convergence)

- Select appropriate vertical and/or horizontal datum and basis of bearing.
- Determine levels of precision and orders of accuracy.
- Perform geodetic and plane surveys using conventional methods and geodetic and/or plane surveys using GPS methods.
- Perform astronomic measurements.
- Perform record, as-built, ALTA/ACSM, hydrographic, photogrammetric control, boundary, route, right-of-way, topographic, flood plain surveys, construction, and condominium surveys.
- Perform trigonometric and differential leveling.
- Produce survey data using photogrammetric methods.
- Compute, analyze and adjust survey data.
- Convert survey data to appropriate datum.
- Prepare work sheets for analysis of surveys.
- Determine locations of boundary lines and encumbrances.
- Develop and/or provide data for LIS/GIS.

17. Measurement Analysis and Data Adjustment (6%)

(analysis of error sources; error propagation; control network analysis; blunder trapping and elimination; least squares adjustment; calculation of uncertainty of position; accuracy standards)

- Determine levels of precision and orders of accuracy.
- Perform geodetic and plane surveys using conventional methods and geodetic and/or plane surveys using GPS methods.
- Perform astronomic measurements.
- Perform record, as-built, ALTA/ACSM, hydrographic, photogrammetric control, boundary, route, right-of-way, topographic, flood plain surveys, construction, and condominium surveys.
- Perform trigonometric and differential leveling.
- Compute, analyze and adjust survey data.
- Reconcile survey and record data.
- Convert survey data to appropriate datum.
- Prepare work sheets for analysis of surveys.
- Determine locations of boundary lines and encumbrances.

- Determine and prepare lot and street patterns for land division.
- Design horizontal and vertical alignment for roads within a subdivision.
- Develop and/or provide data for LIS/GIS.

18. Geographic Information System Concepts (4%)

- (spatial data storage and retrieval and analysis systems; relational database systems; spatial objects; attribute value measurement; data definitions; schemas; metadata concepts; coding standards; GIS analysis of polygons and networks; buffering; overlay; spatial data accuracy standards)
- Utilize computer-aided drafting systems.
- Perpetuate and/or establish monuments and their records.
- Prepare and file records of surveys.
- Prepare survey maps, plats and reports.
- Develop and/or provide data for LIS/GIS.

19. Land Development Principles (5%)

(soil classifications and properties; hydrology and hydraulics; land planning and practices and laws controlling land use; drainage systems; construction methods; geometric and physical aspects of site analysis and design of land subdivisions; street alignment calculations; application of subdivision standards to platting of land)

20. Survey Planning, Processes, and Procedures (6%)

(techniques for planning and conducting surveys including boundary surveys, control surveys, hydrographic surveys, topographic surveys, route surveys, aerial surveys, construction surveys; issues related to professional liability, ethics and courtesy)

- Evaluate project elements to define scope of work.
- Prepare and negotiate proposals and/or contracts.
- Consult and coordinate with allied professionals and/or regulatory agencies.
- Consult with and advise clients and/or their agents.
- Facilitate regulatory review and approval of project documents and maps.
- Determine and secure entry rights.
- Advise clients regarding boundary uncertainties.
- Testify as an expert witness.
- Review documents with clients and/or attorneys.
- Document potential possession claims.
- Prepare survey maps, plats and reports.
- Develop and/or provide data for LIS/GIS.

Principles and Practice of Land Surveying (PLS)

1. Project Management (12%)

   - Evaluate project elements to define scope of work.

   - Select appropriate vertical and/or horizontal datum and basis of bearings.

   - Determine levels of precision and order of accuracy.

   - Prepare and negotiate proposals and/or contracts.

   - Consult and coordinate with allied professionals and/or regulatory agencies.

   - Consult with and advise clients and/or their agents.

   - Facilitate regulatory review and approval of project documents and maps.

   - Determine and secure entry rights.

2. Research (7%)

   - Research and evaluate evidence from private and public record sources.

   - Research court records and case law.

   - Gather parol evidence.

3. Measurements/Locations (32%)

   - Recover horizontal/vertical control.

   - Identify pertinent physical features, landmarks, and existing monumentation.

   - Calibrate instruments.

   - Perform geodetic and plane surveys using conventional methods and geodetic and/or plane surveys using GPS methods.

   - Perform astronomic measurements.

   - Perform record, as-built, ALTA/ACSM, hydrographic, photogrammetric control, boundary, route, right-of-way, topographic, flood plain, construction, and condominium surveys.

   - Perform trigonometric and differential leveling.

   - Perform field verifications of photogrammetric maps.

   - Produce survey data using photogrammetric methods.

4. Computations/Analysis (17%)

   - Compute, analyze, and adjust survey data.

   - Evaluate parol evidence.

   - Reconcile survey and record data.

   - Compute areas and volumes.

   - Convert survey data to an appropriate datum.

   - Prepare worksheets for analysis of surveys.

   - Utilize computer-aided drafting systems.

5. Legal Principles/Reconciliation (16%)

   - Identify and evaluate field evidence for possession, boundary line discrepancies, and potential adverse possession claims.

   - Identify riparian and/or littoral boundaries.

   - Apply public land and other survey system principles.

   - Evaluate the priority of conflicting title elements.

   - Determine locations of boundary lines and encumbrances.

   - Advise clients regarding boundary uncertainties.

   - Testify as an expert witness.

   - Review documents with clients and/or attorneys.

6. Land Planning and Design (3%)

   - Determine subdivision development requirements and constraints.

   - Determine and prepare lot and street patterns for land division.

   - Design horizontal and vertical alignment for roads within a subdivision.

7. Documentation/Land Information Systems (13%)

   - Perpetuate and/or establish monuments and their records.

   - Prepare sketches and/or preliminary plats.

   - Document potential possession claims.

   - Prepare and file records of surveys.

   - Prepare survey maps, plats, and reports.

   - Develop and/or provide data for LIS/GIS.

Typical Question Format

As previously mentioned, the FLS examination includes 170 basic surveying questions in a multiple-choice format. The questions are typically short, straightforward and designed to test your knowledge of the basic fundamentals of surveying and mapping. A typical question (taken from Cole, George M., *Land Surveyor-in-Training Sample Examination*, Professional Publications, 1994) might be as follows.

The bearing of the tangent into a highway curve with a radius of 1000 ft is N 60° E and the coordinates, in feet, of the point of curvature are N = 1000, E = 1000. What are the coordinates, in feet, for the radius point?

   (A) N = 134.0, E = 1500.0
   (B) N = 144.0, E = 1500.0
   (C) N = 1500.0, E = 134.0
   (D) N = 1500.0, E = 144.0

The PLS examination is designed to test your ability to apply surveying fundamentals to typical problems encountered in surveying practice. For example, a series of land descriptions from deeds might be provided, followed by a series of multiple-choice questions requiring you to analyze the descriptions and indicate how you would establish certain boundaries, corner positions, or treat encroachments.

Scoring of Examinations

Neither the FLS nor the PLS examination is graded on a curve since a certain minimum competency must be demonstrated to safeguard the public welfare. Nevertheless, it is recognized that the tests may vary slightly in difficulty, depending upon the questions selected for a particular examination. Therefore, questions are reviewed by committees of practicing land surveyors before the examinations. These committees evaluate the difficulty of each question in order to develop a "standard of minimum competency," or recommended passing score for each examination. However, the individual state boards have the authority to determine the passing score in their respective states. In the grading process, credit is given for each correct answer and no points are deducted for incorrect answers. The sum of the correct answers is scaled so that the grade of 70 reflects the standard minimum competency.

Use of Calculators and Computers in the Examinations

Obviously, a calculator is needed for the examinations. Although rules may vary with individual state boards, calculators are usually allowed as long as they are battery-operated, silent, nonprinting, and do not have a QWERTY keyboard. Laptop and palmtop computers are not permitted. Contact your state board for rules specific to your state. In some states, programmable or preprogrammed calculators may not be used. For most problems, a $15 scientific calculator will suffice as long as it has trigonometric functions, pi, square root, x^2, logarithms, and standard deviation. In addition, it is helpful to have functions for converting degrees, minutes and seconds to decimal degrees; and economic analysis functions.

Reference Material Permitted in the Examinations

The FLS examination is "closed book." No reference material of any kind may be used. However, each portion of the examination contains a collection of pertinent reference formulas that may be used during the examination. These include formulas for triangle solutions, horizontal and vertical curves, statistics, state plane coordinates, earthwork, tape correction, astronomy, photogrammetry, and stadia. In addition, some conversion factors are provided. However, the formulas and conversion factors included are not necessarily all-inclusive. Some formulas and conversion factors other than those provided may be necessary to complete the examination. Therefore, a well-prepared examinee should know many of the basic formulas.

The PLS examination is "open book." Therefore, you may use textbooks, handbooks, and bound reference material, although you may not share books with other examinees in the room. Generally, you can use personal notes bound in a three-ring notebook. However, loose paper and scratch pads are not permitted. Since specific requirements may vary with individual state boards, you should contact your state board for rules specific to your state. A few states do not allow the use of published collections of solved problems. Those states typically maintain a list of such banned books.

Cheating and Examination Subversion

The proctors for these examinations are well-trained to insure that the regulations regarding use of reference material and other types of cheating do not occur. Obviously, you should not talk to other examinees in the room during the examination, nor should you pass notes back and forth. Typically, the number of people released to use the restrooms at any given time during the examination may be restricted to prevent discussion.

The NCEES regularly reuses good problems that have appeared on previous exams. Therefore, examination security is a serious issue with NCEES, which goes to great lengths to prevent copying of problems. You may not keep your examination booklet, enter text of problems into your calculator, or copy problems into your own material.

The proctors are especially concerned about exam subversion, which generally means any activity that might invalidate the examination or the examination process. The most common form of exam subversion involves trying to copy exam problems for future use.

Preparation for and Taking the Examinations

Plan Your Approach

Preparation for the licensure examinations should be considered a long-term project and should be carefully planned. As currently structured, the examinations are both comprehensive and fast-paced. Rapid recall, discipline, stamina, and mastery of all areas to be covered are essential in order to succeed on the examinations. Development of these qualities may require months of preparation in addition to the years of academic study and work practice necessary to qualify for the examinations. Therefore, it is important to plan your preparation for the examinations as you would for a large surveying and mapping project. A rigid schedule for your review should be established and adhered to.

It is suggested that preparation follow these steps.

(1) Review the listing of subject areas contained in the previous section to gain insight into the nature and content of the examinations.

(2) Go through each chapter of *this* book and answer the practice problems at the end of each chapter.

It is a good idea to prepare a concise outline as you work through each area, for future reference. This review should be on a rigorous schedule to help you develop the discipline and stamina necessary to do well on the examinations.

(3) For any areas in which you are not comfortable, read additional reference material as you work through each chapter. It is a good idea to tab any reference books at pages where frequently used or hard-to-find information is located. This speeds up the process of obtaining such information when needed during the examinations as well as during the review process. For problem areas, you may consider taking continuing education courses in those topics.

(4) To make your review realistic, take a sample examination, such as that available in *Land Surveyor-in-Training Sample Examination* (available from Professional Publications, Inc.) to evaluate your readiness for the examinations.

(5) Work on any weak areas detected by the sample examination.

(6) Conduct a final review of your notes.

Learning to use your time wisely is one of the most important things that you can do during your review. You will undoubtedly encounter review problems that end up taking much longer than you expected. In some instances, you may cause your own delays by spending too much time looking through books for things that you need. At other times, the problems will just entail too much work. Learning to recognize such situations will help you make intelligent decisions about such problems during the examinations.

Additional Reference Material

You will find that this book is an excellent starting point for preparing for the examinations. However, it may be helpful to supplement the material in this book with additional references, especially in areas in which you are uncomfortable. There are countless texts available that cover the various topics in depth. Below are listed several personal favorites which offer coverage of the areas to be tested on the examinations.

Legal Principles

Brown, Robillard and Wilson. *Boundary Control and Legal Principles*, John Wiley & Sons.

Brown, Robillard and Wilson. *Evidence and Procedures for Boundary Location*, John Wiley & Sons.

Bureau of Land Management. *Manual of Instruction for Surveys of Public Lands*, Government Printing Office.

Cole, George M. *Water Boundaries*, John Wiley & Sons.

Wattles, W.C. *Land Survey Descriptions*, Gordon Wattles Publications.

Measurement and Computation Theory and Practice

Bureau of Land Management. *Manual of Instruction for Surveys of Public Lands*, Government Printing Office.

Cole, George M. *Water Boundaries*, John Wiley & Sons.

Davis, Foote, Anderson and Mikhail. *Surveying Theory and Practice*, McGraw-Hill.

Hickerson, Thomas F. *Route Location and Design*, McGraw-Hill.

Wolf and Ghilani. *Adjustment Computations*, John Wiley & Sons.

Geodesy (including GPS) and Survey Astronomy

Buckner, R.B. *A Manual on Astronomic and Grid North*, Landmark Enterprises.

Smith, James R. *Introduction to Geodesy*, John Wiley & Sons.

Van Sickle, Jan. *GPS for Land Surveyors*, Ann Arbor Press.

Geographic Information Systems and Photogrammetry

Clarke, Keith C. *Getting Started with Geographic Information Systems*, Prentice Hall.

Wolf & Dewitt. *Elements of Photogrammetry (with Applications in GIS)*, McGraw-Hill.

Land Development

Colley, Barbara C. *Practical Manual of Land Development*, McGraw-Hill.

Business Law, Management, Economics, and Finance

Denny, Milton E. *Surveyors and Engineers Small Business Handbook*, CED Technical Services.

Lindeburg, Michael. *Engineering Economic Analysis*, Professional Publications.

Practice Questions

Cole, George M. *Land Surveyor-in-Training Sample Examination*, Professional Publications.

NCEES. *Fundamentals of Land Surveying Sample Questions and Solutions.*

NCEES. *Principles and Practice of Land Surveying Sample Questions and Solutions.*

Last-Minute Preparation

A week or so before the exam, you should conduct an intensive review of the outlines prepared during your review. However, do not attempt to cram during the last night before the exam.

During the last week or so before the examination, make arrangements for child care and transportation. Since the examination does not always start or end at the designated time, make sure such arrangements are flexible. If it is convenient, visit the examination site to locate the building, parking areas, examination rooms, and restrooms.

You should always take a backup calculator to the examinations. If your spare calculator is not the same type as your primary one, spend some time familiarizing yourself with its operation. Make sure that you have the correct size of replacement batteries for both calculators. In addition, you should prepare a kit of items and reference material to be taken to the examination.

Take the day before the examination off from work to relax. If you live a considerable distance from the examination site, consider getting a hotel room in which to spend the night. Calculate your wake-up time and set two alarms. Select and lay out your clothing and breakfast items, and make sure that you have gas in your car and money in your wallet.

Taking the Examination

What to Take to the Examination

In addition to the review process, another important aspect of preparation for the examination is selection of the materials to be taken to the examination. Even for the FLS examination, which as previously mentioned is closed book, there are a number of documents, tools, and personal comfort items that should be taken as your "exam kit." The following is a suggested list.

- letter admitting you to the examination
- photo identification (such as driver's license)
- eye glasses
- primary calculator
- spare calculator
- spare batteries for calculators
- several sharpened pencils or mechanical pencils
- erasers
- ruler and protractor
- unobtrusive snacks or candies
- travel pack of tissues
- handkerchief
- headache remedy
- several dollars in miscellaneous change
- light jacket or sweater
- wristwatch with alarm

For the PLS examination, which as previously mentioned is open book, you will also need to take reference material and books. You actually will not use many books during the examination. However, you do not know in advance which ones you will need. (That is why many examinees show up with boxes and boxes of books.) The examination is very fast paced. You will not have time to thumb through unfamiliar texts, searching for critical information. So take only books with which you are intimately familiar, and tab important formulas, charts, tables, and the like, or copy them into a well-organized notebook. This book and five or six other references should be sufficient for most of the problems you encounter. As a final touch, take along the morning newspaper to read while waiting for the exam to begin. Two items that you should *not* take to the examination are your pager and cellular phone. At the very least, be sure to turn them off before the examination.

What to Do at the Examination

Arrive at least 30 minutes before the examination is scheduled to begin. This will allow you to find a convenient parking place, get your materials into the examination room, find a good seat and calm down. Be prepared, though, to find that the examination room is not open or ready at the designated time. Once you have arranged your materials on your table, take out your morning newspaper and look cool.

All of the procedures typically associated with timed, proctored, computer-graded assessment tests will be in effect when you take your licensure examinations. The proctors will distribute the examination booklets and answer sheet. However, you should not open the booklets until instructed to do so.

Listen carefully to everything the proctors say. Do not ask the proctors any technical questions. Even if knowledgeable in engineering, they are not permitted to answer your questions. They will guide you through the process of putting your name and other biographical information on the material. Time for these and other instructions and for initializing the answer sheets is not part of the timed period.

The common suggestions to "use number 2 pencils, completely fill the bubbles, and erase completely" apply here. Actually, mechanical pencils with HB lead can also be used, but pens and felt-tip markers should not be used. Make sure that all of your responses on the answer sheet are dark to insure proper credit.

All of the questions on the exam are worth the same number of points, so it is a good idea to answer all of the questions that you can within a reasonable time before attempting to solve problems that will take a disproportionate amount of time. If time allows, you can go back to those difficult problems after you have answered all of the "easy" questions.

Many points are lost due to carelessness. Therefore, it is a good idea to read each question carefully twice before solving. Check to make sure that you used all of the given data and made the appropriate conversion of units. While the examination questions are not "tricky," you may find that answers using commonly made mistakes are represented among the available answer choices. Thus, just because there is an answer matching your results does not mean that you have obtained the correct results.

Both the FLS and PLS examinations are multiple choice. Credit is given for correct answers, but no credit is deducted for wrong answers. Therefore, it is in your best interest to answer every question. It is a good idea to set your wristwatch alarm for five minutes before the end of each session and use that remaining time to guess at any remaining unsolved multiple-choice problems. You will be successful with about 25% of your guesses, and those points will more than make up for the few points you might earn by working during the last five minutes.

If you finish the examination early and there are still more than 30 minutes remaining, you will be permitted to leave the room. If you finish less than 30 minutes before the end of the examination, you may be required to remain until the end in consideration of the people who are still working. You will not be permitted to keep your examination booklet for later review. When you leave, you must return it to the proctors.

After the Examination

People react quite differently to the examination experience. Some people are energized and need to unwind by talking with other examinees, describing every detail of their experience and dissecting every examination question. However, most people are completely exhausted and need a lot of quiet space and a hot tub to soak and sulk. Since everyone who took the examination has seen it, you will not be violating your "oath of silence" if you talk about the details with other examinees. It is difficult not to ask how someone else approached a problem that had you completely stumped. However, it is also very disquieting to think you did well on a problem, only to have someone else tell you where you went wrong.

Waiting for the results of the examination is its own form of mental torture. There is no predictable pattern to the release of the results. The examination results are not released by NCEES to all states simultaneously. They are not released alphabetically by state or examinee name. The people who failed are not notified first or last. Your co-worker might receive his or her notification today, and you might have to wait another three weeks. It all depends on when the entire process is complete. Some states have to have the results approved at a board meeting. Some prepare certificates before sending out notifications. Some states are more highly automated that others. The number of examinees also varies by state, as do numerous other factors. Therefore you just have to wait patiently.

To find out which states have released their results check in regularly with the Exam Forum on PPI's website (www.ppi2pass.com). You will find many other anxious examinees there.

Conclusion

Now that you know all there is to know about the examinations and about how to prepare for them, the rest is up to you. Plan your approach, and get to work. The very best of luck to you!

1 Arithmetic and Measurements

1. COMMON FRACTIONS

A *common fraction* is the division of one number by another. In the fraction $\frac{2}{3}$, the number above the line is known as the *numerator*, and the number below the line is known as the *denominator*. The line between the numbers indicates division.

A fraction whose numerator is smaller than its denominator is known as a *proper fraction*. A fraction whose numerator is equal to or larger than its denominator is known as an *improper fraction*.

2. CHANGING THE FORM OF FRACTIONS

If the numerator and the denominator of a fraction are multiplied by the same number, the value of the fraction is not changed. Thus, if the numerator and denominator of the fraction $\frac{1}{2}$ are each multiplied by 2, the

fraction becomes $\frac{2}{4}$; if the numerator and denominator of $\frac{2}{3}$ are each multiplied by 3, the fraction becomes $\frac{6}{9}$. The fractions $\frac{1}{2}$ and $\frac{2}{4}$ are known as *equivalent fractions*, as are $\frac{2}{3}$ and $\frac{6}{9}$.

The numerator and denominator of a fraction may be divided by the same number without changing the value of the fraction. Thus, $\frac{3}{6}$ is equivalent to $\frac{1}{2}$ and $\frac{2}{8}$ is equivalent to $\frac{1}{4}$. In each case the numerator and denominator have been divided by 2. If the numerator and denominator have no common divisor except 1, the fraction is said to be in its *lowest terms*. Examples of fractions in their lowest terms are $\frac{1}{2}$, $\frac{2}{9}$, $\frac{3}{8}$, $\frac{4}{11}$, and $\frac{5}{16}$.

Example 1.1

(a) $\dfrac{1}{4} \times \dfrac{2}{2}$

(b) $\dfrac{1}{3} \times \dfrac{3}{3}$

(c) $\dfrac{7}{8} \times \dfrac{4}{4}$

Solution

(a) $\dfrac{2}{8}$ $\left[\frac{1}{4} \text{ in lowest terms}\right]$

(b) $\dfrac{3}{9}$ $\left[\frac{1}{3} \text{ in lowest terms}\right]$

(c) $\dfrac{28}{32}$ $\left[\frac{7}{8} \text{ in lowest terms}\right]$

3. MIXED NUMBERS

A number that consists of a whole number and a fraction, such as $5\frac{3}{8}$, is known as a *mixed number*. An improper fraction can be changed to a whole number or to a whole number and a fraction. A mixed number can be changed to an improper fraction.

To change an improper fraction to a mixed number, divide the numerator by the denominator.

To change a mixed number to an improper fraction, multiply the whole number by the denominator of the fraction and add the improper fraction and the proper fraction.

Example 1.2

(a) Change $\frac{15}{4}$ to a mixed number.

(b) Change $5\frac{3}{8}$ to an improper fraction.

Solution

(a) $\dfrac{15}{4} = 15 \div 4 = 3\frac{3}{4}$

(b) $5\frac{3}{8} = \dfrac{5}{1} + \dfrac{3}{8} = \dfrac{5 \times 8}{1 \times 8} + \dfrac{3}{8} = \dfrac{40}{8} + \dfrac{3}{8} = \dfrac{43}{8}$

Example 1.3

Change each of the following fractions to an equivalent fraction having the higher denominator.

(a) $\dfrac{1}{2} = \dfrac{x}{8}$

(b) $\dfrac{1}{4} = \dfrac{x}{32}$

(c) $\dfrac{5}{16} = \dfrac{x}{64}$

(d) $\dfrac{7}{12} = \dfrac{x}{72}$

Solution

(a) $\dfrac{1}{2} = \dfrac{4}{8}$

(b) $\dfrac{1}{4} = \dfrac{8}{32}$

(c) $\dfrac{5}{16} = \dfrac{20}{64}$

(d) $\dfrac{7}{12} = \dfrac{42}{72}$

Example 1.4

Reduce the following fractions to lowest terms.

(a) $\dfrac{3}{9}$

(b) $\dfrac{12}{16}$

(c) $\dfrac{15}{35}$

(d) $\dfrac{9}{24}$

Solution

(a) $\dfrac{3}{9} = \dfrac{1}{3}$

(b) $\dfrac{12}{16} = \dfrac{3}{4}$

(c) $\dfrac{15}{35} = \dfrac{3}{7}$

(d) $\dfrac{9}{24} = \dfrac{3}{8}$

Example 1.5

Change the following improper fractions to equivalent mixed numbers reduced to the lowest terms.

(a) $\dfrac{37}{7}$

(b) $\dfrac{21}{8}$

(c) $\dfrac{51}{16}$

(d) $\dfrac{455}{64}$

Solution

(a) $\dfrac{37}{7} = 5\frac{2}{7}$

(b) $\dfrac{21}{8} = 2\frac{5}{8}$

(c) $\dfrac{51}{16} = 3\frac{3}{16}$

(d) $\dfrac{455}{64} = 7\frac{7}{64}$

Example 1.6

Change the following numbers to improper fractions.

(a) $3\frac{3}{5}$

(b) $3\frac{5}{16}$

(c) $4\frac{5}{7}$

(d) $7\frac{8}{9}$

Solution

(a) $3\frac{3}{5} = \dfrac{18}{5}$

(b) $3\frac{5}{16} = \dfrac{53}{16}$

(c) $4\frac{5}{7} = \dfrac{33}{7}$

(d) $7\frac{8}{9} = \dfrac{71}{9}$

4. ADDITION AND SUBTRACTION OF FRACTIONS

Before fractions can be added or subtracted, they must have the same denominator. The fractions $\frac{1}{2}$, $\frac{2}{3}$, and $\frac{3}{4}$ cannot be added directly, but if they are changed to the equivalent fractions $\frac{6}{12}$, $\frac{8}{12}$, and $\frac{9}{12}$, they can be added.

5. PRIME FACTORS

When $12 = 3 \times 4$ is written, 12 is factored into a product of two numbers. The numbers 3 and 4 are factors of 12. Also, 6 and 2 are factors of 12, and 12 and 1 are factors of 12.

A *prime factor*, or *prime number*, is a number that is divisible only by 1 and itself. The prime numbers are 2, 3, 5, 7, 11, 13, and so on.

The prime factors of numbers can be found by dividing the number by the lowest prime, 2; dividing the result by 2 until it is not divisible by 2; then dividing by the next higher prime until the quotient is 1.

Example 1.7

Find the prime factors of each of the following numbers.

(a) 252

(b) 1575

Solution

(a) $2 \overline{)\,252}$
$\quad 2 \overline{)\,126}$
$\quad 3 \overline{)\,\;63}$
$\quad 3 \overline{)\,\;21}$
$\quad 7 \overline{)\,\;\;7}$
$\qquad\quad 1$

$252 = 2 \times 2 \times 3 \times 3 \times 7$

(b) $3 \overline{)\,1575}$
$\quad 3 \overline{)\,\;525}$
$\quad 3 \overline{)\,\;175}$
$\quad 5 \overline{)\,\;\;35}$
$\quad 7 \overline{)\,\;\;\;7}$
$\qquad\quad 1$

$1575 = 3 \times 3 \times 5 \times 5 \times 7$

6. LEAST COMMON MULTIPLE

The *least common multiple* (LCM) of two or more numbers is the least number that has each of the numbers as a factor. The least common multiple of two or more numbers is found by finding the prime factors of each number and the product of all the prime factors using each prime the greatest number of times it appears in either number.

Example 1.8

Find the LCM of 12, 15, and 18.

Solution

$12 = 2 \times 2 \times 3$

$15 = 3 \times 5$

$18 = 2 \times 3 \times 3$

The LCM is $2 \times 2 \times 3 \times 3 \times 5 = 180$

7. LEAST COMMON DENOMINATOR

The *least common denominator* (LCD) of two or more fractions is the least common multiple of the denominators. It is the smallest number that can be exactly divided by the denominators.

To add or subtract fractions with different denominators, first find the LCD for the fractions and change the fractions to equivalent ones using the LCD. Then add the numerators and place the sum over the common denominator.

Example 1.9

Add the following fractions.

(a) $\dfrac{3}{4} + \dfrac{2}{3} + \dfrac{5}{9}$

(b) $\dfrac{11}{16} + \dfrac{12}{14} + \dfrac{13}{21}$

Solution

(a) $4 = 2 \times 2$
$\quad 3 = 3$
$\quad 9 = 3 \times 3$

$\text{LCD} = 2 \times 2 \times 3 \times 3 = 36$

$\dfrac{3}{4} + \dfrac{2}{3} + \dfrac{5}{9} = \dfrac{27}{36} + \dfrac{24}{36} + \dfrac{20}{36} = \dfrac{71}{36} = 1\dfrac{35}{36}$

(b) $16 = 2 \times 2 \times 2 \times 2$
$\quad 14 = 2 \times 7$
$\quad 21 = 3 \times 7$

$\text{LCD} = 2 \times 2 \times 2 \times 2 \times 3 \times 7 = 336$

$\dfrac{11}{16} + \dfrac{12}{14} + \dfrac{13}{21} = \dfrac{231}{336} + \dfrac{288}{336} + \dfrac{208}{336} = \dfrac{727}{336} = 2\dfrac{55}{336}$

8. MULTIPLICATION OF FRACTIONS

The *product of fractions* is the product of the numerators over the product of the denominators.

Example 1.10

(a) $\dfrac{3}{8} \times \dfrac{2}{3} \times \dfrac{1}{4}$

(b) $2\tfrac{7}{8} \times 3\tfrac{2}{3} \times 418 = \dfrac{23}{8} \times \dfrac{11}{3} \times \dfrac{33}{8}$

Solution

(a) $\dfrac{6}{96} = \dfrac{1}{16}$

(b) $\dfrac{8349}{192} = 43\tfrac{93}{192}$

9. DIVISION OF FRACTIONS

To divide a number by a fraction, to divide a fraction by a number, or to divide a fraction by a fraction, invert the second number and multiply by the first. Note that when a number is divided by a proper fraction, the quotient will be larger than the number.

Example 1.11

(a) $4 \div \dfrac{3}{8}$

(b) $\dfrac{3}{4} \div 5$

(c) $\dfrac{3}{4} \div \dfrac{2}{3}$

Solution

(a) $\dfrac{4}{1} \times \dfrac{8}{3} = \dfrac{32}{3} = 10\tfrac{2}{3}$

(b) $\dfrac{3}{4} \div \dfrac{5}{1} = \dfrac{3}{4} \times \dfrac{1}{5} = \dfrac{3}{20}$

(c) $\dfrac{3}{4} \times \dfrac{3}{2} = \dfrac{9}{8} = 1\tfrac{1}{8}$

10. CANCELLATION

A fraction containing the product of several numbers in the numerator and the product of several numbers in the denominator may be simplified by performing as many divisions as possible by cancellation.

Example 1.12

Simplify the following by cancellation.

(a) $\dfrac{(15)(6)}{(12)(5)}$

(b) $\dfrac{(6)(10)(8)}{(4)(3)(5)}$

(c) $\dfrac{(16)(15)(14)}{(30)(28)(8)}$

(d) $\dfrac{(15)(7)(8)(9)}{(21)(30)(5)}$

Solution

(a) $\dfrac{(\overset{3}{\cancel{15}})(\overset{1}{6})}{(\underset{2}{\cancel{12}})(\underset{1}{\cancel{5}})} = \dfrac{3}{2} = 1\tfrac{1}{2}$

(b) $\dfrac{(\overset{2}{6})(\overset{2}{\cancel{10}})(\overset{2}{8})}{(\underset{1}{4})(\underset{1}{3})(\underset{1}{\cancel{5}})} = 8$

(c) $\dfrac{(\overset{\overset{1}{\cancel{2}}}{\cancel{16}})(\overset{1}{\cancel{15}})(\overset{1}{\cancel{14}})}{(\underset{\underset{1}{\cancel{2}}}{\cancel{30}})(\underset{2}{\cancel{28}})(\underset{1}{8})} = \dfrac{1}{2}$

(d) $\dfrac{(\overset{1}{\cancel{15}})(\overset{1}{\cancel{7}})(\overset{4}{8})(\overset{3}{\cancel{9}})}{(\underset{\underset{1}{\cancel{3}}}{\cancel{21}})(\underset{\underset{1}{\cancel{2}}}{\cancel{30}})(\underset{}{5})} = \dfrac{12}{5} = 2\tfrac{2}{5}$

11. READING DECIMAL FRACTIONS

A *decimal fraction* is a fraction whose denominator is 10 or some power of 10. The fraction $\frac{5}{10}$ is written as 0.5; $\frac{75}{100}$ is written as 0.75. The period preceding 75 is the *decimal point*. The number 25,376.4921 is read as "twenty-five thousand, three hundred seventy-six and four thousand nine hundred twenty-one ten-thousandths." The "and" is used for the decimal point. Decimal terminology for this number is illustrated in Fig. 1.1.

A decimal fraction that is greater than one is called a *mixed decimal*. An *integer* (a whole number such as 325) is understood to have a decimal point to the right of the units digit (325.).

Figure 1.1 *Decimal Terminology*

Example 1.13

Write the following numbers using words.

(a) 388.152

(b) 72.006

(c) 0.00005

(d) 326.2200

Solution

(a) three hundred eighty-eight and one hundred fifty-two thousandths

(b) seventy-two and six thousandths

(c) zero and five hundred thousandths

(d) three hundred twenty-six and two thousand two hundred ten-thousandths

12. MULTIPLYING AND DIVIDING DECIMAL FRACTIONS

Multiplying a number by 10 moves the decimal point in the number one place to the *right*. Multiplying by 100 moves the decimal point two places to the right. In general, multiplying a number by 10 or a multiple of 10 moves the decimal to the right a number of places equivalent to the number of zeros in the multiplier.

Dividing a number by 10 or a multiple of 10 moves the decimal point to the *left* a number of places equivalent to the number of zeros in the divisor.

Multiplying a number by 0.1 is the same as dividing by 10 and moves the decimal point to the *left*.

Dividing a number by 0.1 is the same as multiplying by 10 and moves the decimal point to the *right*.

Example 1.14

(a) 537.26×10

(b) 421.38×100

(c) 6.7382×1000

Solution

(a) 5372.6

(b) 42138

(c) 6738.2

Example 1.15

(a) $537.26 \div 10$

(b) $421.38 \div 100$

(c) $6.7382 \div 1000$

Solution

(a) 53.726

(b) 4.2138

(c) 0.0067382

Example 1.16

(a) $(4762)(0.1) = 476 \times \dfrac{1}{10}$

(b) $(378)(0.01) = 378 \times \dfrac{1}{100}$

(c) $(9843)(0.0001) = 9843 \times \dfrac{1}{10,000}$

Solution

(a) 476.2

(b) 3.78

(c) 0.9843

Example 1.17

(a) $\dfrac{537.26}{0.1}$

(b) $\dfrac{4221.38}{0.01}$

Solution

(a) $\dfrac{537.26}{\frac{1}{10}} = 537.26 \times \dfrac{10}{1} = 5372.6$

(b) $\dfrac{421.38}{\frac{1}{100}} = 421.38 \times \dfrac{100}{1} = 42{,}138$

13. PERCENT

The word "percent" comes from a Latin word meaning *by the hundred*. The symbol "%" is used for the word *percent*. A fraction having 100 as a denominator expresses a percent. Thus, $\frac{6}{100}$ is the same as 6%, as is 0.06.

14. CHANGING A DECIMAL FRACTION TO A PERCENT

To change a decimal fraction to a percent, move the decimal point two places to the right and place the percent sign after the number.

Example 1.18

$$0.005 = 0.5\%$$
$$0.07 = 7\%$$
$$0.25 = 25\%$$
$$8.39 = 839\%$$

15. CHANGING A PERCENT TO A DECIMAL FRACTION

To change a percent to a decimal fraction, remove the percent sign and move the decimal point two places to the left.

Example 1.19

$$0.7\% = 0.007$$
$$4\% = 0.04$$
$$75\% = 0.75$$
$$150\% = 1.50$$

16. FINDING A PERCENT OF A NUMBER

To find 25% of a number, multiply 25% times the number. To find a percent of a number, change the percent to a decimal fraction and multiply the number by this decimal fraction.

Example 1.20

Find the percent of the following numbers as indicated.

(a) 1% of 7823

(b) 10% of 156

(c) 35% of 200

(d) 68% of 300

Solution

(a) 1% of $7823 = (0.01)(7823) = 78.23$

(b) 10% of $156 = (0.10)(156) = 15.6$

(c) 35% of $200 = (0.35)(200) = 70$

(d) 68% of $300 = (0.68)(300) = 204$

17. FINDING WHAT PERCENT ONE NUMBER IS OF ANOTHER

To find what percent one number is of another, divide the first by the second and express as a decimal fraction. Then move the decimal point two places to the right and add the percent sign.

Example 1.21

(a) 15 is what percent of 750?

(b) What percent of 64 is 16?

Solution

(a) $\dfrac{15}{750} = 0.02 = 2\%$

(b) $\dfrac{16}{64} = 0.25 = 25\%$

18. FINDING A NUMBER WHEN A PERCENT OF THE NUMBER IS KNOWN

If 30 is 75% of some number, to find that number, divide 30 by 75%.

$$\frac{30}{75\%} = \frac{30}{0.75} = 40$$

By converting percent to a common fraction, many problems can be solved more rapidly.

Knowing that $75\% = 0.75 = \frac{3}{4}$, divide 30 by $\frac{3}{4}$.

$$\frac{30}{\frac{3}{4}} = \left(\frac{30}{1}\right)\left(\frac{4}{3}\right) = 40$$

By relating the *common-fraction equivalent* to certain percentages in solving problems involving percent, the solution becomes much simpler. The first step in using this method is to memorize the following equivalents.

$$\frac{1}{100} = 1\% \qquad \frac{1}{3} = 33\tfrac{1}{3}\%$$
$$\frac{1}{10} = 10\% \qquad \frac{1}{2} = 50\%$$
$$\frac{1}{8} = 12.5\% \qquad \frac{2}{3} = 66\tfrac{2}{3}\%$$
$$\frac{1}{4} = 25\% \qquad \frac{3}{4} = 75\%$$
$$\frac{3}{8} = 37.5\% \qquad 1 = 100\%$$

Example 1.22

(a) 25% of what number is 10?

(b) 40 is 80% of what number?

Solution

(a) $\dfrac{10}{25\%} = \dfrac{10}{0.25} = 40$

(b) $\dfrac{40}{80\%} = \dfrac{40}{0.80} = 50$

Example 1.23

(a) 6 is 50% of what number?

(b) 12 is $12\tfrac{1}{2}\%$ of what number?

(c) 12 is 25% of what number?

(d) 45 is $33\tfrac{1}{3}\%$ of what number?

(e) 120 is 75% of what number?

(f) 6.5 is 25% of what number?

(g) $\dfrac{3}{8}$ is $37\tfrac{1}{2}\%$ of what number?

(h) $\dfrac{2}{3}$ is 75% of what number?

Solution

(a) $\dfrac{6}{50\%} = \dfrac{6}{\frac{1}{2}} = \left(\dfrac{6}{1}\right)\left(\dfrac{2}{1}\right) = 12$

(b) $\dfrac{12}{12\frac{1}{2}\%} = \dfrac{12}{\frac{1}{8}} = \left(\dfrac{12}{1}\right)\left(\dfrac{8}{1}\right) = 96$

(c) $\dfrac{12}{25\%} = \left(\dfrac{12}{1}\right)\left(\dfrac{4}{1}\right) = 48$

(d) $\dfrac{45}{33\frac{1}{3}\%} = \dfrac{45}{\frac{1}{3}} = \left(\dfrac{45}{1}\right)\left(\dfrac{3}{1}\right) = 135$

(e) $\dfrac{120}{75\%} = \dfrac{120}{\frac{3}{4}} = \left(\dfrac{120}{1}\right)\left(\dfrac{4}{3}\right) = 160$

(f) $\dfrac{6.5}{25\%} = \dfrac{6.5}{\frac{1}{4}} = \left(\dfrac{6.5}{1}\right)\left(\dfrac{4}{1}\right) = 26$

(g) $\dfrac{\frac{3}{8}}{37.5\%} = \dfrac{3}{8} \div \dfrac{3}{8} = \dfrac{3}{8} \times \dfrac{8}{3} = 1$

(h) $\dfrac{\frac{2}{3}}{75\%} = \dfrac{2}{3} \div \dfrac{3}{4} = \dfrac{2}{3} \times \dfrac{4}{3} = \dfrac{8}{9}$

19. ROUNDING OFF NUMBERS

The term *in round numbers* is often used. A crowd at a football game might be estimated at 50,000, meaning that it is between 40,000 and 60,000. Or, a certain university might be said to have 12,000 students enrolled, meaning that there are between 11,000 and 13,000 students. The exact figure is often unimportant in conveying an idea of size or quantity.

In like manner, the use of *approximate numbers—rounded-off numbers*—can make locating the decimal point in computations relatively simple.

For instance, the product of 310 and 21 is approximately 6000 because (300)(20) is equal to 6000. The number 310 has been rounded to 300 and 21 has been rounded to 20. For simplicity (3)(2) is 6, and because there are three zeros in the approximation (300)(20) three zeros are added to the right of 6 to get 6000. Thus, the approximate product is estimated. This system is one of those used in finding the decimal point when approximate answers are needed quickly.

In rounding off a number so that it contains a certain number of significant digits, the following procedure should be observed.

- If the digit to be dropped is less than 5, no change is made in the number retained. Thus, 58.463 becomes 58.46.

- If the digit to be dropped is greater than 5, the digit preceding it is increased by 1. Thus, 58.467 becomes 58.47.

- If the digit to be dropped is exactly 5, the digit preceding it is increased by 1 if it is odd but left unchanged if it is even. Thus, 3.55 becomes 3.6, and 15.45 becomes 15.4.

The first two procedures are universally accepted; however, the last procedure is not always followed. Some computers prefer to take the next higher number when the digit to be dropped is 5.

Example 1.24

Round off each of the following numbers to the nearest tenth.

(a) 2.749

(b) 2.751

(c) 3.66

(d) 5.35

(e) 5.45

(f) 6.74

(g) 7.049

(h) 7.051

Solution

(a) 2.7

(b) 2.8

(c) 3.7

(d) 5.4

(e) 5.4

(f) 6.7

(g) 7.0

(h) 7.1

Example 1.25

Round off each of the following numbers to the nearest hundredth.

(a) 1.006

(b) 2.355

(c) 3.015

(d) 5.295

(e) 6.114

(f) 7.465

(g) 8.344

(h) 9.455

Solution

(a) 1.01

(b) 2.36

(c) 3.02

(d) 5.30

(e) 6.11

(f) 7.46

(g) 8.34

(h) 9.46

Example 1.26

Round off each of the following numbers to the nearest hundred.

(a) 1990

(b) 8501

(c) 9552

(d) 25,962

(e) 28,232

(f) 32,008

(g) 41,500

(h) 60,499

Solution

(a) 2000

(b) 8500

(c) 9600

(d) 26,000

(e) 28,200

(f) 32,000

(g) 41,500

(h) 60,500

20. EXACT AND APPROXIMATE NUMBERS

Many beginning students in engineering technology experience some confusion in dealing with *approximate numbers*. Mathematics, to these students, has dealt largely with exact numbers. Counting numbers are exact numbers. The number of students in a classroom is an exact number. The number of inches in a foot is an exact number. Thus, when students are asked to multiply (2.3)(12.78), they find the result to be 29.394. But if the numbers 2.3 and 12.78 represent measured quantities, the result 29 might better represent the actual value. It is difficult for the students to know "how many places to carry the answer."

A *measured quantity* is an approximate number; it can never be an exact number. Consider measuring the distance between two points with two different tapes. The first tape is graduated into feet and tenths of a foot. The second tape is graduated into feet, tenths, and hundredths of a foot.

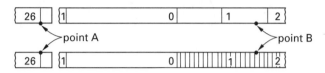

Figure 1.2 *Measured Quantity*

Suppose the distance is recorded as 26.18 ft using the first tape. This means that the first *three* digits are exact or positive digits, but that the last digit is estimated. The last digit, 8, has been determined by estimating that one of the points lies along the tape between the 1 and 2 tenth marks, and it is estimated that it lies 8 tenths of the way between the 1 and the 2.

Using the other tape, graduated into hundredths, the distance is recorded as 26.185. The first *four* digits are exact or positive numbers and the last digit is estimated. Thus, it can be seen that the number of digits recorded for a measurement is an indication of the accuracy of the measurement. There will always be an indeterminable amount of error in a measured quantity, and an exact number will never be reached.

21. SIGNIFICANT DIGITS

The accuracy of computed solutions can be no greater than the accuracy of the data used in the computations. A method commonly used to indicate the degree of accuracy of computations is the use of *significant digits*. The number of decimal places in a number does *not* indicate the number of significant digits in the number.

All the digits in a number are significant except zero, which may or may not be significant.

A zero is significant if it is used as a placeholder following a decimal, as in 0.004, 0.062, and 0.0002.

A zero is not significant if it occurs at the end of a measured number unless information is available that indicates that it is. For example, if a distance is recorded as 260 ft, the zero is significant if the distance is measured to the nearest foot but not significant if measured to the nearest 10 ft.

Zeros following a decimal point are usually considered significant. For example, a measurement recorded as 216.00 ft would indicate that the zeros are significant.

A zero is always significant when it occurs between two other digits as in 306.

In general, the following rules can be applied.

rule 1: Initial zeros are not significant, as in 0.0065.

rule 2: Final zeros are not significant unless they follow a decimal point, as in 22.00.

rule 3: Zeros between two other digits are significant, as in 30,605.

Example 1.27

Write the number of significant digits in each of the following numbers.

(a) 0.0006

(b) 0.0020

(c) 0.0301

(d) 0.036

(e) 16.50

(f) 300

(g) 369.5

(h) 603

Solution

(a) 1

(b) 2

(c) 3

(d) 2

(e) 4

(f) 1

(g) 4

(h) 3

Example 1.28

Round off each of the following numbers to three significant digits.

(a) 0.00606

(b) 0.05078

(c) 2.002

(d) 2.465

(e) 4.255

(f) 36,374

(g) 286,501

(h) 325,496

Solution

(a) 0.00606

(b) 0.0508

(c) 2.00

(d) 2.46

(e) 4.26

(f) 36,400

(g) 287, 000

(h) 325,000

Example 1.29

Round off each of the following numbers to one significant digit.

(a) 0.0056

(b) 0.0719

(c) 0.449

(d) 3.448

(e) 16.32

(f) 251

(g) 56,325

(h) 100,788

Solution

(a) 0.006

(b) 0.07

(c) 0.4

(d) 3

(e) 20

(f) 300

(g) 60,000

(h) 100,000

22. COMPUTATIONS WITH APPROXIMATE DATA

Suppose a rectangle is measured as 14.2 ft long and 12.6 ft wide. Students not familiar with computations using measured quantities might be tempted to say that the area of the rectangle is (14.2 ft)(12.6 ft) = 178.92 ft^2, but they would indicate false accuracy. The result should be rounded off to the same number of significant digits as in the measured quantities. A more accurate measurement is 179 ft^2.

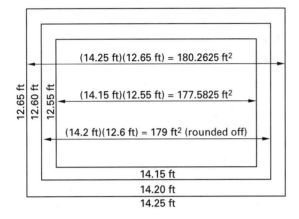

Figure 1.3 *Computations with Approximate Data*

The length was measured to the nearest tenth of a foot and indicates that the exact distance lies between 14.15 ft and 14.25 ft. Likewise, the exact distance for the width would lie between 12.55 ft and 12.65 ft. Using the lower limits of the measurements the area is (14.15 ft)(12.55 ft) = 177.5825 ft^2, and using the upper limits the area is (14.25 ft^2)(12.55 ft^2) = 180.2625 ft^2. The more accurate result lies somewhere between the two measurements.

In multiplying (or dividing) two approximate numbers, the product (or quotient) should not contain more significant digits than the number containing the fewer significant digits.

In adding numbers, the accuracy of values is governed by the number of places of digits rather than the number of significant digits. However, when two quantities are added together that were measured to a different degree of accuracy, the accuracy of the result cannot be greater than that of the quantity of least accuracy. For instance, in summing up earthwork, a quantity of 26.5 cubic yards added to a quantity of 156 cubic yards would give a result of 182 cubic yards and not 182.5 cubic yards.

Example 1.30

Multiply the following approximate numbers.

$$(246.34)(3.12)$$

Solution

$$(246.34)(3.12) = 768.5808$$

$$\approx 769 \quad \text{[three significant digits]}$$

23. SCIENTIFIC NOTATION

Scientists often work with very large or very small numbers. For instance, the distance from the earth to the sun is 93,000,000 mi, and the coefficient of linear expansion of steel is 0.0000065. In *scientific notation* these numbers are written 9.3×10^7 and 6.5×10^{-6}.

A number expressed in scientific notation is expressed as the product of two numbers, the first number having a value between 1 and 10 and the second number being a power of 10. Thus, 6,500,000 becomes 6.5×10^6 and 0.0000078 becomes 7.8×10^{-6}.

Example 1.31

Express 128,000 in scientific notation.

Solution

Place a caret to the right of the first significant digit as follows: $1_\wedge 28000$. Starting at the caret, count the number of places to the decimal point. Counting 5 places to the right to the decimal point, the number 1.28000 must be multiplied by 10^5 to equal 128,000. Therefore, $128{,}000 = 1.28 \times 10^5$.

Example 1.32

Express 0.000433 in scientific notation.

Solution

Place a caret to the right of the first significant digit as follows: $0.0004_\wedge 33$. Starting at the caret, count four places to the left to the decimal point. Then 4.33 must be multiplied by 10^{-4} to equal 0.000433. Therefore, $0.000433 = 4.33 \times 10^{-4}$ in scientific notation.

Where a number is expressed in scientific notation, the number of digits in the first factor should equal the number of significant digits of the number. For example, the number 4.23×10^6 has three significant digits.

Example 1.33

Express each of the following numbers in scientific notation.

(a) 0.00042

(b) 0.563

(c) 7.73

(d) 55,200

(e) 856,000

(f) 1,500,000

Solution

(a) 4.2×10^{-4}

(b) 5.63×10^{-1}

(c) 7.73×10^0

(d) 5.52×10^4

(e) 8.56×10^5

(f) 1.5×10^6

24. SQUARES AND SQUARE ROOTS

To say "3 squared equals 9" means that 3 multiplied by itself equals 9.

To say "the square root of 9 is 3" means that 3 is the number that produces 9 when multiplied by itself. The symbol $\sqrt{}$ is used to represent "the square root of." The symbol is known as the *radical sign*, or simply the *radical*. Thus,

$$(1)^2 = 1, \text{ and } \sqrt{1} = 1$$
$$(2)^2 = 4, \text{ and } \sqrt{4} = 2$$
$$(3)^2 = 9, \text{ and } \sqrt{9} = 3$$

$$(4)^2 = 16, \text{ and } \sqrt{16} = 4$$
$$(5)^2 = 25, \text{ and } \sqrt{25} = 5$$
$$(6)^2 = 36, \text{ and } \sqrt{36} = 6$$
$$(7)^2 = 49, \text{ and } \sqrt{49} = 7$$
$$(8)^2 = 64, \text{ and } \sqrt{64} = 8$$
$$(9)^2 = 81, \text{ and } \sqrt{81} = 9$$
$$(10)^2 = 100, \text{ and } \sqrt{100} = 10$$

25. THE PYTHAGOREAN THEOREM

The Greek mathematician *Pythagoras*, who lived some 500 years before Christ, is given credit for a very important theorem that is widely used today. The *Pythagorean theorem* states: In a right triangle, the square of the hypotenuse is equal to the sum of the squares of the other two sides.

If the base is called b, the altitude a, and the hypotenuse c, then

$$c^2 = a^2 + b^2$$
$$a^2 = c^2 - b^2$$
$$b^2 = c^2 - a^2$$
$$c = \sqrt{a^2 + b^2}$$
$$a = \sqrt{c^2 - b^2}$$
$$b = \sqrt{c^2 - a^2}$$

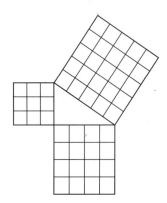

Figure 1.4 *Pythagorean Theorem*

Example 1.34

Find the hypotenuse, c, of the triangle shown.

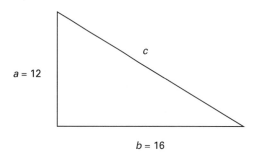

Solution

$$c^2 = a^2 + b^2$$
$$c = \sqrt{a^2 + b^2} = \sqrt{(12)^2 + (6)^2}$$
$$= \sqrt{144 + 256} = \sqrt{400} = 20$$

Example 1.35

A city lot forms a right triangle with the hypotenuse 100 ft and one side 60 ft. Find the other side.

Solution

$$b = \sqrt{c^2 - a^2}$$
$$= \sqrt{(100 \text{ ft})^2 - (60 \text{ ft})^2} = 80 \text{ ft}$$

26. RATIO AND PROPORTION

A *ratio* is a comparison of two quantities. It is the quotient of the first quantity divided by the second. The ratio of 1 to 2 is $\frac{1}{2}$. The ratio of 2 to 1 is $\frac{2}{1}$. A ratio can be stated in four ways: the ratio of 2 to 4, 2:4, $2 \div 4$, or $\frac{2}{4}$. Thus, a ratio is a fraction and can be treated as a fraction.

A *proportion* is a statement of equality between two ratios. Since $\frac{2}{4}$ and $\frac{1}{2}$ have the same value, the equality $\frac{2}{4} = \frac{1}{2}$ is a proportion. This can also be written as 2:4 = 1:2. In the proportion $a:b = c:d$, the terms a, b, c and d are *terms* of the proportion. The first and last terms are referred to as the *extremes*, and the two middle terms are referred to as the *means*.

In a proportion, the product of the extremes is equal to the product of the means. In the proportion $a:b = c:d$, $ad = bc$. In the proportion 3:4 = 6:8, $3 \times 8 = 4 \times 6$. In the proportion $1/2 = 5/10$, $2 \times 5 = 1 \times 10$. (This is sometimes called *cross-multiplying*.)

In the proportion $x{:}3 = 4{:}6$, $6x = 12$ and $x = 2$.

Example 1.36

Find the value of x.

(a) $x{:}4 = 2{:}8$.

(b) $2{:}x = 3{:}9$

(c) $3{:}6 = x{:}2$

(d) $\dfrac{4}{12} = \dfrac{x}{6}$

(e) $\dfrac{9}{6} = \dfrac{12}{x}$

Solution

(a) $x{:}4 = 2{:}8$; $8x = 8$; $x = 1$

(b) $2{:}x = 3{:}9$; $x = 18$; $x = 6$

(c) $3{:}6 = x{:}2$; $6x = 6$; $x = 1$

(d) $\dfrac{4}{12} = \dfrac{x}{6}$; $12x = 24$; $x = 2$

(e) $\dfrac{9}{6} = \dfrac{12}{x}$; $9x = 72$; $x = 8$

PRACTICE PROBLEMS

1. Change the given fraction to an equivalent fraction having the higher denominator indicated.

Examples: (1) $\dfrac{2}{5} = \dfrac{10}{25}$;   (2) $\dfrac{2}{3} = \dfrac{6}{9}$

(a) $\dfrac{1}{4} = \dfrac{x}{16}$ (b) $\dfrac{1}{6} = \dfrac{x}{24}$ (c) $\dfrac{3}{5} = \dfrac{x}{15}$ (d) $\dfrac{3}{4} = \dfrac{x}{16}$

(e) $\dfrac{3}{8} = \dfrac{x}{32}$ (f) $\dfrac{1}{3} = \dfrac{x}{15}$ (g) $\dfrac{2}{3} = \dfrac{x}{18}$ (h) $\dfrac{2}{7} = \dfrac{x}{21}$

(i) $\dfrac{5}{8} = \dfrac{x}{24}$ (j) $\dfrac{8}{3} = \dfrac{x}{9}$ (k) $\dfrac{7}{8} = \dfrac{x}{32}$ (l) $\dfrac{1}{5} = \dfrac{x}{25}$

(m) $\dfrac{4}{5} = \dfrac{x}{30}$ (n) $\dfrac{5}{8} = \dfrac{x}{32}$ (o) $\dfrac{1}{6} = \dfrac{x}{30}$ (p) $\dfrac{3}{9} = \dfrac{x}{81}$

(q) $\dfrac{2}{9} = \dfrac{x}{45}$ (r) $\dfrac{3}{16} = \dfrac{x}{48}$ (s) $\dfrac{5}{12} = \dfrac{x}{72}$ (t) $\dfrac{3}{7} = \dfrac{x}{56}$

2. Reduce the following fractions to lowest terms.

Examples: (1) $\dfrac{6}{8} = \dfrac{3}{4}$;   (2) $\dfrac{8}{12} = \dfrac{2}{3}$

(a) $\dfrac{2}{8}$ (b) $\dfrac{12}{16}$ (c) $\dfrac{21}{42}$ (d) $\dfrac{4}{10}$

(e) $\dfrac{12}{32}$ (f) $\dfrac{27}{45}$ (g) $\dfrac{6}{54}$ (h) $\dfrac{9}{16}$

(i) $\dfrac{30}{45}$ (j) $\dfrac{18}{54}$ (k) $\dfrac{36}{72}$ (l) $\dfrac{48}{64}$

(m) $\dfrac{14}{28}$ (n) $\dfrac{9}{27}$ (o) $\dfrac{12}{16}$ (p) $\dfrac{24}{32}$

(q) $\dfrac{54}{81}$ (r) $\dfrac{17}{51}$ (s) $\dfrac{27}{81}$ (t) $\dfrac{25}{35}$

3. Change the following improper fractions to equivalent mixed numbers reduced to lowest terms.

Examples: (1) $\dfrac{16}{3} = 5\tfrac{1}{3}$;   (2) $\dfrac{8}{6} = 1\tfrac{2}{6} = 1\tfrac{1}{3}$

(a) $\dfrac{15}{12}$ (b) $\dfrac{25}{4}$ (c) $\dfrac{19}{3}$ (d) $\dfrac{48}{15}$

(e) $\dfrac{30}{9}$ (f) $\dfrac{15}{4}$ (g) $\dfrac{50}{7}$ (h) $\dfrac{71}{12}$

(i) $\dfrac{77}{44}$ (j) $\dfrac{51}{12}$ (k) $\dfrac{64}{15}$ (l) $\dfrac{71}{56}$

(m) $\dfrac{71}{12}$ (n) $\dfrac{43}{5}$ (o) $\dfrac{71}{9}$ (p) $\dfrac{70}{8}$

(q) $\dfrac{77}{16}$ (r) $\dfrac{84}{13}$ (s) $\dfrac{78}{5}$ (t) $\dfrac{87}{15}$

4. Change the following mixed numbers to improper fractions.

Examples: (1) $4\tfrac{2}{3} = \dfrac{14}{3}$;   (2) $3\tfrac{3}{8} = \dfrac{27}{8}$

(a) $2\tfrac{1}{2}$ (b) $4\tfrac{7}{8}$ (c) $2\tfrac{4}{5}$ (d) $3\tfrac{1}{8}$

(e) $2\tfrac{5}{8}$ (f) $5\tfrac{3}{7}$ (g) $3\tfrac{7}{9}$ (h) $5\tfrac{2}{7}$

(i) $3\tfrac{5}{9}$ (j) $5\tfrac{3}{8}$ (k) $7\tfrac{8}{9}$ (l) $4\tfrac{5}{6}$

(m) $8\tfrac{3}{5}$ (n) $3\tfrac{4}{7}$ (o) $8\tfrac{1}{3}$ (p) $5\tfrac{2}{7}$

(q) $9\tfrac{1}{8}$ (r) $3\tfrac{3}{16}$ (s) $4\tfrac{7}{9}$ (t) $7\tfrac{7}{64}$

5. Find the prime factors.

Examples:

(1)
$$
\begin{array}{r}
2\overline{)180} \\
2\overline{)\ 90} \\
3\overline{)\ 45} \\
3\overline{)\ 15} \\
5\overline{)\ \ 5} \\
\overline{1}
\end{array}
$$

(2)
$$
\begin{array}{r}
2\overline{)2310} \\
3\overline{)1155} \\
5\overline{)\ 385} \\
7\overline{)\ \ 77} \\
11\overline{)\ \ 11} \\
\overline{1}
\end{array}
$$

$180 = 2 \times 2 \times 3 \times 3 \times 5$ $\qquad$ $2310 = 2 \times 3 \times 5 \times 7 \times 11$

(a) 24 $\qquad$ (b) 210 $\qquad$ (c) 126

(d) 1323 $\qquad$ (e) 1125 $\qquad$ (f) 3234

(g) 5775 $\qquad$ (h) 4851 $\qquad$ (i) 15,015

6. Find the least common multiple (LCM) of the following sets of numbers.

Example: 18, 24, 45

Solution: $18 = 2 \times 3 \times 3$

$24 = 2 \times 2 \times 2 \times 3$

$45 = 3 \times 3 \times 5$

$\text{LCM} = 2 \times 2 \times 2 \times 3 \times 3 \times 5 = 360$

(a) 4, 6, 12 $\qquad$ (b) 10, 15, 20

(c) 24, 35, 60 $\qquad$ (d) 16, 45, 75

(e) 45, 54, 75 $\qquad$ (f) 30, 42, 60

7. Find the least common denominator (LCD) of the following fractions and add or subtract as indicated. Express the answers in lowest terms.

Examples:

(1) $\dfrac{3}{8} + \dfrac{5}{8} + \dfrac{7}{8} = \dfrac{15}{8} = 1\dfrac{7}{8}$

(2) $\dfrac{1}{4} + \dfrac{1}{6} + \dfrac{1}{9} = \dfrac{9}{36} + \dfrac{6}{36} + \dfrac{4}{36} = \dfrac{19}{36}$

(a) $\dfrac{1}{2} + \dfrac{2}{3} + \dfrac{3}{5}$

(b) $\dfrac{3}{8} + \dfrac{1}{4} + \dfrac{2}{3}$

(c) $\dfrac{9}{16} + \dfrac{5}{12} - \dfrac{1}{3}$

(d) $\dfrac{1}{4} + \dfrac{3}{8} + \dfrac{2}{3} + \dfrac{5}{6}$

(e) $\dfrac{5}{8} + \dfrac{3}{16} + \dfrac{3}{4} - \dfrac{5}{6}$

(f) $6\dfrac{2}{3} + 4\dfrac{5}{8}$

(g) $5\dfrac{7}{8} - 3\dfrac{1}{9}$

(h) $5\dfrac{4}{9} + 2\dfrac{7}{12} - 3\dfrac{1}{8}$

(i) $2\dfrac{2}{3} + 3\dfrac{5}{16} - 1\dfrac{1}{16}$

(j) $2\dfrac{5}{8} + 3\dfrac{1}{4} - \dfrac{15}{16}$

8. Multiply the following and express the answers in lowest terms.

Examples:

(1) $\dfrac{2}{3} \times \dfrac{1}{4} = \dfrac{2}{12} = \dfrac{1}{6}$

(2) $\dfrac{3}{5} \times \dfrac{1}{3} \times \dfrac{15}{16} = \dfrac{3}{16}$

(a) $\dfrac{3}{8} \times \dfrac{2}{3}$

(b) $\dfrac{3}{5} \times \dfrac{2}{3} \times \dfrac{3}{8}$

(c) $\dfrac{3}{4} \times \dfrac{2}{5} \times \dfrac{1}{2}$

(d) $\dfrac{3}{8} \times \dfrac{1}{3} \times \dfrac{3}{4}$

(e) $\dfrac{2}{3} \times \dfrac{4}{5} \times \dfrac{1}{4} \times \dfrac{5}{8} \times \dfrac{9}{16}$

(f) $\dfrac{3}{5} \times \dfrac{7}{9} \times \dfrac{20}{21} \times \dfrac{3}{8} \times \dfrac{1}{12}$

(g) $5\dfrac{3}{5} \times 2\dfrac{3}{4} \times 3\dfrac{3}{7}$

(h) $3\dfrac{1}{4} \times 2\dfrac{2}{5} \times 4\dfrac{2}{3}$

(i) $1\dfrac{15}{16} \times 2\dfrac{5}{8} \times 3\dfrac{1}{2}$

(j) $2\dfrac{4}{9} \times 1\dfrac{8}{9} \times 3\dfrac{1}{3}$

9. Divide the following and show the answers in lowest terms.

Examples:

(1) $3 \div \dfrac{2}{3} = 3 \times \dfrac{3}{2} = 4\dfrac{1}{2}$

(2) $\dfrac{2}{3} \div 4 = \dfrac{2}{3} \times \dfrac{1}{4} = \dfrac{1}{6}$

(a) $3 \div \dfrac{3}{5}$

(b) $5 \div \dfrac{5}{6}$

(c) $3 \div \dfrac{5}{8}$

(d) $\dfrac{3}{4} \div \dfrac{5}{8}$

(e) $\dfrac{5}{6} \div \dfrac{2}{3}$

(f) $\dfrac{3}{8} \div 5$

(g) $\dfrac{3}{8} \div 3$

(h) $\dfrac{1}{8} \div 5$

(i) $\dfrac{4}{5} \div \dfrac{1}{4}$

(j) $\dfrac{2}{3} \div \dfrac{3}{4}$

(c) 1.0066

(d) 0.400

(e) 0.20200

(f) 1.0001

(g) 2037.566

(h) 46,388.07

(i) 2000.55

(j) 10.001

10. Simplify the following by cancellation.

Examples:

(1) $\dfrac{(\overset{1}{\cancel{5}})(\overset{2}{\cancel{12}})}{(\underset{1}{\cancel{6}})(\underset{3}{\cancel{15}})} = \dfrac{2}{3}$

(2) $\dfrac{(\overset{1}{\cancel{15}})(\overset{3}{\cancel{12}})(\overset{1}{\cancel{4}})}{(\underset{1}{\cancel{4}})(\underset{2}{\cancel{30}})(\underset{2}{\cancel{8}})} = \dfrac{3}{4}$

(a) $\dfrac{(6)(10)}{(5)(3)}$

(b) $\dfrac{(6)(15)(8)}{(3)(4)(5)}$

(c) $\dfrac{(9)(12)(15)}{(3)(8)(5)}$

(d) $\dfrac{(21)(32)(45)}{(7)(8)(15)}$

(e) $\dfrac{(14)(15)(24)}{(30)(28)(8)}$

(f) $\dfrac{(27)(45)(64)}{(15)(9)(16)}$

(g) $\dfrac{(15)(7)(8)(9)}{(21)(15)(5)}$

(h) $\dfrac{(72)(24)(15)(18)}{(8)(12)(5)(9)}$

12. Find the following products or quotients as indicated.

Examples:

(1) $537.26 \times 100 = 53,726$

(2) $\dfrac{6666}{0.02} = \dfrac{3333}{0.01} = 333,300$

(a) 12×10

(b) 23.16×100

(c) $\dfrac{0.3695}{100}$

(d) $\dfrac{183,914}{0.002}$

(e) $\dfrac{9823}{0.01}$

(f) $\dfrac{0.3695}{0.001}$

(g) $\dfrac{274.767}{0.00001}$

(h) $\dfrac{6.37}{10,000}$

(i) $\dfrac{44.444}{0.04}$

(j) $\dfrac{0.032}{0.002}$

(k) $\dfrac{808.0}{0.004}$

(l) $\dfrac{0.003}{0.03}$

(m) 0.0063×0.2

(n) 0.272×0.002

(o) 32.42×0.002

(p) 0.0058×1000

(q) 11.11×0.01

(r) 0.000412×10

(s) $22.2 \div 0.0002$

(t) $0.000033 \div 0.03$

11. Write the following numbers in words.

Examples:

(1) 476.232: four hundred seventy-six and two hundred thirty-two thousandths

(2) 0.0521: five hundred twenty-one ten-thousandths

(a) 46.32

(b) 132.036

13. Change the following to percents.

Examples: (1) $0.06 = 6\%$; (2) $5.45 = 545\%$

(a) 0.04

(b) 0.15

(c) 0.004

(d) 1.37

(e) 0.0002

(f) 0.275

(g) $2\dfrac{1}{2}$

(h) $1\dfrac{1}{4}$

(i) $1\dfrac{1}{10}$

14. Change the following numbers from percents to decimal fractions.

Example: 0.6% = 0.006

(a) 13%   (b) 0.007%   (c) 125%

(d) $33\frac{1}{3}\%$ (e) 250%   (f) 0.5%

15. Change the following fractions from common fractions to percents.

Example: $\frac{1}{2} = 50\%$

(a) $\frac{1}{3}$ (b) $1\frac{1}{2}$ (c) $2\frac{2}{3}$

(d) $1\frac{1}{4}$ (e) $\frac{3}{4}$ (f) $\frac{2}{3}$

(g) $1\frac{3}{8}$ (h) $2\frac{1}{8}$ (i) $3\frac{1}{10}$

16. Find the following percents.

Examples: (1) 50 is 50% of 100;   (2) 6 is 25% of 24

(a) 6 is $x\%$ of 12

(b) 20 is $x\%$ of 80

(c) 60 is $x\%$ of 90

(d) 8 is $x\%$ of 32

(e) 16 is $x\%$ of 64

(f) 5 is $x\%$ of 40

(g) 44 is $x\%$ of 22

(h) 36 is $x\%$ of 48

(i) 50 is $x\%$ of 20

(j) 60 is $x\%$ of 40

17. Find the following numbers.

Example: 70 is 35% of 200 (70 ÷ 0.35 = 200)

(a) 8 is 50% of x

(b) 15 is 25% of x

(c) 60 is $66\frac{2}{3}\%$ of x

(d) 30 is 75% of x

(e) 8 is $12\frac{1}{2}\%$ of x

(f) 12 is 2% of x

(g) $\frac{2}{3}$ is 75% of x

(h) $\frac{1}{2}$ is $66\frac{2}{3}\%$ of x

(i) $\frac{1}{8}$ is $12\frac{1}{2}\%$ of x

(j) $\frac{3}{4}$ is 25% of x

18. Round off each of the following numbers to the nearest tenth.

Examples: (1) 2.56 = <u>2.6</u>;   (2) 4.62 = <u>4.6</u>;
           (3) 6.55 = <u>6.6</u>

(a) 2.09   (b) 2.48   (c) 3.429

(d) 3.651   (e) 4.272   (f) 5.552

(g) 6.77   (h) 8.009   (i) 8.039

19. Round off each of the following numbers to the nearest hundredth.

(a) 1.455   (b) 2.295   (c) 2.599

(d) 3.001   (e) 4.009   (f) 5.455

(g) 5.486   (h) 5.494   (i) 7.365

20. Round off each of the following numbers to the nearest hundred.

(a) 5962   (b) 8951   (c) 9950

(d) 11,449   (e) 11,990   (f) 15,966

(g) 21,009   (h) 21,500   (i) 22,009

(j) 30,499   (k) 45,249   (l) 50,499

21. Write the number of significant digits in each of the following numbers.

Examples: (1) 2.506 = <u>4</u>;   (2) 0.0032 = <u>2</u>;
           (3) 2.00 = <u>3</u>

(a) 0.002   (b) 0.046   (c) 2.407

(d) 3.00   (e) 50.00   (f) 245.0

(g) 348.4   (h) 5651   (i) 7000

22. Round off each of the following numbers to three significant digits.

(a) 0.03606   (b) 3.585   (c) 12.448

(d) 24,100   (e) 323,202   (f) 447,501

23. Round off each of the following numbers to one significant digit.

(a) 0.076   (b) 0.451   (c) 1.449

(d) 47.2   (e) 1389   (f) 200,501

24. Express each of the following numbers in scientific notation.

(a) 0.00004   (b) 0.00015   (c) 0.0045

(d) 0.011   (e) 0.549   (f) 848

(g) 10,800   (h) 30,376   (i) 132,000

25. Find the missing sides of the following right triangles.

Example: Given $a = 6$ and $c = 10$, find b.

Solution: $b = \sqrt{c^2 - a^2} = \sqrt{10^2 - 6^2} = 8$

(a) Given $a = 3$ and $b = 4$, find c.

(b) Given $a = 15$ and $c = 17$, find b.

(c) Given $b = 9$ and $c = 15$, find a.

(d) Given $a = 12$ and $c = 13$, find b.

(e) Given $b = 7$ and $c = 25$, find a.

(f) A guy wire is attached to a TV-antenna tower 120 ft above ground. It is anchored 50 ft from the tower on level ground. What is the length of the guy wire?

(g) The foundation for a building is to be 80 ft by 120 ft. The builder, in staking the building, wants to check to see if it is square. Compute the length of the diagonal.

(h) Crossroads is 14 mi due east of Pumpkin Hollow; Ghostown is 27 mi due south of Crossroads. How far is Pumpkin Hollow from Ghostown? (Calculate to the nearest mile.)

26. Find the value of x that makes the following ratios correct.

Examples:

(1) $x{:}4 = 3{:}6$, $6x = 12$, $x = 2$

(2) $\dfrac{x}{3} = \dfrac{2}{6}$, $6x = 6$, $x = 1$

(a) $x{:}4 = 12{:}144$ (b) $x{:}3 = 18{:}6$

(c) $x{:}6 = 12{:}72$ (d) $9{:}x = 27{:}3$

(e) $3{:}x = 4{:}24$ (f) $9{:}6 = 15{:}x$

(g) $12{:}144 = 3{:}x$ (h) $x{:}y = a{:}b$

(i) $x{:}0.5 = 0.8{:}0.4$ (j) $\dfrac{12}{x} = \dfrac{16}{4}$

(k) $\dfrac{x}{8} = \dfrac{12}{6}$ (l) $\dfrac{2.5}{7.5} = \dfrac{3}{x}$

SOLUTIONS

1. (a) $\dfrac{4}{16}$ (b) $\dfrac{4}{24}$ (c) $\dfrac{9}{15}$ (d) $\dfrac{12}{16}$

(e) $\dfrac{12}{32}$ (f) $\dfrac{5}{15}$ (g) $\dfrac{12}{18}$ (h) $\dfrac{6}{21}$

(i) $\dfrac{15}{24}$ (j) $\dfrac{24}{9}$ (k) $\dfrac{28}{32}$ (l) $\dfrac{5}{25}$

(m) $\dfrac{24}{30}$ (n) $\dfrac{20}{32}$ (o) $\dfrac{5}{30}$ (p) $\dfrac{27}{81}$

(q) $\dfrac{10}{45}$ (r) $\dfrac{9}{48}$ (s) $\dfrac{30}{72}$ (t) $\dfrac{24}{56}$

2. (a) $\dfrac{1}{4}$ (b) $\dfrac{3}{4}$ (c) $\dfrac{1}{2}$ (d) $\dfrac{2}{5}$

(e) $\dfrac{3}{8}$ (f) $\dfrac{3}{5}$ (g) $\dfrac{1}{9}$ (h) $\dfrac{9}{16}$

(i) $\dfrac{2}{3}$ (j) $\dfrac{1}{3}$ (k) $\dfrac{1}{2}$ (l) $\dfrac{3}{4}$

(m) $\dfrac{1}{2}$ (n) $\dfrac{1}{3}$ (o) $\dfrac{3}{4}$ (p) $\dfrac{3}{4}$

(q) $\dfrac{2}{3}$ (r) $\dfrac{1}{3}$ (s) $\dfrac{1}{3}$ (t) $\dfrac{5}{7}$

3. (a) $1\frac{1}{4}$ (b) $6\frac{1}{4}$ (c) $6\frac{1}{3}$ (d) $3\frac{1}{5}$

(e) $3\frac{1}{3}$ (f) $3\frac{3}{4}$ (g) $7\frac{1}{7}$ (h) $5\frac{11}{12}$

(i) $1\frac{3}{4}$ (j) $4\frac{1}{4}$ (k) $4\frac{4}{15}$ (l) $1\frac{15}{16}$

(m) $5\frac{11}{12}$ (n) $8\frac{3}{5}$ (o) $7\frac{8}{9}$ (p) $8\frac{3}{4}$

(q) $4\frac{13}{16}$ (r) $6\frac{6}{13}$ (s) $15\frac{3}{5}$ (t) $5\frac{4}{5}$

4. (a) $\dfrac{5}{2}$ (b) $\dfrac{39}{8}$ (c) $\dfrac{14}{5}$ (d) $\dfrac{25}{8}$

(e) $\dfrac{21}{8}$ (f) $\dfrac{38}{7}$ (g) $\dfrac{34}{9}$ (h) $\dfrac{37}{7}$

(i) $\dfrac{32}{9}$ (j) $\dfrac{43}{8}$ (k) $\dfrac{71}{9}$ (l) $\dfrac{29}{6}$

(m) $\dfrac{43}{5}$ (n) $\dfrac{25}{7}$ (o) $\dfrac{25}{3}$ (p) $\dfrac{37}{7}$

(q) $\dfrac{73}{8}$ (r) $\dfrac{51}{16}$ (s) $\dfrac{43}{9}$ (t) $\dfrac{455}{64}$

5. (a) $2\underline{)\,24}$

$2\underline{)\,12}$

$2\underline{)\,6}$

$3\underline{)\,3}$

1

$24 = 2 \times 2 \times 2 \times 3$

ARITHMETIC AND MEASUREMENTS 1-17

Arithmetic

(b) $2\overline{)\,210}$
$3\overline{)\,105}$
$5\overline{)\,35}$
$7\overline{)\,7}$
1

$210 = 2 \times 3 \times 5 \times 7$

(c) $2\overline{)\,126}$
$3\overline{)\,63}$
$3\overline{)\,21}$
$7\overline{)\,7}$
1

$126 = 2 \times 3 \times 3 \times 7$

(d) $3\overline{)\,1323}$
$3\overline{)\,441}$
$3\overline{)\,147}$
$7\overline{)\,49}$
$7\overline{)\,7}$
1

$1323 = 3 \times 3 \times 3 \times 7 \times 7$

(e) $3\overline{)\,1125}$
$3\overline{)\,375}$
$5\overline{)\,125}$
$5\overline{)\,25}$
$5\overline{)\,5}$
1

$1125 = 3 \times 3 \times 5 \times 5 \times 5$

(f) $2\overline{)\,3234}$
$3\overline{)\,1617}$
$7\overline{)\,539}$
$7\overline{)\,77}$
$11\overline{)\,11}$
1

$3234 = 2 \times 3 \times 7 \times 7 \times 11$

(g) $3\overline{)\,5775}$
$5\overline{)\,1925}$
$5\overline{)\,385}$
$7\overline{)\,77}$
$11\overline{)\,11}$
1

$5775 = 3 \times 5 \times 5 \times 7 \times 11$

(h) $3\overline{)\,4851}$
$3\overline{)\,1617}$
$7\overline{)\,539}$
$7\overline{)\,77}$
$11\overline{)\,11}$
1

$4851 = 3 \times 3 \times 7 \times 7 \times 11$

(i) $3\overline{)\,15,015}$
$5\overline{)\,5005}$
$7\overline{)\,1001}$
$11\overline{)\,143}$
$13\overline{)\,13}$
1

$15,015 = 3 \times 5 \times 7 \times 11 \times 13$

6. (a) $4 = 2 \times 2$
$6 = 2 \times 3$
$12 = 2 \times 2 \times 3$
$\text{LCM} = 2 \times 2 \times 3 = 12$

(b) $10 = 2 \times 5$
$15 = 3 \times 5$
$20 = 2 \times 2 \times 5$
$\text{LCM} = 2 \times 2 \times 3 \times 5 = 60$

(c) $24 = 2 \times 2 \times 2 \times 3$
$35 = 5 \times 7$
$60 = 2 \times 2 \times 3 \times 5$
$\text{LCM} = 2 \times 2 \times 2 \times 3 \times 5 \times 7 = 840$

(d) $16 = 2 \times 2 \times 2 \times 2$
$45 = 3 \times 3 \times 5$
$75 = 3 \times 5 \times 5$
$\text{LCM} = 2 \times 2 \times 2 \times 2 \times 3 \times 3 \times 5 \times 5 = 3600$

(e) $45 = 3 \times 3 \times 5$
$54 = 2 \times 3 \times 3 \times 3$
$75 = 3 \times 5 \times 5$
$\text{LCM} = 2 \times 3 \times 3 \times 3 \times 5 \times 5 = 1350$

(f) $30 = 2 \times 3 \times 5$
$42 = 2 \times 3 \times 7$
$60 = 2 \times 2 \times 3 \times 5$
$\text{LCM} = 2 \times 2 \times 3 \times 5 \times 7 = 420$

PROFESSIONAL PUBLICATIONS, INC.

7. (a) $\dfrac{15}{30} + \dfrac{20}{30} + \dfrac{18}{30} = \dfrac{53}{30} = 1\frac{23}{30}$

(b) $\dfrac{9}{24} + \dfrac{6}{24} + \dfrac{16}{24} = \dfrac{31}{24} = 1\frac{7}{24}$

(c) $\dfrac{27}{48} + \dfrac{20}{48} - \dfrac{16}{48} = \dfrac{31}{48}$

(d) $\dfrac{6}{24} + \dfrac{9}{24} + \dfrac{16}{24} + \dfrac{20}{24} = \dfrac{51}{24} = 2\frac{1}{8}$

(e) $\dfrac{30}{48} + \dfrac{9}{48} + \dfrac{36}{48} - \dfrac{40}{48} = \dfrac{35}{48}$

(f) $6 + 4 + \dfrac{16}{24} + \dfrac{15}{24} = 11\frac{7}{24}$

(g) $5 - 3 + \dfrac{6}{72} - \dfrac{8}{72} = 2\frac{55}{72}$

(h) $5 + 2 - 3 + \dfrac{32}{72} + \dfrac{42}{72} - \dfrac{9}{72} = 4\frac{65}{72}$

(i) $2 + 3 - 1 + \dfrac{32}{48} + \dfrac{15}{48} - \dfrac{3}{48} = 4\frac{11}{12}$

(j) $2 + 3 + \dfrac{10}{16} + \dfrac{4}{16} - \dfrac{15}{16} = 4\frac{15}{16}$

8. (a) $\dfrac{6}{24} = \dfrac{1}{4}$

(b) $\dfrac{18}{120} = \dfrac{3}{20}$

(c) $\dfrac{6}{40} = \dfrac{3}{20}$

(d) $\dfrac{9}{96} = \dfrac{3}{32}$

(e) $\dfrac{360}{7680} = \dfrac{3}{64}$

(f) $\dfrac{1260}{90,720} = \dfrac{1}{72}$

(g) $\left(\dfrac{28}{5}\right)\left(\dfrac{11}{4}\right)\left(\dfrac{24}{7}\right) = \dfrac{7392}{140} = 52\frac{4}{5}$

(h) $\left(\dfrac{13}{4}\right)\left(\dfrac{12}{5}\right)\left(\dfrac{14}{3}\right) = \dfrac{2184}{60} = 36\frac{2}{5}$

(i) $\left(\dfrac{31}{16}\right)\left(\dfrac{21}{8}\right)\left(\dfrac{7}{2}\right) = \dfrac{4557}{256} = 17\frac{205}{256}$

(j) $\left(\dfrac{22}{9}\right)\left(\dfrac{17}{9}\right)\left(\dfrac{10}{3}\right) = \dfrac{3740}{243} = 15\frac{95}{243}$

9. (a) $(3)\left(\dfrac{5}{3}\right) = 5$

(b) $(5)\left(\dfrac{6}{5}\right) = 6$

(c) $(3)\left(\dfrac{8}{5}\right) = \dfrac{24}{5} = 4\frac{4}{5}$

(d) $\left(\dfrac{3}{4}\right)\left(\dfrac{8}{5}\right) = \dfrac{24}{20} = 1\frac{1}{5}$

(e) $\left(\dfrac{5}{6}\right)\left(\dfrac{3}{2}\right) = \dfrac{15}{12} = 1\frac{1}{4}$

(f) $\left(\dfrac{3}{8}\right)\left(\dfrac{1}{5}\right) = \dfrac{3}{40}$

(g) $\left(\dfrac{3}{8}\right)\left(\dfrac{1}{3}\right) = \dfrac{3}{24} = \dfrac{1}{8}$

(h) $\left(\dfrac{1}{8}\right)\left(\dfrac{1}{5}\right) = \dfrac{1}{40}$

(i) $\left(\dfrac{4}{5}\right)\left(\dfrac{4}{1}\right) = \dfrac{16}{5} = 3\frac{1}{5}$

(j) $\left(\dfrac{2}{3}\right)\left(\dfrac{4}{3}\right) = \dfrac{8}{9}$

10. (a) $\dfrac{(6)(10)}{(5)(3)} = \dfrac{(2)(2)}{(1)(1)} = 4$

(b) $\dfrac{(6)(15)(8)}{(3)(4)(5)} = \dfrac{(2)(3)(2)}{(1)(1)(1)} = 12$

(c) $\dfrac{(9)(12)(15)}{(3)(8)(5)} = \dfrac{(3)(3)(3)}{(1)(2)(1)} = 13\frac{1}{2}$

(d) $\dfrac{(21)(32)(45)}{(7)(8)(15)} = \dfrac{(3)(4)(9)}{(1)(1)(3)} = 36$

(e) $\dfrac{(14)(15)(24)}{(30)(28)(8)} = \dfrac{(1)(1)(3)}{(2)(2)(1)} = \dfrac{3}{4}$

(f) $\dfrac{(27)(45)(64)}{(15)(9)(16)} = \dfrac{(3)(3)(4)}{(1)(1)(1)} = 36$

(g) $\dfrac{(1)(1)(8)(9)}{(3)(1)(5)} = \dfrac{(1)(1)(8)(3)}{(1)(1)(5)} = 4\frac{4}{5}$

(h) $\dfrac{(72)(24)(15)(18)}{(8)(12)(5)(9)} = \dfrac{(9)(2)(3)(2)}{(1)(1)(1)(1)} = 108$

11. (a) forty-six and thirty-two hundredths

(b) one hundred thirty-two and thirty-six thousandths

(c) one and sixty-six ten-thousandths

(d) zero and four hundred thousandths

(e) zero and twenty thousand two hundred hundred-thousandths

(f) one and one ten-thousandth

(g) two thousand thirty-seven and five hundred sixty-six thousandths

(h) forty-six thousand three hundred eighty-eight and seven hundredths

(i) two thousand and fifty-five hundredths

(j) ten and one thousandth

12. (a) 120   (b) 2316
(c) 0.003695   (d) 431,957,000
(e) 982,300   (f) 369.5
(g) 27,476,700   (h) 0.000637
(i) 1111.1   (j) 16
(k) 202,000   (l) 0.1
(m) 0.00126   (n) 0.000544
(o) 0.06484   (p) 5.8
(q) 0.1111   (r) 0.00412
(s) 111,000   (t) 0.0011

13. (a) 4%   (b) 15%   (c) 0.4%
(d) 137%   (e) 0.02%   (f) 27.5%
(g) 250%   (h) 125%   (i) 110%

14. (a) 0.13   (b) 0.00007   (c) 1.25
(d) 0.333   (e) 2.50   (f) 0.005

15. (a) $33\frac{1}{3}\%$ (b) 150%   (c) $266\frac{2}{3}\%$
(d) 125%   (e) 75%   (f) $66\frac{2}{3}\%$
(g) 137.5%   (h) 212.5%   (i) 310%

16. (a) 50%   (b) 25%   (c) $66\frac{2}{3}\%$
(d) 25%   (e) 25%   (f) $12\frac{1}{2}\%$
(g) 200%   (h) 75%   (i) 250%
(j) 150%

17. (a) 16   (b) 60   (c) 90   (d) 40
(e) 64   (f) 600   (g) $\frac{8}{9}$ (h) $\frac{3}{4}$
(i) 1   (j) 3

18. (a) 2.1   (b) 2.5   (c) 3.4   (d) 3.7
(e) 4.3   (f) 5.6   (g) 6.8   (h) 8.0
(i) 8.0

19. (a) 1.46   (b) 2.30   (c) 2.60
(d) 3.00   (e) 4.01   (f) 5.46
(g) 5.49   (h) 5.49   (i) 7.36

20. (a) 6000   (b) 9000   (c) 10,000
(d) 11,400   (e) 12,000   (f) 16,000
(g) 21,000   (h) 21,500   (i) 22,000
(j) 30,500   (k) 45,200   (l) 50,500

21. (a) 1   (b) 2   (c) 4   (d) 3   (e) 4
(f) 4   (g) 4   (h) 4   (i) 1

22. (a) 0.0361   (b) 3.58   (c) 12.4
(d) 24,100   (e) 323,000   (f) 448,000

23. (a) 0.08   (b) 0.5   (c) 1
(d) 50   (e) 1000   (f) 200,000

24. (a) 4×10^{-5} (b) 1.5×10^{-4}
(c) 4.5×10^{-3} (d) 1.1×10^{-2}
(e) 5.49×10^{-1} (f) 8.48×10^2
(g) 1.08×10^4 (h) 3.0376×10^4
(i) 1.32×10^5

25. (a) $c = \sqrt{(3)^2+(4)^2} = 5$
(b) $b = \sqrt{(17)^2-(15)^2} = 8$
(c) $a = \sqrt{(15)^2-(9)^2} = 12$
(d) $b = \sqrt{(13)^2-(12)^2} = 5$
(e) $a = \sqrt{(25)^2-(7)^2} = 24$
(f) length $= \sqrt{(120)^2+(50)^2} = 130$ ft
(g) diagonal $= \sqrt{(120)^2+(80)^2} = 144.22$ ft
(h) distance $= \sqrt{(14)^2+(27)^2} = 30$ mi

26. (a) $144x = 48$, $x = \frac{1}{3}$ (b) $6x = 54$, $x = 9$
(c) $72x = 72$, $x = 1$ (d) $27x = 27$, $x = 1$
(e) $4x = 72$, $x = 18$ (f) $9x = 90$, $x = 10$
(g) $12x = 432$, $x = 36$ (h) $bx = ay$, $x = \frac{ay}{b}$
(i) $0.4x = 0.4$, $x = 1$ (j) $16x = 48$, $x = 3$
(k) $6x = 96$, $x = 16$ (l) $2.5x = 22.5$, $x = 9$

2 Geometry

1. DEFINITION

Geometry is a branch of mathematics that deals with the measurement, properties, and relationships of points, lines, angles, surfaces, and solids. The word "geometry" is derived from the words *geo*, meaning earth, and *metro*, meaning measure.

2. HISTORY

The Egyptians first used geometry in about 2500 B.C. because the seasonal overflowing of the Nile made it necessary to reestablish boundaries so that taxes could be levied and collected. In about 500 B.C. the Greeks began to develop information received from the Egyptians into the branch of mathematics we now know as geometry. By the 4th century A.D. they had developed arithmetic and geometry into separate branches of mathematical science.

3. POINTS AND LINES

Points and lines are undefined elements of geometry, yet everyone has some understanding of these terms. A *point* is understood to have no length, width, or thickness, and it indicates a location. A point is usually shown on paper as a small dot and is named with a capital letter such as A.

A *line* is considered to have length but not width or thickness. A line connecting two points is said to be a *straight line* if it does not curve. A straight line is usually designated by two points that it connects, such as AB. A curved line is a line of which no part is straight.

4. PARALLEL LINES

Two lines that lie in the same plane and do not intersect are called *parallel lines*.

5. ANGLE

There are many definitions of an *angle*. In geometry, an angle may be defined as the space between two lines diverging from a common point called the *vertex*. In trigonometry, an angle may be defined as the amount of rotation to bring one line into coincidence with another. In surveying, an angle may be defined as the difference in direction of two intersecting lines.

6. MEASURE OF ANGLES

The most common measure of an angle is the *degree*. It is defined as $1/360$ of a complete angle or turn. A circle can be divided into 360 equal arcs. If radii connect each end of these small arcs, the angle formed by the two radii measures 1 degree.

For closer measurement, the degree is divided into 60 equal parts, with one part measuring 1 minute. And for even closer measurement, the 1 minute angle is divided into 60 equal parts, with each part measuring 1 second.

The symbols used for degrees, minutes, and seconds are: degrees (°), minutes (′), and seconds (″). As an example, 36 degrees, 24 minutes, and 52 seconds is written as $36°24'52''$.

A protractor can be used to measure angles on paper; a transit and theodolite measure angles in the field.

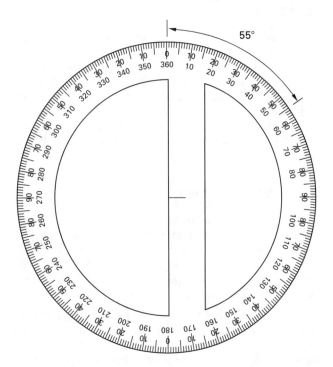

Figure 2.1 *Angle Measurements*

7. ACUTE ANGLE

An *acute angle* is an angle of less than 90°.

8. RIGHT ANGLE

A *right angle* is an angle of 90°.

9. OBTUSE ANGLE

An *obtuse angle* is an angle of more than 90° and less than 180°.

10. STRAIGHT ANGLE

A *straight angle* is an angle of 180°.

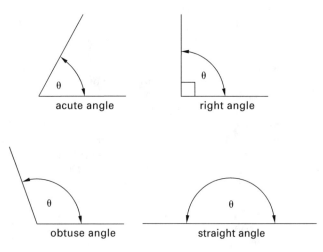

Figure 2.2 *Angles*

11. COMPLEMENTARY ANGLES

Two angles are said to be *complementary* if their sum is 90°.

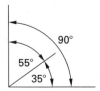

Figure 2.3 *Complementary Angles*

12. SUPPLEMENTARY ANGLES

Two angles are said to be *supplementary* if their sum is 180°.

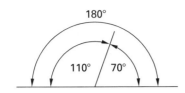

Figure 2.4 *Supplementary Angles*

13. TRANSVERSAL

A line that cuts two or more lines is called a *transversal*.

14. ALTERNATE INTERIOR ANGLES

When two parallel lines are cut by a transversal, the *alternate interior angles* are equal.

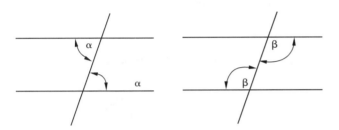

Figure 2.5 *Alternate Interior Angles*

15. ALTERNATE EXTERIOR ANGLES

When two parallel lines are cut by a transversal, the *alternate exterior angles* are equal.

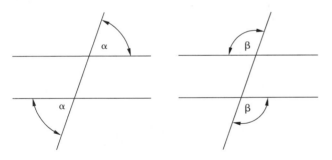

Figure 2.6 *Alternate Exterior Angles*

16. ADDING AND SUBTRACTING ANGLES

The surveying technician is often called on to add the measurements of two angles or to find the difference in the measurements of two angles. The procedures often involve "borrowing" 1 degree and converting it to 60 minutes and "borrowing" 1 minute and converting it to 60 seconds.

Example 2.1

Add $24°40'$ and $16°30'$.

Solution

When the number of minutes in the sum is 60 or more, $60'$ are subtracted from the minutes column and $1°$ is added to the degree column. Likewise, when the number of seconds in the sum is 60 or more, $60''$ are subtracted from the seconds column and $1'$ is added to the minutes column.

$$\begin{array}{r} 24°40' \\ + 16°30' \\ \hline 40°70' = 41°10' \end{array}$$

Example 2.2

Add $3°46'32''$ and $14°22'44''$.

Solution

$$\begin{array}{r} 32°46'32'' \\ + 14°22'44'' \\ \hline 46°68'76'' = 46°69'16'' = 47°09'16'' \end{array}$$

Example 2.3

Find the difference between $60°12'$ and $40°32'$.

Solution

Before finding the difference, convert $60°12'$ to its equivalent $59°72'$. "Borrow" $1°$ from $60°$ leaving $59°$ and add its equivalent, $60'$ to $12'$, making $72'$.

$$\begin{array}{rr} 60°12' & 59°72' \\ -40°32' & -40°32' \\ & \hline & 19°40' \end{array}$$

Example 2.4

Find the difference between $96°08'14''$ and $52°33'50''$.

Solution

One degree was borrowed from the degree column, leaving $95°$ and making the minutes column $68'$. One minute was borrowed from the minutes column, leaving $67'$ and making the seconds column $74''$.

When a number of angle measurements are added, as is common in surveying, each column (degrees, minutes, and seconds) is added separately and recorded. If the sum of either the minutes column or the seconds column, or both, is 60 or more, the same procedure is followed.

$$\begin{array}{rr} 96°08'14'' & 95°67\ 74'' \\ -52°33'50'' & -52°33'50'' \\ & \hline & 43°34'24'' \end{array}$$

Geometry I

Example 2.5

Add the following angle measurements.

$$93°18'22''$$
$$65°13'8''$$
$$218°19'30''$$
$$67°05'20''$$
$$96°04'50'$$

Solution

$$
\begin{array}{rrr}
93° & 18' & 22'' \\
65° & 13' & 08'' \\
218° & 19' & 30'' \\
67° & 05' & 20'' \\
96° & 04' & 50'' \\
\hline
539° & 59' & 130'' \\
+ & 02' & 120'' \\
\hline
539° & 61' & 10'' \\
+1° & -60' & \\
\hline
540° & 01' & 10'' \\
\end{array}
$$

17. AVERAGE OF SEVERAL MEASUREMENTS OF AN ANGLE

Surveyors often measure angles by repetition. An angle of $36°30'30''$ might have read on the first reading as $36°30'$, but after turning the angle six times, the accumulated angle may have been read $216°03'00''$. Dividing this by six gives $36°30'30''$, which is closer to the true measurement.

Example 2.6

An angle was doubled, and the accumulated reading was $84°26'$. What was the angle?

Solution

$$84°26' \div 2 = 42°13'$$

Example 2.7

An angle that was doubled read $314°13'$. What was the closest value for the single angle?

Solution

$$314°13' \div 2 = 157°06'30''$$

Example 2.8

An angle reads $318°03'$ after having been turned six times. What was the average for the single angle?

Solution

$$318°03' = 318°00'180'' \div 6 = 53°00'30''$$

18. CHANGING DEGREES AND MINUTES TO DEGREES AND DECIMALS OF A DEGREE

In some situations and in some tables of trigonometric functions, angles are expressed in degrees and decimals of a degree. To change degrees and minutes to degrees and decimals of a degree, first express the minutes as a common fraction with a denominator of 60, and then convert the common fraction to a decimal fraction. Add this fraction to the degrees.

Example 2.9

Change $73°15'$ to degrees and decimals of a degree.

Solution
$$73°15' = 73\tfrac{15}{60}^{°} = 73.25°$$

19. CHANGING DEGREES, MINUTES, AND SECONDS TO DEGREES AND DECIMALS OF A DEGREE

To change degrees, minutes, and seconds to degrees and decimals of a degree, first convert degrees, minutes, and seconds to degrees, minutes, and decimals of a minute; then convert degrees, minutes, and decimals of a minute to degrees and decimals of a degree.

Example 2.10

Change $46°24'36''$ to degrees and decimals of a degree.

Solution
$$46°24'36'' = 46°24\tfrac{36}{60}' = 46°24.6' = 46\tfrac{24.6}{60}^{°} = 46.41°$$

20. CHANGING DEGREES AND DECIMALS OF A DEGREE TO DEGREES, MINUTES, AND SECONDS

To change degrees and decimals of a degree to degrees, minutes, and seconds, multiply the decimal fraction by 60 and add the product (in minutes and decimals of a minute) to the degrees. Then multiply the decimal fraction in minutes by 60 and add the product (in seconds) to the degrees and minutes. The decimal fraction will be left as such.

Example 2.11

Change 36.12345° to degrees, minutes and seconds.

Solution

$$(0.12345°)(60) = 7.407'$$
$$(0.407')(60) = 24.42''$$
$$36.12345° = 36°07'24.42''$$

21. POLYGON

A closed figure bounded by straight lines lying in the same plane is known as a *polygon*.

The sum of the interior angles of a closed polygon is equal to

$$(n-2)(180°)$$

n is the number of sides. Thus, the sum of the interior angles of a triangle is 180°, of a rectangle 360°, of a five-sided figure 540°, and so on.

22. TRIANGLE

A polygon of three sides is known as a *triangle*.

23. RIGHT TRIANGLE

A *right triangle* is a triangle that has one right angle (90°).

24. ISOSCELES TRIANGLE

An *isosceles triangle* is a triangle that has two equal sides and two equal angles.

25. EQUILATERAL TRIANGLE

An *equilateral triangle* is a triangle that has three equal sides and three equal angles.

26. OBLIQUE TRIANGLE

An *oblique triangle* is a triangle that has no right angle and no two sides equal. It is also known as a *scalene triangle*.

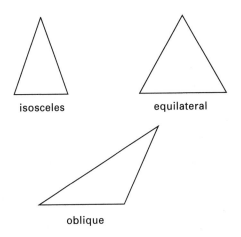

isosceles   equilateral

oblique

Figure 2.7 Triangles

27. CONGRUENT TRIANGLES

Two triangles are *congruent* if their corresponding sides and corresponding angles are equal.

28. SIMILAR TRIANGLES

Two triangles are *similar* if their corresponding angles are equal and their corresponding sides are proportional.

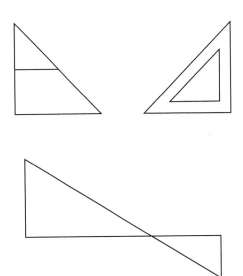

Figure 2.8 Similar Triangles

29. RECTANGLE

A *rectangle* is a four-sided polygon whose angles are right angles. A *square* is a rectangle whose sides are equal.

Figure 2.9 *Rectangles*

30. TRAPEZOID

A *trapezoid* is a four-sided polygon that has two parallel sides and two nonparallel sides.

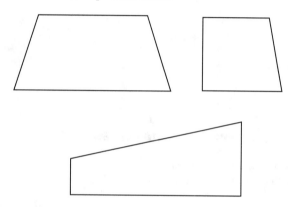

Figure 2.10 *Trapezoids*

31. CIRCLE

A *circle* is a closed plane curve, all points on which are equidistant from a point within called the *center*.

32. RADIUS

The distance from the center of the circle to any point on the circle is called the *radius* of the circle.

33. DIAMETER

The distance across the circle through the center is called the *diameter*. One diameter is two radii.

34. CHORD

A straight line between points on a circle is called a *chord*. The length of the chord is designated LC.

35. SECANT

A *secant* of a circle is a line that intersects the circle at two points.

36. TANGENT

A *tangent* of a circle is a line that touches the circle at only one point.

37. ARC

Any part of a circle is called an *arc*.

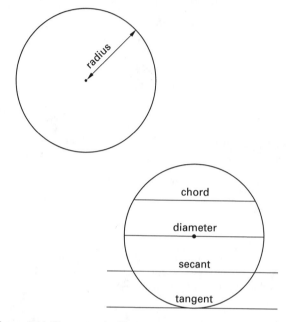

Figure 2.11 *Elements of a Circle*

38. SEMICIRCLE

An arc equal to one-half the circumference of a circle is a *semicircle*.

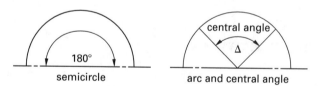

Figure 2.12 *Semicircle and Central Angle*

39. CENTRAL ANGLE

A *central angle* is an angle formed by two radii. The Greek letter Δ (delta) is often used to denote a central angle. A central angle has the same number of degrees as the arc it intercepts. A $60°$ central angle intercepts a $60°$ arc, and so on. Thus, a central angle is measured by its intercepted arc.

40. SECTOR

A figure bounded by an arc of a circle and two radii of the circle is called a *sector* of the circle.

41. SEGMENT

A figure bounded by a chord and an arc of a circle is called a *segment* of a circle.

42. CONCENTRIC CIRCLES

Two circles of different radius but with the same center are called *concentric circles*.

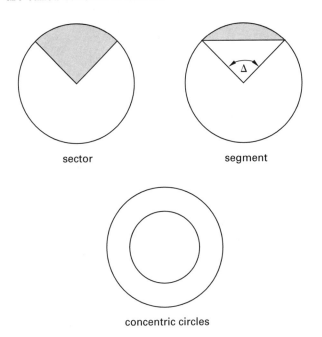

sector   segment

concentric circles

Figure 2.13 *Concentric Circles, Sector, and Segment of Circles*

43. RADIUS PERPENDICULAR TO TANGENT

The radius of a circle is perpendicular to a tangent to the circle at the point of *tangency*.

44. RADIUS AS PERPENDICULAR BISECTOR OF A CHORD

The perpendicular *bisector* of a chord passes through the center of the circle.

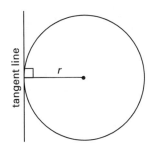

Figure 2.14 *Tangents and Chords*

45. TANGENTS TO CIRCLE FROM OUTSIDE POINT

Tangents to a circle from an outside point are equal.

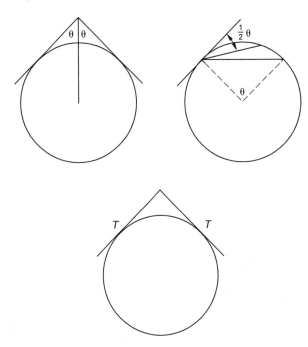

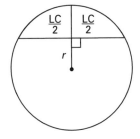

Figure 2.15 *Tangents to a Circle*

46. LINE FROM CENTER OF CIRCLE TO OUTSIDE POINT

A line from the center of a circle to an outside point bisects the angle between the tangents from the point to the circle.

47. ANGLE FORMED BY TANGENT AND CHORD

The angle formed by a tangent and a chord is equal to one-half its intercepted arc.

48. ANGLE FORMED BY TWO CHORDS

The angle formed by two chords is equal to one-half its intercepted arc.

49. SOLID GEOMETRY

Figures shown thus far are plane figures and are included in the study of *plane geometry*. *Solid geometry* is the study of figures of three dimensions such as cubes, cones, pyramids, and spheres.

50. POLYHEDRON

A *polyhedron* is any solid formed by plane surfaces.

51. PRISM

A *prism* is a polyhedron with parallel edges and parallel bases.

52. RIGHT PRISM

A prism with edges perpendicular to the bases is known as a *right prism*. A cylinder is a right prism with circular bases.

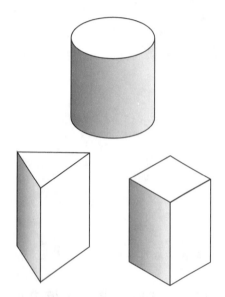

Figure 2.16 *Right Prisms*

53. PYRAMID

A *pyramid* is a polyhedron having for its base a polygon and for its faces, triangles with a common vertex.

The altitude of a pyramid is the perpendicular distance from the vertex to the base.

A *right pyramid* is a pyramid in which the base is a regular polygon, and a line from the vertex to the center of the polygon is perpendicular to the polygon.

The slant height of a right pyramid is the altitude of one of the lateral faces.

54. FRUSTUM OF A PYRAMID

A *frustum of a pyramid* is that part left after cutting off the top part with a plane parallel to the base.

55. CONE

A *cone* is a polyhedron with a circular base and with sides that taper evenly up to a vertex.

The altitude of a cone is the perpendicular distance from the vertex to the base.

A *right circular cone* is a cone in which a line from the vertex to the center of the base is perpendicular to the base.

The slant height of a cone is the distance from the vertex to the base measured along the surface.

56. FRUSTUM OF A PYRAMID AND FRUSTUM OF A CONE

The *frustum of a right circular cone* is the part left after cutting off the top part with a plane parallel to the base.

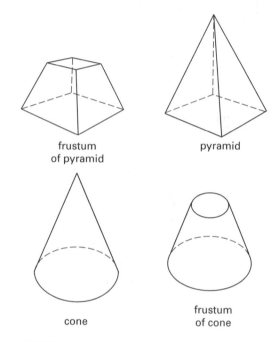

frustum
of pyramid   pyramid

cone   frustum
of cone

Figure 2.17 *Frustums*

57. CONSTRUCTION OF GEOMETRIC FIGURES

Geometric figures may be constructed by using a compass and straight-edge.

To bisect a given angle A, construct an arc intersecting the sides of the angle; then, at these intersections, construct arcs of equal radii. A line from the vertex A to the intersection of these two arcs bisects the angle A.

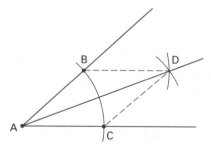

Figure 2.18 *Bisecting an Angle*

To construct a line perpendicular to a given line, w, through a given point, P, on the line, bisect the straight angle at P.

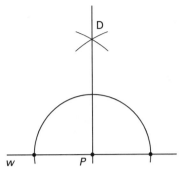

Figure 2.19 *Constructing a Perpendicular*

To construct the perpendicular bisector of a line AB, use A as a radius point and construct an arc with a radius more than half of AB. Then, use B as a radius point and construct an arc with the same radius. A line through the intersections of these arcs is the perpendicular bisector of line AB.

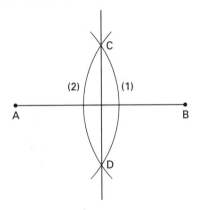

Figure 2.20 *Constructing a Perpendicular Bisector*

To circumscribe a circle about a triangle, construct the perpendicular bisectors of two sides of the triangle. Their intersection is at the center of the circle. The radius is the distance from the circle's center to any vertex of the triangle.

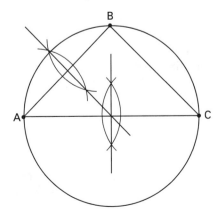

Figure 2.21 *Circumscribing a Circle about a Triangle*

To locate the center of the circle, select three points on the circle and connect them with two chords; then, erect perpendicular bisectors to the chords. Their intersection is at the center of the circle.

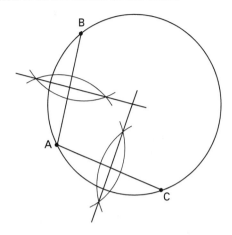

Figure 2.22 *Locating a Circle's Center*

To inscribe a circle in a triangle, construct the bisectors of two of the angles of the triangle. Their intersection is at the center of the circle.

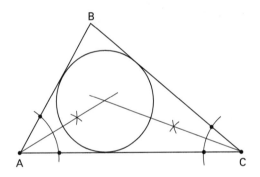

Figure 2.23 *Inscribing a Circle in a Triangle*

PRACTICE PROBLEMS

1. Add the following angles.

Example:

$$21°41'12''$$
$$11°32'54''$$
$$\overline{32°73'66''} = 33°14'06''$$

Solution

(a)	$46°27'$ $+22°24$	(b)	$56°24'$ $+33°26'$	(c)	$35°52'$ $+47°39'$
(d)	$21°46'52''$ $+40°25'26''$	(e)	$46°19'22''$ $+35°51'40''$	(f)	$13°49'58''$ $+12°21'32''$

2. Find the average of the following angles that were doubled in the field with the accumulated value shown.

Example:

$$\frac{2)311°17'20'' = 2)310°76'80''}{155°38'40''}$$

(a) 237°27'17''

(b) 329°47'16''

3. Find the average of the following angle that was repeated six times in the field with the accumulated value shown.

$$390°13'24''$$

4. Change the following to degrees and decimals of a degree.

Example:

$$36°14'52'' = 36°14\frac{52'}{60} = 36°14.8667' = 36\frac{14.8667°}{60}$$
$$= 36.2478°$$

(a) 16°24'30''

(b) 24°30'

(c) 36°45'

(d) 68°44'05''

(e) 69°11'

(f) 118°55'11''

(g) 127°17'23''

(h) 173°32'56''

(i) 186°08'34''

(j) 223°37'48''

5. Change the following to degrees, minutes, and seconds. Show each step.

Example:

$$142.276843° = 142° + (60)(0.276843)' = 142°16.61058'$$
$$= 142°16' + (60)(0.61058)'' = 142°16'37''$$

(a) 68.176°

(b) 96.564722°

(c) 145.882222°

(d) 221.347778°

(e) 303.107778°

6. The interior angles of polygons of 5, 6, and 7 sides were measured. Find the sum of the angles for each and indicate the error of measurement.

	pt	angle		pt	angle		pt	angle
(a)	A	83°23'	(b)	A	96°34'	(c)	A	98°08'05''
	B	105°27'		B	111°42'		B	149°16'12''
	C	158°31'		C	183°12'		C	134°12'55''
	D	53°19'		D	88°57'		D	93°20'10''
	E	139°18'		E	139°21'		E	152°39'47''
				F	100°18'		F	174°32'50''
							G	97°51'11''
	error			error			error	

7. Write the missing word in each of the following sentences.

(a) Two lines that lie in the same plane and do not intersect are called _____ lines.

(b) An _____ angle is an angle of less than 90°.

(c) A _____ angle is an angle of 90°.

(d) An _____ angle is an angle of more than 90° and less than 180°.

(e) A _____ angle is an angle of 180°.

(f) Two angles are said to be _____ if their sum is 90°.

(g) Two angles are said to be _____ if their sum is 180°.

(h) A line that cuts two or more lines is called a _____ .

(i) When two parallel lines are cut by a transversal, the alternate _____ angles are equal and the alternate _____ angles are equal.

(j) A polygon of three sides is known as a _____ .

(k) A _____ triangle is a triangle that has one right angle.

(l) An _____ triangle is a triangle that has two equal sides and two equal angles.

(m) An _____ triangle is a triangle that has three equal sides and three equal angles.

(n) An _____ triangle is a triangle that has no right angle and no two sides equal.

(o) Two triangles are _____ if their corresponding sides and corresponding angles are equal.

(p) Two triangles are _____ if their corresponding angles are equal and their corresponding sides are proportional.

(q) A _____ is a four-sided polygon that has two parallel sides and two nonparallel sides.

(r) A _____ is a closed plane curve, all points on which are equidistant from a point within called the center.

(s) The distance from the center of the circle to any point on the circle is called the _____

(t) The distance across a circle through the center is called the _____ .

(u) A straight line between two points on a circle is called a _____ .

(v) A _____ of a circle is a line that intersects the circle at two points.

(w) A _____ is an angle formed by two radii.

(x) A _____ of a circle is a line that touches the circle at only one point.

(y) A figure bounded by an arc of a circle and two radii of a circle is called a _____ of a circle.

(z) A figure bounded by a chord and an arc of a circle is called a _____ of a circle.

(aa) Two circles of different radii but having the same center are called _____ circles.

(bb) A _____ is a polyhedron with parallel edges and parallel bases.

(cc) A prism with edges perpendicular to the base is known as a _____ prism.

SOLUTIONS

1. (a)
$$46°27'$$
$$+24°24'$$
$$\overline{68°51'}$$

(b)
$$56°24'$$
$$+33°26'$$
$$\overline{89°50'}$$

(c)
$$35°52'$$
$$+47°39'$$
$$\overline{82°91'}$$
$$= 83°31'$$

(d)
$$21°46'52''$$
$$+40°25'26''$$
$$\overline{61°71'78''}$$
$$= 62°12'18''$$

(e)
$$46°19'22''$$
$$+35°51'40''$$
$$\overline{81°70'62''}$$
$$= 82°11'02''$$

(f)
$$13°49'58''$$
$$+12°21'32''$$
$$\overline{25°70'90''}$$
$$= 26°11'30''$$

2. (a) $2\overline{)237°27'17''} = 2\overline{)236°86'77''}$
$$118°43'38.5''$$
$$= \quad 118°43'38''$$

(b) $2\overline{)329°47'16''} = 2\overline{)328°106'76''}$
$$164°53'38''$$
$$= \quad 164°53'38''$$

3. (a) $6\overline{)390°13'24''} = 2\overline{)390°12'84''}$
$$65°02'14''$$
$$= \quad 65°02'14''$$

4. (a) 16.4083°   (b) 24.50°   (c) 36.75°
(d) 68.7347°   (e) 69.18°   (f) 118.9197°
(g) 127.2897°   (h) 173.5489°   (i) 186.1428°
(j) 223.6300°

5. (a) 68°10'34''   (b) 96°33'53''   (c) 145°52'56''
(d) 221°20'52''   (e) 303°06'28''

6.

pt	angle	pt	angle	pt	angle
(a) A	83° 23'	(b) A	96° 34'	(c) A	98° 08' 05''
B	105° 27'	B	111° 42'	B	149° 16' 12''
C	158° 31'	C	183° 12'	C	134° 12' 55''
D	53° 19'	D	88° 57'	D	93° 20' 10''
E	139° 18'	E	139° 21'	E	152° 39' 47''
	538°118'	F	100° 18'	F	174° 32' 50''
	539° 58'		717°184'	G	97° 51' 11''
	error = 02'		720° 04'		897°178'190''
			error = 04'		900° 01' 10''
					error = 01' 10''

7.
(a) parallel   (b) acute
(c) right   (d) obtuse
(e) straight   (f) complementary
(g) supplementary (h) transversal
(i) interior, exterior (j) triangle
(k) right   (l) isosceles
(m) equilateral   (n) oblique
(o) congruent   (p) similar
(q) trapezoid   (r) circle
(s) radius   (t) diameter
(u) chord   (v) secant
(w) central   (x) tangent
(y) sector   (z) segment
(aa) concentric   (bb) prism
(cc) right

3 Systems of Weights and Measures

1. THE ENGLISH SYSTEM

American colonists from England brought the system of weights and measures in use at the time in England, called the *English system*.

The English system was originally based on standards determined by parts of the body, such as the foot, the hand, and the thumb. Several hundred years ago, the English king proclaimed the length of the English inch to be the length of three barley corn grains laid end to end. Later, of course, more sophisticated methods were used for standardization, but the system had no uniform conversion factors.

During the Colonial period, the English system was well standardized (12 inches per foot, 3 feet per yard, $5\frac{1}{2}$ yards per rod, and 16 ounces per pound), but such was not the case in the rest of Europe. There was such a wide variety of weights and measures in use that commerce was difficult.

2. THE METRIC SYSTEM

This situation prompted the National Assembly of France to enact a decree in 1790 that directed the French Academy of Sciences to find standards for all weights and measures. The French Academy was supposed to work with the Royal Society of London, but the English did not participate so the French proceeded alone. The result was the *metric system*, which used the base ten in coverting units of measure.

The metric system spread rapidly in the 19th century. In 1872, France called an international meeting that was attended by 26 nations, including the United States, to further refine the system.

The meeting resulted in the establishment of the *International Bureau of Weights and Measures*, and in 1960 an extensive revision and simplification resulted in the *International System of Units*, which is in use in most countries today.

3. THE SI SYSTEM

The International System of Units—officially abbreviated SI—uses the base ten in expressing multiples and submultiples just as the metric system has always done. The six base units of measurement are as follows.

Table 3.1 Six Base Units of Measurement

length	meter	m
time	second	s
mass	kilogram	kg
temperature	kelvin	K
electric current	ampere	A
luminous intensity	candela	cd

The *meter* is defined as 1,650,763.73 wave lengths in a vacuum of the orange-red line of the spectrum of krypton-86. The SI unit of measure is the square meter, m^2. Land is measured by the *hectare* (10,000 square meters). The SI unit of volume is the cubic meter, m^3. Fluid volume is measured by the liter (0.001 cubic meter).

The *second* is defined as the duration of 9,192,631,770 cycles of the radiation associated with a specified transition of the atom.

The standard for the *kilogram* is a cylinder of platinum-iridium alloy kept by the International Bureau of Weights and Measures in Paris. A duplicate is in the custody of the National Bureau of Standards, Washington, D.C.

To make a conversion in the metric system (SI system), the decimal is moved to the right or left just as in working with decimal fractions. To facilitate use of the system, names are given to the various powers of ten.

Table 3.2 Names of Powers of Ten

milli	one thousandth of	0.001
centi	one hundredth of	0.01
deci	one tenth of	0.1
deka	ten times	10
hecto	one hundred times	100
kilo	one thousand times	1000

Thus, centimeter means one hundredth of a meter, and kilometer means one thousand meters.

To express meters in centimeters, move the decimal two places to the right.

To express meters in kilometers, move the decimal three places to the left.

Prefixes and symbols for all SI units are as shown in Table 3.3.

Table 3.3 *Prefixes and Symbols for SI Units*

multiples and submultiples		prefixes	symbols
$1,000,000,000,000 = 10^{12}$		tera	T
$1,000,000,000 = 10^{9}$		giga	G
$1,000,000 = 10^{6}$		mega	M
$1000 = 10^{3}$		kilo	k
$100 = 10^{2}$		hecto	h
$10 = 10$		deka	da
$0.1 = 10^{-1}$		deci	d
$0.01 = 10^{-2}$		centi	c
$0.001 = 10^{-3}$		milli	m
$0.000\,001 = 10^{-6}$		micro	μ
$0.000\,000\,001 = 10^{-9}$		nano	n
$0.000\,000\,000\,001 = 10^{-12}$		pico	p
$0.000\,000\,000\,000\,001 = 10^{-15}$		femto	f
$0.000\,000\,000\,000\,000\,000\,001 = 10^{-18}$		atto	a

Table 3.4 *The English System of Weights and Measures*

linear measure	square measure
1 ft = 12 in	$1\ \text{ft}^2 = 144\ \text{in}^2$
1 yd = 3 ft	$1\ \text{yd}^2 = 9\ \text{ft}^2$
$16\frac{1}{2}$ ft = 1 rod	$1\ \text{ac} = 43{,}560\ \text{ft}^2$
5280 ft = 1 mi	$1\ \text{mi}^2 = 640\ \text{ac}$

cubic measure	weight
$1728\ \text{in}^3 = 1\ \text{ft}^3$	1 gal of water = 8.33 lbm
$27\ \text{ft}^3 = 1\ \text{yd}^3$	$1\ \text{ft}^3$ of water = 62.5 lbm
$231\ \text{in}^3 = 1\ \text{gal}$	1 kip = 1000 lbm
$1\ \text{ft}^3 = 7.5\ \text{gal}$	1 ton = 2000 lbm

Table 3.5 *United States Rectangular Survey System Measures*

1 m	= 39.37 in exactly
1 m	= 39.37/12
	= 3.2808333 U.S. survey ft
1 U.S. survey ft	= 12/39.37
	= 0.3048006 m
1 international ft	= 0.3048 m
1 Gunter's chain	= 100 links
	= 66 ft
1 Gunter's link	= 7.92 in
1 Gunter's chain	= 4 rods
	= 4 poles
1 mi	= 80 chains
$10\ \text{chains}^2$	$= 43{,}560\ \text{ft}^2$
	= 1 ac

4. CONVERSION OF INCHES TO DECIMALS OF A FOOT

Engineering plans usually show dimensions of structures in feet and inches, while elevations are established in feet and decimals of a foot. The surveying technician's job is to make the necessary conversions to establish finished elevations. Construction stakes are usually set to the nearest hundredth of a foot for concrete, asphalt, pipe flow-lines, and so on. For earthwork, stakes are set to the nearest tenth of a foot.

The key to conversion is shown in Table 3.6. The values of 1 in and $^1/_8$ in are important parts of the key.

Table 3.6 *Conversion of Inches to Decimals of a Foot*

1 in	= 0.08 ft
	$\left(1\ \text{in} = \frac{1}{12}\ \text{ft} = 1 \div 12 = 0.083\dots\text{ft}\right)$
$\frac{1}{8}$ in	= 0.01 ft
	$\left(\frac{1}{8}\ \text{in} = 1\ \text{in} \div 8 \div 12 = 0.01\dots\text{ft}\right)$
2 in	= 0.17 ft
	$(1\ \text{in} + 1\ \text{in} = 0.166\dots\text{ft})$
3 in	= 0.25 ft
	$\left(\frac{3}{12}\ \text{ft} = \frac{1}{4}\ \text{ft} = 0.250\ \text{ft}\right)$
4 in	= 0.33 ft
	$\left(\frac{4}{12}\ \text{ft} = \frac{1}{3}\ \text{ft} = 0.333\dots\text{ft}\right)$
5 in	= 0.42 ft
	$(4\ \text{in} + 1\ \text{in} = 0.333 + 0.083 = 0.416\ \text{ft})$
6 in	= 0.50 ft
	$\left(\frac{6}{12}\ \text{ft} = \frac{1}{2}\ \text{ft} = 0.500\ \text{ft}\right)$
7 in	= 0.58 ft
	$(6\ \text{in} + 1\ \text{in} = 0.500 + 0.083 = 0.583\ \text{ft})$
8 in	= 0.67 ft
	$\left(\frac{8}{12}\ \text{ft} = \frac{2}{3}\ \text{ft} = 0.666\dots\text{ft}\right)$
9 in	= 0.75 ft
	$\left(\frac{9}{12}\ \text{ft} = \frac{3}{4}\ \text{ft} = 0.750\ \text{ft}\right)$
10 in	= 0.83 ft
	$(9\ \text{in} + 1\ \text{in} = 0.750 + 0.083 = 0.833\ \text{ft})$
11 in	= 0.92 ft
	$(10\ \text{in} + 1\ \text{in} = 0.833 + 0.083 = 0.916\ \text{ft})$
12 in	= 1.00 ft

Conversions can be made mentally by using the following steps.

step 1: First memorize:

$$6\ \text{in} = 0.50\ \text{ft}$$
$$3\ \text{in} = 0.25\ \text{ft}$$
$$9\ \text{in} = 0.75\ \text{ft}$$

step 2: Next memorize:

$$4\ \text{in} = 0.33\ \text{ft}$$
$$8\ \text{in} = 0.67\ \text{ft}$$

step 3: Then memorize:

$$1 \text{ in} = 0.08 \text{ ft}$$
$$\tfrac{1}{8} \text{ in} = 0.01 \text{ ft}$$

step 4: In converting measurements expressed in feet, inches, and fractions of an inch to feet and decimals of a foot, convert the inches and fractions of an inch separately to decimals of a foot, then add the three parts. (In some cases subtraction can be used.)

Example 3.1

Convert the following measurements to feet and decimals of a foot.

(a) 1 ft 4 in

(b) 2 ft $8\tfrac{7}{8}$ in

(c) 5 ft $11\tfrac{1}{2}$ in

(d) 7 ft $5\tfrac{3}{4}$ in

(e) 11 ft $9\tfrac{1}{8}$ in

Solution

(a) 1 ft $\quad = 1.00$ ft
  4 in $\quad = 0.33$ ft
  ——————
  1 ft 4 in $= 1.33$ ft

(b) 2 ft $\quad = 2.00$ ft
  8 in $\tfrac{7}{8}$ in $= 0.74$ ft
  ——————
  2 ft $8\tfrac{7}{8}$ in $= 2.74$ ft

(c) 6 ft 0 in $\quad = \quad 6.00$ ft
  -0 ft $\tfrac{1}{2}$ in $= -0.04$ ft
  ——————
  5 ft $11\tfrac{1}{2}$ in $= \quad 5.96$ ft

(d) 7 ft $\quad = 7.00$ ft
  5 in $\quad = 0.42$ ft
  $\tfrac{3}{4}$ in $\quad = 0.06$ ft
  ——————
  7 ft $5\tfrac{3}{4}$ in $= 7.48$ ft

(e) 11 ft $\quad = 11.00$ ft
  9 in $\quad = \quad 0.75$ ft
  $\tfrac{1}{8}$ in $\quad = \quad 0.01$ ft
  ——————
  11 ft $9\tfrac{1}{8}$ in $= 11.76$ ft

5. CONVERSION OF DECIMALS OF A FOOT TO INCHES

In converting measurements expressed in feet and decimals of a foot to feet, inches, and fractions of an inch, mentally recall the "decimal-to-foot value" for the full inch that is nearest to and less than the given measurement. Then convert the remainder, which will be in hundredths of a foot, to a fraction that is expressed in eighths of an inch. Or, recall the "decimal of a foot value" for the full inch that is nearest to and more than

the given measurement and subtract the given measurement from it. Remember that conversions are made only to the nearest $\tfrac{1}{8}$ inch in this procedure.

Example 3.2

(a) 3.72 ft

(b) 3.79 ft

(c) 5.65 ft

(d) 6.34 ft

Solution

(a) 3.00 ft $\quad = 3$ ft 0 in
  0.79 ft $= 0.75$ ft $+ 0.04$ ft $= 0$ ft $9\tfrac{1}{2}$ in
  ——————
  3.79 ft $\quad = 3$ ft $9\tfrac{1}{2}$ in

(b) 6.00 ft $\quad = 6$ ft 0 in
  0.34 ft $= 0.33$ ft $+ 0.01$ ft $= 0$ ft $4\tfrac{1}{8}$ in
  ——————
  6.34 ft $\quad = 6$ ft $4\tfrac{1}{8}$ in

(c) 5.00 ft $\quad = 5$ ft 0 in
  0.65 ft $= 0.67$ ft $- 0.02$ ft $= 0$ ft $7\tfrac{3}{4}$ in
  ——————
  5.65 ft $\quad = 5$ ft $7\tfrac{3}{4}$ in

(d) 3.00 ft $\quad = 3$ ft 0 in
  0.72 ft $= 0.75$ ft $- 0.03$ ft $= 0$ ft $8\tfrac{5}{8}$ in
  ——————
  3.72 ft $\quad = 3$ ft $8\tfrac{5}{8}$ in

PRACTICE PROBLEMS

1. Write the missing word or number in each of the following sentences.

(a) 4 yd = _____ in

(b) 288 in^2 = _____ ft^2

(c) 54 ft^3 = _____ yd^3

(d) 3 acres = _____ ft^2

(e) 0.5 ft^2 = _____ in^2

(f) 2 gal = _____ in^3

(g) 15 gal = _____ ft^3

(h) 3 gal = _____ lbm (water)

(i) 250 lbm = _____ ft^3 (water)

(j) 3000 lbm = _____ ton

(k) kilo means _____ times.

(l) centi means _____ of.

(m) deci means _____ of.

(n) milli means _____ of.

(o) hecto means _____ times.

(p) deka means _____ times.

2. Convert the following feet and inches to feet and decimals of a foot (to two decimals only).

Example: 3 ft $9\frac{3}{8}$ in = 3.78 ft

(a) 1 ft $10\frac{1}{8}$ in

(b) 2 ft $6\frac{1}{2}$ in

(c) 2 ft $6\frac{3}{4}$ in

(d) 3 ft $7\frac{5}{8}$ in

(e) 4 ft $3\frac{3}{8}$ in

(f) 4 ft $6\frac{1}{8}$ in

(g) 4 ft $9\frac{3}{4}$ in

(h) 5 ft $0\frac{1}{4}$ in

(i) 5 ft $4\frac{3}{4}$ in

(j) 5 ft $10\frac{5}{8}$ in

(k) 6 ft $3\frac{1}{2}$ in

(l) 6 ft $4\frac{7}{8}$ in

(m) 7 ft $2\frac{1}{2}$ in

(n) 7 ft $2\frac{7}{8}$ in

(o) 8 ft $7\frac{1}{8}$ in

(p) 8 ft $8\frac{5}{8}$ in

(q) 9 ft $4\frac{1}{8}$ in

(r) 9 ft $8\frac{1}{2}$ in

(s) 10 ft $5\frac{1}{4}$ in

(t) 10 ft $11\frac{1}{4}$ in

3. Convert the following feet and decimals to feet and inches.

(a) 0.36 ft

(b) 1.35 ft

(c) 2.69 ft

(d) 2.94 ft

(e) 3.52 ft

(f) 3.87 ft

(g) 4.76 ft

(h) 4.79 ft

(i) 4.83 ft

(j) 5.06 ft

(k) 5.60 ft

(l) 6.08 ft

(m) 6.16 ft

(n) 6.25 ft

(o) 6.67 ft

(p) 7.81 ft

(q) 8.21 ft

(r) 8.72 ft

(s) 9.23 ft

(t) 9.27 ft

SOLUTIONS

1. (a) 144 in (b) 2 ft^2
(c) 2 yd^3 (d) 130,680 ft^2
(e) 72 in^2 (f) 462 in^3
(g) 2 ft^3 (h) 25 lbm (water)
(i) 4 ft^3 (water) (j) 1.5 ton
(k) 1000 (l) 0.01
(m) 0.1 (n) 0.001
(o) 100 (p) 10

2. (a) 1.84 ft (b) 2.54 ft (c) 2.56 ft
(d) 3.63 ft (e) 4.28 ft (f) 4.51 ft
(g) 4.81 ft (h) 5.02 ft (i) 5.39 ft
(j) 5.88 ft (k) 6.29 ft (l) 6.40 ft
(m) 7.21 ft (n) 7.24 ft (o) 8.59 ft
(p) 8.72 ft (q) 9.34 ft (r) 9.71 ft
(s) 10.44 ft (t) 10.94 ft

3. (a) 0 ft $4\frac{3}{8}$ in (b) 1 ft $4\frac{1}{4}$ in
(c) 2 ft $8\frac{1}{4}$ in (d) 2 ft $11\frac{1}{4}$ in
(e) 3 ft $6\frac{1}{4}$ in (f) 3 ft $10\frac{1}{2}$ in
(g) 4 ft $9\frac{1}{8}$ in (h) 4 ft $9\frac{1}{2}$
(i) 4 ft 10 in (j) 5 ft $0\frac{3}{4}$ in
(k) 5 ft $7\frac{1}{4}$ in (l) 6 ft 1 in
(m) 6 ft $1\frac{7}{8}$ in (n) 6 ft 3 in
(o) 6 ft 8 in (p) 7 ft $9\frac{3}{4}$ in
(q) 8 ft $2\frac{1}{2}$ in (r) 8 ft $8\frac{5}{8}$ in
(s) 9 ft $2\frac{3}{4}$ in (t) 9 ft $3\frac{1}{4}$ in

4 Perimeter and Circumference

1. DEFINITION

The sum of the lengths of the sides of a polygon is called the *perimeter* of the polygon.

Example 4.1

Find the perimeter of the right triangle shown.

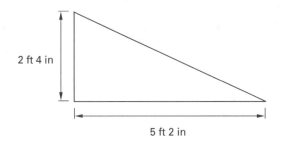

Solution

$$\text{side} = \sqrt{\left(5.17 \text{ ft}\right)^2 + \left(2.33 \text{ ft}\right)^2}$$
$$= 5.67 \text{ ft} = 5 \text{ ft } 8 \text{ in}$$
$$\text{perimeter} = 5 \text{ ft } 2 \text{ in} + 2 \text{ ft } 4 \text{ in} + 5 \text{ ft } 8 \text{ in}$$
$$= 13 \text{ ft } 2 \text{ in}$$

Example 4.2

Find the perimeter of the isosceles trapezoid shown.

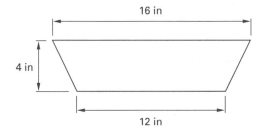

Solution

$$\text{side} = \sqrt{(4 \text{ in})^2 + (2 \text{ in})^2} = 4.5 \text{ in}$$
$$\text{perimeter} = 12 \text{ in} + 4.5 \text{ in} + 16 \text{ in} + 4.5 \text{ in} = 37 \text{ in}$$

2. CIRCUMFERENCE OF A CIRCLE

The *circumference* of a circle is the distance around the circle. It contains 360°. Regardless of the size, the circumference of any circle is always approximately 3.14 times the length of the diameter. The exact ratio of the circumference to the diameter is the number π. It is an irrational number, but its value is usually considered to be 3.1416 or 3.14, depending on the accuracy desired, based on the accuracy of the measurement of the diameter. Thus, the circumference of any circle is

$$C = \pi D \qquad \qquad 4.1$$

C is the circumference and D is the diameter of the circle. Because the diameter is twice the radius,

$$C = 2\pi R \qquad \qquad 4.2$$

C is the circumference and R is the radius of the circle.

Example 4.3

Find the circumference of a 10 ft diameter circle.

Solution

$$C = \pi D = \pi(10 \text{ ft}) = 31.42 \text{ ft}$$
$$= 31 \text{ ft } 5 \text{ in}$$

Example 4.4

Find the circumference of a circle that has a radius of 21 ft 9 in.

Solution

$$C = 2\pi R = 2\pi(21.75 \text{ ft}) = 136.66 \text{ ft}$$
$$= 136 \text{ ft } 8 \text{ in}$$

Perimeter

Example 4.5

Find the outside circumference of a concrete pipe with an inside diameter of 36 in and a wall thickness of 3 in.

Solution

$$C = \pi D = \pi(42 \text{ in}) = 132 \text{ in}$$
$$= 11 \text{ ft } 0 \text{ in}$$

Example 4.6

Find the diameter of a tank that measures 62 ft 10 in around.

Solution

$$C = \pi D$$
$$D = \frac{C}{\pi} = \frac{62.83 \text{ ft}}{\pi} = 20 \text{ ft } 0 \text{ in}$$

3. LENGTH OF AN ARC OF A CIRCLE

The *length of an arc* of a circle is proportional to its central angle. A central angle of 90° (one-fourth of 360°) subtends an arc that is one-fourth the circumference in length. Thus, an arc whose central angle is 45° is $(\frac{45°}{360°})C$ in length.

Example 4.7

Find the length of an arc of a 100 ft circle that has a central angle of 36°.

Solution

$$\text{arc} = \frac{(36°)(100 \text{ ft})\pi}{360°} = \frac{100\pi}{10} = 31 \text{ ft}$$

Example 4.8

Find the perimeter of a right triangle with a base of 9 in and an altitude of 12 in.

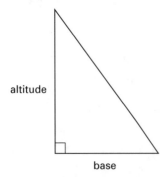

Solution

$$\text{side} = \sqrt{(9 \text{ in})^2 + (12 \text{ in})^2} = 15 \text{ in}$$
$$\text{perimeter} = 9 \text{ in} + 12 \text{ in} + 15 \text{ in} = 36 \text{ in}$$

PRACTICE PROBLEMS

1. Find the perimeter of the floor of a room 18 ft by 22 ft.

2. Find the perimeter of a right triangle with a base of 12 in and an altitude of 5 in.

3. Find the perimeter of an isosceles triangle with a base of 12 in and an altitude of 8 in. (Note: For an isosceles triangle, a line from the vertex perpendicular to the base bisects the base.)

4. Find the circumference of a 10 in circle.

5. Find the circumference of a circle with a radius of 7 in.

6. Find the length of an arc of a circle of 24 in radius that has a central angle of 60°.

7. What is the diameter of a cylindrical tank that measures 47.10 ft around the outside?

8. Find the length of an arc of a circle of 16 in diameter that has a central angle of 45°.

9. Find the diameter of a tree that measures 3 ft $1\frac{3}{4}$ in around.

10. Find the perimeter of each of the following figures. Show computations for the length of any side needed that is not dimensioned.

(a)

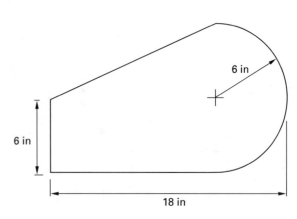

(b)

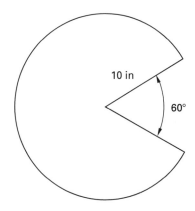

10 in

60°

11. Find the length of the following steel reinforcing bars.

(a)

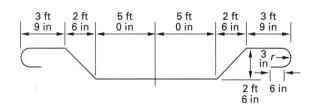

| 3 ft 9 in | 2 ft 6 in | 5 ft 0 in | | 5 ft 0 in | 2 ft 6 in | 3 ft 9 in |

3 in r→

2 ft 6 in 6 in

(b)

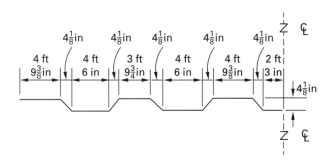

SOLUTIONS

1. $(2)(22 \text{ ft}) + (2)(18 \text{ ft}) = 80 \text{ ft}$

2.
$$\text{side} = \sqrt{(5 \text{ in})^2 + (12 \text{ in})^2} = 13 \text{ in}$$
$$\text{perimeter} = 5 \text{ in} + 12 \text{ in} + 13 \text{ in} = 30 \text{ in}$$

3.
$$\text{side} = \sqrt{(8 \text{ in})^2 + (6 \text{ in})^2} = 10 \text{ in}$$
$$\text{perimeter} = 10 \text{ in} + 10 \text{ in} + 12 \text{ in} = 32 \text{ in}$$

4. $\text{circumference} = (\pi)(10 \text{ in}) = 31.4 \text{ in}$

5. $\text{circumference} = (2)(\pi)(7 \text{ in}) = 44 \text{ in}$

6. $\text{arc} = \left(\dfrac{1}{6}\right)(2)(\pi)(24 \text{ in}) = 25 \text{ in}$

7. $\text{diameter} = \dfrac{47.10 \text{ ft}}{\pi} = 15.0 \text{ ft}$

8. $\text{arc} = \left(\dfrac{1}{8}\right)(\pi)(16 \text{ in}) = 6.3 \text{ in}$

9. $\text{diameter} = \dfrac{3.14 \text{ ft}}{\pi} = 1 \text{ ft}$

10. (a) $\text{arc} = (\pi)(6 \text{ in}) = 19 \text{ in}$
$$\text{side} = \sqrt{(6 \text{ in})^2 + (12 \text{ in})^2} = 13 \text{ in}$$
$$\text{perimeter} = 19 \text{ in} + 13 \text{ in} + 6 \text{ in} + 12 \text{ in} = 50 \text{ in}$$

(b) $\text{arc} = (2)\left(\dfrac{5}{6}\right)(\pi)(10 \text{ in}) = 52 \text{ in}$
$$\text{side} = 10 \text{ in} + 10 \text{ in} = 20 \text{ in}$$
$$\text{perimeter} = 52 \text{ in} + 10 \text{ in} + 10 \text{ in} = 72$$

11. (a) $\text{side} = \sqrt{(2.5 \text{ ft})^2 + (2.5 \text{ ft})^2} = 3.54 \text{ ft}$
$$\text{arc} = \left(\dfrac{1}{2}\right)(\pi)(0.25 \text{ ft}) = 0.78 \text{ ft}$$
$$\text{length} = (2)(0.50 \text{ ft} + 0.78 \text{ ft} + 3.50 \text{ ft}$$
$$+ 3.54 \text{ ft} + 5.00 \text{ ft})$$
$$= 26.64 \text{ ft}$$

(b) $\text{side} = \sqrt{(0.34 \text{ ft})^2 + (0.34 \text{ ft})^2} = 0.48 \text{ ft}$
$$\text{length} = 4.78 \text{ ft} + 0.48 \text{ ft} + 4.50 \text{ ft} + 0.48 \text{ ft}$$
$$+ 3.81 \text{ ft} + 0.48 \text{ ft} + 4.50 \text{ ft} + 0.48 \text{ ft}$$
$$+ 4.78 \text{ ft} + 0.48 \text{ ft} + 2.25 \text{ ft}$$
$$= 27.02 \text{ ft}$$

Perimeter

5 Area

Nomenclature
A area
b base
D diameter
h altitude
r radius
s half the perimeter

1. DEFINITION

Area is defined as the surface within a set of lines. Thus, the area of a triangle is the surface within the three sides; the area of a circle is the surface within the circumference.

Area is measured in square units: square inches (in^2), square feet (ft^2), square miles (mi^2), and so on.

A square inch is a square, each side of which is 1 in in length. The rectangle shown in Fig. 5.1 has an area of 6 in^2. It could be exactly covered by six squares, each 1 in on a side.

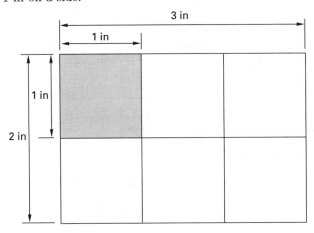

Figure 5.1 *Square Inch*

2. AREA OF A RECTANGLE

The area of a rectangle is equal to the product of the length and the width.

Example 5.1

Find the area of the floor of a room that is 20.25 ft long and 16.33 ft wide.

Solution

$$
\begin{aligned}
\text{area} &= (\text{length})(\text{width}) \\
&= (20.25 \text{ ft})(16.33 \text{ ft}) \\
&= 330.7 \text{ ft}^2
\end{aligned}
$$

Example 5.2

Find the area of the walls of a room that is 8.0 ft high if the length of the room is 20.0 ft and the width is 15.0 ft.

Solution

$$
\begin{aligned}
\text{area of two walls} &= (2)(8 \text{ ft})(20 \text{ ft}) = 320 \text{ ft}^2 \\
\text{area of two walls} &= (2)(8 \text{ ft})(15 \text{ ft}) = \underline{240 \text{ ft}^2} \\
\text{total area} &= 560 \text{ ft}^2
\end{aligned}
$$

3. AREA OF A TRIANGLE

The area of a triangle is expressed in terms of its base and its altitude. Any side of a triangle can be called the *base*. The *vertex* of a triangle is the vertex opposite the base. The *altitude* of a triangle is the perpendicular distance from the vertex to the base.

The area of any triangle is equal to one-half the product of the base and the altitude.

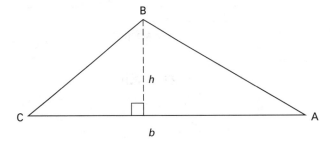

Figure 5.2 *Area of a Triangle*

In the triangle ABC,

$$A = \tfrac{1}{2}bh \qquad \qquad 5.1$$

$$A = \text{area}$$
$$b = \text{base}$$
$$h = \text{altitude}$$

Example 5.3

Find the area of a triangle with a base of 12 in and an altitude of 4 in.

Solution

$$A = \frac{bh}{2} = \frac{(12 \text{ in})(4 \text{ in})}{2} = 24 \text{ in}^2$$

4. AREA OF A RIGHT TRIANGLE

The area of a right triangle is equal to one-half the product of the base and the altitude.

In Fig. 5.3, a rectangle has a side b and a side a. The area of the rectangle is ab. If the rectangle is cut into two equal triangles as shown by the dashed line, then a represents the altitude of a right triangle and b represents the base of the right triangle; the area of each triangle is $\tfrac{1}{2}ab$.

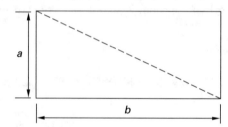

Figure 5.3 *Area of a Right Triangle*

Example 5.4

Find the area of the right triangle shown.

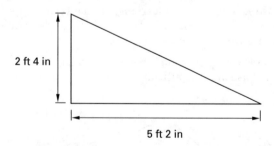

Solution

$$\text{base} = 5 \text{ ft } 2 \text{ in} = 5.16 \text{ ft}$$
$$\text{altitude} = 2 \text{ ft } 4 \text{ in} = 2.33 \text{ ft}$$
$$\text{area} = \frac{ab}{2} = \frac{(2.33 \text{ ft})(5.16 \text{ ft})}{2}$$
$$= 6.01 \text{ ft}^2$$

5. AREA OF A TRIANGLE WITH KNOWN SIDES

If the lengths of the three sides of a triangle are known, the area of the triangle can be found from Eq. 5.2.

$$A = \sqrt{s(s-a)(s-b)(s-c)} \qquad \qquad 5.2$$

$$A = \text{area}$$
$$s = \text{half the perimeter}$$
$$a, \ b, \text{ and } c = \text{lengths of each of the sides}$$

Example 5.5

Find the area of a triangle with sides 32 ft, 46 ft, and 68 ft.

Solution

$$A = \sqrt{s(s-a)(s-b)(s-c)}$$
$$= \sqrt{(73 \text{ ft})(73 \text{ ft} - 32 \text{ ft})(73 \text{ ft} - 46 \text{ ft})(73 \text{ ft} - 68 \text{ ft})}$$
$$= 636 \text{ ft}^2$$

6. AREA OF A TRAPEZOID

The area of a trapezoid is equal to the average width times the altitude. This may be expressed in another way: the area of a trapezoid is equal to one-half the sum of the bases times the altitude.

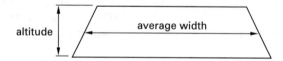

Figure 5.4 *Area of a Trapezoid*

Example 5.6

Find the area of the trapezoid shown.

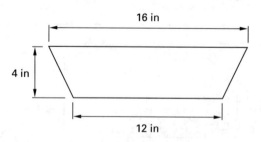

Solution

$$\text{area} = \tfrac{1}{2} \text{ (sum of bases) (altitude)}$$
$$= \left(\frac{16 \text{ in} + 12 \text{ in}}{2} \right) (4 \text{ in})$$
$$= 56 \text{ in}^2$$

Example 5.7

A swimming pool is 4 ft deep at one end, 8 ft deep at the other end, and 100 ft long. Find the area of the trapezoidal section through the long axis.

Solution

$$A = \frac{(4 \text{ ft} + 8 \text{ ft})(100 \text{ ft})}{2} = 600 \text{ ft}^2$$

Example 5.8

A drainage ditch has a trapezoidal cross section with a bottom width of 6 ft, a top width of 24 ft, and a depth of 4 ft. Find the area of the cross section.

Solution

$$A = \frac{(6 \text{ ft} + 24 \text{ ft})(4 \text{ ft})}{2} = 60 \text{ ft}^2$$

7. AREA OF A CIRCLE

The relation between the circumference of a circle and its radius is that the circumference is always 2π times the radius.

The area of a circle is always π times the square of its radius, or

$$A = \pi r^2 \qquad\qquad 5.3$$

$$A = \text{the area of any circle}$$

$$r = \text{the radius of that circle}$$

Since the radius of a circle is equal to half the diameter, the area of a circle can be expressed as

$$A = \pi \left(\frac{D}{2}\right)^2 = \left(\frac{\pi}{4}\right) D^2 \qquad\qquad 5.4$$

Example 5.9

Find the area of a 12.0 in circle.

Solution

$$A = \pi r^2 = \pi(6 \text{ in})^2 = \pi(36 \text{ in}^2)$$
$$= 113 \text{ in}^2$$

Example 5.10

Find the area of a 10.0 in circle.

Solution

$$A = \left(\frac{\pi}{4}\right) D^2 = \frac{\pi(10.0 \text{ in})^2}{4} = \frac{\pi(100 \text{ in}^2)}{4}$$
$$= 78.5 \text{ in}^2$$

Example 5.11

Find the area of an 11.0 in circle.

Solution

$$A = \left(\frac{\pi}{4}\right) D^2 = \left(\frac{\pi}{4}\right) (11.0 \text{ in})^2$$
$$= 95.0 \text{ in}^2$$

8. AREA OF A SECTOR OF A CIRCLE

The area of a sector of a circle is a fractional part of the area of the circle. The central angle of the sector is a measure of the fraction. A sector whose central angle is 90° is one-fourth of a circle because 90° is one-fourth of 360°.

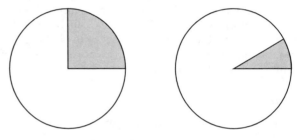

Figure 5.5 Area of a Sector of a Circle

Example 5.12

Find the area of a 30° sector in a 6 in circle.

Solution

$$A = \left(\frac{30°}{360°}\right) \left(\frac{\pi}{4}\right) D^2$$
$$= \left(\frac{30°}{360°}\right) \left(\frac{\pi}{4}\right) (36 \text{ in}^2)$$
$$= 2.4 \text{ in}^2$$

9. AREA OF A SEGMENT OF A CIRCLE

A segment of a circle is bounded by an arc and a straight line that connects the ends of the arc. The area of a segment is found by subtracting the area of the triangle formed by the chord and the two radii to its end points from the area of the sector formed by the two radii and the arc.

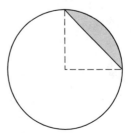

Figure 5.6 Area of a Segment of a Circle

Example 5.13

Find the area of the segment whose arc subtends an angle of 90° in a circle with an 8 in radius.

Solution

$$A = \left(\frac{90°}{360°} \right) \pi \, (8 \text{ in})^2 - \frac{(8 \text{ in})\,(8 \text{ in})}{2}$$

$$= 50 \text{ in}^2 - 32 \text{ in}^2$$

$$= 18 \text{ in}^2$$

10. COMPOSITE AREAS

Irregularly shaped areas can sometimes be divided into components that consist of geometric figures, the areas of which can be found. Total area can be found by adding the areas of the components. In some cases it may be appropriate to subtract the areas of geometric figures in order to find the net area desired.

Example 5.14

Find the area of the following figure.

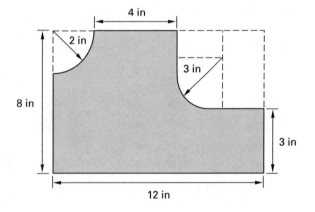

Solution

$$\text{area} = (12 \text{ in})(8 \text{ in}) - \frac{\pi(2 \text{ in})^2}{4} - (2 \text{ in})(3 \text{ in})$$

$$- \frac{\pi(3 \text{ in})^2}{4} - (3 \text{ in})(5 \text{ in})$$

$$= 65 \text{ in}^2$$

PRACTICE PROBLEMS

1. Find the area of a right triangle with a base of 12 in and an altitude of 8 in.

2. Find the number of square feet of wallboard needed to cover the walls and ceiling of a room 24 ft long, 16 ft wide, and 8 ft high. Find the number of 4 ft by 8 ft sheets needed.

3. Find the cross-sectional area of a ditch of trapezoidal cross section with a top width of 28 ft, a bottom width of 4 ft, and a depth of 6 ft.

4. Find the cross-sectional area of a highway fill of trapezoidal cross section with a top width of 44 ft, a base width of 92 ft, and a height of 8 ft.

5. Find the area of a circle that has a 20 ft radius.

6. Find the area of a 10 ft diameter circle.

7. Find the area of a 60° sector of a 6 in circle.

8. Find the area of the segment whose arc subtends an angle of 90° in a 12 ft circle (12 ft diameter).

9. Find the area of a triangle with sides of 18 ft, 12 ft, and 10 ft.

10. Divide the following figures into component parts, then find the total area by either adding the areas of the component parts or by subtracting areas from a larger area that includes the area shown.

Example:

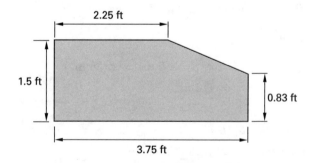

Solution:

$$A = (3.75 \text{ ft})(1.5 \text{ ft}) - \left(\frac{1}{2} \right) (1.50 \text{ ft})(0.67 \text{ ft})$$

$$= 5.1 \text{ ft}^2$$

(a)

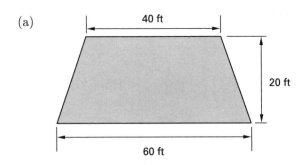

(b)

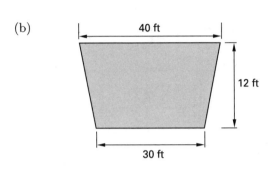

(c)

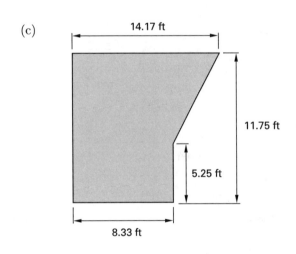

(d)

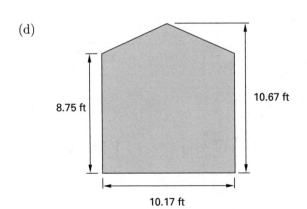

(e)

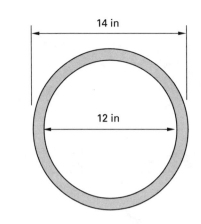

(f)

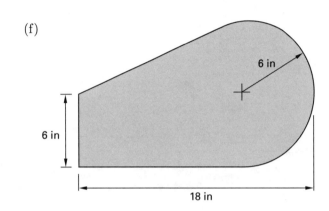

(g)

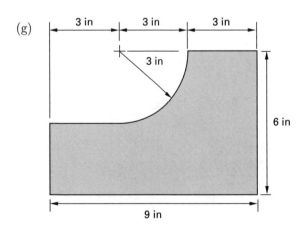

(h)

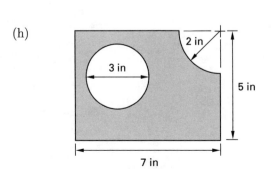

Area

(i)

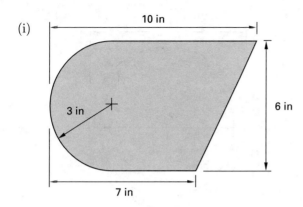

10 in

3 in

6 in

7 in

SOLUTIONS

1. The area is

$$A = \left(\frac{1}{2}\right)(12 \text{ in})(8 \text{ in}) = \boxed{48 \text{ in}^2}$$

2. The area is

$$A = (2)(24 \text{ ft})(8 \text{ ft}) + (2)(16 \text{ ft})(8 \text{ ft})$$
$$+ (24 \text{ ft})(16 \text{ ft}) = 1024 \text{ ft}^2$$

$$\text{no. of sheets} = \frac{1024 \text{ ft}^2}{32 \dfrac{\text{ft}^2}{\text{sheet}}}$$

$$= \boxed{32 \text{ sheets}}$$

3. The area is

$$A = \left(\frac{28 \text{ ft} + 4 \text{ ft}}{2}\right)(6 \text{ ft}) = \boxed{96 \text{ ft}^2}$$

4. The area is

$$A = \left(\frac{44 \text{ ft} + 92 \text{ ft}}{2}\right)(8 \text{ ft}) = \boxed{544 \text{ ft}^2}$$

5. The area is

$$A = \pi(20 \text{ ft})^2 = \boxed{1256 \text{ ft}^2}$$

6. The area is

$$A = \left(\frac{\pi}{4}\right)(10 \text{ ft})^2 = \boxed{78.5 \text{ ft}^2}$$

7. The area is

$$A = \left(\frac{1}{6}\right)\left(\frac{\pi}{4}\right)(6 \text{ in})^2 = \boxed{4.7 \text{ in}^2}$$

8. The area is

$$A = \left(\frac{1}{4}\right)\left(\frac{\pi}{4}\right)(12 \text{ ft})^2 - \left(\frac{1}{2}\right)(6 \text{ ft})(6 \text{ ft})$$

$$= \boxed{10.3 \text{ ft}^2}$$

9. The area is

$$A = \sqrt{\begin{array}{c}(20 \text{ ft})(20 \text{ ft} - 18 \text{ ft}) \\ \times \ (20 \text{ ft} - 12 \text{ ft})(20 \text{ ft} - 10 \text{ ft})\end{array}}$$

$$= \boxed{57 \text{ ft}^2}$$

10. The area is

(a) $A = \left(\dfrac{40 \text{ ft} + 60 \text{ ft}}{2}\right)(20 \text{ ft}) = \boxed{1000 \text{ ft}^2}$

(b) $A = \left(\dfrac{30 \text{ ft} + 40 \text{ ft}}{2}\right)(12) = \boxed{420 \text{ ft}^2}$

(c) $A = (8.33 \text{ ft})(11.75 \text{ ft}) + \left(\dfrac{1}{2}\right)(5.84 \text{ ft})(6.50 \text{ ft})$

$$= \boxed{116.9 \text{ ft}^2}$$

(d) $A = (10.17 \text{ ft})(8.75 \text{ ft}) + \left(\dfrac{1}{2}\right)(1.92 \text{ ft})(10.17 \text{ ft})$

$$= \boxed{99 \text{ ft}^2}$$

(e) $A = \left(\dfrac{\pi}{4}\right)\left((14 \text{ in})^2 - (12 \text{ in})^2\right) = \boxed{41 \text{ in}^2}$

(f) $A = (6 \text{ in})(12 \text{ in}) + \left(\dfrac{1}{2}\right)(6 \text{ in})(12 \text{ in})$

$$+ \left(\dfrac{1}{2}\right)\pi(6 \text{ in})^2$$

$$= \boxed{165 \text{ in}^2}$$

(g) $A = (9 \text{ in})(6 \text{ in}) - (3 \text{ in})(3 \text{ in}) - \left(\dfrac{1}{4}\right)\pi(3 \text{ in})^2$

$$= \boxed{38 \text{ in}^2}$$

(h) $A = (7 \text{ in})(5 \text{ in}) - \left(\dfrac{\pi}{4}\right)(3 \text{ in})^2 - \left(\dfrac{1}{4}\right)\pi(2 \text{ in})^2$

$$= \boxed{25 \text{ in}^2}$$

(i) $A = (6 \text{ in})(4 \text{ in}) + \left(\dfrac{1}{2}\right)\pi(3 \text{ in})^2$

$$+ \left(\dfrac{1}{2}\right)(3 \text{ in})(6 \text{ in})$$

$$= \boxed{47 \text{ in}^2}$$

Area

6 Volume

Nomenclature
A area
h altitude
r radius
V volume

Subscripts
i inside
o outside

1. DEFINITION

Volume is defined as the amount of substance occupying a certain space. It is measured in cubic units. The block shown in Fig. 6.1 has a volume of 6 cubic inches (6 in^3). One cubic inch is a cube that measures 1 in on each edge.

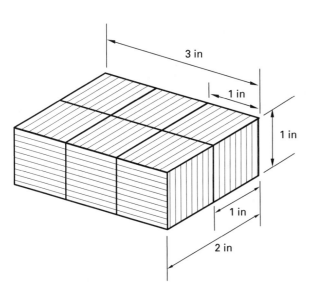

Figure 6.1 *Volume of a Block*

2. VOLUME OF RIGHT PRISMS AND CYLINDERS

The volume of a right prism or cylinder is the product of the area of the base and the altitude. Expressed as a formula,

$$V = Ah \qquad\qquad 6.1$$

V is volume in cubic units, A is area in square units, and h is altitude in linear units.

Example 6.1

Find the volume of a rectangular prism with a base of 8 in by 6 in and an altitude of 10 in.

Solution

$$V = Ah = (8 \text{ in})(6 \text{ in})(10 \text{ in}) = 480 \text{ in}^3$$

Example 6.2

Find the volume of a triangular prism with a triangular base that has sides 3 in, 4 in, and 5 in, and with an 8 in altitude.

Solution

$$V = Ah = \left(\frac{1}{2}\right)(3 \text{ in})(4 \text{ in})(8 \text{ in}) = 48 \text{ in}^3$$

Example 6.3

Find the number of cubic yards of dirt in 500 ft of a highway fill of trapezoidal cross section with a bottom base of 112 ft, a top base of 40 ft, and a height of 12 ft.

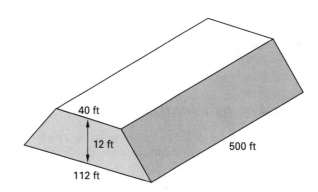

Solution

$$V = \frac{(112 \text{ ft} + 40 \text{ ft})(12 \text{ ft})(500 \text{ ft})}{(2)\left(27 \dfrac{\text{ft}^3}{\text{yd}^3}\right)} = 16{,}889 \text{ yd}^3$$

Example 6.4

Find the volume of the shell of a hollow cylinder that has an outside diameter of 8 in, an inside diameter of 6 in, and a height of 5 in.

Solution

$$V = Ah$$

$$= (\text{outside area} - \text{inside area})h$$

$$= \left(\frac{\pi}{4}\right)(D_o^2 - D_i^2)h$$

$$= \left(\frac{\pi}{4}\right)\left((8 \text{ in})^2 - (6 \text{ in})^2\right)(5 \text{ in})$$

$$= 110 \text{ in}^3$$

3. VOLUME OF CONE

The volume of a right circular cone is equal to one-third the product of the area of its base and its altitude.

$$V = \tfrac{1}{3}\pi r^2 h \qquad\qquad 6.2$$

Example 6.5

Find the volume of a cone that is 6 in high with a base of 4 in in diameter.

Solution

$$V = \tfrac{1}{3}\pi r^2 h = \tfrac{1}{3}(\pi)(2 \text{ in})^2(6 \text{ in})$$

$$= 25 \text{ in}^3$$

4. VOLUME OF PYRAMID

The volume of a pyramid is equal to one-third the product of the area of its base and its altitude.

$$V = \tfrac{1}{3}Ah \qquad\qquad 6.3$$

5. VOLUME OF SPHERE

The volume of a sphere is equal to $\frac{4}{3}\pi r^3$.

$$V = \tfrac{4}{3}\pi r^3 \qquad\qquad 6.4$$

PRACTICE PROBLEMS

(Note: When dimensions are predominantly in feet but one dimension is in inches, convert inches to feet by using a common fraction: 3 in $= \frac{1}{4}$ ft, 4 in $= \frac{1}{3}$ ft, and 6 in $= \frac{1}{2}$ ft. The denominators can be used for cancellation.)

1. Find the volume of a rectangular right prism with a base of 3 ft by 4 ft and an altitude of 6 ft.

2. Find the volume of a triangular right prism with base sides of 9 in, 12 in, and 15 in, and an altitude of 10 in.

3. Find the volume of a cylinder with a base with a diameter of 10 in and an altitude of 8 in.

4. Find the number of cubic feet of concrete (to the nearest tenth) in a pipe of 8 in inside diameter, 2 ft wall thickness, and 30 in length.

5. Find the necessary height of a cylindrical tank 6 ft in diameter if its volume is to be 226 ft^3 (to the nearest tenth).

6. Find the volume of a right prism with an altitude of 10 in and a base which enscribes an isosceles triangle with a side length of 8 in and an altitude of 3 in.

7. Find the number of cubic yards of dirt in 810 ft of a highway fill of trapezoidal cross section with a base at the bottom of 120 ft, a base at the top of 80 ft, and an 8 ft height of fill.

8. How many cubic yards of concrete are needed to pour a parking area 30 ft long, 27 ft wide, and 4 in thick?

SOLUTIONS

1. The volume is

$$V = LWh = (3 \text{ ft})(4 \text{ ft})(6 \text{ ft}) = \boxed{72 \text{ ft}^3}$$

2. The volume is

$$\left(\frac{1}{2}\right)(9 \text{ in})(12 \text{ in})(10 \text{ in}) = \boxed{540 \text{ in}^3}$$

3. The volume is

$$\left(\frac{\pi}{4}\right)(10 \text{ in})^2(8 \text{ in}) = \boxed{628 \text{ in}^3}$$

4. The volume is

$$\frac{\left(\frac{\pi}{4}\right)\left((12 \text{ in})^2 - (8 \text{ in})^2\right)(30 \text{ in})}{1728 \ \dfrac{\text{in}^3}{\text{ft}^3}} = \boxed{1.1 \text{ ft}^3}$$

5. The height is

$$h = \frac{V}{D} = \frac{226 \text{ ft}^3}{\left(\dfrac{\pi}{4}\right)(6 \text{ ft})^2} = \boxed{8 \text{ ft}}$$

6. The volume is

$$\left(\frac{1}{2}\right)(8 \text{ in})(3 \text{ in})(10 \text{ in}) = \boxed{120 \text{ in}^3}$$

7. The volume is

$$\frac{\left(\dfrac{120 \text{ ft} + 80 \text{ ft}}{2}\right)(8 \text{ ft})(810 \text{ ft})}{27 \ \dfrac{\text{ft}^3}{\text{yd}^3}} = \boxed{24{,}000 \text{ yd}^3}$$

8. The volume is

$$\frac{(30 \text{ ft})(27 \text{ ft})(1 \text{ ft})}{\left(27 \ \dfrac{\text{ft}^3}{\text{yd}^3}\right)(3)} = \boxed{10 \text{ yd}^3}$$

Volume

7 Dimensional Equations

1. MEASUREMENT

A *measurement* consists of a number that expresses quantity and a unit of measure. The surveyor and the surveying technician are intricately involved in measurements and in converting measurements expressed in one unit of measure to an equivalent in another unit of measure.

In converting values from one unit of measure to another, it is just as important to find the correct unit of measure as it is to find the correct quantity.

2. DEFINITION OF DIMENSIONAL EQUATION

A *dimensional equation* is one that contains units of measure but does not contain the corresponding numerical values. For example, to express in cubic yards the volume of a dump truck bed with dimensions of 6 ft by 8 ft by 4 ft, arithmetically multiply $(6)(8)(4)$ to find the volume in cubic feet (192 ft^3). Since there are 27 ft^3 in a cubic yard, divide 192 by 27 to find that the bed has a volume of 7 yd^3. This operation is written as

$$\frac{(6 \text{ ft})(8 \text{ ft})(4 \text{ ft})}{27 \dfrac{\text{ft}^3}{\text{yd}^3}} = 7 \text{ yd}^3$$

The dimensional equation that corresponds to this is

$$\frac{(\text{ft})(\text{ft})(\text{ft})}{\dfrac{\text{ft}^3}{\text{yd}^3}} = \frac{\dfrac{\text{ft}^3}{1}}{\dfrac{\text{ft}^3}{\text{yd}^3}} = \left(\frac{\text{ft}^3}{1}\right)\left(\frac{\text{yd}^3}{\text{ft}^3}\right) = \text{yd}^3$$

Including the numbers in the equation,

$$\frac{\left(\dfrac{6}{1} \text{ ft}\right)\left(\dfrac{8}{1} \text{ ft}\right)\left(\dfrac{4}{1} \text{ ft}\right)}{\dfrac{27 \text{ ft}^3}{1 \text{ yd}^3}} = \left(\frac{192 \text{ ft}^3}{1}\right)\left(\frac{1}{27} \frac{\text{yd}^3}{\text{ft}^3}\right)$$

$$= 7 \text{ yd}^3$$

Example 7.1

Write a dimensional equation for finding the area in acres of a rectangular tract of land 300 ft by 200 ft. Include the measured quantities in the equation.

Solution

$$
\begin{aligned}
A &= \frac{\left(\dfrac{300}{1} \text{ ft}\right)\left(\dfrac{200}{1} \text{ ft}\right)}{\dfrac{43{,}560 \text{ ft}^2}{1 \text{ ac}}} \\
&= \left(\frac{60{,}000}{1} \text{ ft}^2\right)\left(\frac{1}{43{,}560} \frac{\text{ac}}{\text{ft}^2}\right) \\
&= 1.4 \text{ ac}
\end{aligned}
$$

Example 7.2

Write a dimensional equation for finding the velocity in feet per second of a vehicle traveling 36 mi/hr. Include the measured quantities in the equation.

Solution

$$
\begin{aligned}
v &= \frac{\left(\dfrac{36 \text{ mi}}{1 \text{ hr}}\right)\left(\dfrac{5280 \text{ ft}}{1 \text{ mi}}\right)}{\dfrac{3600 \text{ sec}}{1 \text{ hr}}} \\
&= \left(\frac{36 \text{ mi}}{1 \text{ hr}}\right)\left(\frac{5280 \text{ ft}}{1 \text{ mi}}\right)\left(\frac{1 \text{ hr}}{3600 \text{ sec}}\right) \\
&= 53 \text{ ft/sec}
\end{aligned}
$$

Example 7.3

Write a dimensional equation for finding the mass of water, in tons, in a fully rectangular tank that is 10 ft long, 8 ft wide, and 6 ft deep. Include the measured quantities in the equation.

Solution

$$m = \frac{\left(\frac{10}{1}\ \text{ft}\right)\left(\frac{8}{1}\ \text{ft}\right)\left(\frac{6}{1}\ \text{ft}\right)\left(\frac{62.5}{1}\ \frac{\text{lbm}}{\text{ft}^3}\right)}{\frac{2000\ \text{lbm}}{1\ \text{ton}}}$$

$$= \frac{\left(\frac{480}{1}\ \text{ft}^3\right)\left(\frac{62.5}{1}\ \frac{\text{lbm}}{\text{ft}^3}\right)}{\frac{2000\ \text{lbm}}{1\ \text{ton}}}$$

$$= \left(\frac{480}{1}\ \text{ft}^3\right)\left(\frac{62.5}{1}\ \frac{\text{lbm}}{\text{ft}^3}\right)\left(\frac{1}{2000}\ \frac{\text{ton}}{\text{lbm}}\right)$$

$$= 15\ \text{tons}$$

3. FORM FOR PROBLEM SOLVING

The use of common conversion factors often makes it unnecessary to set up a dimensional equation for solving problems involving several measured quantities. However, setting up a single equation that includes the numbers expressing quantity but not units of measure (similar to the dimensional equation) is advantageous. It allows cancellation and is easily followed by someone whose task is to check its accuracy. Each number that expresses quantity should be shown in the equation. For example, to find the area of a 10 in circle do not simply write $A = (\pi/4)D^2 = 78.5\ \text{in}^2$. Write out $A = (\pi/4)(10\ \text{in})^2 = 78.5\ \text{in}^2$. For solutions that involve unfamiliar formulas, it is good practice to write the formula and then substitute the measured quantities.

Example 7.4

What is the cost of concrete, delivered to a site at $36 per cubic yard, for a parking lot 100 ft long, 54 ft wide, and 4 in thick?

Solution

(Note: Where length and width are measured in feet and thickness in inches, use a common fraction of a foot as the thickness measurement. Four inches is *exactly* one-third of a foot but *approximately* 0.33 of a foot.)

$$\text{cost} = \frac{(100\ \text{ft})(54\ \text{ft})(\$36)}{(27\ \text{ft})(3\ \text{ft})} = \$2400$$

Example 7.5

What is the cost of filling a rectangular tank, 100 ft long, 40 ft wide, and 10 ft deep, with water at $0.06 per 1000 gallons? (There are 7.5 gallons per cubic foot.)

Solution

$$\text{cost} = \frac{(100\ \text{ft})(40\ \text{ft})(10\ \text{ft})\left(7.5\ \frac{\text{gal}}{\text{ft}^3}\right)(\$0.60)}{1000}$$

$$= \$180.00$$

PRACTICE PROBLEMS

1. Write a dimensional equation to convert the given quantities to an equivalent quantity in the unit of measure indicated.

Examples:

(1) Convert 48 in to feet.

$$\frac{48\ \text{in}}{12\ \frac{\text{in}}{\text{ft}}} = (48\ \text{in})\left(\frac{1\ \text{ft}}{12\ \text{in}}\right) = 4\ \text{ft}$$

(2) Convert 3 ft to inches.

$$(3\ \text{ft})\left(12\ \frac{\text{in}}{\text{ft}}\right) = 36\ \text{in}$$

(3) Convert 72 ft^2 to square yards.

$$\frac{72\ \text{ft}^2}{9\ \frac{\text{ft}^2}{\text{yd}^2}} = (72\ \text{ft}^2)\left(\frac{1\ \text{yd}^2}{9\ \text{ft}^2}\right) = 8\ \text{yd}^2$$

(a) Convert 588 ft to yards.

(b) Convert 121 yd to feet.

(c) Convert 2 mi to feet.

(d) Convert 4 ft^2 to square inches.

(e) Convert 432 in^2 to square feet.

(f) Convert 5 yd^2 to square feet.

(g) Convert 81 ft^2 to square yards.

(h) Convert 2 ac to square feet.

(i) Convert 21,780 ft^2 to acres.

(j) Convert 3 ft^3 to cubic inches.

(k) Convert 3456 in^3 to cubic feet.

(l) Convert 5 yd^3 to cubic feet.

(m) Convert 135 ft^3 to cubic yards.

(n) Convert 3 gal to cubic inches.

(o) Convert 693 in^3 to gallons.

(p) Convert 4 gal of water to pounds.

(q) Convert 25 lbm of water to gallons.

(r) Convert 87,120 ft^2 to acres.

(s) Convert 1320 ft to miles.

(t) Convert 7 yd^2 to square feet.

2. Find the required quantities by including the given quantities within a dimensional equation.

Examples:

(1) Find the number of square yards in a driveway 20 ft wide and 54 ft long.

$$A = \frac{(20 \text{ ft}) (54 \text{ ft})}{9 \, \frac{\text{ft}^2}{\text{yd}^2}}$$

$$= (20 \text{ ft}) (54 \text{ ft}) \left(\frac{1 \text{ yd}^2}{9 \text{ ft}^2} \right)$$

$$= 120 \text{ yd}^2$$

(2) Find the cost of a concrete sidewalk 4 ft wide, 81 ft long, and 4 in thick at \$30 per cubic yard.

$$\text{cost} = \frac{(4 \text{ ft})(81 \text{ ft}) \left(\dfrac{4 \text{ in}}{12 \, \frac{\text{in}}{\text{ft}}} \right)}{\left(27 \, \dfrac{\text{ft}^3}{\text{yd}^3} \right) \left(\dfrac{\$30}{1 \text{ yd}^3} \right)}$$

$$= (4 \text{ ft}) (81 \text{ ft}) \left(\frac{1}{3} \text{ ft} \right) \left(\frac{1 \text{ ft}}{12 \text{ in}} \right) \left(\frac{1 \text{ yd}^3}{27 \text{ ft}^3} \right) \left(\frac{\$30}{1 \text{ yd}^3} \right)$$

$$= \$120$$

(a) How many acres are in a rectangular plot 545 ft long and 400 ft wide?

(b) What is the weight, in tons, of the water in a tank containing 2000 gal?

(c) Find the velocity, in feet per second, of a vehicle traveling 72 miles per hour.

(d) What is the weight of water, in tons, in a full cylindrical tank of 10 ft diameter and 10 ft height?

(e) What is the cost of excavation of a ditch of rectangular cross section 3 ft wide, 4 ft deep, and 324 ft long at \$0.30 per cubic yard?

3. Solve each of the following problems by writing an equation in the form of a dimensional equation.

Example: Find the mass of water in a full rectangular tank that is 8 ft long, 5 ft wide, and 5 ft deep.

$$m = \frac{(8 \text{ ft})(5 \text{ ft})(5 \text{ ft}) \left(62.5 \, \dfrac{\text{lbm}}{\text{ft}^3} \right)}{2000 \, \dfrac{\text{lbm}}{\text{ton}}}$$

$$= 6.25 \text{ tons}$$

(a) A 6 ft by 3 ft concrete box culvert, 54 ft long, is to be constructed. Walls, footing, and deck are 6 in thick. How many cubic yards of concrete are required?

(Disregard wing walls.) Note: Culvert dimensions refer to waterway openings. The horizontal dimension is 6 ft, and the vertical dimension is 3 ft.

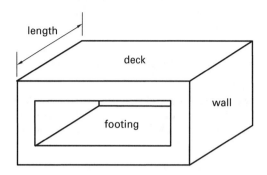

(b) The cross-sectional view of a concrete curb and gutter to be used in street paving is shown. How many lineal feet of curb and gutter can be poured with 1 yd^3 of concrete?

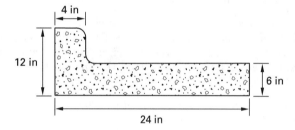

(c) A contractor is to be paid for sprinkling water in units of 1000 gallons. The empty weight of his water truck is 11,808 lbm. Loaded with water, the truck weighs 28,468 lbm. How many thousand gallons of water does the truck hold?

(d) A canal is to be excavated to a trapezoidal cross section, 30 ft at the top and 6 ft at the bottom with a 5 ft depth. What will be the cost of excavation at \$0.50 per cubic yard if the length is 540 ft?

(e) A drainage ditch has a 4 ft flat bottom, 12 ft top width, and average depth of 2 ft through 162 ft of level ground. How many cubic yards of earth were excavated?

(f) How high must a cylindrical tank, 10 ft in diameter, be in order to have a capacity of 3000 gal? (Calculate to the nearest tenth of a foot.)

(g) A parking space is 100 ft by 81 ft. What is the cost of paving this area at \$9.00 per square yard?

(h) A building lot has an area of 3840 ft^2. How deep is the lot if it is 32 ft wide?

(i) An electric power line is to be built from one city to another. One city is 16 mi due north and 12 mi due west of the other. (a) What length of wire is needed to connect the two cities? (b) If the wire weighs 50 lbm per 100 ft, what weight of wire is needed?

(j) A triangular piece of land has one side 320 yd long running north and south and another ¼ mi long at right angles. A second piece of land, rectangular in shape and 250 yd on one side, has the same acreage as the triangular piece. Which piece of land would require the most fence to enclose?

(k) A 24 in shaft was drilled 54 ft deep and filled with concrete as part of a bridge pier. How many cubic yards of concrete were poured?

(l) A swimming pool 100 ft long, 50 ft wide, 2 ft deep at the shallow end, and 10 ft deep at the deep end is to be filled with water. What is the cost of the water at $0.20 per 1000 gallons?

(m) A cylindrical piece of cheese, 16 in in diameter and 8 in high, weighs 24 lbm. If a 30° sector is cut from it, (a) what is the cost of the sector at $1.00 per pound, and (b) how many cubic inches of cheese are in the sector?

(n) A rectangular concrete tank, 11 ft long, 6 ft wide, and 4 ft 6 in high (outside) is ¾ full of water. The walls and floor of the tank are 6 in thick. How many gallons of water are in the tank?

(o) A piece of property to be purchased for highway right-of-way is bounded by an arc of a circle and a chord of that circle. The radius of the circle is 500 ft, and the central angle formed by radii to the ends of the chord is 90°. Find the area of the segment.

(p) A cylindrical water tank contains 60,000 gal of water when the water is 5 ft deep. What is the diameter of the tank?

(q) A lot 150 ft in depth and 100 ft wide is to be leveled for building construction. The fill at the front is 1.4 ft and at the rear is 2.2 ft. How many cubic yards of dirt will be required to make the fill? (Disregard shrinkage of soil.)

SOLUTIONS

1. (a) $\dfrac{588 \text{ ft}}{3\,\dfrac{\text{ft}}{\text{yd}}} = \boxed{196 \text{ yd}}$

(b) $(121 \text{ yd})\left(3\,\dfrac{\text{ft}}{\text{yd}}\right) = \boxed{363 \text{ ft}}$

(c) $(2 \text{ mi})\left(5280\,\dfrac{\text{ft}}{\text{mi}}\right) = \boxed{10{,}560 \text{ ft}}$

(d) $(4 \text{ ft}^2)\left(144\,\dfrac{\text{in}^2}{\text{ft}^2}\right) = \boxed{576 \text{ in}^2}$

(e) $\dfrac{432 \text{ in}^2}{144\,\dfrac{\text{in}^2}{\text{ft}^2}} = \boxed{3 \text{ ft}^2}$

(f) $(5 \text{ yd}^2)\left(9\,\dfrac{\text{ft}^2}{\text{yd}^2}\right) = \boxed{45 \text{ ft}^2}$

(g) $\dfrac{81 \text{ ft}^2}{9\,\dfrac{\text{ft}^2}{\text{yd}^2}} = \boxed{9 \text{ yd}^2}$

(h) $(2 \text{ ac})\left(43{,}560\,\dfrac{\text{ft}^2}{\text{ac}}\right) = \boxed{87{,}120 \text{ ft}^2}$

(i) $\dfrac{21{,}780 \text{ ft}^2}{43{,}560\,\dfrac{\text{ft}^2}{\text{ac}}} = \boxed{0.5 \text{ ac}}$

(j) $(3 \text{ ft}^3)\left(1728\,\dfrac{\text{in}^3}{\text{ft}^3}\right) = \boxed{5184 \text{ in}^3}$

(k) $\dfrac{3456 \text{ in}^3}{1728\,\dfrac{\text{in}^3}{\text{ft}^3}} = \boxed{2 \text{ ft}^3}$

(l) $(5 \text{ yd}^3)\left(27\,\dfrac{\text{ft}^3}{\text{yd}^3}\right) = \boxed{135 \text{ ft}^3}$

(m) $\dfrac{135 \text{ ft}^3}{27 \dfrac{\text{ft}^3}{\text{yd}^3}} = \boxed{5 \text{ yd}^3}$

(n) $(3 \text{ gal})\left(231 \dfrac{\text{in}^3}{\text{gal}}\right) = \boxed{693 \text{ in}^3}$

(o) $\dfrac{693 \text{ in}^3}{231 \dfrac{\text{in}^3}{\text{gal}}} = \boxed{3 \text{ gal}}$

(p) $(4 \text{ gal})\left(8\tfrac{1}{3} \dfrac{\text{lbm}}{\text{gal}}\right) = \boxed{33\tfrac{1}{3} \text{ lbm}}$

(q) $\dfrac{25 \text{ lbm}}{8\tfrac{1}{3} \dfrac{\text{lbm}}{\text{gal}}} = \boxed{3 \text{ gal}}$

(r) $\dfrac{87,120 \text{ ft}^2}{43,560 \dfrac{\text{ft}^2}{\text{ac}}} = \boxed{2 \text{ ac}}$

(s) $\dfrac{1320 \text{ ft}}{5380 \dfrac{\text{ft}}{\text{mi}}} = \boxed{\tfrac{1}{4} \text{ mi}}$

(t) $(7 \text{ yd}^2)\left(9 \dfrac{\text{ft}^2}{\text{yd}^2}\right) = \boxed{63 \text{ ft}^2}$

2. (a) The area is

$$A = \dfrac{(545 \text{ ft})(400 \text{ ft})}{43,560 \dfrac{\text{ft}^2}{\text{ac}}} = \boxed{5 \text{ ac}}$$

(b) The weight is

$$W = \dfrac{(2000 \text{ gal})\left(8.33 \dfrac{\text{lb}}{\text{gal}}\right)}{2000 \dfrac{\text{lbm}}{\text{ton}}} = \boxed{8.33 \text{ ton}}$$

(c) The velocity is

$$v = \dfrac{\left(72 \dfrac{\text{mi}}{\text{hr}}\right)\left(5280 \dfrac{\text{ft}}{\text{mi}}\right)}{3600 \dfrac{\text{sec}}{\text{hr}}} = \boxed{106 \text{ ft/sec}}$$

(d) The weight is

$$W = \dfrac{(0.785)(10 \text{ ft})^2(10 \text{ ft})\left(62.5 \dfrac{\text{lbm}}{\text{ft}^3}\right)}{2000 \dfrac{\text{lbm}}{\text{ton}}} = \boxed{24.5 \text{ ton}}$$

(e) The cost is

$$\dfrac{(3 \text{ ft})(4 \text{ ft})(324 \text{ ft})\left(\dfrac{\$0.30}{\text{yd}^3}\right)}{27 \dfrac{\text{ft}^3}{\text{yd}^3}} = \boxed{\$43.20}$$

3. (a) The volume is

$$V = \dfrac{\big((7 \text{ ft})(4 \text{ ft}) - (6 \text{ ft})(3 \text{ ft})\big)(54 \text{ ft})}{27 \dfrac{\text{ft}^3}{\text{yd}^3}} = \boxed{20 \text{ yd}^3}$$

(b) The length is

$$L = \dfrac{27 \dfrac{\text{ft}^3}{\text{yd}^3}}{(1.0 \text{ ft})(0.5 \text{ ft}) + (1.5 \text{ ft})(0.5 \text{ ft})} = \boxed{21.6 \text{ ft}}$$

(c) The number of gallons is

$$\dfrac{28,468 \text{ lbm} - 11,808 \text{ lbm}}{\left(8.33 \dfrac{\text{lbm}}{\text{gal}}\right)(1000)} = \boxed{2000 \text{ gal}}$$

(d) The cost is

$$\text{cost} = \dfrac{\left(\dfrac{30 \text{ ft} + 6 \text{ ft}}{2}\right)(5 \text{ ft})(540 \text{ ft})\left(\dfrac{\$0.50}{1 \text{ yd}^3}\right)}{27 \dfrac{\text{ft}^3}{\text{yd}^3}}$$

$$= \boxed{\$900.00}$$

(e) The volume is

$$V = \dfrac{\left(\dfrac{12 \text{ ft} + 4 \text{ ft}}{2}\right)(2 \text{ ft})(162 \text{ ft})}{27 \dfrac{\text{ft}^3}{\text{yd}^3}} = \boxed{96 \text{ yd}^3}$$

(f) The height is

$$h = \dfrac{3000 \text{ gal}}{\left(7.5 \dfrac{\text{gal}}{\text{ft}^3}\right)\left(\dfrac{\pi}{4}\right)(10 \text{ ft})^2} = \boxed{5.1 \text{ ft}}$$

Dim Eqs

(g) The cost is

$$\text{cost} = \frac{(100 \text{ ft})(81 \text{ ft})\left(\dfrac{\$9.00}{1 \text{ yd}^2}\right)}{9 \dfrac{\text{ft}^2}{\text{yd}^2}} = \boxed{\$8100.00}$$

(h) The depth is

$$d = \frac{3840 \text{ ft}^2}{32 \text{ ft}} = \boxed{120 \text{ ft}}$$

(i) (a) The length is

$$L = \sqrt{(16 \text{ mi})^2 + (12 \text{ mi})^2} = \boxed{20 \text{ mi}}$$

(b) The weight is

$$W = \frac{(20 \text{ mi})\left(5280 \dfrac{\text{ft}}{\text{mi}}\right)(50 \text{ lbm})}{(100 \text{ ft})\left(2000 \dfrac{\text{lbm}}{\text{ton}}\right)} = \boxed{26.4 \text{ ton}}$$

(j) The area of the triangle is

$$\left(\frac{1}{2}\right)(320 \text{ yd})\left(3 \dfrac{\text{ft}}{\text{yd}}\right)(1320 \text{ ft}) = \boxed{633,600 \text{ ft}^2}$$

The perimeter of the triangle is

$$\sqrt{(960)^2 + (1320 \text{ ft})^2} + 960 \text{ ft} + 1320 \text{ ft} = 3912 \text{ ft}$$

The side of the rectangle is

$$\frac{633,600 \text{ ft}^2}{(250 \text{ yd})\left(3 \dfrac{\text{ft}}{\text{yd}}\right)} = 845 \text{ ft}$$

The perimeter of the rectangle is

$$(2)\left(3 \dfrac{\text{ft}}{\text{yd}}\right)(250 \text{ yd}) + (2)(845 \text{ ft}) = 3190 \text{ ft}$$

$$\boxed{\text{The triangular piece would require the most fence.}}$$

(k) The volume is

$$V = \frac{(3.14)(1 \text{ ft})^2(54 \text{ ft})}{27 \dfrac{\text{ft}^3}{\text{yd}^3}} = \boxed{6.3 \text{ yd}^3}$$

(l) The cost is

$$\text{cost} = \frac{\left(\dfrac{2 \text{ ft} + 10 \text{ ft}}{2}\right)(50 \text{ ft})(100 \text{ ft})}{1000 \dfrac{\text{gal}}{1000 \text{ gal}}}$$
$$\times \frac{\left(7.5 \dfrac{\text{gal}}{\text{ft}^3}\right)\left(0.20 \dfrac{\$}{1000 \text{ gal}}\right)}{}$$

$$= \boxed{\$45.00}$$

(m) (a) The cost is

$$\left(\frac{30°}{360°}\right)(24 \text{ lbm})\left(\frac{\$1.00}{1 \text{ lbm}}\right) = \$2.00$$

(b) The volume is

$$V = \left(\frac{1}{12}\right)\pi(16 \text{ in})^2(8 \text{ in}) = 134 \text{ in}^3$$

(n) The volume is

$$V = \left(\frac{3}{4}\right)(10 \text{ ft})(5 \text{ ft})(4 \text{ ft})\left(7.5 \dfrac{\text{gal}}{\text{ft}^3}\right) = \boxed{1125 \text{ gal}}$$

(o) The area is

$$A = \left(\frac{\pi}{4}\right)(1000 \text{ ft})^2 - \left(\frac{1}{2}\right)(500 \text{ ft})(500 \text{ ft})$$

$$= \boxed{71,250 \text{ ft}^2}$$

(p) The diameter is

$$D = 2\sqrt{\frac{60,000 \text{ gal}}{\left(7.5 \dfrac{\text{gal}}{\text{ft}^3}\right)(5 \text{ ft})\pi}} = \boxed{45 \text{ ft}}$$

(q) The volume is

$$V = \frac{\left(\dfrac{1.4 \text{ ft} + 2.2 \text{ ft}}{2}\right)(150 \text{ ft})(100 \text{ ft})}{27 \dfrac{\text{ft}^3}{\text{yd}^3}} = \boxed{1000 \text{ yd}^3}$$

8 Signed Numbers

1. POSITIVE AND NEGATIVE NUMBERS

In arithmetic the symbol for plus (+) indicates that something is to be added; the symbol for minus (−) indicates that something is to be subtracted. These same symbols are also used to show the values of numbers. The numbers are called *signed numbers*.

In algebra the plus sign + before a number indicates that the number is a *positive* number; the sign − before a number indicates that the number is a *negative* number. If there is no sign before a number, it is considered to be a positive number.

Positive numbers are greater than zero; negative numbers are less than zero. Zero is neither positive nor negative.

The relative value of numbers can be shown by a graduated horizontal line as shown in Fig. 8.1.

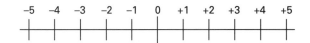

Figure 8.1 *Relative Value of Numbers*

Numbers to the right of zero are positive; numbers to the left of zero are negative. Values of numbers increase from left to right. For instance, in Fig. 8.1, −4 is less than −3 and −3 is less than −2. This statement can be simplified by use of symbols: $-4 < -3 < -2$. The symbol < means "less than," and the symbol > means "greater than." $5 > 3 > 1$ means "5 is greater than 3, and 3 is greater than 1."

2. ABSOLUTE VALUE

The *absolute value* of a number is the numerical value of a number, the value of the number without regard to its sign. The absolute value of −5 is 5; the absolute value of +5 is 5. The symbol for the absolute value of −5 is $|-5|$.

3. ADDITION OF SIGNED NUMBERS

To add signed numbers, the following three rules should be remembered.

rule 1: To add two or more numbers with like signs, find the sum of their absolute values and prefix the common sign to the sum.

rule 2: To add two numbers with unlike signs, find the difference between their absolute values and prefix the sign of the number having the greater absolute value.

rule 3: To add more than two numbers with unlike signs, add the positive numbers and negative numbers separately and use the two sums as in *rule 2*.

Example 8.1

(a) +9
   +6

(b) −8
   −5

(c) $(+9) + (+6) =$

(d) $(-8) + (-5) =$

Solution

(a) +15

(b) −13

(c) +15

(d) −13

Example 8.2

(a) +9
   −6

(b) −8
   +5

(c) $(+9) + (-6) =$

(d) $(-8) + (+5) =$

Solution

(a) $+3$

(b) -3

(c) $+3$

(d) -3

Example 8.3

(a) -3
$+6$
$+4$
$\underline{-2}$

(b) $+5$
-8
-3
$\underline{-2}$

(c) -7
$+4$
$+5$
$\underline{-2}$

Solution

(a) $+5$

(b) -8

(c) 0

4. SUBTRACTION OF SIGNED NUMBERS

rule: To subtract two signed numbers, change the sign of the subtrahend and add it to the minuend.

Subtracting $+3$ from $+10$ is the same as adding -3 to $+10$.

(a) $(+10) - (+3) = 7$

(b) $(+10) - (-3) = 13$

(c) $(+10) + (+3) = 7$

(d) $(+10) + (+3) = 13$

Example 8.4

In the following examples, subtract the bottom number from the top number.

(a) $+12$
$\underline{+3}$

(b) $+10$
$\underline{-4}$

(c) -18
$\underline{+6}$

(d) -16
$\underline{-7}$

Solution

(a) $+12$
$(-)\underline{+3}$
$+9$

(b) $+10$
$(+)\underline{-4}$
$+14$

(c) -18
$(-)\underline{+6}$
-24

(d) -16
$(+)\underline{-7}$
-9

5. HORIZONTAL ADDITION AND SUBTRACTION

Rules established for addition and subtraction of signed numbers apply to horizontal addition and subtraction.

When parentheses are preceded by a positive sign, they can be removed and the sign of the term enclosed will remain unchanged, but when parentheses are preceded by a negative sign, they can be removed only if the sign of the term enclosed is changed. This is in accordance with the rule for subtraction of signed numbers.

Example 8.5

Combine the following numbers into a single number.

(a) $(-8) + (-3) - (+2)$

(b) $(+12) - (-4) + (-3)$

(c) $(+16) + (-2) - (-3) - (+5)$

(d) $(+8) - (+1) + (-2) - (-5)$

Solution

(a) $-8 - 3 - 2 = -13$

(b) $12 + 4 - 3 = 13$

(c) $16 - 2 + 3 - 5 = 12$

(d) $8 - 1 - 2 + 5 = 10$

Example 8.6

(a) $(+5) + (-2) - (+4) - (-6)$

(b) $(-7) - (+3) + (-5) - (-2)$

(c) $(8) - (3) + (-7) - (6)$

(d) $(-15) - (-12) + (-8) - (5)$

(e) $(27) - (36) - (-45) + (-15)$

(f) $(18) + (12) - (-21) - (16)$

Solution

(a) $5 - 2 - 4 + 6 = 5$

(b) $-7 - 3 - 5 + 2 = -13$

(c) $8 - 3 - 7 - 6 = -8$

(d) $-15 + 12 - 8 - 5 = -16$

(e) $27 - 36 + 45 - 15 = 21$

(f) $18 + 12 + 21 - 16 = 35$

6. MULTIPLICATION OF SIGNED NUMBERS

The *product* of signed numbers is found by multiplying the absolute value of the numbers, as in arithmetic, and prefixing the sign of the product according to the following rule.

rule: The product of two numbers with like signs is a positive number. The product of two numbers with unlike signs is a negative number.

The product of two or more signed numbers is positive if there is an even number of negative factors and is negative if there is an odd number of negative factors.

Example 8.7

(a) $(-3)(+4) =$

(b) $(-3)(-4) =$

(c) $(6)(-2)(3) =$

(d) $(4)(-5)(-3) =$

(e) $(5)(-4)(-3) =$

(f) $(-3)(4)(-5)(-6) =$

Solution

(a) -12

(b) 12

(c) -36

(d) 60

(e) 60

(f) -360

7. DIVISION OF SIGNED NUMBERS

Division is the inverse of multiplication. The quotient of signed numbers is found by dividing the absolute value of the numbers as in arithmetic and prefixing the sign of the quotient according to the following rule.

rule: The quotient of two numbers with like signs is a positive number. The quotient of two numbers with unlike signs is a negative number.

Example 8.8

(a) $\dfrac{(-6)(-2)}{(3)(4)} =$

(b) $\dfrac{(-5)(-2)(21)}{(4)(15)(-7)} =$

(c) $\dfrac{(21)(-36)(-48)}{(-7)(12)(24)} =$

(d) $\dfrac{(16)(-24)(0)}{(12)(48)(18)} =$

(e) $\dfrac{(-15)(12)(-18)}{(9)(-5)(4)(36)} =$

(f) $\dfrac{(-30)(28)(-64)(72)}{(-45)(14)(-16)(-12)} =$

Solution

(a) 1

(b) $-\dfrac{1}{2}$

(c) -18

(d) 0

(e) $-\dfrac{1}{2}$

(f) -32

PRACTICE PROBLEMS

1. Add the following numbers algebraically.

Examples:

(1) $\begin{array}{r} +8 \\ +7 \\ \hline 15 \end{array}$ (3) $\begin{array}{r} -12 \\ +8 \\ \hline -4 \end{array}$ (5) $\begin{array}{r} -4 \\ +5 \\ +3 \\ -2 \\ -7 \\ \hline -5 \end{array}$ (6) $\begin{array}{r} -12 \\ -14 \\ +16 \\ -11 \\ +19 \\ \hline -2 \end{array}$

(2) $\begin{array}{r} -9 \\ -7 \\ \hline -16 \end{array}$ (4) $\begin{array}{r} +5 \\ -12 \\ \hline -7 \end{array}$

(7) $(+7) + (-4) + (-3) + (-6) + (+9)$
$= 7 - 4 - 3 - 6 + 9 = 3$

(a) $+12$ -22 $+35$ -19 -51 -76
    $+27$ $+13$ -18 -26 $+51$ $\underline{0}$

(b)

-18	22	14	-21	13	90
20	-12	16	-32	11	85
12	-13	-10	18	-44	-25
14	31	-12	-47	27	-75
-10	-14	20	36	-18	-30
$\underline{-17}$	$\underline{25}$	$\underline{-15}$	$\underline{12}$	$\underline{-14}$	$\underline{45}$

(c) $(-8) + (-9) + (+5) + (-2)$

(d) $(+9) + (+5) + (-12) + (-10)$

(e) $(+21) + (-12) + (-13) + (+30)$

(f) $(-17) + (-13) + (+11) + (-10)$

(g) $(+13) + (+12) + (+20) + (-16)$

(h) $(+18) + (-16) + (-11) + (-10)$

(i) $(-15) + (+20) + (-12) + (+12)$

(j) $(+13) + (+11) + (-44) + (+27)$

2. In the following problems, subtract the bottom number from the top number.

Examples:

(1) $+8$ (2) $+9$ (3) -12 (4) -10
   $(-)\ +5$ $(+)\ -3$ $(+)\ -8$ $(-)\ +3$
     $\overline{+3}$ $\overline{+12}$ $\overline{-4}$ $\overline{-13}$

(a) $+8$ -12 -15 $+21$ $+32$ -28
   $\underline{-3}$ $\underline{+6}$ $\underline{-5}$ $\underline{-9}$ $\underline{+18}$ $\underline{-32}$

(b) 47 65 -38 18 -62 -27
   $\underline{56}$ $\underline{-35}$ $\underline{24}$ $\underline{18}$ $\underline{0}$ $\underline{-27}$

3. Combine each of the following numbers into a single number.

Example:

$$(+6) - (+3) - (-4) + (-5) + (+8)$$
$$= 6 - 3 + 4 - 5 + 8 = 10$$

(a) $(-6) + (-3) - (+4) - (-5)$

(b) $(12) - (+3) - (-4) + (-8)$

(c) $(9) - (4) + (-3) - (11)$

(d) $(-8) + (12) - (-8) - (10)$

(e) $(15) - (13) - (17) - (-14)$

(f) $(-28) - (36) + (32) + (-12)$

(g) $(57) - (43) - (68) + (22)$

(h) $(125) - (100) + (55) - (40)$

4. Multiply the following numbers.

Examples:

(1) $(-6)(+3) = -18$ (2) $(-4)(-5) = 20$

(3) $(5)(-8)(-2) = 80$ (4) $(-6)(-5)(-3) = -90$

(a) $(-4)(5)(3)(2)$

(b) $(6)(-4)(-2)(3)$

(c) $(3)(2)(4)(-5)(6)$

(d) $(2)(3)(-5)(-4)(6)$

5. Perform the following indicated operations. (Note: An odd number of negative factors will produce a negative result. First determine the sign of the result, then perform cancellation without regard to sign.)

Examples:

(1) $\dfrac{(-6)(2)}{(-3)(4)} = 1$ (2) $\dfrac{(4)(-3)}{(-2)(-6)} = -1$

(3) $\dfrac{(8)(15)(-6)}{(5)(-4)(2)} = 18$ (4) $\dfrac{(16)(-18)(24)}{(27)(-4)(-8)} = -8$

(a) $\dfrac{(-36)(21)(-48)}{(12)(-7)(24)}$ (d) $\dfrac{(28)(-3)(-48)}{(12)(-18)(14)}$

(b) $\dfrac{(24)(-16)(0)}{(38)(-14)(-12)}$ (e) $\dfrac{(45)(-9)(-20)}{(-27)(-15)(4)}$

(c) $\dfrac{(-28)(45)(21)}{(-15)(14)(-7)}$ (f) $\dfrac{(11)(-16)(15)}{(-5)(2)(3)}$

SOLUTIONS

1. (a)

+12	−22	+35	−19	−51	−76
+27	+13	−18	−26	+51	0
$\boxed{39}$	$\boxed{-9}$	$\boxed{17}$	$\boxed{-45}$	$\boxed{0}$	$\boxed{-76}$

(b)

−18	22	14	−21	13	90
20	−12	16	−32	11	85
12	−13	−10	18	−44	−25
14	31	−12	−47	27	−75
−10	−14	20	36	−18	−30
−17	25	−15	12	−14	45
$\boxed{1}$	$\boxed{39}$	$\boxed{13}$	$\boxed{-34}$	$\boxed{-25}$	$\boxed{90}$

(c) $(-8) + (-9) + (+5) + (-2) =$

$$-8 - 9 + 5 - 2 = \boxed{-14}$$

(d) $(+9) + (+5) + (-12) + (-10) =$

$$9 + 5 - 12 - 10 = \boxed{-8}$$

(e) $(+21) + (-12) + (-13) + (+30) =$

$$21 - 12 - 13 + 30 = \boxed{26}$$

(f) $(-17) + (-13) + (+11) + (-10) =$

$$-17 - 13 + 11 - 10 = \boxed{-29}$$

(g) $(+13) + (+12) + (+20) + (-16) =$

$$13 + 12 + 20 - 16 = \boxed{29}$$

(h) $(+18) + (-16) + (-11) + (-10) =$

$$18 - 16 - 11 - 10 = \boxed{-19}$$

(i) $(-15) + (+20) + (-12) + (+12) =$

$$-15 + 20 - 12 + 12 = \boxed{5}$$

(j) $(+13) + (+11) + (-44) + (+27) =$

$$13 + 11 - 44 + 27 = \boxed{7}$$

2. (a)

+8	−12	−15
(−) −3	(−) +6	(−) −5
$\boxed{11}$	$\boxed{-18}$	$\boxed{-10}$

+21	+32	−28
(−) −9	(−) +18	(−) −32
$\boxed{30}$	$\boxed{14}$	$\boxed{4}$

(b)

+47	+65	−38
(−) +56	(−) −35	(−) +24
$\boxed{-9}$	$\boxed{100}$	$\boxed{-62}$

+18	−62	−27
(−) +18	(−) 0	(−) −27
$\boxed{0}$	$\boxed{-62}$	$\boxed{0}$

3. (a) $(-6) + (-3) - (+4) - (-5) =$

$$-6 - 3 - 4 + 5 = \boxed{-8}$$

(b) $(12) - (+3) - (-4) + (-8) =$

$$12 - 3 + 4 - 8 = \boxed{5}$$

(c) $(9) - (4) + (-3) - (11) =$

$$9 - 4 - 3 - 11 = \boxed{-9}$$

(d) $(-8) + (12) - (-8) - (10) =$

$$-8 + 12 + 8 - 10 = \boxed{2}$$

(e) $(15) - (13) - (17) - (-14) =$

$$15 - 13 - 17 + 14 = \boxed{-1}$$

(f) $(-28) - (36) + (32) + (-12) =$

$$-28 - 36 + 32 - 12 = \boxed{-44}$$

(g) $(57) - (43) - (68) + (22) =$

$$57 - 43 - 68 + 22 = \boxed{-32}$$

(h) $(125) - (100) + (55) - (40) =$

$$125 - 100 + 55 - 40 = \boxed{40}$$

4. (a) $\quad (-4)(5)(3)(2) = \boxed{-120}$

(b) $\quad (6)(-4)(-2)(3) = \boxed{144}$

(c) $\quad (3)(2)(4)(-5)(6) = \boxed{-720}$

(d) $\quad (2)(3)(-5)(-4)(6) = \boxed{720}$

5. (a) $\quad \dfrac{(-36)(21)(-48)}{(12)(-7)(24)} = \boxed{-18}$

(b) $\quad \dfrac{(24)(-16)(0)}{(38)(-14)(-12)} = \boxed{0}$

(c) $\quad \dfrac{(-28)(45)(21)}{(-15)(14)(-7)} = \boxed{-18}$

(d) $\quad \dfrac{(28)(-3)(-48)}{(12)(-18)(14)} = \boxed{-1\tfrac{1}{3}}$

(e) $\quad \dfrac{(45)(-9)(-20)}{(-27)(-15)(4)} = \boxed{5}$

(f) $\quad \dfrac{(11)(-16)(15)}{(-5)(2)(3)} = \boxed{88}$

9 Algebra

1. LITERAL NUMBERS

Letters of the alphabet used to represent numbers are called *literal numbers*. The letters a, b, c, x, y, and z are commonly used to represent a number, but any letter of the alphabet can be used. Literal numbers are called *general numbers* because they do not represent a specific number. For example, the area of a certain room that is 12 ft long and 10 ft wide is equal to the product 12×10. To express the area of any rectangle, it can be written as $A = LW$, where A is area in square measure, L is length in linear measure, and W is width in linear measure.

2. USING LITERAL NUMBERS

Addition, subtraction, multiplication, and division using literal numbers are performed in the same manner as in arithmetic. If a and b represent any two numbers, their sum is $a + b$; their difference is $a - b$; their product is $a \times b$, $a \cdot b$, $(a)(b)$, or ab; and their quotient is $a \div b$ or a/b. Expressing the product as $a \cdot b$, or ab prevents confusing the letter x with the multiplication sign $\times$.

3. DEFINITIONS

A *term* is an algebraic expression not separated within itself by a plus or minus sign, such as $4xy$, $5x^2y$, or $2ab^2c$.

A *product* implies multiplication. The term $5xy$ means $5 \times x \times y$.

A *monomial* consists of one term, such as $4a^2b$; a *binomial* consists of two terms, such as $3x^2 - 5xy$; a *trinomial* consists of three terms, and a *polynomial* consists of any number of terms more than one.

The quantities multiplied together to form a product are called the *factors* of the product. The factors of $3xy$ are 3, x, and y.

The numerical factor in a monomial is known as the *numerical coefficient* or simply the *coefficient*. The coefficient of $3x^2y$ is 3, and the coefficient of $-5ab$ is -5.

Like or *similar terms* are terms that have the same literal factors. Terms that do not have the same literal factors are called *unlike* or *dissimilar* terms. $3x^2y$ and $5x^2y$ are like terms; $5x^2y$ and $5x^2y^2$ are unlike terms.

4. HORIZONTAL ADDITION AND SUBTRACTION OF MONOMIALS

Horizontal addition and subtraction of monomials is carried out according to the rules of addition and subtraction of signed numbers. Like terms can be combined; unlike terms cannot be combined.

Example 9.1

(a) $11ab + 4ab - 10ab - 8ab =$

(b) $5x^3 + 2x^2y - 8x^2y + 2x^3 =$

(c) $2ab^2 - 3ab + 2 - 4ab^2 + 3ab =$

Solution

(a) $-3ab$

(b) $7x^3 - 6x^2y$

(c) $-2ab^2 + 2$

5. EXPONENTS

As shown previously, 3 squared equals 9, meaning that 3 multiplied by itself equals 9. This can be written as $3 \cdot 3 = 9$ or $3^2 = 9$. It is also true that 2 cubed equals 8, meaning that $2 \cdot 2 \cdot 2 = 8$. This can be written as $2^3 = 8$. The small number 3 is known as the *exponent*; the number 2 is known as the *base*. 4^6 means that six 4's are to be multiplied. The exponent is the power to which a number is to be raised. If a number does not have an exponent, it is of the first power and the exponent is considered to be 1. To raise a literal number to a power, use an exponent. Thus, x^5 means $x \cdot x \cdot x \cdot x \cdot x$ and is read "x raised to the fifth power" or "x to the fifth."

6. EXPONENTS USED IN MULTIPLICATION

Multiplication in algebra follows the same rules, or laws, as multiplication in arithmetic. Exponents simplify the process. To multiply $x^3 \cdot x^4$,

$$x^3 \cdot x^4 = (x \cdot x \cdot x)(x \cdot x \cdot x \cdot x) = x^7$$

However, it is much easier to say that

$$x^3 \cdot x^4 = x^{3+4} = x^7$$

This can be expressed in the form of a rule.

rule: The exponent of the product of powers with the same base is the sum of the exponents of the factors.

The rule holds for either positive or negative exponents, and the base may be an arithmetic number. Thus,

$$(2)^5(2)^{-2} = (2)^3 = 8$$

The rule for multiplication holds only for the product of powers of the same base; however, multiplication involving powers of more than one base may be simplified. For example,

$$(x^3y^4z^2)(x^2y^{-2}z) = x^5y^2z^3$$
$$(2)^4(3)^3(4)^4 = 16 \cdot 27 \cdot 16 = 6912$$

7. EXPONENTS USED IN DIVISION

Division is the inverse of multiplication, so it is logical to conclude that in division the exponent of the divisor is subtracted from the exponent of the dividend.

rule: The exponent of the quotient of two powers with the same base is the difference of the exponent of the dividend and the exponent of the divisor.

Example 9.2

Divide the following numbers.

(a) $\dfrac{x^5}{x^2}$

(b) $\dfrac{(3)^5}{(3)^{5-2}}$

(c) $a^2 \div a^{-3}$

(d) $\dfrac{x^3}{x^5}$

Solution

(a) $\dfrac{x \cdot x \cdot x \cdot x \cdot x}{x \cdot x} = x \cdot x \cdot x = x^3$

(b) $3^{5-2} = (3)^3 = 27$

(c) $a^{2+3} = a^5$

(d) $x^{3-5} = x^{-2}$

8. EXPONENT OF THE POWER OF A POWER

The exponent of a power is the product of the exponents of the powers.

Example 9.3

(a) $(x^3)^2$

(b) $(-5x^2y^3)^2$

(c) $\dfrac{a^3(b^2)^4}{(ab^3)^2}$

Solution

(a) $(x \cdot x \cdot x) \cdot (x \cdot x \cdot x) = x^{3 \cdot 2} = x^6$

(b) $(-5)^2 x^{2 \cdot 2} y^{3 \cdot 2} = 25x^4y^6$

(c) $\dfrac{a^3 b^8}{a^2 b^6} = ab^2$

9. ZERO POWER

Any number other than zero raised to the zero power is equal to 1. This applies to arithmetic numbers and to literal numbers, as well as to a polynomial enclosed in parentheses.

Example 9.4

(a) $\dfrac{(2)^3}{(2)^3}$

(b) $\dfrac{x^3}{x^3}$

Solution

(a) $\dfrac{2 \cdot 2 \cdot 2}{2 \cdot 2 \cdot 2} = (2)^{3-3} = (2)^0 = 1$

(b) $x^{3-3} = x^0 = 1$

10. NEGATIVE EXPONENTS

A number with a negative exponent is equal to 1 divided by the number with the sign of the exponent changed to positive.

Example 9.5

(a) $\dfrac{(2)^2}{(2)^4}$

(b) $\dfrac{(10)^3}{(10)^4}$

(c) $\dfrac{x^{-2}y^3}{x^3 y^{-2}}$

(d) $\dfrac{a^{-3}}{b^{-2}}$

Solution

(a) $\dfrac{4}{16} = \dfrac{1}{4}$

(b) $10^{-1} = \dfrac{1}{10}$

(c) $\dfrac{y^3 \cdot y^2}{x^2 \cdot x^3} = \dfrac{y^5}{x^5}$

(d) $\dfrac{b^2}{a^3}$

Example 9.6

(a) $5 \cdot 5 \cdot 5$

(b) $a^3 \cdot a^5$

(c) $(4)^3 (4)^2$

(d) $(3x)(4x)$

(e) $x^3 \div x^{-2}$

(f) $(2)^2 (3)^2 (4)^2$

(g) $(ab^2)^3$

(h) $(-3x)^5$

(i) $(-2xy^2)^2$

(j) $\dfrac{x^3 \cdot x^4}{x^2}$

(k) $\dfrac{(3)^7 (3)^2 \cdot 3}{3^5}$

(l) $\dfrac{64a^3 b^3}{4ab^2}$

(m) $\dfrac{5x^5 y^5}{45x^6 y^6}$

(n) $\dfrac{(-2x^2 y^2)^3}{(2xy^2)^2}$

(o) $\dfrac{(-3ab^2)^3}{(12ab^2)^2}$

(p) $(5)^{-2}(5)^3$

(q) $a^3 \cdot a^{-1}$

(r) $(10)^0$

(s) $3a^0$

(t) $x^0 y^2$

(u) $(4-2)^0$

(v) $\dfrac{x^{-5} y^3}{xy}$

(w) $\dfrac{a^{-4} b^5 c^{-6}}{ab^{-2} c^3}$

(x) $\dfrac{x^3 y^{-2}}{xy^3}$

Solution

(a) 125

(b) a^8

(c) $(4)^5 = 1024$

(d) $12x^2$

(e) x^5

(f) $4 \cdot 9 \cdot 16 = 576$

(g) $a^3 b^6$

(h) $-243x^5$

(i) $4x^2 y^4$

(j) x^5

(k) $(3)^5$

(l) $16a^2 b$

(m) $\dfrac{1}{9xy}$

(n) $-2x^4 y^2$

(o) $\dfrac{-27b^4}{144a}$

(p) 5

(q) a^2

(r) 1

(s) 3

(t) y^2

(u) 1

(v) $\dfrac{y^2}{x^6}$

(w) $\dfrac{b^7}{a^5 c^9}$

(x) $\dfrac{x^2}{y^5}$

11. MULTIPLYING A MONOMIAL AND A POLYNOMIAL

When a polynomial is multiplied by a monomial, each term of the polynomial must be multiplied by the monomial.

Example 9.7

$$-4x(2x^2 - 3x - 4)$$

Solution

$$(-4x)(2x^2) - (-4x)(3x) - (-4x)(4)$$
$$= -8x^3 + 12x^2 + 16x$$

12. MULTIPLYING BINOMIALS OR TRINOMIALS

In multiplying two binomials, each term of the multiplicand must be multiplied by each term of the multiplier.

Example 9.8

Multiply the following numbers.

(a)
$$\begin{array}{r} 2x - 3y \\ 3x + 2y \end{array}$$

(b)
$$\begin{array}{r} x^2 + 4x - 3 \\ 3x - 2 \end{array}$$

Solution

(a)
$$\begin{array}{r} 2x - 3y \\ 3x + 2y \\ \hline 6x^2 - 9xy \\ + 4xy - 6y^2 \\ \hline 6x^2 - 5xy - 6y^2 \end{array}$$

(b)
$$\begin{array}{r} x^2 + 4x - 3 \\ 3x - 2 \\ \hline 3x^3 + 12x^2 - 9x \\ - 2x^2 - 8x + 6 \\ \hline 3x^3 + 10x^2 - 17x + 6 \end{array}$$

In both solutions, the factors are arranged in a manner similar to multiplication in arithmetic with the multiplier placed under the multiplicand. Multiplication is performed from right to left.

Using Ex. 9.8., the steps in the operation are as follows.

step 1: The left term in the multiplier is multiplied by the left term in the multiplicand, and the product is placed under the first column: $(3x)(2x) = 6x^2$.

step 2: The left term in the multiplier is multiplied by the next term (from left to right) in the multiplicand, and the product is placed under the second column: $(3x)(-3y) = -9xy$.

step 3: The second term in the multiplier is multiplied by the first term in the multiplicand, and the product is placed under $-9xy$ because it is a similar term: $(2y)(2x) = +4xy$.

step 4: The second term in the multiplier is multiplied by the second term in the multiplicand, and the product is placed in the third column: $(2y)(-3y) = -6y^2$.

step 5: Similar terms are combined to give the product: $6x^2 - 5xy - 6y^2$.

Example 9.9

(a) $3xy(x^2 - xz + z^2)$

(b) $2ab^2(a^2 + 2ab - 3b^2)$

(c) $-3x^2y(xy - 2xy^2 - 3y^3)$

(d) $(x + 3)(x + 4)$

(e) $(x - 2)(x^2 + 4x - 8)$

Solution

(a) $3x^3y - 3x^3yz + 3xyz^2$

(b) $2a^3b^2 + 4a^2b^3 - 6ab^4$

(c) $-3x^3y^2 + 6x^3y^3 + 9x^2y^4$

(d) $x^2 + 7x + 12$

(e) $x^3 + 2x^2 - 16x + 16$

13. DIVISION OF A POLYNOMIAL BY A MONOMIAL

To divide a polynomial by a monomial, each term in the dividend must be divided by the divisor.

Example 9.10

Divide the following numbers.

$$\frac{6ax^3y^3 - 8a^2x^2y^2 + 4ax^3y}{2axy}$$

Solution

$$\frac{6ax^3y^3 - 8a^2x^2y^2 + 4ax^3y}{2axy}$$

$$= \frac{6ax^3y^3}{2axy} - \frac{8a^2x^2y^2}{2axy} + \frac{4ax^3y}{2axy}$$

$$= 3x^2y^2 - 4axy + 2x^2$$

14. DIVISION OF POLYNOMIAL BY A POLYNOMIAL

To perform division of a polynomial by a polynomial, the terms in the dividend and the divisor should be arranged in order of decreasing power. It is performed much like division in arithmetic.

Example 9.11

Divide the following number.

$$(15x^2 - x - 6) \div (5x + 3)$$

Solution

$$\begin{array}{r} 3x - 2 \\ 5x+3\overline{\smash{\big)}15x^2 - x - 6} \\ \underline{15x^2 + 9x} \\ -10x - 6 \\ \underline{-10x - 6} \end{array}$$

The steps in the procedure are as follows.

step 1: Divide the first term in the dividend by the first term in the divisor, $15x^2 \div 5x = 3x$, and place $3x$ over the dividend.

step 2: Multiply the divisor by the first term in the quotient: $3x(5x + 3) = 15x^2 + 9x$. Place these two terms under similar terms in the dividend and subtract: $(15x^2 - x) - (15x^2 + 9x) = -10x$.

step 3: Bring down the next term in the dividend to form the new dividend: $-10x - 6$.

step 4: Divide the first term in the new dividend by the first term in the divisor, $-10x \div 5x = -2$, and place -2 as the second term in the quotient.

step 5: Multiply the divisor by the second term in the quotient: $(-2)(5x+3) = -10x-6$. Place these two terms under similar terms in the dividend and subtract: $(-10x-6)-(-10x-6) = 0$.

If there had been a remainder of lower order than the first term in the divisor, it would be written over the divisor, as in arithmetic, to represent the remainder. Because there was no remainder, $(5x + 3)(3x - 2)$ are factors of $15x^2 - x - 6$.

Example 9.12

Perform the indicated divisions.

(a) $\dfrac{36a^3 - 24a^2 - 12a}{2a}$

(b) $\dfrac{21a^3b^2 - 14a^2b^3 + 7ab}{7ab}$

(c) $\dfrac{15a^4b^3c^2 + 20a^3b^2c^3 - 25a^2b^2c^3}{5a^2b^2c}$

(d) $\dfrac{9x^5y^4z^3 - 18x^3y^2z^2 + 3x^4y^3z^2}{3x^3y^2z}$

Solution

(a) $18a^2 - 12a - 6$

(b) $3a^2b - 2ab^2 + 1$

(c) $3a^2bc + 4ac^2 - 5c^2$

(d) $3x^2y^2z^2 - 6z + xyz$

15. FACTORING

Factoring is the reverse of multiplication. The product of $8 \cdot 9$ is 72. The factors of 72 are 8 and 9, as well as 12 and 6. The prime factors of 72 are $2 \cdot 2 \cdot 2 \cdot 3 \cdot 3$.

16. FACTORING A POLYNOMIAL CONTAINING A COMMON MONOMIAL

The first step in factoring a polynomial is to factor a common monomial, if one exists. The other factor is found by dividing each term of the polynomial by the common factor and writing the result as the product of the common factor and the quotient.

Example 9.13

Factor $20x^2 + 12x$.

Solution

The greatest monomial factor in the binomial is $4x$.

$$20x^2 + 12x = 4x(5x + 3)$$

Example 9.14

Factor $xy^2z^3 - x^2y^3z^4$.

Solution

$$xy^2z^3 - x^2y^3z^4 = xy^2z^3(1 - xyz)$$

17. FACTORING A TRINOMIAL THAT IS A PERFECT SQUARE

The number 4 is a *perfect square* because both factors of 4 are the same. Likewise, 9, 16, and 25 are perfect squares.

A trinomial is also a perfect square if both factors are the same. The product of $(x+2)(x+2)$ is $x^2 + 4x + 4$, which is a perfect square. It can be said that $x^2 + 4x + 4 = (x+2)^2$.

To factor a trinomial that is a perfect square, it is necessary to recognize a trinomial that is a perfect square. Examine the following equation.

$$x^2 + 4x + 4 = (x+2)(x+2) = (x+2)^2$$

In the factor $(x+2)$,

(1) x is the square root of the first term in the trinomial.

(2) 2 is the square root of the third term in the trinomial.

(3) The product of these two terms is $2x$, which is one half the middle term of the trinomial.

(4) The sign between the two terms is the same as the sign of the middle term of the trinomial.

For a trinomial to be a perfect square, the following must be true.

(1) The first and third terms must be perfect squares.

(2) The middle term must be twice the product of the square roots of the first and third terms.

(3) The sign of the middle term can be plus or minus, but the sign of the second term of the factors must be the same as the sign of the middle term of the trinomial.

Examine the following trinomial and its factors.

$$x^2 - 14x + 49 = (x-7)^2$$

(1) Are the first and third terms perfect squares? Yes.

(2) Is the square root of the first term of the trinomial equal to the first term of the factors? Yes: $\sqrt{x^2} = x$.

(3) Is the square root of the third term of the trinomial equal to the second term of the factors? Yes: $\sqrt{49} = 7$.

(4) Is the middle term of the trinomial twice the product of the terms of the factors? Yes: $14x = (2)(7x)$.

(5) Is the sign of the second term in the factors the same as the sign of the middle term of the trinomial? Yes.

Therefore, the trinomial $x^2 - 14x + 49$ is a perfect square.

Example 9.15

(a) $25a^2 - 20a + 4$

(b) $9 - 24x + 16x^2$

(c) $4a^2 - 4a + 1$

(d) $x^2 - 6x + 9$

Solution

(a) $(5a - 2)^2$

(b) $(3 - 4x)^2$

(c) $(2a - 1)^2$

(d) $(x - 3)^2$

18. FACTORING THE DIFFERENCE BETWEEN TWO SQUARES

The product of $(a+b)(a-b)$ is $a^2 - b^2$.

$$
\begin{array}{r}
a + b \\
a - b \\
\hline
a^2 + ab \\
- ab - b^2 \\
\hline
a^2 - b^2
\end{array}
$$

Notice that the factors are the same except for the algebraic sign. Also notice that the product $(a^2 - b^2)$ is equal to the square of the first term of the factors minus the square of the second term. There is no middle term in the product because the sum of $+ab$ and $-ab$ is zero.

The factors of the difference between two squares are the product of the square root of the first term plus the square root of the second term times the square root of the first term minus the square root of the second term.

Example 9.16

(a) $x^2 - 25$

(b) $9x^2 - 16y^2$

(c) $36x^2 - 81y^2$

(d) $1 - 64a^2$

Solution

(a) $(x+5)(x-5)$

(b) $(3x+4y)(3x-4y)$

(c) $(9)(4x^2-9y^2) = (9)(2x+3y)(2x-3y)$

(d) $(1+8a)(1-8a)$

19. FACTORING A TRINOMIAL OF THE FORM $Ax^2 + Bx + C$

The factors of a perfect square, such as $a^2+2ab+b^2$, or the difference between two squares, such as a^2-b^2, are apparent as soon as the type of polynomial is identified. However, some polynomials, such as $2x^2-5x-12$, do not conform to either type.

Because factoring is the opposite of multiplication, start with the assumption that the factors of $2x^2-5x-12 = (2x+3)(x-4)$ and multiply the two factors.

$$2x+3$$
$$x-4$$
$$\overline{2x^2+3x}$$
$$-8x-12$$
$$\overline{2x^2-5x-12}$$

The first term in the trinomial $2x^2-5x-12$ is found by multiplying the first terms of the factors: $x(2x) = 2x^2$, and the third term of the trinomial is found by multiplying the second terms of the factors: $(-4)(3) = -12$. The middle term of the trinomial is the algebraic sum of the cross products: $x(3) + (-4)(2x) = -5x$.

Now factor $2x^2-x-15$.

The factors are not immediately apparent, so set up two pairs of parentheses to contain the factors and insert trial numbers.

$$2x^2-x-15 = (\quad)(\quad)$$

The product of the first terms of the factors must equal $2x^2$. The only possibility is $x(2x)$. But which of these two terms will be placed in the first set of parentheses and which in the second?

$$2x^2-x-15 = (x\quad)(2x\quad)$$

Because the middle and third terms of the trinomial are negative, one of the signs in the parentheses must be positive and one negative. (If both signs were negative, the sign of the third term in the trinomial would be positive.) But which will be positive and which negative?

$$2x^2-x-15 = (x-\quad)(2x+\quad)$$

The product of the second terms of the factors must be -15, so insert 5 in the first parentheses and 3 in the second. The product of these two terms is -15, but the algebraic sum of the cross products gives $-10x$, so try

$$2x^2-x-15 = (x-3)(2x+5)$$

The product of these two factors is $(2x^2-x-15)$, so the trinomial is factored.

With practice, various combinations of numbers within the parentheses can be tried by performing mental multiplication and addition.

Example 9.17

Factor the following polynomials.

(a) x^2+6x+5

(b) x^2-4x+3

(c) $a^2-9a+14$

(d) $5a^2+11a+6$

(e) $9x^2-13x+4$

Solution

(a) $(x+5)(x+1)$

(b) $(x-3)(x-1)$

(c) $(a-7)(a-2)$

(d) $(5a+6)(a+1)$

(e) $(9x-4)(x-1)$

Example 9.18

Factor the following polynomials.

(a) $8x^4-4x^3$

(b) a^2-16

(c) x^2+6x+9

(d) a^4-16

(e) $5a^2-30a+45$

(f) $16x^2-24x+9$

(g) $36a^2-81b^2$

(h) x^2-5x+6

(i) $a^2+4a-12$

(j) x^3-10x^2+9x

Solution

(a) $4x^3(2x-1)$

(b) $(a+4)(a-4)$

(c) $(x+3)^2$

(d) $(a^2+4)(a+2)(a-2)$

(e) $(5)(a-3)^2$

(f) $(4x - 3)^2$

(g) $(9)(2a + 3b)(2a - 3b)$

(h) $(x - 3)(x - 2)$

(i) $(a + 6)(a - 2)$

(j) $x(x - 9)(x - 1)$

20. EQUATIONS

Probably the most important concept in algebra is the *equation*. It is a means of solving problems in science, engineering, and everyday life.

An equation is a statement that two quantities are equal. The equal sign ($=$) separates the two quantities. The terms on the left of the equal sign are known as the *left member*; the terms on the right of the equal sign are known as the *right member*.

21. CONDITIONAL EQUATIONS

The statement "$2x + 4 = 10$" is true if, and only if, x is equal to 3. In other words, the equation is true on the condition that x is equal to 3. Therefore, 3 satisfies the equation.

It is important to be able to determine an unknown quantity in an equation.

22. ROOT OF AN EQUATION

In the equation $2x + 4 = 10$, 3 is the solution of the equation, or the *root* of the equation, and 3 is the only root of the equation.

23. SOLVING AN EQUATION

Solving an equation simply means finding the value of the literal number in the equation, or finding the root of the equation. The root of the equation $2x + 4 = 10$ can be found by inspection, but some equations are not so easily solved, so certain principles of algebra are necessary. These principles, or truths, are known as *axioms*.

24. AXIOMS

axiom 1: The same quantity may be added to both sides of an equation without altering the truth of the statement.

axiom 2: The same quantity may be subtracted from both sides of an equation without altering the truth of the statement.

axiom 3: Both sides of the equation may be multiplied by the same quantity, other than zero, without altering the truth of the statement.

axiom 4: Both sides of an equation may be divided by the same quantity, other than zero, without altering the truth of the statement.

axiom 5: The square root (or any root) of each side of an equation may be taken without altering the truth of the statement.

Example 9.19

Using axiom 1, solve the following equation.

$$x - 3 = 12$$

Solution

$$x - 3 = 12$$
Adding: $x - 3 + 3 = 12 + 3$
$$x = 15$$

Example 9.20

Using axiom 2, solve the following equation.

$$x + 6 = 12$$

Solution

$$x + 6 = 12$$
Subtracting: $x + 6 - 6 = 12 - 6$
$$x = 6$$

Example 9.21

Using axioms 3 and 4, solve the following equation.

$$4 = \frac{12}{x}$$

Solution

$$4 = \frac{12}{x}$$
Multiplying: $x(4) = \frac{x(12)}{x}$
$$4x = 12$$
Dividing: $\frac{4x}{4} = \frac{12}{4}$
$$x = 3$$

Example 9.22

Using axiom 5, solve the following equation.

$$x^2 = 16$$

Solution

$$x^2 = 16$$
Taking the square root: $\sqrt{x^2} = \sqrt{16}$
$$x = \pm 4$$

25. TRANSPOSING

Note in Ex. 9.19 (axiom 1) that when 3 was added to both sides of the eqation, -3 disappeared from the left side, leaving only x on that side. Also, in Ex. 9.20 (axiom 2) when 6 was subtracted from both sides of the equation, $+6$ disappeared from the left side, leaving only x. Putting the unknown quantity x on the left side with no other quantity can be accomplished by a shortcut method known as *transposing*. It is not an axiom but gives the same results as were obtained using axiom 1 and axiom 2.

Transposing means that any term may be moved from one side of an equation to the other side if its sign is changed.

Example 9.23

Solve the following equation.

$$3x - 2 = x + 6$$

Solution

$$3x - 2 = x + 6$$

Transposing: $\quad 3x - x = 6 + 2$

$$2x = 8$$

Dividing: $\quad\quad\quad x = 4$

Example 9.24

Solve the following equation.

$$4x + 2 - 3x - 6 = 5x + 4$$

Solution

$$4x + 2 - 3x - 6 = 5x + 4$$

Transposing: $\quad 4x - 3x - 5x = 4 - 2 + 6$

$$-4x = 8$$

Dividing by -4: $\quad\quad x = -2$

26. PARENTHESES

If a quantity within parentheses, brackets, or braces is preceded by a plus sign $(+)$, the parentheses may be removed without changing the sign of the terms within the parentheses.

If a quantity within parentheses is preceded by a minus sign $(-)$, the parentheses may be removed if the sign of each term within the parentheses is changed.

Example 9.25

Solve the following equation.

$$7x - (2x - 4) - (2)(4x - 2) + (3x - 3)$$
$$= 8 - (x - 5) - (3)(x + 4)$$

Solution

Removing parentheses:

$$7x - 2x + 4 - 8x + 4 + 3x - 3$$
$$= 8 - x + 5 - 3x - 12$$

Transposing: $\quad 7x - 2x - 8x + 3x + x + 3x$
$$= 8 + 5 - 12 - 4 - 4 + 3$$

Combining: $\quad\quad 4x = -4$

Dividing: $\quad\quad\quad x = -1$

27. FRACTIONAL EQUATIONS

To solve fractional equations, the first step is to get rid of the denominators. This is done by multiplying both sides of the equation by the lowest common denominator (LCD). Then, after the equation is cleared of fractions, the equation is solved using the axioms.

Example 9.26

Solve the following equation.

$$\frac{3x}{4} + 5 = 2x - \frac{1}{3}$$

Solution

To solve equations with fractions, it is convenient to write any term without a denominator with a denominator of 1.

$$\frac{3x}{4} + \frac{5}{1} = \frac{2x}{1} - \frac{1}{3}$$

The LCD is 12.

Multiplying by the LCD:

$$\left(\frac{12}{1}\right)\left(\frac{3x}{4}\right) + \left(\frac{12}{1}\right)\left(\frac{5}{1}\right)$$
$$= \left(\frac{12}{1}\right)\left(\frac{2x}{1}\right) - \left(\frac{12}{1}\right)\left(\frac{1}{3}\right)$$
$$9x + 60 = 24x - 4$$

Transposing: $\quad -15x = -64$

$$x = \frac{64}{15}$$

Example 9.27

Solve the following equations.

(a) $3x + 5 = 14$

(b) $5x - 8 = x + 16$

(c) $4x + (3)(2x - 5) = 7 + (5)(3x - 7)$

(d) $(2x - 3)(x + 4) - 23 = (x + 7)(2x)$

(e) $(2)(x+3)(x-4) = 6 + 2x(x-5)$

(f) $\dfrac{x-5}{x-3} - \dfrac{x+6}{x^2+x-12} = \dfrac{x-2}{x+4}$

Solution

(a)
$$3x + 5 = 14$$
$$3x = 14 - 5$$
$$3x = 9$$
$$x = 3$$

(b)
$$5x - 8 = x + 16$$
$$5x - x = 16 + 8$$
$$4x = 24$$
$$x = 6$$

(c)
$$4x + (3)(2x - 5) = 7 + (5)(3x - 7)$$
$$4x + 6x - 15 = 7 + 15x - 35$$
$$4x + 6x - 15x = 7 - 35 + 15$$
$$-5x = -13$$
$$x = \frac{13}{5}$$

(d)
$$(2x - 3)(x + 4) - 23 = (x + 7)(2x)$$
$$2x^2 + 5x - 12 - 23 = 2x^2 + 14x$$
$$5x - 14x = 35$$
$$-9x = 35$$
$$x = -\frac{35}{9}$$

(e)
$$(2)(x + 3)(x - 4) = 6 + 2x(x - 5)$$
$$(2)(x^2 - x - 12) = 6 + 2x^2 - 10x$$
$$2x^2 - 2x - 24 = 6 + 2x^2 - 10x$$
$$-2x + 10x = 6 + 24$$
$$8x = 30$$
$$x = \frac{30}{8} = \frac{15}{4}$$

(f)
$$\frac{x-5}{x-3} - \frac{x+6}{x^2+x-12} = \frac{x-2}{x+4}$$
$$(x + 4)(x - 5) - (x + 6) = (x - 3)(x - 2)$$
$$x^2 - x - 20 - x - 6 = x^2 - 5x + 6$$
$$-x - x + 5x = 6 + 20 + 6$$
$$3x = 32$$
$$x = \frac{32}{3}$$

28. LITERAL EQUATIONS AND FORMULAS

Literal equations are equations in which some, or all, of the quantities are literal numbers.

Literal equations are solved in the same way as any other equation, by using the axioms mentioned previously.

For example, if the length and width of a room are known and the area is needed, the formula $A = LW$ is in proper form, but if the length and area are known and the width is needed, the formula can be written as $W = A/L$.

Example 9.28

From the formula for the area of a rectangle, $A = LW$, find the formula for the width of a rectangle.

Solution
$$A = LW$$
Transposing: $\quad LW = A$

Dividing by L: $\quad W = \dfrac{A}{L}$

Example 9.29

Solve the following formula for C.

$$F = \frac{9}{5}C + 32$$

Solution
$$\frac{F}{1} = \frac{9C}{5} + \frac{32}{1}$$

Multiplying by the LCD (5):
$$v5F = 9C + (5)(32)$$

Transposing: $\quad 9C = 5F - (5)(32)$
$$= (5)(F - 32)$$

Dividing by 9: $\quad C = \dfrac{5(F - 32)}{9}$

Or, $\quad C = \dfrac{5}{9}(F - 32)$

29. QUADRATIC EQUATIONS

Equations that have been solved to this point have been first-degree equations known as *linear equations*.

A first-degree equation is an equation that contains only the first power of x. A *linear equation* is a first-degree equation of one or more variables, such as $2x + 3 = 9$ or $3x - 2y = 8$.

The general form of the linear equation is

$$Ax + By + C = 0 \qquad 9.1$$

A represents the coefficient of x, B represents the coefficient of y, and C represents any number that does not contain x. The C term is called the *constant*. In the equation $2x + 3 = 9$, the coefficient of y is zero (0) and the constant is 3.

A quadratic equation contains a term in the second degree x but no higher, such as $2x^2 - 3x + 4 = 0$.

The general form of the quadratic equation is

$$Ax^2 + Bx + C = 0 \qquad \textit{9.2}$$

When B is zero, the equation contains no term in x and is known as a *pure quadratic*, such as $3x^2 - 12 = 0$.

The root of a quadratic equation is any number that satisfies the equation. There are two roots for a quadratic equation.

30. SOLVING A PURE QUADRATIC EQUATION

In solving a pure quadratic equation, the term including x^2 is isolated on the left side of the equation and the other term is on the right. If the coefficient of x^2 is not 1, both sides of the equation are divided by this coefficient according to axiom 4, Sec. 24. The value of x is found by taking the square root of each side of the equation, according to axiom 5, Sec. 24.

Example 9.30

Solve the following equation.

$$3x^2 - 12 = 0$$

Solution
$$3x^2 - 12 = 0$$
$$3x^2 = 12$$
$$x^2 = 4$$
$$\sqrt{x^2} = \sqrt{4}$$
$$x = +2 \text{ or } -2$$

31. SOLVING A QUADRATIC EQUATION BY FACTORING

Factoring the left side of the equation $x^2 - 4x + 3 = 0$ gives
$$(x - 3)(x - 1) = 0$$

For this statement to be true—that is, for the product of two factors to be zero—one of the two factors must be zero or both must be zero. Consider two numbers, a and b, whose product is equal to zero.

$$ab = 0$$

For this statement to be true, either a must be zero, b must be zero, or both must be zero. Setting $x = 3$ in the equation $(x - 3)(x - 1) = 0$ gives

$$(3 - 3)(3 - 1) = 0$$

This is a true statement. Therefore, the roots of the equation $x^2 - 4x + 3 = 0$ are 3 and 1. For proof, substitute these numbers in the equation.

$$(3)^2 - (4)(3) + 3 = 0$$
$$9 - 12 + 3 = 0$$

and

$$(1)^2 - (4)(1) + 3 = 0$$
$$1 - 4 + 3 = 0$$

To simplify the operation, let each of the factors equal zero and solve for x.

$$(x - 3) = 0$$
$$x = 3$$
$$(x - 1) = 0$$
$$x = 1$$

Example 9.31

Solve the following equation.

$$2x^2 - x - 15 = 0$$

Solution
$$2x^2 - x - 15 = 0$$
$$(x - 3)(2x + 5) = 0$$
$$x - 3 = 0$$
$$x = 3$$
$$2x + 5 = 0$$
$$x = -\frac{5}{2}$$

Example 9.32

Solve the following equation.

$$2x^2 = 3x$$

Solution
$$2x^2 = 3x$$
$$2x^2 - 3x = 0$$
$$x(2x - 3) = 0$$
$$x = 0$$
$$2x - 3 = 0$$
$$2x = 3$$
$$x = \frac{3}{2}$$

32. SOLVING A QUADRATIC EQUATION BY COMPLETING THE SQUARE

The left side of the equation $x^2 + 4x + 4 = 16$ is a perfect square because the factors are $(x+2)(x+2) = (x+2)^2$. The right side of the equation is also a perfect square: $\sqrt{16} = \pm 4$. The values of x can be found by taking the square root of each side of the equation.

$$\sqrt{(x+2)^2} = \sqrt{16}$$
$$x + 2 = \pm 4$$
$$x = +2 \text{ or } -6$$

Quadratic equations that are not perfect squares can be solved by a method known as *completing the square*. The method involves using the five axioms found in Sec. 24 to make both sides of the equation perfect squares and then solving. The method is lengthy and will not be explained here.

33. SOLVING A QUADRATIC EQUATION BY FORMULA

In the general form of the quadratic equation $Ax^2 + Bx + C = 0$, A is the coefficient of x^2, B is the coefficient of x, and C is the constant. As mentioned, if $B = 0$, there will be no middle term Bx. In the equation $3x^2 - 5x - 8 = 0$, $A = 3$, $B = -5$, and $C = -8$.

The quadratic formula is

$$x = \frac{-B + \sqrt{B^2 - 4AC}}{2A} \qquad 9.3(a)$$

$$x = \frac{-B - \sqrt{B^2 - 4AC}}{2A} \qquad 9.3(b)$$

It is usually written as

$$x = \frac{-B \pm \sqrt{B^2 - 4AC}}{2A} \qquad 9.3(c)$$

Example 9.33

Solve the following equation by formula.

$$4x^2 - 5x - 6 = 0$$

Solution

$4x^2 - 5x - 6 = 0$: $A = 4$, $B = -5$, $C = -6$

$$x = \frac{-(-5) + \sqrt{(-5)^2 - (4)(4)(-6)}}{(2)(4)}$$

$$= \frac{5 + \sqrt{25 + 96}}{8}$$

$$= \frac{5 + \sqrt{121}}{8}$$

$$x = 2$$

or

$$x = \frac{-(-5) - \sqrt{(-5)^2 - (4)(4)(-6)}}{(2)(4)}$$

$$= \frac{5 - \sqrt{25 + 96}}{8}$$

$$= \frac{5 - \sqrt{121}}{8}$$

$$x = -\frac{3}{4}$$

In using the formula, the equation must be arranged in the form $Ax^2 + Bx + C = 0$ with A being positive. If the equation is not in this form, it must be rearranged.

Example 9.34

Arrange the equations in the form $Ax^2 + Bx + C = 0$ for solution by formula and write the values of A, B, and C.

(a) $3x^2 = 4x + 2$

(b) $-4x^2 + 2x = 6$

(c) $x^2 = 4 - x$

(d) $x^2 - 7 = 0$

Solution

(a) $3x^2 - 4x - 2 = 0$: $A = 3$, $B = -4$, $C = -2$

(b) $4x^2 - 2x + 6 = 0$: $A = 4$, $B = -2$, $C = 6$

(c) $x^2 + x - 4 = 0$: $A = 1$, $B = 1$, $C = -4$

(d) $x^2 - 7 = 0$: $A = 1$, $B = 0$, $C = -7$

PRACTICE PROBLEMS

1. Identify each of the following as a monomial, binomial, trinomial, or polynomial. (Note: *Mono* means one, *bi* means two, *tri* means three, and *poly* means many.)

(a) $x^2 + 2xy + y^2$

(b) $4a^2b^2c^2$

(c) $3x^2 + 5xy^2$

(d) $2x^3 + 3x^2 + 3x + 8$

2. Write the numerical coefficient of each of the following terms.

(a) $3x^2y$

(b) $6x^3$

(c) $4abc$

(d) $x^2y^2z^2$

Algebra

3. Combine like terms into one algebraic expression.

Example:

$3x^3 + 5xy + 2x^3 + x^2 + 2xy + 3x^2 = 5x^3 + 4x^2 + 7xy$

(a) $2x^2 + xy + x^2 + 3xy + 4x^2$

(b) $x^2 + x^2y + xy^2 + 2x^2 + x^2y^2$

(c) $3ab^2 + 2ab - 4ab^2 - 2ab + 5$

(d) $x^2 + y^2 + 2 - y^2 - y^2 + 4x^2 - 4$

(e) $x^3 + 2x^2 + 4 - 2x^3 - 5x^2 - x^3$

(f) $3x^3 + x^2y - 6x^2y + 2x + 4 - 6x^3$

(g) $x^2 + 3xy - x^3 + 3x^2 - 5xy + x^3$

(h) $x^2y^2 - 2xy^2 + 2x^2y^2 + 3xy^2 - 4x^2y + x^2y^2$

(i) $a^2b^2c - ab^2c + 2a^2b^2c + 2ab^2c + 4ab^2c$

(j) $2a - 3 + 11a + 5 - 4a - 9$

4. Find the indicated products and quotients.

(a) $(2)^4$

(b) $(x^3)(x^4)$

(c) $(-3)^4$

(d) $-(-4)^2$

(e) $(2a)(3a)$

(f) $(-4x^2)(2x)$

(g) $(x^2y^3)(xy)$

(h) $(3xy^2)(-4x^2y)$

(i) $(2)^3(3^2)(4)^2$

(j) $(-4x^3)(-2x^2)$

(k) $\dfrac{x^8}{x^5}$

(l) $\dfrac{a^2}{a^{-3}}$

(m) $\dfrac{(x^3)(x^4)}{x^5}$

(n) $\dfrac{(2)^2(2)^3(2)}{(2)^5}$

(o) $\dfrac{15x^4y^3}{5x^3y^2}$

(p) $\dfrac{-64a^3b^2}{4ab^2}$

(q) $(x^2y)^3$

(r) $(-2ab^2c^3)^2$

(s) $\dfrac{(4xy^2)^4}{(-8x^2y)^2}$

(t) $\dfrac{a^3(b^2)^4}{(ab^3)^2}$

(u) $\dfrac{(3)^4}{(3)^4}$

(v) $\dfrac{x^4}{x^4}$

(w) $\dfrac{(2)^2}{(2)^5}$

(x) $\dfrac{a^{-2}b^3}{a^3b^{-2}}$

(y) $(4-2)^{-2}$

(z) $(a+b)^0$

5. Multiply the following terms.

(a) $2xy(x^2 - yz + z^2)$

(b) $3a^2b(a^2 - 2ab + 2b^2)$

(c) $5x^2(6x^2 - 3x + 4)$

(d) $-3xy(2x^2 - xy + 2y^2)$

(e) $x + 3$
$x + 2$

(f) $a - 4$
$a + 2$

(g) $a^2 + 2a + 6$
$a - 3$

(h) $x^2 + 3x - 2$
$3x - 2$

(i) $a^3 - 3a^2 + 2a - 8$
$5a + 4$

(j) $x^2 - 2x - 3$
$\overline{x^2 + x + 4}$

6. Divide the following terms.

(a) $\dfrac{8x^4 + 6x^3 + 4x^2 + 2x + 2}{2}$

(b) $\dfrac{24a^4 - 16a^3 - 12a}{4a}$

(c) $\dfrac{14a^2b^3 - 21a^2b^2 - 28ab}{7ab}$

(d) $\dfrac{18x^4y^3z^2 - 6x^3yz^2 + 3x^2y^2z}{3x^2yz}$

(e) $\dfrac{24a^5bc^3 - 12a^3bc^2 - 18a^4bc^4}{6a^3bc^2}$

(f) $\dfrac{25xy^3z^5 - 20x^3y^2z^2 - 15x^4y^3z^3}{5xy^2z^2}$

(g) $x + 3 \overline{)\, x^2 + 8x + 15}$

(h) $x + 2 \overline{)\, 2x^2 + 7x + 6}$

(i) $x + 5 \overline{)\, x^2 + 8x + 15}$

(j) $x + 4 \overline{)\, 3x^2 + 10x - 6}$

7. Factor the following terms.

(a) $12x^3 - 3x^2$

(b) $x^3y - 3x^2y$

(c) $a^2 + 2ab + b^2$

(d) $a^2 - b^2$

(e) $4a^2 - 8a + 4$

(f) $x^2 - 16$

(g) $x^3 + 2x^2y + xy^2$

(h) $25a^2 - 49$

(i) $3a^2 - 18a + 27$

(j) $a^4 - 81$

(k) $x^2 - 4x - 12$

(l) $a^2 - 4a - 5$

(m) $y^2 - 5y + 6$

(n) $x^2 + 7x + 12$

(o) $2x^2 - 5x - 12$

8. Solve the following equations.

Example:

$10x - (3)(x+3)(2x-5) = 3 - (2)(4x-7)(x+1) - x(3-2x)$

$10x - 6x^2 - 3x + 45 = 3 - 8x^2 + 6x + 14 - 3x + 2x^2$

$10x - 6x^2 - 3x + 8x^2 - 2x^2 - 6x + 3x = 3 + 14 - 45;$

$4x = -28; \ x = -7$

(a) $3x + 4 = 6$

(b) $5a + 6 = a - 10$

(c) $15 - 2x = 1 - 5x$

(d) $9x - 12 = 7x - 11$

(e) $7x + 7 - x = 2x - 8$

(f) $5a - 7 - 4a - 8 + 8a - 15 = 0$

(g) $6 + 2x - 3 - 5x = x - 5 - 2x + 7$

(h) $12x - (4x - 6) = 3x - (9x - 27)$

(i) $5x - (x+3)(x-4) + 3 = 7 - x(x-7)$

9. Solve the following equations.

Examples:

(1)
$$\frac{x}{3} + \frac{x}{4} = \frac{7}{2}$$
$$\text{LCD} = 12$$
$$\frac{12x}{3} + \frac{12x}{4} = \frac{(12)(7)}{2}$$
$$4x + 3x = 42; \ 7x = 42; \ x = 6$$

(2)
$$\frac{2x+2}{3} - \frac{3x-1}{4} = 1$$
$$\text{LCD} = 12$$
$$\frac{(12)(2x+2)}{3} - \frac{(12)(3x-1)}{4} = (12)(1)$$
$$8x + 8 - 9x + 3 = 12; \ -x = 1; \ x = -1$$

(3)
$$\frac{x-4}{x-2} - \frac{x}{x+2} = \frac{8-x}{x^2-4}$$
$$\text{LCD} = x^2 - 4$$
$$(x+2)(x-4) - x(x-2) = 8 - x$$
$$x^2 - 2x - 8 - x^2 + 2x = 8 - x; \ x = 16$$

(a) $x + 3 = \dfrac{3x}{4} - \dfrac{x}{2}$

(b) $\dfrac{3x}{2} - x = \dfrac{x}{3} + 12$

(c) $\dfrac{3}{x+4} - \dfrac{4}{x-4} = \dfrac{x}{x^2-16}$

(d) $\dfrac{2}{x-1} - \dfrac{4}{x-3} = \dfrac{6}{x^2-4x+3}$

10. Solve each of the following formulas for the letter indicated at the right.

(a) $C = \pi D$ $\qquad\qquad$ (D)

(b) $C = 2\pi r$ $\qquad\qquad$ (r)

(c) $A = \pi r^2$ $\qquad\qquad$ (r)

(d) $A = \dfrac{\pi D^2}{4}$ $\qquad\qquad$ (D)

(e) $V = \pi r^2 h$ $\qquad\qquad$ (h)

(f) $V = \pi r^2 h$ $\qquad\qquad$ (r)

(g) $A = \dfrac{\pi}{4} D^2$ $\qquad\qquad$ (D)

(h) $A - P = Prt$ $\qquad\qquad$ (P)

(i) $I = \dfrac{E}{R+r}$ $\qquad\qquad$ (R)

(j) $\dfrac{a}{a+b} = \dfrac{c}{c+d}$ $\qquad$ (b)

11. Write each of the following equations in the form $Ax^2 + Bx + C = 0$.

(a) $7x^2 = 4x - 3$

(b) $x^2 = 4 - x$

12. Write the value of A, B, and C in each of the following equations.

(a) $2x^2 + 2x + 5 = 0$

(b) $5x^2 - 9x = 0$

13. Solve the following equations by factoring.

(a) $3x^2 + 7x + 2 = 0$

(b) $4x^2 = 5x + 6$

(c) $3x^2 = 5x$

14. Solve the following equation by formula.

$$3x^2 + 7x + 2 = 0$$

SOLUTIONS

1. (a) $\boxed{\text{trinomial}}$ (b) $\boxed{\text{monomial}}$

(c) $\boxed{\text{binomial}}$ (d) $\boxed{\text{polynomial}}$

2. (a) $\boxed{3}$ (b) $\boxed{6}$ (b) $\boxed{4}$ (b) $\boxed{1}$

3. (a) $\boxed{7x^2 + 4xy}$

(b) $\boxed{3x^2 + x^2y^2 + x^2y + xy^2}$

(c) $\boxed{-ab^2 + 5}$

(d) $\boxed{5x^2 - y^2 - 2}$

(e) $\boxed{-2x^3 - 3x^2 + 4}$

(f) $\boxed{-3x^2 - 5x^2y + 2x + 4}$

(g) $\boxed{4x^2 - 2xy}$

(h) $\boxed{4x^2y^2 + xy^2 - 4x^2y}$

(i) $\boxed{3a^2b^2c + ab^2c}$

(j) $\boxed{9a - 7}$

4. (a) $\boxed{16}$ (b) $\boxed{x^7}$ (c) $\boxed{81}$

(d) $\boxed{-16}$ (e) $\boxed{6a^2}$ (f) $\boxed{-8x^3}$

(g) $\boxed{x^3y^4}$ (h) $\boxed{-12x^3y^3}$ (i) $\boxed{1152}$

(j) $\boxed{8x^5}$ (k) $\boxed{x^3}$ (l) $\boxed{a^5}$

(m) $\boxed{x^2}$ (n) $\boxed{2}$ (o) $\boxed{3xy}$

(p) $\boxed{-16a^2}$ (q) $\boxed{x^6y^3}$ (r) $\boxed{4a^2b^4c^6}$

(s) $\boxed{4y^6}$ (t) $\boxed{ab^2}$ (u) $\boxed{1}$

(v) $\boxed{1}$ (w) $\boxed{\dfrac{1}{8}}$ (x) $\boxed{\dfrac{b^5}{a^5}}$

(y) $\boxed{\dfrac{1}{4}}$ (z) $\boxed{1}$

5. (a) $\boxed{2x^3y - 2xy^2z + 2xyz^2}$

(b) $\boxed{3a^4b - 6a^3b^2 + 6a^2b^3}$

(c) $\boxed{30x^4 - 15x^3 + 20x^2}$

(d) $\boxed{-6x^3 + 3x^2y^2 - 6xy^3}$

(e)
$$
\begin{array}{r}
x\ +3 \\
\times\ x\ +2 \\
\hline
x^2 + 3x \\
2x + 6 \\
\hline
\boxed{x^2 + 5x + 6}
\end{array}
$$

(f)
$$
\begin{array}{r}
a\ -4 \\
\times\ a\ +2 \\
\hline
\boxed{a^2 - 4a} \\
2a - 8 \\
\hline
a^2 - 2a - 8
\end{array}
$$

(g)
$$
\begin{array}{r}
a^2 + 2a\ +6 \\
\times\ \ a\ -3 \\
\hline
a^3 + 2a^2 + 6a \\
-3a^2 - 6a - 18 \\
\hline
\boxed{a^3 -\ a^2 - 18}
\end{array}
$$

(h)
$$
\begin{array}{r}
x^2 + 3x\ -\ 2 \\
\times\ \ 3x\ -\ 2 \\
\hline
3x^3 + 9x^2 -\ 6x \\
-2x^2 -\ 6x + 4 \\
\hline
\boxed{3x^3 + 7x^2 - 12x + 4}
\end{array}
$$

(i)
$$
\begin{array}{r}
a^3 -\ 3a^2 +\ 2a\ -\ 8 \\
\times\ \ 5a\ +\ 4 \\
\hline
5a^4 - 15a^3 + 10a^2 - 40a \\
4a^3 - 12a^2 +\ 8a - 32 \\
\hline
\boxed{5a^4 - 11a^3 -\ 2a^2 - 32a - 32}
\end{array}
$$

(j)
$$
\begin{array}{r}
x^2 - 2x\ -\ 3 \\
\times\ x^2 +\ x\ +4 \\
\hline
x^4 - 2x^3 - 3x^2 \\
x^4 - 2x^3 - 2x^2 -\ 3x \\
4x^2 -\ 8x - 12 \\
\hline
\boxed{x^4 -\ x^3 -\ x^2 - 11x - 12}
\end{array}
$$

6. (a) $\boxed{4x^4 + 3x^3 + 2x^2 + x + 1}$

(b) $\boxed{6a^3 - 4a^2 - 3}$

(c) $\boxed{2ab^2 - 3ab - 4}$

(d) $\boxed{6x^2y^2z - 2xy + y}$

(e) $\boxed{4a^2c - 2 - 3ac^2}$

(f) $\boxed{5yz^3 - 4x^2 - 3x^3yz}$

(g)
$$
\begin{array}{r}
\boxed{x\ +5} \\
x+3\overline{)x^2 + 8x + 15} \\
\underline{x^2 + 3x} \\
5x + 15 \\
\underline{5x + 15}
\end{array}
$$

(h)
$$
\begin{array}{r}
\boxed{2x\ +3} \\
x+2\overline{)2x^2 + 7x + 6} \\
\underline{2x^2 + 4x} \\
3x + 6 \\
\underline{3x + 6}
\end{array}
$$

(i)
$$
\begin{array}{r}
\boxed{x\ +3} \\
x+5\overline{)x^2 + 8x + 15} \\
\underline{x^2 + 5x} \\
3x + 15 \\
\underline{3x + 15}
\end{array}
$$

(j)
$$
\begin{array}{r}
\boxed{3x\ -\ 2} \\
x+4\overline{)3x^2 + 10x\ -6} \\
\underline{3x^2 + 12x} \\
-2x\ -\ 6 \\
\underline{-\ 2x - 8} \\
2
\end{array}
$$

7. (a) $\boxed{3x^2(4x - 1)}$

(b) $\boxed{x^2y(x - 3)}$

(c) $\boxed{(a + b)^2}$

(d) $\boxed{(a + b)(a - b)}$

(e) $\boxed{(4)(a - 1)^2}$

(f) $\boxed{(x + 4)(x - 4)}$

(g) $\boxed{x(x + y)^2}$

(h) $\boxed{(5a + 7)(5a - 7)}$

(i) $\boxed{(3)(a - 3)^2}$

(j) $(a^2 + 9)(a^2 - 9) = \boxed{(a^2 + 9)(a + 3)(a - 3)}$

(k) $\boxed{(x + 2)(x - 6)}$

(l) $\boxed{(a + 1)(a - 5)}$

(m) $\boxed{(y - 2)(y - 3)}$

(n) $\boxed{(x + 3)(x + 4)}$

(o) $\boxed{(2x + 3)(x - 4)}$

8. (a) $3x + 4 = 6;\ \ 3x = 2;\ \ x = \boxed{\dfrac{2}{3}}$

(b) $5a + 6 = a - 10;\ \ 4a = -16;\ \ a = \boxed{-4}$

(c) $15 - 2x = 1 - 5x;\ \ 3x = -14;\ \ x = \boxed{-\dfrac{14}{3}}$

(d) $9x - 12 = 7x - 11;\ \ 2x = 1;\ \ x = \boxed{\dfrac{1}{2}}$

(e) $7x + 7 - x = 2x - 8;\ \ 4x = -15;\ \ x = \boxed{-\dfrac{15}{4}}$

(f) $5a - 7 - 4a - 8 + 8a - 15 = 0;$
$$9a = 30;\ \ a = \boxed{\dfrac{10}{3}}$$

(g) $6 + 2x - 3 - 5x = x - 5 - 2x + 7;$
$$-2x = -1;\ \ x = \boxed{\dfrac{1}{2}}$$

(h) $12x - (4x - 6) = 3x - (9x - 27);$
$$12x - 4x + 6 = 3x - 9x + 27;$$
$$14x = 21;\ \ x = \boxed{\dfrac{3}{2}}$$

(i) $5x - (x+3)(x-4) + 3 = 7 - x(x-7);$
$$5x - x^2 + x + 12 + 3 = 7 - x^2 + 7x;$$
$$x = \boxed{8}$$

9. (a) $x + 3 = \dfrac{3x}{4} - \dfrac{x}{2};\quad 4x + 12 = 3x - 2x;$
$$3x = 12;\ x = \boxed{-4}$$

(b) $\dfrac{3x}{2} - x = \dfrac{x}{3} + 12;\quad 9x - 6x = 2x + 72;$
$$x = \boxed{72}$$

(c) $\dfrac{3}{x+4} - \dfrac{4}{x-4} = \dfrac{x}{x^2 - 16};$
$$(3)(x-4) - (4)(x+4) = x;$$
$$3x - 12 - 4x - 16 = x;$$
$$-2x = 28;$$
$$x = \boxed{-14}$$

(d) $\dfrac{2}{x-1} - \dfrac{4}{x-3} = \dfrac{6}{x^2 - 4x + 3};$
$$(2)(x-3) - (4)(x-1) = 6;\quad 2x - 6 - 4x + 4 = 6;$$
$$-2x = 8;\ \boxed{x = -4}$$

10. (a) $D = \boxed{\dfrac{C}{\pi}}$ (b) $r = \boxed{\dfrac{C}{2\pi}}$

(c) $r = \boxed{\sqrt{\dfrac{A}{\pi}}}$ (d) $D = \boxed{\sqrt{\dfrac{4A}{\pi}}}$

(e) $h = \boxed{\dfrac{V}{\pi r^2}}$ (f) $r = \boxed{\sqrt{\dfrac{V}{\pi h}}}$

(g) $D = \boxed{\sqrt{\dfrac{A}{\frac{\pi}{4}}}}$ (h) $P = A - Prt = \boxed{\dfrac{A}{1 + rt}}$

(i) $R = \boxed{\dfrac{E}{I} - r}$ (j) $b = \boxed{\dfrac{ad}{c}}$

11. (a) $\boxed{7x^2 - 4x + 3 = 0}$ (b) $\boxed{x^2 + x - 4 = 0}$

12. (a) $A = 2,\ B = 2,\ C = 5$
(b) $A = 5,\ B = -9,\ C = 0$

13. (a) $(3x+1)(x+2) = 0$
$$x = -\dfrac{1}{3}$$
$$x = \boxed{-2}$$

(b) $4x^2 - 5x - 6 = 0$
$$(4x+3)(x-2) = 0$$
$$x = -\dfrac{3}{4}$$
$$x = \boxed{2}$$

(c) $3x^2 - 5x = 0$
$$x(3x - 5) = 0$$
$$x = 0$$
$$x = \boxed{\dfrac{5}{3}}$$

14. $A = \boxed{3};\ B = \boxed{7};\ C = 2$
$$x = \dfrac{-7 \pm \sqrt{(7)^2 - (4)(3)(2)}}{(2)(3)} = \dfrac{-7 \pm \sqrt{49 - 24}}{6}$$
$$= \dfrac{-7 \pm \sqrt{25}}{6}$$
$$x = \dfrac{-7 + 5}{6} = \boxed{-\dfrac{1}{3}}$$
$$x = \dfrac{-7 - 5}{6} = \boxed{-2}$$

10 The Rectangular Coordinate System

1. DIRECTED LINE

Suppose a surveyor establishes a west-east line through a point on a monument called the *origin*. Moving east from the origin, the surveyor marks off points 1 ft apart, labeling them $+1, +2, +3, \ldots$. Then, moving west from the origin, the surveyor marks off points 1 ft apart, labeling them $-1, -2, -3, \ldots$.

The surveyor then establishes a *directed line*, as shown in Fig. 10.1. From left to right (west to east), the direction is positive and numbers increase in value. From right to left (east to west), the direction is negative and numbers decrease in value.

Figure 10.1 *Directed Line*

A directed line, on paper, can be called an *axis*. A horizontal (west-east) line is called the *x-axis*. A point on the axis associated with a particular number is called the *graph* of the number, and the number is called the *coordinate* of the point. This coordinate is the directed distance from the origin to the point. Point A in Fig. 10.1 has the coordinate -4, which is the distance from 0 to A, not the distance from A to 0. Point B has the coordinate $+3$, which is the distance from 0 to B, not B to 0.

$$\text{distance AB} = 3 - (-4) = 7$$
$$\text{distance BA} = -4 - (+3) = -7$$

The word "distance" is used loosely in this text. A more appropriate term for distance on a directed line would be "the measure of travel in a specified direction." The actual length from A to B, or B to A, is 7. This actual length, which disregards the negative sign, is known as the *absolute value* or AB or BA. Symbolically, $|-7| = 7$. The enclosure indicates absolute value.

In general, if P_1 and P_2 are any two points on the x-axis with coordinates x_1 and x_2, then

$$P_1 P_2 = x_2 - x_1 \qquad 10.1$$
$$P_2 P_1 = x_1 - x_2 \qquad 10.2$$

The surveyor can also establish a south-north line through the same point on the monument and mark points on the line at 1 ft intervals from the origin in a northerly direction, labeling them $+1, +2, +3, \ldots$. Points at 1 ft intervals on the line in a southerly direction are labeled $-1, -2, -3, \ldots$.

Figure 10.2 shows a vertical line through the origin representing this south-north line. It is a directed line that is positive from south to north and is called the *y-axis*.

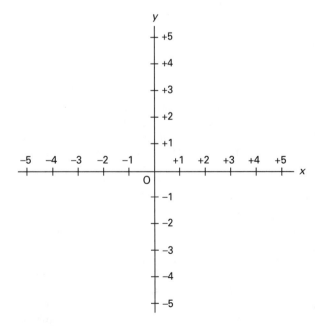

Figure 10.2 *x-y Axis*

If P_1 and P_2 are any two points on the y-axis with coordinates y_1 and y_2, then

$$P_1P_2 = y_2 - y_1 \qquad 10.3$$
$$P_2P_1 = y_1 - y_2 \qquad 10.4$$

2. THE RECTANGULAR COORDINATE SYSTEM

The French mathematician René Descartes (17th century) devised the *rectangular coordinate system*, sometimes called the *Cartesian plane*. This system uses an ordered pair of coordinates to locate a point. The ordered pair of coordinates are the x-coordinate and the y-coordinate of a point, enclosed in parentheses with the x-coordinate always written first, followed by a comma and the y-coordinate.

The Cartesian plane consists of an x- (horizontal) and a y- (vertical) axis, which are directed lines as shown in Fig. 10.3.

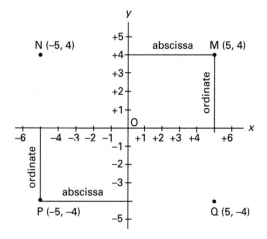

Figure 10.3 *Abscissas and Ordinates*

Point M has the coordinates $(5, 4)$. The horizontal distance 5, from the y-axis to M, is known as the *abscissa*. The vertical distance 4, from the x-axis to M, is known as the *ordinate*. The abscissa and ordinate are measured from the axis to point and not from point to the axis, which is in accordance with distances on a directed line. In Fig. 10.3, point N has an abscissa of -5 and an ordinate of $+4$; point P has an abscissa of -5 and an ordinate of -4; and point Q has an abscissa of $+5$ and an ordinate of -4.

The x- and y-axes divide the plane into four parts, numbered in a counterclockwise direction as shown in Fig. 10.4. Signs of the coordinates of points in each quadrant are also shown in Fig. 10.4. In quadrant I, x is positive and y is positive; in quadrant II, x is negative and y is positive; in quadrant III, x is negative and y is negative; and in quadrant IV, x is positive and y is negative.

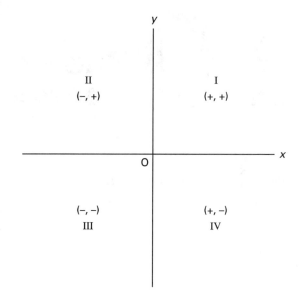

Figure 10.4 *Signs of the Quadrants*

Example 10.1

Determine the coordinates of the points shown.

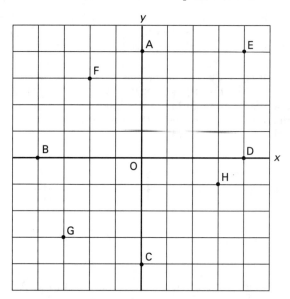

Solution

A: $(0, 4)$ B: $(-4, 0)$
C: $(0, -4)$ D: $(4, 0)$
E: $(4, 4)$ F: $(-2, 3)$
G: $(-3, -3)$ H: $(3, -1)$

3. DISTANCE FORMULA

A formula for finding the distance between any two points in a rectangular coordinate system can be derived from the *Pythagorean theorem*: in a right triangle, the square of the length of the hypotenuse equals the sum of the squares of the lengths of the other two sides.

If c represents the length of the hypotenuse and a and b are the lengths of the other two sides,

$$c^2 = a^2 + b^2 \qquad 10.5$$

Taking the square root of both sides of the equation,

$$c = \sqrt{a^2 + b^2} \qquad 10.6$$

In Fig. 10.5, P_1 and P_2 represent any two points in a rectangular coordinate system with the coordinates (x_1, y_1) and (x_2, y_2). If a horizontal line is passed through P_1 and a vertical line is passed through P_2, they will intersect at Q, forming a right triangle, with the line P_1P_2 being the hypotenuse. The x-coordinate of Q will be the same as the x-coordinate of P_2: x_2; and the y-coordinate of Q will be the same as the y-coordinate of P_1: y_1.

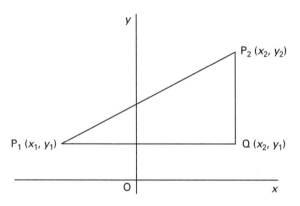

Figure 10.5 *Distance Between Points*

By the Pythagorean theorem,

$$(P_1P_2)^2 = (P_1Q)^2 + (QP_2)^2 \qquad 10.7$$
$$P_1P_2 = \sqrt{(P_1Q)^2 + (QP_2)^2}$$
$$= \sqrt{(x_2 - x_1)^2 + (y_2 - y_1)^2} \qquad 10.8$$

It is also true that

$$P_1P_2 = \sqrt{(x_1 - x_2)^2 + (y_1 - y_2)^2} \qquad 10.9$$

Example 10.2

Find the distance between $P_1(-3, -1)$ and $P_2(5, 5)$.

Solution

$$P_1P_2 = \sqrt{(x_2 - x_1)^2 + (y_2 - y_1)^2}$$
$$= \sqrt{\left(5 - (-3)\right)^2 + \left(5 - (-1)\right)^2}$$
$$= \sqrt{(5 + 3)^2 + (5 + 1)^2}$$
$$= 10$$

The distance r from the origin to any point $P(x, y)$ is called the *radius*. This distance is always positive.

From the Pythagorean theorem,

$$r = \sqrt{x^2 + y^2} \qquad 10.10$$

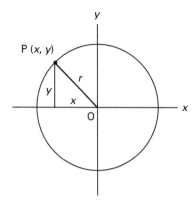

Figure 10.6 *Radius*

Example 10.3

Find the distance from the origin to the point $P(-3, 4)$.

Solution

$$\mathbf{r} = \sqrt{(-3)^2 + (4)^2} = 5$$

4. MIDPOINT OF A LINE

In Fig. 10.7, point M is the midpoint of the line PQ.

The x-coordinate of M is the distance from the y-axis to M. This is equal to the distance from the y-axis to L, which is the average of the x-coordinates of P and Q.

$$\frac{6 + (-2)}{2} = 2$$

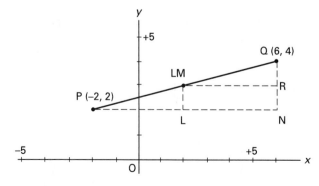

Figure 10.7 *Midpoint of a Line*

The y-coordinate of M is the distance from the x-axis to M. This is equal to the distance from the x-axis to R, which is the average of the y-coordinates of P and Q.

$$\frac{4+2}{2} = 3$$

From this it can be seen that the midpoint M of a line P_1P_2 has the coordinates

$$M\left(\frac{x_1 + x_2}{2}, \frac{y_1 + y_2}{2}\right) \qquad 10.11$$

Example 10.4

Find the midpoint M of line with endpoints $P_1\,(-2, 3)$ and $P_2\,(-8, -3)$.

Solution

$$M\left(\frac{x_1 + x_2}{2}, \frac{y_1 + y_2}{2}\right) = M\left(\frac{-2-8}{2}, \frac{3-3}{2}\right)$$
$$= M\,(-5, -0)$$

PRACTICE PROBLEMS

1. (a) In the rectangular coordinate system, plot the points and connect with lines in the order PQRSTP.

P $(12, 16)$

Q $(-14, 18)$

R $(-12, 1)$

S $(-14, -17)$

T $(14, -14)$

(b) Find the length of each line segment (PQ, QR, etc.) and the perimeter.

2. Find the radii for circles centered at the origin $(0,0)$ and with the points on the circles given.

(a) $(3,4)$

(b) $(-3,3)$

(c) $(-5,-12)$

(d) $(5,-5)$

SOLUTIONS

1. (a)

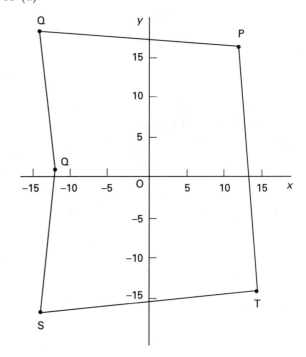

(b) $PQ = \sqrt{(-14-12)^2 + (18-16)^2} = \boxed{26}$

$QR = \sqrt{\left(-12-(-14)\right)^2 + (1-18)^2} = \boxed{17}$

$RS = \sqrt{\left(-14-(-12)\right)^2 + (-17-1)^2} = \boxed{18}$

$ST = \sqrt{\left(14-(-14)\right)^2 + \left(-14-(-17)\right)^2} = \boxed{28}$

$TP = \sqrt{(12-14)^2 + \left(16-(-14)\right)^2} = \boxed{30}$

perimeter $= 26 + 17 + 18 + 28 + 30 = \boxed{119}$

2. (a)

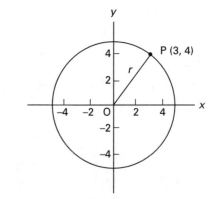

$$r = \sqrt{(3)^2 + (4)^2} = \boxed{5}$$

(b)

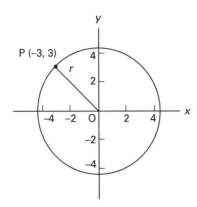

$$r = \sqrt{(-3)^2 + (3)^2} = \boxed{4.2}$$

(d)

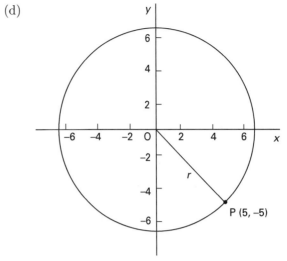

$$r = \sqrt{(5)^2 + (-5)^2} = 7.1$$

(c)

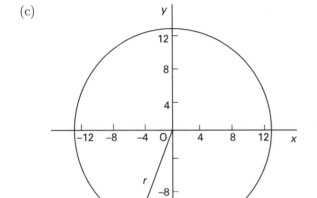

$$r = \sqrt{(-5)^2 + (-12)^2} = 13$$

Coordinates

11 Trigonometry for Surveyors

1. DEFINITION OF AN ANGLE

In trigonometry, an *angle* is considered to be the measure of the rotation of a *ray* (line) from one position to another in a counterclockwise direction.

2. STANDARD POSITION OF AN ANGLE

An angle is in *standard position* when its vertex is at the origin and the initial side coincides with the positive x-axis of a rectangular coordinate system.

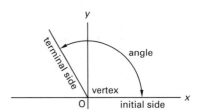

Figure 11.1 Angle Terminology

3. QUADRANTS

The coordinate axes divide the plane into four *quadrants* designated I, II, III, and IV, as shown in Fig. 11.2. An angle is in one of the quadrants when its *terminal side* is in that quadrant.

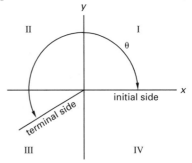

Figure 11.2 Angle Quadrants

4. TRIGONOMETRIC FUNCTIONS OF ANY ANGLE

For any angle in standard position, the six trigonometric functions are given by Eqs. 11.1 through 11.6. Note that the abbreviations for sine, cosine, tangent, cosecant, secant, and cotangent are used in equations.

$$\sin \theta = \frac{y}{r} \qquad\qquad 11.1$$

$$\cos \theta = \frac{x}{r} \qquad\qquad 11.2$$

$$\tan \theta = \frac{y}{x} \qquad\qquad 11.3$$

$$\csc \theta = \frac{r}{y} \qquad\qquad 11.4$$

$$\sec \theta = \frac{r}{x} \qquad \text{11.5}$$

$$\cot \theta = \frac{x}{y} \qquad \text{11.6}$$

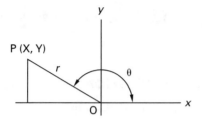

Figure 11.3 *Functions of an Angle*

Because x, y, and r represent measured lengths, the six trigonometric functions are actually ratios of two numbers—that is, they are ratios of the length of one side of a triangle to the length of another side. A *ratio* is a comparison of two numbers. The ratio 2 to 3 can be written $^2/_3$, which is a fraction. Therefore, a trigonometric function is a number.

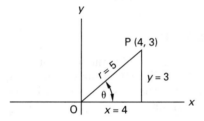

Figure 11.4 *Functions Related to Sides*

In Fig. 11.4,

$$\sin \theta = \frac{y}{r} = \frac{3}{5} = 0.60$$

$$\cos \theta = \frac{x}{r} = \frac{4}{5} = 0.80$$

$$\tan \theta = \frac{y}{x} = \frac{3}{4} = 0.75$$

The value of a trigonometric function of an angle θ depends on the value of θ. As θ changes, $\sin \theta$ changes so that $\sin \theta$ is a *function* of θ.

5. RECIPROCAL OF A NUMBER

The *reciprocal* of a number is 1 divided by the number. Thus, the reciprocal of 3 is $^1/_3$, and the reciprocal of $^2/_3$ is $^3/_2$.

6. RECIPROCAL OF A TRIGONOMETRIC FUNCTION

Trigonometric functions are ratios of numbers and can be treated as such. Therefore, the reciprocal of $\sin \theta = 1 \sin /\theta$. It follows that if $\sin \theta = y/r$, the reciprocal is r/y.

$$\sin \theta = \frac{1}{\csc \theta} \qquad \text{11.7}$$

$$\cos \theta = \frac{1}{\sec \theta} \qquad \text{11.8}$$

$$\tan \theta = \frac{1}{\cot \theta} \qquad \text{11.9}$$

$$\csc \theta = \frac{1}{\sin \theta} \qquad \text{11.10}$$

$$\sec \theta = \frac{1}{\cos \theta} \qquad \text{11.11}$$

$$\cot \theta = \frac{1}{\tan \theta} \qquad \text{11.12}$$

Therefore, $\sin \theta$ and $\csc \theta$ are reciprocals, $\cos \theta$ and $\sec \theta$ are reciprocals, and $\tan \theta$ and $\cot \theta$ are reciprocals.

From the relation between a function and its reciprocal,

$$(\sin \theta)(\csc \theta) = \left(\frac{y}{r}\right)\left(\frac{r}{y}\right) = 1 \qquad \text{11.13}$$

$$(\cos \theta)(\sec \theta) = \left(\frac{x}{r}\right)\left(\frac{r}{x}\right) = 1 \qquad \text{11.14}$$

$$(\tan \theta)(\cot \theta) = \left(\frac{y}{x}\right)\left(\frac{x}{y}\right) = 1 \qquad \text{11.15}$$

7. ALGEBRAIC SIGN OF TRIGONOMETRIC FUNCTIONS

An angle in standard position is considered to be in the quadrant in which its terminal side lies; therefore, the values of x and y have algebraic signs. The value of r is always considered to be positive. It follows then, that functions expressed as ratios of positive and negative numbers will have positive and negative values. The algrebraic sign of any function can be determined by memorizing the terms shown in Fig. 11.5.

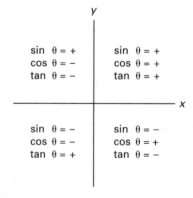

Figure 11.5 *Signs of the Natural Functions*

8. VALUES OF TRIGONOMETRIC FUNCTIONS OF QUADRANTAL ANGLES

The *quadrantal angles* are the angles that are common to two quadrants. They are 0°, 90°, 180°, 270°, and 360°. In Fig. 11.6, where $r = 5$, points P_1, P_2, P_3, and P_4 will have the coordinates shown.

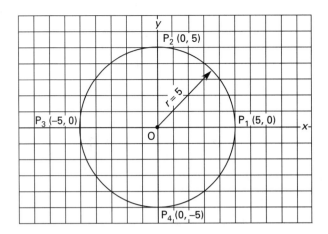

Figure 11.6 *Quadrantal Angles*

If the terminal side of θ_1 ($0°$ or $360°$) passes through point P_1 $(5, 0)$, θ_2 ($90°$) passes through point P_2 $(0, 5)$, θ_3 ($180°$) passes through point P_3 $(-5, 0)$, and θ_4 ($270°$) passes through P_4 $(0, -5)$, then

$$\sin 0° = \frac{y}{r} = \frac{0}{5} = 0$$

$$\sin 90° = \frac{y}{r} = \frac{5}{5} = +1$$

$$\sin 180° = \frac{y}{r} = \frac{0}{5} = 0$$

$$\sin 270° = \frac{y}{r} = \frac{-5}{5} = -1$$

$$\sin 360° = \frac{y}{r} = \frac{0}{5} = 0$$

As θ increases from $0°$ to $90°$, $\sin \theta$ increases from 0 to $+1$; as θ increases from $90°$ to $180°$, $\sin \theta$ decreases from $+1$ to 0; as θ increases from $180°$ to $270°$, $\sin \theta$ decreases from 0 to -1; and as θ increases from $270°$ to $360°$, $\sin \theta$ increases from -1 to 0.

An important fact that can be considered in the solution of right angles is that, except for the quadrantal angles, $\sin \theta$ always has a numerical value of less than 1.

Considering the cosine function for the angles in Fig. 11.6,

$$\cos 0° = \frac{x}{r} = \frac{5}{5} = +1$$

$$\cos 90° = \frac{x}{r} = \frac{0}{5} = 0$$

$$\cos 180° = \frac{x}{r} = \frac{-5}{5} = -1$$

$$\cos 270° = \frac{x}{r} = \frac{0}{5} = 0$$

$$\cos 360° = \frac{x}{r} = \frac{5}{5} = +1$$

Except for the quadrantal angles, $\cos \theta$ also has a numerical value less than 1.

In Fig. 11.6, the tangent function varies as follows.

$$\tan 0° = \frac{y}{x} = \frac{0}{5} = 0$$

$$\tan 90° = \frac{y}{x} = \frac{5}{0} = \infty$$

$$\tan 180° = \frac{y}{x} = \frac{0}{-5} = 0$$

$$\tan 270° = \frac{y}{x} = \frac{-5}{0} = \infty$$

$$\tan 360° = \frac{y}{x} = \frac{0}{5} = 0$$

The symbol ∞ is sometimes interpreted to mean infinity; however, $\tan 90°$ does not exist because a number cannot be divided by 0.

Table 11.1 *Summary of Functions of Quadrantal Angles*

angle	sin	cos	tan	cot	sec	csc
$0°$	0	1	0	∞	1	∞
$90°$	1	0	∞	0	∞	1
$180°$	0	-1	0	∞	-1	∞
$270°$	-1	0	∞	0	∞	-1
$360°$	0	1	0	∞	1	∞

9. TRIGONOMETRIC FUNCTIONS OF AN ACUTE ANGLE

All angles in quadrant I are *acute angles* and, therefore, positive. In dealing with acute angles only, it is more convenient to consider them as part of a right triangle and express the trigonometric functions in terms of the sides of a triangle that are given the names *opposite*, *adjacent*, and *hypotenuse* as they are shown in Fig. 11.7.

$$\sin \theta = \frac{\text{opposite}}{\text{hypotenuse}} \qquad 11.16$$

$$\cos \theta = \frac{\text{adjacent}}{\text{hypotenuse}} \qquad 11.17$$

$$\tan \theta = \frac{\text{opposite}}{\text{adjacent}} \qquad 11.18$$

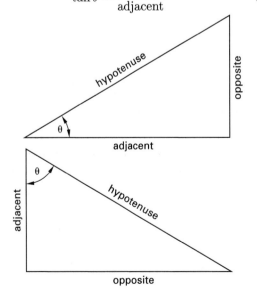

Figure 11.7 *Sides of a Triangle*

Trigonometry

10. COFUNCTIONS

Any function of an acute angle is equal to the *cofunction* of its *complementary angle*. The sine and cosine are cofunctions, as are the tangent and the cotangent.

$$\sin 30° = \cos 60° \qquad \textit{11.19}$$

$$\tan 20° = \cot 70° \qquad \textit{11.20}$$

11. TRIGONOMETRIC FUNCTIONS OF 30°, 45°, AND 60°

Consider an equilateral triangle with sides of length 2 ft. The bisector of any of the 60° angles will bisect the opposite side and form a right triangle with acute angles of 30° and 60° as shown in Fig. 11.8.

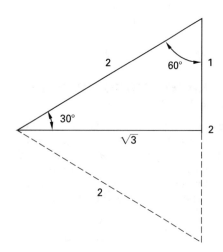

Figure 11.8 *A 30-60-90 Triangle*

In Fig. 11.8,

$$\sin 30° = \frac{1}{2} = \cos 60° = 0.500$$

$$\sin 60° = \frac{\sqrt{3}}{2} = \cos 30° = 0.866\ldots$$

$$\tan 30° = \frac{1}{\sqrt{3}} = \cot 60° = 0.577\ldots$$

Consider an isosceles right triangle with two of the sides equal to 1 as shown in Fig. 11.9. The angles opposite the sides will be 45°. Then,

$$\sin 45° = \frac{1}{\sqrt{2}} = \frac{\sqrt{2}}{2} = 0.707\ldots$$

$$\cos 45° = \frac{1}{\sqrt{2}} = \frac{\sqrt{2}}{2} = 0.707\ldots$$

$$\tan 45° = \frac{1}{1} = 1 = 1.000$$

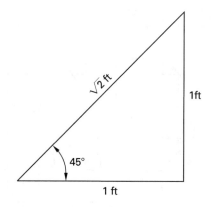

Figure 11.9 *A 45-45-90 Triangle*

12. TABLE OF VALUES OF TRIGONOMETRIC FUNCTIONS

In Secs. 7 and 10, certain trigonometric functions of angles of 0°, 30°, 45°, 60°, 90°, 180°, 270°, and 360° were computed. Many years ago, mathematicians compiled tables showing values of trigonometric functions expressed in decimal form. Some tables show values for each degree of an angle, some for each degree and minute of an angle, and some for each degree, minute, and second of an angle. Values of functions are computed to four decimal places in some tables, and up to ten decimal places in others.

The use of handheld calculators has essentially eliminated the need for tables of trigonometric functions. However, the technique of interpolation is still required in surveying work.

13. INTERPOLATION

Interpolation has been defined as finding an intermediate term in a sequence. Tables often give the sine and cosine functions for each minute of an angle, but to determine the sine and cosine functions of an angle measured in degrees, minutes, and seconds, interpolation is necessary. The process can be best explained with examples.

Example 11.1

Find $\sin 52°15'24''$ if $\sin 52°15' = 0.7906896$ and $\sin 52°16' = 0.7908676$.

Solution

The angle $52°15'24''$ is $24/60 = 0.4$ of the way from $52°15'00''$ to $52°16'00''$. The difference in the sine function of these two angles is $0.7908676 - 0.7906896 = 0.0001780$, remembering that as the angle increases, the sine increases. Because the angle $52°15'24''$ is 0.4 of the way between the other two angles, the sine of $52°15'24''$ is also 0.4 of the way between the sines of the other two angles.

$$\sin 52°15'24'' = 0.7906896 + (0.4)(0.0001780)$$

$$= 0.7907608$$

Example 11.2

Find $\cos 37°25'48''$ given that $\cos 37°25' = 0.7942379$ and $\cos 37°26' = 0.7940611$.

Solution

As θ increases $\cos\theta$ decreases, so the cosine of $37°25'48''$ will be less than the cosine of $37°25'00''$.

$$\frac{48}{60} = 0.8$$

$$0.7942379 - 0.7940611 = 0.0001768$$

$$\cos 37°25'48'' = 0.7942379$$
$$- (0.8)(0.0001768)$$
$$= 0.7940965$$

Example 11.3

Find θ if $\cos\theta = 0.8047643$.

Solution

The cosine of $36°24'00''$, 0.8048938, and the cosine of $36°25'00''$, 0.8047211, are just greater than and just less than 0.8047643. So θ is more than $36°24'00''$ and less than $36°25'00''$. The cosine of θ is $1295/1727 = 0.75$ of the way between 0.8048938 and 0.8047211, so θ is 0.75 of the way between $36°24'00''$ and $36°25'00''$.

$$\arccos 0.8047643 = 36°24'00'' + (0.75)(60'')$$
$$= 36°24'45''$$

Example 11.4

Find $\tan 44°17'06''$.

Solution

$$\tan 44°17'00'' = 0.9752914$$
$$\tan 44°18'00'' = 0.9758591$$
$$\tan 44°17'06'' = 0.9752914$$
$$+ \left(\frac{6}{60}\right)(0.9758591 - 0.9752914)$$
$$= 0.9753482$$

Example 11.5

Find θ if $\cot\theta = 1.0967405$.

Solution

$$\cot 42°21' = 1.0970609$$
$$1.0970609 - 1.0967405 = 0.0003204$$
$$\cot 42°22' = 1.0964201$$

$$1.0970609 - 1.0964201 = 0.0006408$$
$$\text{arccot } 1.0967405 = 42°21'00''$$
$$+ \left(\frac{0.0003204}{0.0006408}\right)(60'')$$
$$= 42°21'30''$$

14. BEARING OF A LINE

The *bearing of a line* is the acute horizontal angle between the meridian (north line) and the line.

15. ANGLE OF ELEVATION AND ANGLE OF DEPRESSION

The *angle of elevation* is the vertical angle between the horizontal and a line rotated upward from the horizontal. It is positive. The *angle of depression* is the vertical angle between the horizontal and a line rotated downward from the horizontal. It is negative.

16. SOLUTION OF RIGHT TRIANGLES

Solving a right triangle means finding the value of its three angles and the length of each of its sides. To solve a right triangle, two of these values must be known. Either two sides must be known, or one side and an acute angle must be known.

If an acute angle of a right triangle is known, the other acute angle is the complement of it, since the sum of the interior angles of a triangle equals $180°$.

Example 11.6

Solve the right triangle ABC having angle $A = 23°30'$ and side $a = 400$ ft. (Note: It is customary to use capital letters to name the vertices of a triangle and the corresponding lower case letter to name the side opposite each of the vertices.)

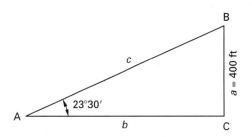

Solution

$$B = 90° - 23°30' = 66°30'$$

In solving for c, the side opposite the known angle A is the known side a, and side c is the hypotenuse. So, select the sine function for use in the solution.

$$\sin A = \frac{\text{opposite}}{\text{hypotenuse}}$$

$$\sin 23°30' = \frac{400}{c}$$

To remove c from the denominator, multiply both sides of the equation by c.

$$c \sin 23°30' = \frac{400c}{c}$$

$$c \sin 23°30' = 400$$

To isolate c, divide both sides of the equation by $\sin 23°30'$.

$$\frac{c \sin 23°30'}{\sin 23°30'} = \frac{400}{\sin 23°30'}$$

$$c = \frac{400}{\sin 23°30'}$$

From trigonometric tables or a calculator,

$$\sin 23°30' = 0.3987$$

$$c = \frac{400}{0.3987} = 1000 \text{ ft}$$

b is the side adjacent to the known angle A, so select the tangent function.

$$\tan A = \frac{\text{opposite}}{\text{adjacent}}$$

$$\tan 23°30' = \frac{400}{b}$$

Multiplying both sides of the equation by b and dividing both sides by $\tan 23°30'$,

$$b = \frac{400}{\tan 23°30'}$$

$$b = \frac{400}{0.4348} = 920 \text{ ft}$$

Example 11.7

Solve the right triangle EFG having angle $E = 40°$ and the hypotenuse, side $g = 200$ ft.

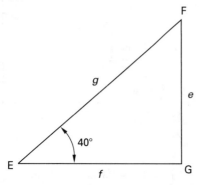

Solution

$$F = 90° - 40° = 50°$$

To solve for e, the sine function is selected.

$$\sin E = \frac{\text{opposite}}{\text{hypotenuse}}$$

$$\sin 40° = \frac{e}{200 \text{ ft}}$$

To isolate e, multiply both sides of the equation by 200 ft.

$$(200 \text{ ft}) \sin 40° = \left(\frac{200 \text{ ft}}{200 \text{ ft}} \right) e$$

$$e = 200 \sin 40°$$

From trigonometric tables or a calculator,

$$\sin 40° = 0.6428$$

$$e = (200 \text{ ft})(0.6428) = 130 \text{ ft}$$

To solve for f, select the cosine function.

$$\cos E = \frac{\text{adjacent}}{\text{hypotenuse}}$$

$$\cos 40° = \frac{f}{200 \text{ ft}}$$

Multiplying both sides by 200 ft,

$$f = (200 \text{ ft})(\cos 40°) = (200 \text{ ft})(0.7660)$$

$$= 150 \text{ ft}$$

17. ALTERNATE SOLUTION METHODS FOR RIGHT ANGLES

Since there are many different relationships between the sides and angles of right triangles, it is not surprising that triangle problems may have more than a single correct solution method. Some solutions are simpler than others. Practice is required to be able to select the simplest solution procedure.

Additional examples are provided to illustrate various solution procedures.

Example 11.8

Solve the right triangle ABC having angle $A = 23°30'$ and side $a = 400$ ft. (Refer to the illustration in Ex. 11.6).

Solution

$$B = 90° - 23°30' = 66°30'$$

The known side is opposite the known angle. To find the hypotenuse, select the sine function.

$$c = \frac{400 \text{ ft}}{\sin 23°30'} = 1000 \text{ ft}$$

To solve for b, select the tangent function.

$$b = \frac{400 \text{ ft}}{\tan 23°30'} = 920 \text{ ft}$$

Example 11.9

Solve the right triangle EFG in Ex. 11.7 having angle $E = 40°$ and side $g = 200$ ft.

Solution

$$F = 90° - 40° = 50°$$

The hypotenuse, which is the longest side, is known; therefore, to find the other sides, the hypotenuse must be multiplied by the sine or cosine function.

To solve for e, use the sine function.

$$e = (200 \text{ ft})(\sin 40°) = 130 \text{ ft}$$

To solve for f, use the cosine function.

$$f = (200 \text{ ft})(\cos 40°) = 150 \text{ ft}$$

Example 11.10

Solve the right triangle ABC.

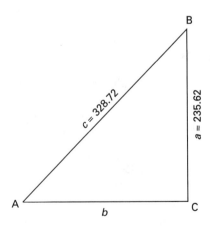

Solution

Find the solution from the inverse sine function on a calculator.

$$\sin A = \frac{235.62}{328.72} = 0.7167802$$

$$A = \arcsin 0.7167802 = 45°47'21''$$

$$B = 90°00'00'' - 45°47'21'' = 44°12'39''$$

$$b = \sqrt{(328.72)^2 - (235.62)^2} = 229.22$$

Example 11.11

To measure the width of a river, surveyors establish points A and B on the west bank and find the distance between them to be 90.0 ft. They set up a transit on point B and establish point C on the east bank so that BC is at right angles to AB. They then measure the angle at A between AB and AC and find it to be 68°20′. How wide is the river?

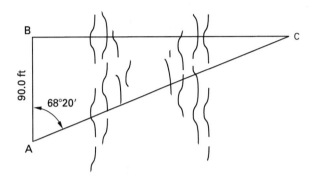

Solution

Select the tangent function.

$$BC = (90.0 \text{ ft})(\tan 68°20')$$
$$= (90.0 \text{ ft})(2.517) = 226.5 \text{ ft}$$

Example 11.12

San Angelo is due west of Waco, and Arlington is 100 mi due north of Waco. The bearing of Arlington from San Angelo is N 66°30′ E. (a) How far is San Angelo from Arlington? (b) How far is San Angelo from Waco?

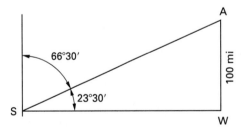

Solution

The triangle formed by the three cities is a right triangle. The angle at S is the complement of the bearing angle. In solving for SA, select the sine function.

$$SA = \frac{100 \text{ mi}}{\sin 23°30'} = 250 \text{ mi}$$

In solving for SW, select the tangent function.

$$SW = \frac{100 \text{ mi}}{\tan 23°30'} = 230 \text{ mi}$$

18. RELATED ANGLES

Trigonometric tables give values for acute angles only. Larger angles must be expressed in terms of an acute angle that has the same values for the functions. These angles are known as *related angles* or *reference angles*.

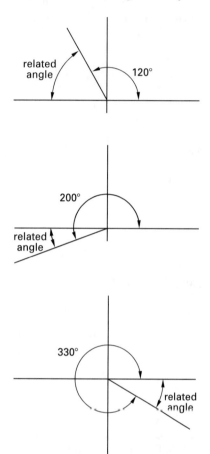

Figure 11.10 *Related Angles*

The related angle of θ is the positive acute angle between the x-axis and the terminal side of the angle. For example, in Fig. 11.10, the related angle of 120° is 60°, the related angle of 200° is 20°, and the related angle of 330° is 30°.

Regardless of which quadrant an angle lies in, the numerical value of any one of the six trigonometric functions is the same as that of the related angle. But the algebraic sign of any function depends on the quadrant in which the angle lies.

As illustrated in Fig. 11.2 and Sec. 3, an angle is a third-quadrant angle if its terminal side lies in the third quadrant. Likewise, an angle is in one of the other quadrants if its terminal side lies in that quadrant. Figure 11.5 gives the signs of the natural functions for each quadrant.

As solving oblique triangles involves first- and second-quadrant angles only, it is important to remember that the sine function will always be positive and the cosine function will be negative for second-quadrant angles (90° to 180°).

Example 11.13

Find the sine, cosine, and tangent of 150°.

Solution

The related angle is $180° - 150° = 30°$. The angle is less than 180° and more than 90°; therefore, it is a second-quadrant angle. The sign of the sine function is positive, the sign of the cosine function is negative, and the sign of the tangent function is negative.

$$\sin 150° = +0.500$$
$$\cos 150° = -0.866\ldots$$
$$\tan 150° = -0.577\ldots$$

Example 11.14

Find the sine, cosine, and tangent of 315°.

Solution

The related angle is $360° = 45°$.

$$\sin 315° = -0.707\ldots$$
$$\cos 315° = +0.707\ldots$$
$$\tan 315° = -1.000$$

19. SINE CURVE

The variations in the value of $\sin \theta$ can be shown by plotting θ as the abscissa and $\sin \theta$ as the ordinate on a system of coordinate axes. This is done in Fig. 11.11 as a *sine curve*.

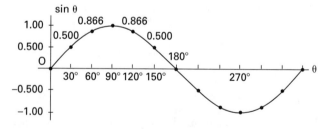

Figure 11.11 *Sine Curve*

20. COSINE CURVE

The *cosine* curve has the same shape as the sine curve, but is offset 90°.

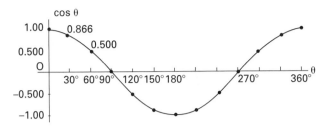

Figure 11.12 *Cosine Curve*

21. OBLIQUE TRIANGLES

An *oblique triangle* is a triangle that does not contain a right angle. All of the angles in an oblique triangle may be acute, or there may be one obtuse angle and two acute angles.

As with right triangles, the three angles in an oblique triangle are identified with capital letters, and most often the letters A, B, and C are used. The sides of the triangle are often identified with small (lower case) letters, with side a opposite angle A, side b opposite angle B, and side c opposite angle C. However, in surveys, angles can also be identified with letters other than A, B, and C and sometimes are identified with numbers. Sides can be identified with two capital letters.

The three angles and three sides of any triangle make up the six parts of the triangle. An oblique triangle can be solved if three of its parts, at least one of which is a side, are known. However, the solution is not as simple as the solution of a right triangle. Oblique triangles can be solved by forming two right triangles within the oblique triangle, but the task is made easier by the use of formulas. The most important formulas are the *law of sines* and the *law of cosines*. The choice of which of the two laws to use to solve a particular triangle depends on which three parts of the triangle are known.

If one side and two angles of a triangle are known, the triangle can be solved by the law of sines. This case is represented by the abbreviation SAA (side, angle, angle).

Example 11.15

What abbreviation represents the known triangle parameters?

Given: $a = 32.16$, $B = 64°20'$, $C = 50°20'$

Solution

If two sides and the angle opposite one of them are known, the triangle can also be solved by the law of sines. This case is represented by the abbreviation SSA (side, side, angle).

Example 11.16

What abbreviation represents the known triangle parameters?

Given: $a = 251.5$, $b = 647.3$, $A = 22°20'$

Solution

If two sides and the angle included between the two sides are known, the triangle cannot be solved by the law of sines alone. However, it can be solved by the law of cosines and the law of sines together. This case is represented by the abbreviation SAS (side, angle, side). After the third side is found by the law of cosines, the second angle can be found by the law of sines.

If three sides of a triangle are known, the triangle can be solved by the law of cosines. This case is represented by the abbreviation SSS (side, side, side).

22. LAW OF SINES

In any triangle, the sides are proportional to the sines of the opposite angles. Equations 11.21 and 11.22 show the *law of sines*.

$$\frac{a}{\sin A} = \frac{b}{\sin B} = \frac{c}{\sin C} \qquad \text{11.21}$$

$$\frac{\sin A}{a} = \frac{\sin B}{b} = \frac{\sin C}{c} \qquad \text{11.22}$$

23. SAA CASE

In solving the SAA case, the law of sines can be expressed as Eqs. 11.23 through 11.25.

$$a = (\sin A)\left(\frac{b}{\sin B}\right)$$

$$= (\sin A)\left(\frac{c}{\sin C}\right) \qquad \text{11.23}$$

$$b = (\sin B)\left(\frac{a}{\sin A}\right)$$

$$= (\sin B)\left(\frac{c}{\sin C}\right) \qquad \text{11.24}$$

$$c = (\sin C)\left(\frac{a}{\sin A}\right)$$

$$= (\sin C)\left(\frac{b}{\sin B}\right) \qquad \text{11.25}$$

In the case with two angles known, the third angle can be readily found, so there will always be a known side opposite a known angle.

Example 11.17

Solve the triangle ABC with $C = 83°$, $B = 61°$, and $c = 150$ ft.

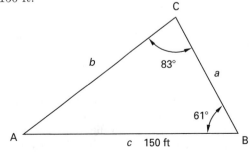

Solution

$$A = 180° - (83° + 61°) = 36°$$

$$a = (\sin A)\left(\frac{c}{\sin C}\right)$$

$$= (\sin 36°)\left(\frac{150 \text{ ft}}{\sin 83°}\right)$$

$$= 89 \text{ ft}$$

$$b = (\sin B)\left(\frac{c}{\sin C}\right)$$

$$= (\sin 61°)\left(\frac{150 \text{ ft}}{\sin 83°}\right)$$

$$= 132 \text{ ft}$$

Example 11.18

Solve the triangle EFG shown.

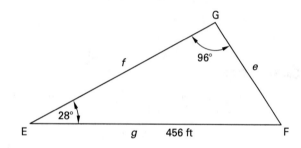

Solution

$$F = 180° - (28° + 96°) = 56°$$

$$e = (\sin 28°)\left(\frac{456 \text{ ft}}{\sin 96°}\right) = 215 \text{ ft}$$

$$f = (\sin 56°)\left(\frac{456 \text{ ft}}{\sin 96°}\right) = 380 \text{ ft}$$

24. SSA CASE

If two sides and an angle opposite one of them are known, the law of sines can be expressed as Eq. 11.26.

$$\sin A = \frac{a \sin B}{b} \qquad 11.26$$

Example 11.19

Solve the triangle ABC with $A = 36°$, $a = 50$ in, and $b = 70$ in.

In using the law of sines to solve triangles in which two sides and the angle opposite one of them are given, it is possible to construct two different triangles from the given information. For example, triangles ABC and

AB'C can both be constructed from the given information.

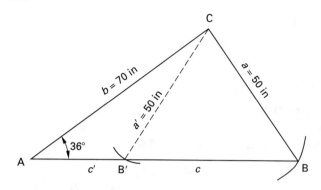

Solution for triangle ABC

$$\sin B = \frac{(70 \text{ in})(\sin 36°)}{50 \text{ in}} = 0.8229 \text{ in}$$

$$B = 55°$$

$$C = 180° - (36° + 55°) = 89°$$

$$c = (\sin 89°)\left(\frac{50 \text{ in}}{\sin 36°}\right) = 85 \text{ in}$$

Solution for triangle AB'C

Angle B could be the related angle to 55°. Then,

$$B' = 125°$$

$$C = 180° - (36° + 125°) = 19°$$

$$c' = (\sin 19°)\left(\frac{50 \text{ in}}{\sin 36°}\right) = 28 \text{ in}$$

25. LAW OF COSINES

Equations 11.25 through 11.27 are alternate forms of the law of cosines. Referring to Fig. 11.13,

$$a^2 = b^2 + c^2 - 2bc \cos A \qquad 11.27$$
$$b^2 = c^2 + a^2 - 2ca \cos B \qquad 11.28$$
$$c^2 = a^2 + b^2 - 2ab \cos C \qquad 11.29$$

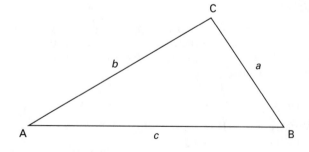

Figure 11.13 *A General Triangle*

Remember that the cosine of an angle greater than 90° and less than 180° has a negative algebraic sign. In substituting the cosines of angles in that range, the negative sign must be included, meaning that the value of $2bc \cos A$ in Eq. 11.27 will be added to the value of $b^2 + c^2$.

$$\cos A = \frac{b^2 + c^2 - a^2}{2bc} \qquad 11.30$$

$$\cos B = \frac{a^2 + c^2 - b^2}{2ac} \qquad 11.31$$

$$\cos C = \frac{a^2 + b^2 - c^2}{2ab} \qquad 11.32$$

26. SAS CASE

The law of cosines can be used when two sides and the included angle of a triangle are known. In this case, the law of cosines can be expressed as Eqs. 11.33 through 11.35.

$$a = \sqrt{b^2 + c^2 - 2bc \cos A} \qquad 11.33$$

$$b = \sqrt{a^2 + c^2 - 2ac \cos B} \qquad 11.34$$

$$c = \sqrt{a^2 + b^2 - 2ab \cos C} \qquad 11.35$$

Example 11.20

Solve the triangle shown.

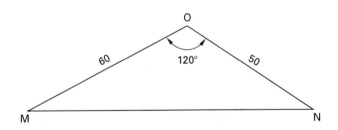

Solution

$$\text{MN} = \sqrt{(60)^2 + (50)^2 - (2)(60)(50)(\cos 120°)}$$

$$= \sqrt{3600 + 2500 - (6000)(0.500)} = 95 \text{ ft}$$

$$\sin M = \frac{(50)(\sin 120°)}{95} = 0.45580 \quad [\text{law of sines}]$$

$$M = 27°$$

$$\sin N = \frac{(60)(\sin 120°)}{95} = 0.54696 \quad [\text{law of sines}]$$

$$N = 33°$$

27. SSS CASE

When three sides of a triangle are known (SSS), the law of cosines can be expressed as

$$\cos A = \frac{b^2 + c^2 - a^2}{2bc} \qquad 11.36$$

$$\cos B = \frac{a^2 + c^2 - b^2}{2ac} \qquad 11.37$$

$$\cos C = \frac{a^2 + b^2 - c^2}{2ab} \qquad 11.38$$

Example 11.21

Solve triangle ABC with sides $a = 3.0$, $b = 5.0$, and $c = 6.0$.

Solution

$$\cos A = \frac{(5.0)^2 + (6.0)^2 - (3.0)^2}{(2)(5.0)(6.0)}$$

$$= \frac{25 + 36 - 9}{60}$$

$$A = 30°$$

$$\cos B = \frac{(3.0)^2 + (6.0)^2 - (5.0)^2}{(2)(3.0)(6.0)}$$

$$= \frac{9 + 36 - 25}{36}$$

$$B = 56°$$

$$\cos C = \frac{(3.0)^2 + (5.0)^2 - (6.0)^2}{(2)(3.0)(5.0)}$$

$$= \frac{9 + 25 - 36}{30} = -0.067$$

$$C = 94°$$

The negative value of $\cos C$ indicates that the angle is greater than 90° and the related angle is 86°.

28. OBLIQUE TRIANGLES USED IN SURVEYING

Example 11.22

Austin is 100 mi S 23° W from Waco. The bearing of Houston from Waco is S 40° E, and the bearing of Austin from Houston is N 76° W. What is the distance from Waco to Houston? What is the distance from Houston to Austin?

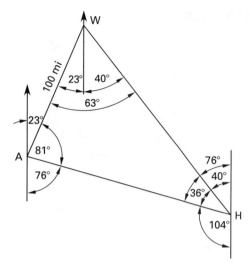

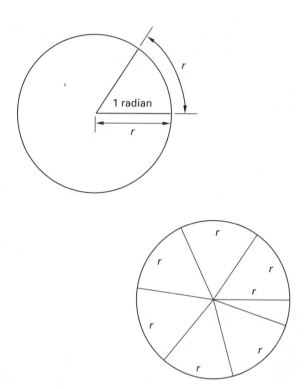

Figure 11.14 *Radians*

Solution

Draw a meridian (north line) and place point W (Waco) on it. From W draw a line that makes an angle of about 23° with the meridian in a southwesterly direction. Using any scale, place point A (Austin) on this line. Also from W draw a line in a southeasterly direction making an angle of about 40° with the meridian. From point A, draw a line in a southeasterly direction making an angle of about 76° with a meridian through A. This line intersects the southeasterly line from W at point H (Houston).

$$\text{WH} = \frac{(100 \text{ mi})\,(\sin 81°)}{\sin 36°} = 170 \text{ mi}$$

$$\text{HA} = \frac{(100 \text{ mi})(\sin 63°)}{\sin 36°} = 150 \text{ mi}$$

29. SELECTION OF LAW TO BE USED

To select the proper law for the solution of triangles, remember the following.

- *law of sines:* one side and two angles (SAA)

   two sides and the angle opposite one (SSA)

- *law of cosines:* two sides and the included angle (SAS)

   three sides (SSS)

30. RADIAN MEASURE

A *radian* is an angle that, when situated as a central angle of a circle, is subtended by an arc whose length is equal to the radius of the circle.

The circumference of a circle is 2π times the length of the radius, r. Therefore, the number of arcs of length r that can be applied to the circumference of a circle is 2π.

$$2\pi \text{ rad} = 360° \qquad \textit{11.39}$$
$$\pi \text{ rad} = 180° \qquad \textit{11.40}$$

Dividing both sides of Eq. 11.40 by π gives

$$1 \text{ rad} = \frac{180°}{\pi} \qquad \textit{11.41}$$

Dividing both sides of Eq. 11.40 by 180° gives

$$1° = \frac{\pi}{180°} \text{ rad} \qquad \textit{11.42}$$

rule: To convert degrees to radians, multiply the number of degrees by $\pi/180°$.

For example,

$$30° \equiv \frac{30°\pi}{180°} = \frac{\pi}{6} \text{ rad}$$

rule: To convert radians to degrees, multiply the number of radians by $180°/\pi$.

For example,

$$2 \text{ rad} = \frac{(2)(180°)}{\pi} = \frac{360°}{\pi}$$

31. LENGTH OF AN ARC OF A CIRCLE

If S equals the length of any arc of a circle, the relationship between the arc length and radius where θ is expressed in radians is

$$S = r\theta \qquad 11.43$$

Example 11.23

What is the length of the arc subtended by a central angle of 30° in a circle of 10 in radius?

Solution

$$S = r\theta = \frac{(10 \text{ in})(30°)\pi}{180°} = 5.2 \text{ in}$$

Example 11.24

What is the radius of the circle on which an arc of 100 ft subtends an angle of 1°?

$$r = \frac{S}{\theta} = \frac{100 \text{ ft}}{\dfrac{(1)\pi}{180°}} = \frac{18,000 \text{ ft}}{\pi} = 5729.58 \text{ ft}$$

32. AREA OF A SECTOR OF A CIRCLE

By geometry, the area of a sector of a circle is equal to one half its arc times the radius of the circle. If S equals the length of the arc, the area = $\frac{1}{2}Sr$.

The length of an arc of a circle equals $r\theta$, where θ is in radians. Therefore, where θ is in radians,

$$\text{area} = \tfrac{1}{2}r^2\theta \qquad 11.44$$

33. AREA OF A SEGMENT OF A CIRCLE

The area of a segment of a circle is found by subtracting the area of the triangle formed by the chord and the radii to the chord's end points from the area of the sector formed by the two radii and the arc of the segment. The area of the triangle equals $\frac{1}{2}ab(\sin C)$, where a and b are radii of the circle and C is the central angle. Therefore, where θ is in radians,

$$A = \tfrac{1}{2}r^2\theta - \tfrac{1}{2}r^2\sin\theta = \tfrac{1}{2}r^2(\theta - \sin\theta) \qquad 11.45$$

Example 11.25

Find the area of the segment with $r = 8$ in and central angle = 45°.

Solution

$$A = \left(\frac{1}{2}\right)(64 \text{ in})(0.7854 \text{ in} - 0.7071 \text{ in}) = 2.5 \text{ in}$$

PRACTICE PROBLEMS

1. Each of the following points lies on the terminal side of an angle θ in standard position. Plot the point, draw the terminal side through the point, and measure the angle with a protractor.

(a) $(12, 6)$

(b) $(-12, 4)$

(c) $(-10, -10)$

(d) $(7, -9)$

2. Write the sin, cos, tan, cot, sec, and csc of angles θ as a common fraction. Show the algebraic sign.

(a)

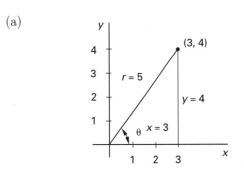

(b)

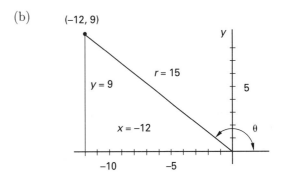

(c)

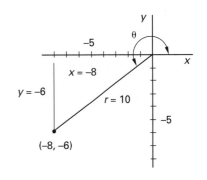

(d)

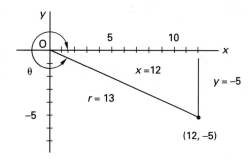

3. Write the reciprocal of each of the following numbers.

Example: $\frac{2}{3}$ $\dfrac{1}{\frac{2}{3}} = \frac{3}{2}$

(a) 3   (b) $\frac{3}{4}$

(c) x (d) y

(e) r (f) $\dfrac{1}{x}$

(g) $\dfrac{1}{y}$ (h) $\dfrac{x}{r}$

(i) $\dfrac{y}{r}$ (j) $\dfrac{y}{x}$

4. Show that $\sin\theta$ and $\csc\theta$, $\cos\theta$ and $\sec\theta$, and $\tan\theta$ and $\cot\theta$ are reciprocals.

Example: $\sin\theta$ $\dfrac{1}{\sin\theta} = \dfrac{1}{\frac{y}{r}} = \dfrac{r}{y} = \csc\theta$

(a) $\cos\theta$

(b) $\tan\theta$

(c) $\cot\theta$

5. Using a calculator and the reciprocal, find the following numbers.

(a) $\sec 60°$

(b) $\csc 30°$

(c) $\cot 45°$

6. Using the following graph, find the sine, cosine, and tangent functions of the angles indicated.

Example:

$$\theta = 150°$$

$$\sin 150° = \frac{5.0}{10} = 0.50$$

$$\cos 150° = \frac{-8.7}{10} = -0.87$$

$$\tan 150° = \frac{5.0}{-8.7} = -0.57$$

(a) $\theta = 30°$

(b) $\theta = 135°$

(c) $\theta = 300°$

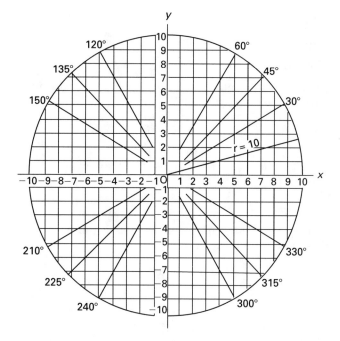

7. Indicate the algebraic sign of the sine, cosine, and tangent functions of each of the following angles.

Example: 30°: sin+
               cos+
               tan+

(a) 185°   (b) 225°

(c) 350°   (d) 190°

(e) 265°   (f) 100°

(g) 89°   (h) 175°

(i) 95°   (j) 275°

(k) 300°   (l) 110°

(m) 85°   (n) 290°

8. Using the information provided, identify the quadrant in which the terminal side of the angle θ lies.

Examples: $\sin\theta = +$, $\cos\theta = -$: II

(a) $\sin\theta = -$, $\cos\theta = +$

(b) $\tan\theta = -$, $\sin\theta = -$

(c) $\tan\theta = +$, $\sin\theta = -$

(d) $\sin\theta = +$, $\cos\theta = +$

(e) $\tan\theta = +$, $\cos\theta = -$

(f) $\sin\theta = +$, $\tan\theta = +$

(g) $\sin\theta = -$, $\tan\theta = -$

9. Indicate the value of the sin, cos, and tan functions of the following quadrantal angles.

Example: $0°$ $\sin = 0$; $\cos = +1$; $\tan = 0$

(a) $90°$

(b) $180°$

(c) $270°$

(d) $360°$

10. Using the figures, write sin, cos, and tan function of each acute angle. Express as a common fraction and as a decimal fraction to two significant digits.

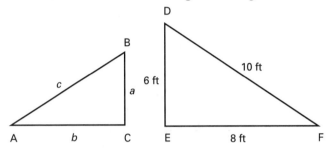

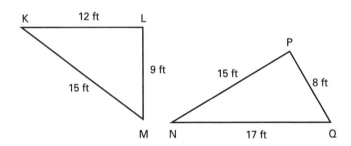

Examples: (1) $\sin A = \dfrac{a}{c}$

(2) $\sin K = \dfrac{9}{15} = 0.60$

(a) $\cos A$ (b) $\tan A$

(c) $\sin B$ (d) $\cos B$

(e) $\tan B$ (f) $\cos K$

(g) $\tan K$ (h) $\sin M$

(i) $\cos M$ (j) $\tan M$

(k) $\sin D$ (l) $\cos D$

(m) $\tan D$ (n) $\sin F$

(o) $\cos F$ (p) $\tan F$

(q) $\sin N$ (r) $\cos N$

(s) $\tan N$ (t) $\sin Q$

(u) $\cos Q$ (v) $\tan Q$

11. Find the trigonometric function for the angle indicated or find the angle for the function indicated.

Example: $\sin 53°13'36'' = 0.8010101$

(a) $\sin 37°29'16''$

(b) $\cos 52°42'51''$

(c) $\cos^{-1} 0.7918605$

(d) $\tan 43°17'28''$

12. For triangle ABC, select the trigonometric function to be used to solve the part of the triangle indicated.

Examples: (1) $A = 36°52'$, $a = 600$ ft, b: tan, c: sin

(2) $a = 300$ ft, $b = 400$ ft, A: tan, B: tan

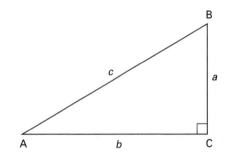

(a) $B = 51°40'$, $a = 650$ ft
    b: _____ c: _____

(b) $A = 46°44'$, $b = 156$ ft
    a: _____ c: _____

(c) $B = 53°21'$, $c = 300$ ft
    a: _____ b: _____

(d) $A = 38°19'$, $c = 700$ ft
    b: _____ c: _____

(e) $a = 600$ ft, $c = 1000$ ft
    A: _____ B: _____

(f) $b = 400$ ft, $c = 500$ ft
    A: _____ B: _____

(g) $B = 55°10'$, $b = 378$ ft
    a: _____ c: _____

(h) $A = 33°40'$, $a = 250$ ft
    b: _____ c: _____

13. Completely solve the right triangle ABC.

Example: $A = 36°52'$, $a = 600$ ft

$$B = 90° - 36°52' = 53°08'$$

$$b = \frac{600 \text{ ft}}{\tan 36°52'} = 800 \text{ ft}$$

$$c = \frac{600 \text{ ft}}{\sin 36°52'} = 1000 \text{ ft}$$

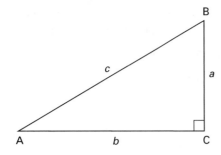

(a) $A = 28°41', b = 540$ ft

(b) $B = 55°13', a = 371$ ft

(c) $B = 61°29', b = 466$ ft

(d) $A = 33°15', c = 263$ ft

(e) $B = 58°55', c = 562$ ft

(f) $a = 300$ ft, $b = 400$ ft

14. Find the height to the nearest inch of a man who casts a shadow 10.0 ft long when the angle of elevation of the sun is $31°18'$.

15. From a point 160 ft from the foot of a flagpole, the angle of elevation to the top of the flagpole is $50°40'$. Find the height of the flagpole.

16. From the top of a tower 500 ft high, the angle of depression to a road intersection is $30°$. How far from the tower is the road intersection?

17. Temple is due south of Fort Worth and 115 mi due east of Brady. The bearing of Brady from Fort Worth is $S45°W$. How far is Brady from Fort Worth?

18. A 20 ft ladder reaches from the ground to the roof of a building. The angle of elevation of the ladder is $60°$. How high is the roof?

19. Completely solve the right triangle ABC, as in Prob. 13.

(a) $B = 41°12'38'', a = 625.18$ ft

(b) $A = 66°22'37'', a = 492.72$ ft

(c) $B = 38°04'48'', c = 585.20$ ft

(d) $A = 22°13'50'', a = 376.26$ ft

(e) $B = 75°35'41'', b = 237.68$ ft

(f) $a = 427.82', b = 396.95$ ft

(g) $b = 445.64', c = 616.38$ ft

20. Write the related (or reference) angle of each angle.

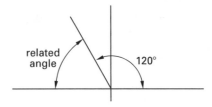

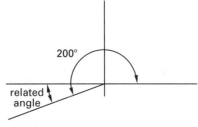

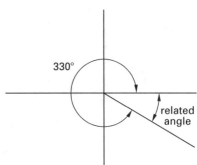

Example: $150°$, $30°$

(a) $260°$	(b) $210°$
(c) $160°$	(d) $220°$
(e) $190°$	(f) $350°$
(g) $330°$	(h) $170°$
(i) $200°$	(j) $300°$
(k) $125°$	(l) $155°$
(m) $110°$	(n) $140°$
(o) $135°$	(p) $100°$
(q) $120°$	(r) $175°$
(s) $95°$	(t) $185°$
(u) $280°$	

21. Solve for the missing sides and angles for the following oblique triangles (SAA).

Example: triangle MNO: $M = 38°48'45''$, $O = 82°23'56''$, MN $= 298.34$ ft

$$N = 180° - (38°48'45'' + 82°23'56'') = 58°47'19''$$

$$NO = \frac{(298.34 \text{ ft})(\sin 38°48'45'')}{\sin 82°23'56''} = 188.65 \text{ ft}$$

$$OM = \frac{(298.34 \text{ ft})(\sin 58°47'19'')}{\sin 82°23'56''} = 257.42 \text{ ft}$$

(a) triangle 1-2-3: $1 = 34°18'24''$, $3 = 62°12'55''$ 1-2 = 1347.77 ft

(b) triangle PQR: $P = 118°34'24''$, $Q = 23°06'54''$, QR = 526.30 ft

(c) triangle 1-2-3: $1 = 82°46'58''$, $2 = 58°54'20''$, 2-3 = 345.43 ft

22. Solve for the missing sides and angles for the following oblique triangles (SSA).

Example: triangle ABC: $A = 76°00'09''$, $a = 256.07$ ft, $b = 172.28$ ft

$$B: \sin B = \frac{(172.28 \text{ ft})(\sin 76°00'09'')}{256.07} \quad B = 40°45'13''$$

$$C = 180° - (76°00'09'' + 40°45'13'') = 63°14'38''$$

$$c = \frac{(\sin 63°14'38'')(256.07 \text{ ft})}{\sin 76°00'09''} = 235.65 \text{ ft}$$

(a) triangle EFG: $E = 25°40'55''$, $e = 646.13$ ft, $f = 1296.20$ ft

(b) triangle ABC: $A = 51°20'14''$, $a = 445.23$ ft, $b = 526.17$ ft (two solutions)

23. Solve for the missing sides and the angles for the following oblique triangles (SAS).

Example:

triangle ABC: $A = 58°33'47''$, $b = 204.38$ ft, $c = 152.15$ ft

$$a = \sqrt{\begin{array}{l}(204.38)^2 + (152.15)^2 \\ - (2)(204.38)(152.15) \cos 58°33'47''\end{array}} = 180.23 \text{ ft}$$

$$B: \sin B = \frac{(204.38 \text{ ft})(\sin 58°33'47'')}{180.23} = 75°21'43''$$

$$C = 180° - (75°21'43'' + 58°33'47'') = 46°04'30''$$

(a) triangle EFG: $G = 95°12'50''$, FG = 146.25 ft, GE = 122.31 ft

(b) triangle KLM: $L = 35°19'16''$, KL = 595.45 ft, LM = 851.78 ft (Hint: The largest angle must be opposite the longest side. The sine function may represent a related angle.)

(c) triangle NOP: $N = 46°07'01''$, NO = 138.38 ft, PN = 165.12 ft

24. Solve for the missing sides and angles for the following oblique triangles (SSS).

Example: triangle ABC: $a = 48.79$ ft, $b = 62.45$ ft, $c = 30.13$ ft

$$A: \cos A = \frac{(62.45)^2 + (30.13)^2 - (48.79)^2}{(2)(62.45)(30.13)} = 49°49'59''$$

$$B: \cos B = \frac{(48.79^2) + (30.13)^2 - (62.45)^2}{(2)(48.79)(30.13)} = 102°00'32''$$

$$C: \cos C = \frac{(48.79)^2 + (62.45)^2 - (30.13)^2}{(2)(48.79)(62.45)} = 28°09'29''$$

$$\text{check } 180°00'00''$$

(a) triangle EFG: EF = 125.83 ft, FG = 171.25 ft, GE = 155.13 ft

(b) triangle MNO: MN = 298.34 ft, NO = 188.65 ft, OM = 257.42 ft

25. Given the parts of an oblique triangle ABC indiacted, select the law applicable for the solution.

Example: a, c, A; law of sines

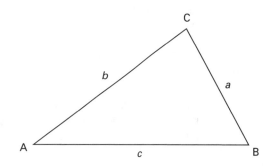

(a) a, b, A

(b) b, c, A

(c) b, A, C

(d) A, B, a

(e) a, c, B

(f) b, c, B

(g) A, a, b

(h) a, b, c

(i) A, C, a

(j) c, b, a

26. Express the following angles in radians using π in each answer.

Example: $30° = \dfrac{\pi}{6}$ rad

(a) $45°$ (b) $120°$ (c) $90°$ (d) $15°$ (e) $270°$

27. Express the following angles in degrees.

Example: π rad $= 180°$

(a) $\dfrac{\pi}{3}$ rad (b) $\dfrac{\pi}{2}$ rad (c) $\dfrac{\pi}{4}$ rad (d) $\dfrac{\pi}{12}$ rad (e) $\dfrac{3\pi}{2}$ rad

28. What is the length of the arc subtended by a central angle of $60°$ in a circle with a 300 ft radius?

29. What is the radius of the circle on which an arc of 300 ft subtends an angle of 12°?

30. Find the radius of a circle on which an arc of 25 in has a central angle of 2.4 rad.

31. Find the length along the equator of an arc subtended by a central angle of 1° if the diameter of the earth at the equator is 7927 mi.

32. The end of a 25 in pedulum swings through a 3.4 in arc. What is the size of the angle through which the pendulum swings?

33. Find the area of a sector of a circle with a radius of 6 in and a central angle of 60°.

34. Find the area of a segment of a circle with a 100 ft radius and a 30° central angle.

35. Using a calculator, find the value of the sine function for each angle indicated, plot on the coordinate system marked sine curve, and connect the points with a curved line. Repeat the procedure for the cosine curve.

SOLUTIONS

1. (a) $\theta = \boxed{27°}$　　　(b) $\theta = \boxed{162°}$

(c) $\theta = \boxed{225°}$　　　(d) $\theta = \boxed{308°}$

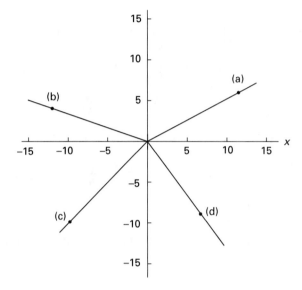

(a)
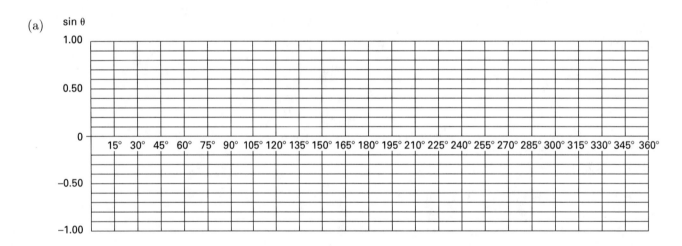
sin θ

(b)
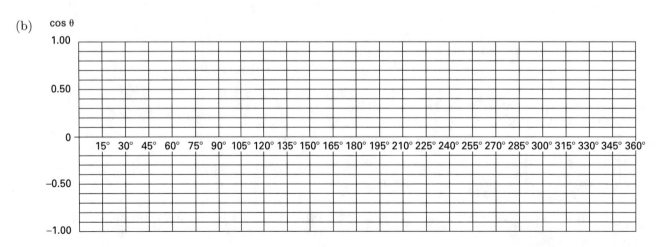
cos θ

2. (a) $\sin \theta = \boxed{\dfrac{4}{5}}$

$\cos \theta = \boxed{\dfrac{3}{5}}$

$\tan \theta = \boxed{\dfrac{4}{3}}$

$\cot \theta = \boxed{\dfrac{3}{4}}$

$\sec \theta = \boxed{\dfrac{5}{3}}$

$\csc \theta = \boxed{\dfrac{5}{4}}$

(b) $\sin \theta = \boxed{\dfrac{9}{15}}$

$\cos \theta = \boxed{-\dfrac{12}{15}}$

$\tan \theta = \boxed{-\dfrac{9}{12}}$

$\cot \theta = \boxed{-\dfrac{12}{9}}$

$\sec \theta = \boxed{-\dfrac{15}{12}}$

$\csc \theta = \boxed{\dfrac{15}{9}}$

(c) $\sin \theta = \boxed{-\dfrac{6}{10}}$

$\cos \theta = \boxed{-\dfrac{8}{10}}$

$\tan \theta = \boxed{\dfrac{6}{8}}$

$\cot \theta = \boxed{\dfrac{8}{6}}$

$\sec \theta = \boxed{-\dfrac{10}{8}}$

$\csc \theta = \boxed{-\dfrac{10}{6}}$

(d) $\sin \theta = \boxed{-\dfrac{5}{13}}$

$\cos \theta = \boxed{\dfrac{12}{13}}$

$\tan \theta = \boxed{-\dfrac{5}{12}}$

$\cot \theta = \boxed{-\dfrac{12}{15}}$

$\sec \theta = \boxed{\dfrac{13}{12}}$

$\csc \theta = \boxed{-\dfrac{13}{5}}$

3. (a) $\boxed{\dfrac{1}{3}}$ (b) $\boxed{\dfrac{4}{3}}$ (c) $\boxed{\dfrac{1}{x}}$ (d) $\boxed{\dfrac{1}{y}}$

(e) $\boxed{\dfrac{1}{r}}$ (f) $\boxed{x}$ (g) $\boxed{y}$ (h) $\boxed{\dfrac{r}{x}}$

(i) $\boxed{\dfrac{r}{y}}$ (j) $\boxed{\dfrac{x}{y}}$

4. (a) $\dfrac{1}{\cos \theta} = \dfrac{1}{\frac{x}{r}} = \dfrac{r}{x} = \boxed{\sec \theta}$

(b) $\dfrac{1}{\tan \theta} = \dfrac{1}{\frac{y}{x}} = \dfrac{x}{y} = \boxed{\cot \theta}$

(c) $\dfrac{1}{\cot \theta} = \dfrac{1}{\frac{x}{y}} = \dfrac{y}{x} = \boxed{\tan \theta}$

5. (a) $\sec 60° = \boxed{2}$

(b) $\csc 30° = \boxed{2}$

(c) $\cot 45° = \boxed{1}$

6. (a) $\sin 30° = \dfrac{5.0}{10} = \boxed{0.5}$

$\cos 30° = \dfrac{8.7}{10} = \boxed{0.87}$

$\tan 30° = \dfrac{5.0}{8.7} = \boxed{0.57}$

(b) $\sin 135° = \dfrac{7.1}{10} = \boxed{0.71}$

$\cos 135° = \dfrac{-7.1}{10} = \boxed{-0.71}$

$\tan 135° = \dfrac{-7.0}{7.0} = \boxed{-1}$

(c) $\sin 300° = \dfrac{-8.7}{10} = \boxed{-0.87}$

$\cos 300° = \dfrac{5.0}{10} = \boxed{0.50}$

$\tan 300° = \dfrac{-8.7}{5.0} = \boxed{-1.7}$

7.

		sin	cos	tan
(a)	185°	−	−	+
(b)	225°	−	−	+
(c)	350°	−	+	−
(d)	190°	−	−	+
(e)	265°	−	−	+
(f)	100°	+	−	−
(g)	89°	+	+	+
(h)	175°	+	−	−
(i)	95°	+	−	−
(j)	275°	−	+	−
(k)	300°	−	+	−
(l)	110°	+	−	−
(m)	85°	+	+	+
(n)	290°	−	+	−

8. (a) $\boxed{\text{IV}}$

(b) $\boxed{\text{IV}}$

(c) $\boxed{\text{III}}$

(d) $\boxed{\text{I}}$

(e) $\boxed{\text{III}}$

(f) $\boxed{\text{I}}$

(g) $\boxed{\text{IV}}$

9. (a) $\sin = \boxed{1}$ $\cos = \boxed{0}$ $\tan = \boxed{\infty}$

(b) $\sin = \boxed{0}$ $\cos = \boxed{-1}$ $\tan = \boxed{0}$

(c) $\sin = \boxed{-1}$ $\cos = \boxed{0}$ $\tan = \boxed{\infty}$

(d) $\sin = \boxed{0}$ $\cos = \boxed{+1}$ $\tan = \boxed{0}$

10. (a) $\cos A = \boxed{\dfrac{b}{c}}$

(b) $\tan A = \boxed{\dfrac{a}{b}}$

(c) $\sin B = \boxed{\dfrac{b}{c}}$

(d) $\cos B = \boxed{\dfrac{a}{c}}$

(e) $\tan B = \boxed{\dfrac{b}{a}}$

(f) $\cos K = \dfrac{12}{15} = \boxed{0.80}$

(g) $\tan K = \dfrac{9}{12} = \boxed{0.75}$

(h) $\sin M = \dfrac{12}{15} = \boxed{0.80}$

(i) $\cos M = \dfrac{9}{15} = \boxed{0.60}$

(j) $\tan M = \dfrac{12}{9} = \boxed{1.33}$

(k) $\sin D = \dfrac{8}{10} = \boxed{0.80}$

(l) $\cos D = \dfrac{6}{10} = \boxed{0.60}$

(m) $\tan D = \dfrac{8}{6} = \boxed{1.33}$

(n) $\sin F = \dfrac{6}{10} = \boxed{0.60}$

(o) $\cos F = \dfrac{8}{10} = \boxed{0.80}$

(p) $\tan F = \dfrac{6}{8} = \boxed{0.75}$

(q) $\sin N = \dfrac{8}{17} = \boxed{0.47}$

(r) $\cos N = \dfrac{15}{17} = \boxed{0.88}$

(s) $\tan N = \dfrac{8}{15} = \boxed{0.53}$

(t) $\sin Q = \dfrac{15}{17} = \boxed{0.88}$

(u) $\cos Q = \dfrac{8}{17} = \boxed{0.47}$

(v) $\tan Q = \dfrac{15}{8} = \boxed{1.88}$

11. (a) $\sin 37°29'00'' = 0.6085306$
$\sin 37°30'00'' = 0.6087614$
$\sin 37°29'16'' = 0.6085306$
$$+ \left(\dfrac{16}{60}\right)(0.0002308)$$
$$= \boxed{0.6085922}$$

(b) $\cos 52°42'00'' = 0.6059884$
$\cos 52°43'00'' = 0.6057570$
$\cos 52°42'51'' = 0.6059884$
$$- \left(\dfrac{51}{60}\right)(0.0002314)$$
$$= \boxed{0.6057917}$$

(c) $\cos 37°38'00'' = 0.7919345$
$\cos 37°39'00'' = 0.7917569$
$\cos^{-1} 0.7918605 = 37°38'00''$
$$+ \left(\dfrac{740}{1776}\right)(60'')$$
$$= \boxed{37°38'25''}$$

(d) $\tan 43°17'00'' = 0.9418033$
$\tan 43°18'00'' = 0.9423523$
$\tan 43°17'28'' = 0.9418033$
$$+ \left(\dfrac{28}{60}\right)(0.0005490)$$
$$= \boxed{0.9420595}$$

12. (a) $b :$ $\boxed{\tan}$ $c :$ $\boxed{\cos}$ (b) $a :$ $\boxed{\tan}$ $c :$ $\boxed{\cos}$

(c) $a :$ $\boxed{\cos}$ $b :$ $\boxed{\sin}$ (d) $a :$ $\boxed{\sin}$ $b :$ $\boxed{\cos}$

(e) $A :$ $\boxed{\sin}$ $B :$ $\boxed{\cos}$ (f) $A :$ $\boxed{\cos}$ $B :$ $\boxed{\sin}$

(g) $a :$ $\boxed{\tan}$ $c :$ $\boxed{\sin}$ (h) $b :$ $\boxed{\tan}$ $c :$ $\boxed{\sin}$

13. (a) $B = 90° - 28°41' = \boxed{61°19'}$

$a = 540 \tan 28°41' = \boxed{295}$

$c = \dfrac{540}{\cos 28°41'} = \boxed{616}$

(b) $A = 90° - 55°13' = 34°47'$
$b = (371 \text{ ft})(\tan 55°13') = 534$
$c = \dfrac{371 \text{ ft}}{\cos 55°13'} = \boxed{650 \text{ ft}}$

(c) $A = 90° - 61°29' = 28°31'$
$a = \dfrac{466 \text{ ft}}{\tan 61°29'} = \boxed{253 \text{ ft}}$
$c = \dfrac{466 \text{ ft}}{\sin 61°29'} = \boxed{530 \text{ ft}}$

(d) $B = 90° - 33°15' = \boxed{56°45'}$

$a = (263 \text{ ft})(\sin 33°15') = \boxed{144 \text{ ft}}$

$b = (263 \text{ ft})(\cos 33°15') = \boxed{220 \text{ ft}}$

(e) $A = 90° - 58°55' = \boxed{31°05'}$

$a = (562 \text{ ft})(\cos 58°55') = \boxed{290 \text{ ft}}$

$b = (562 \text{ ft})(\sin 58°55') = \boxed{481 \text{ ft}}$

Trigonometry

(f) $c = \sqrt{(300)^2 + (400)^2} = \boxed{500}$

$$A = \boxed{36°52'}$$

$$B = \boxed{53°08'}$$

14. $\tan 31°18' = \dfrac{h}{10 \text{ ft}}$

height $= (10 \text{ ft})(\tan 31°18') = \boxed{6 \text{ ft, 1 in}}$

15. $\tan 50°40' = \dfrac{h}{160 \text{ ft}}$

height $= (160 \text{ ft})(\tan 50°40') = \boxed{195 \text{ ft}}$

16. $\tan 30° = \dfrac{500 \text{ ft}}{\tan 30°}$

distance $= \dfrac{500 \text{ ft}}{\tan 30°} = \boxed{866 \text{ ft}}$

17. $\cos 45° = \dfrac{115 \text{ ft}}{d}$

distance $= \dfrac{115 \text{ ft}}{\cos 45°} = \boxed{163 \text{ mi}}$

18. $\sin 60° = \dfrac{h}{20 \text{ ft}}$

height $= (20 \text{ ft})(\sin 60°) = \boxed{17 \text{ ft}}$

19. (a) $A = 90° - 41°12'38'' = \boxed{48°47'22''}$

$b = (625.18 \text{ ft})(\tan 41°12'38'') = \boxed{547.51 \text{ ft}}$

$c = \dfrac{625.18 \text{ ft}}{\cos 41°12'38''} = \boxed{831.03 \text{ ft}}$

(b) $B = 90° - 66°22'37'' = \boxed{23°37'23''}$

$b = \dfrac{492.72 \text{ ft}}{\tan 66°22'37''} = \boxed{215.50 \text{ ft}}$

$c = \dfrac{492.72 \text{ ft}}{\sin 66°22'37''} = \boxed{537.79 \text{ ft}}$

(c) $A = 90° - 38°04'48'' = \boxed{51°55'12''}$

$a = (585.20 \text{ ft})(\cos 38°04'48'') = \boxed{460.64 \text{ ft}}$

$b = (585.20 \text{ ft})(\sin 38°04'48'') = \boxed{360.93 \text{ ft}}$

(d) $B = 90° - 22°13'50'' = \boxed{67°46'10''}$

$b = \dfrac{376.26 \text{ ft}}{\tan 22°13'50''} = \boxed{920.59 \text{ ft}}$

$c = \dfrac{376.26 \text{ ft}}{\sin 22°13'50''} = \boxed{994.52 \text{ ft}}$

(e) $A = 90° - 75°35'41'' = \boxed{14°24'19''}$

$a = \dfrac{237.68 \text{ ft}}{\tan 75°35'41''} = \boxed{61.05 \text{ ft}}$

$c = \dfrac{237.68 \text{ ft}}{\sin 75°35'41''} = \boxed{245.40 \text{ ft}}$

(f) $c = \sqrt{(427.82)^2 + (396.95)^2} = \boxed{583.61 \text{ ft}}$

$\tan A = \dfrac{427.82}{396.95};\ A = \boxed{47°08'37''}$

$\tan B = \dfrac{396.95}{427.82};\ B = \boxed{42°51'23''}$

(g) $a = \sqrt{(616.38)^2 - (445.64)^2} = \boxed{425.83 \text{ ft}}$

$\cos A = \dfrac{445.64}{616.38};\ A = \boxed{43°41'52''}$

$\sin B = \dfrac{445.64}{616.38};\ B = \boxed{46°18'08''}$

20. (a) $\boxed{80°}$ (b) $\boxed{30°}$ (c) $\boxed{20°}$ (d) $\boxed{40°}$

(e) $\boxed{10°}$ (f) $\boxed{10°}$ (g) $\boxed{30°}$ (h) $\boxed{10°}$

(i) $\boxed{20°}$ (j) $\boxed{60°}$ (k) $\boxed{55°}$ (l) $\boxed{25°}$

(m) $\boxed{70°}$ (n) $\boxed{40°}$ (o) $\boxed{45°}$ (p) $\boxed{80°}$

(q) $\boxed{60°}$ (r) $\boxed{5°}$ (s) $\boxed{85°}$ (t) $\boxed{5°}$

(u) $\boxed{80°}$

21. (a) Triangle 1–2–3: $1 = 34°18'24''$;

$3 = \boxed{62°12'55''}$; $1\text{–}2 = \boxed{1347.77 \text{ ft}}$

$2 = 180° - (34°18'24'' + 62°12'55'')$

$= \boxed{83°28'41''}$

$$2\text{–}3 = \frac{(1347.77\text{ ft})(\sin 34°18'24'')}{\sin 62°12'55''}$$

$$= \boxed{858.63\text{ ft}}$$

$$3\text{–}1 = \frac{(1347.77\text{ ft})(\sin 83°28'41'')}{\sin 62°12'55''}$$

$$= \boxed{1513.55\text{ ft}}$$

(b) Triangle PQR: $P = 118°34'24''$;

$$Q = \boxed{23°06'54''}; \quad QR = \boxed{526.30\text{ ft}}$$
$$R = 180° - (118°34'24'' + 23°06'54'')$$
$$= \boxed{38°18'42''}$$

$$PQ = \frac{(526.30\text{ ft})(\sin 38°18'42'')}{\sin 118°34'24''}$$
$$= \boxed{371.52\text{ ft}}$$

$$RP = \frac{(526.30\text{ ft})(\sin 23°06'54'')}{\sin 118°34'24''}$$
$$= \boxed{235.27\text{ ft}}$$

(c) Triangle 1–2–3: $1 = 118°46'58''$;

$$2 = \boxed{58°54'20''}; \quad 2\text{–}3 = \boxed{345.43\text{ ft}}$$
$$3 = 180° - (82°46'58'' + 58°54'20'')$$
$$= \boxed{38°18'42''}$$

$$1\text{–}2 = \frac{(345.43\text{ ft})(\sin 38°18'42'')}{\sin 82°46'58''}$$
$$= \boxed{215.86\text{ ft}}$$

$$3\text{–}1 = \frac{(345.43\text{ ft})(\sin 58°54'20'')}{\sin 82°46'58''}$$
$$= \boxed{298.16\text{ ft}}$$

22. (a) $\sin G = \dfrac{(1296.20\text{ ft})(\sin 25°40'55'')}{646.13}$;

$$G = \boxed{60°23'17''}$$
$$F = 180° - (25°40'55'' + 60°23'17'')$$
$$= \boxed{93°55'48''}$$

$$g = \frac{(646.13\text{ ft})(\sin 93°55'48'')}{\sin 25°40'55''}$$
$$= \boxed{1487.42\text{ ft}}$$

(b) *Solution 1:*

$$\sin B = \frac{(526.17\text{ ft})(\sin 51°20'14'')}{445.23};$$
$$B = \boxed{67°20'13''}$$
$$C = 180° - (51°20'14'' + 67°20'13'')$$
$$= \boxed{61°19'33''}$$

$$c = \frac{(445.23\text{ ft})(\sin 61°19'33'')}{\sin 51°20'14''}$$
$$= \boxed{500.27\text{ ft}}$$

(b) *Solution 2:*

$$B = \boxed{112°39'47''}$$
$$C = 180° - (51°20'14'' + 112°39'47'')$$
$$= \boxed{15°59'59''}$$

$$c = \frac{(445.23\text{ ft})(\sin 15°59'59'')}{\sin 51°20'14''}$$
$$= \boxed{157.16\text{ ft}}$$

23. (a) $EF = \sqrt{\begin{array}{l}(146.25)^2 + (122.31)^2 \\ - (2)(146.25)(122.31)\cos 95°12'50''\end{array}}$

$$= \boxed{199.00\text{ ft}}$$

$$\sin E = \frac{(146.25\text{ ft})(\sin 95°12'50'')}{199.00}$$
$$E = \boxed{47°02'40''}$$
$$F = 180° - (95°12'50'' + 47°02'40'')$$
$$= \boxed{37°44'30''}$$

(b) $MK = \sqrt{\begin{array}{l}(595.45)^2 + (851.78)^2 \\ - (2)(595.45)(851.78)\cos 35°19'16''\end{array}}$

$$= \boxed{502.42\text{ ft}}$$

$$\sin K = \frac{(851.78\text{ ft})(\sin 35°19'16'')}{502.42}$$

Trigonometry

$K = \boxed{101°25'32''}$

$M = 180° - (35°19'16'' + 101°25'32'')$

$= \boxed{43°15'12''}$

(c) $OP = \sqrt{\begin{array}{l}(138.38)^2 + (165.12)^2 \\ - (2)(138.38)(165.12)\cos 46°07'01''\end{array}}$

$= \boxed{121.39}$

$\sin O = \dfrac{(165.12 \text{ ft})(\sin 46°07'01'')}{121.39}$

$O = \boxed{78°38'19''}$

$P = 180° - (46°07'01'' + 78°38'19'')$

$= \boxed{55°14'40''}$

(Note: The largest angle must be opposite the longest side. Hint: The sine function may represent a related angle.)

24. (a) $\cos E = \dfrac{(125.83)^2 + (155.13)^2 - (171.25)^2}{(2)(125.83)(155.13)}$;

$E = \boxed{74°17'18''}$

$\cos F = \dfrac{(125.83)^2 + (171.25)^2 - (155.13)^2}{(2)(125.83)(155.13)}$;

$F = \boxed{60°41'40''}$

$\cos G = \dfrac{(171.25)^2 + (155.13)^2 - (125.83)^2}{(2)(171.25)(155.13)}$;

$G = \boxed{45°01'02''}$

check: $E + F + G = 180°00'00''$

(b) $\cos M = \dfrac{(298.34)^2 + (257.42)^2 - (188.65)^2}{(2)(171.25)(155.13)}$;

$M = \boxed{38°48'46''}$

$\cos N = \dfrac{(298.34)^2 + (188.65)^2 - (257.42)^2}{(2)(298.34)(188.65)}$;

$N = \boxed{58°47'18''}$

$\cos O = \dfrac{(188.65)^2 + (257.42)^2 - (298.34)^2}{(2)(188.65)(257.42)}$;

$O = \boxed{82°23'56''}$

check $180°00'00''$

25. (a) $\boxed{\text{law of sines}}$ (b) $\boxed{\text{law of cosines}}$

(c) $\boxed{\text{law of sines}}$ (d) $\boxed{\text{law of sines}}$

(e) $\boxed{\text{law of cosines}}$ (f) $\boxed{\text{law of sines}}$

(g) $\boxed{\text{law of sines}}$ (h) $\boxed{\text{law of cosines}}$

(i) $\boxed{\text{law of sines}}$ (j) $\boxed{\text{law of cosines}}$

26. (a) $\boxed{\dfrac{\pi}{4} \text{ rad}}$ (b) $\boxed{\dfrac{2\pi}{3} \text{ rad}}$ (c) $\boxed{\dfrac{\pi}{2} \text{ rad}}$

(d) $\boxed{\dfrac{\pi}{12} \text{ rad}}$ (e) $\boxed{\dfrac{3\pi}{2} \text{ rad}}$

27. (a) $\boxed{60°}$ (b) $\boxed{90°}$ (c) $\boxed{45°}$

(d) $\boxed{15°}$ (e) $\boxed{270°}$

28. $S = r\theta = \dfrac{(300 \text{ ft})(60)\pi}{180} = \boxed{314 \text{ ft}}$

29. $r = \dfrac{S}{\theta} = \dfrac{300 \text{ ft}}{\dfrac{12\pi}{180°}} = \boxed{1432 \text{ ft}}$

30. $r = \dfrac{S}{\theta} = \dfrac{25 \text{ in}}{2.4} = \boxed{10.4 \text{ in}}$

31. $S = r\theta = \dfrac{(7927 \text{ mi})\pi}{(2)(180°)} = \boxed{69 \text{ mi}}$

32. $\theta = \dfrac{S}{r} = \dfrac{(3.4 \text{ in})(180°)}{(25 \text{ in})\pi} = \boxed{8°}$

33. area $= \frac{1}{2}r^2\theta = \left(\dfrac{1}{2}\right)(6 \text{ in})^2 \left(\dfrac{\pi}{3}\right) = \boxed{19 \text{ in}^2}$

34. area $= \frac{1}{2}r^2(\theta - \sin\theta)$

$= \left(\dfrac{1}{2}\right)(100 \text{ ft})^2 \left(\dfrac{\pi}{6} - 0.5\right)$

$= 118 \text{ ft}^2$

35. (a)

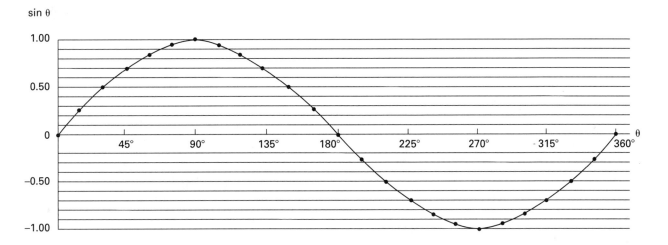

(b)

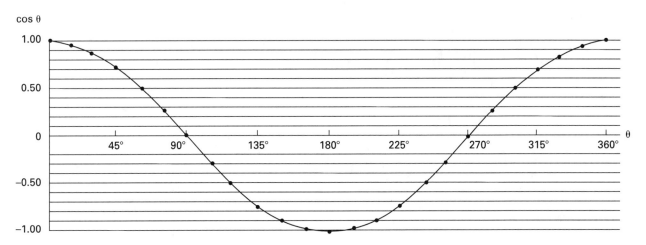

12 Analytic Geometry for Surveyors

Nomenclature

m slope of a line

Symbols

α inclination of a line

1. FIRST-DEGREE EQUATIONS

The statement "four times a number minus three is equal to five" can be expressed in algebraic terms as $4x - 3 = 5$. The letter x represents the unknown number. The number 2 satisfies the equation, and it is called the *root* of the equation.

The statement "the sum of two numbers is eight" can be written in algebraic terms as $x + y = 8$. In this equation there are two unknowns, represented by x and y, and more than one pair of numbers will make the statement true. (If $x = 1$, $y = 7$; if $x = 2$, $y = 6$; if $x = 3$, $y = 5$; etc.) If these pairs of numbers are always expressed in the order of x first and y second, they are called *ordered pairs* and are written symbolically as $(1,7)$, $(2,6)$, $(3,5)$, and so on. Because x and y are of the first power, $x + y = 8$ is an *equation of the first degree*. The equation $x + y = 8$ has two *unknowns*, or two *variables*. The equation $4x - 3 = 5$ is an equation of the first degree with one unknown.

2. GRAPHS OF FIRST-DEGREE EQUATIONS WITH TWO VARIABLES

Consider each ordered pair that satisfies a first-degree equation with two variables (the roots of the equation, or the *solution set* of the equation) to be coordinates of a point. If several of these points are plotted on a rectangular coordinate system and connected with a line, the system will show the *graph of the equation*.

Consider the following equation.

$$2x - 3y = 12$$

To solve the equation—that is, to find ordered pairs that satisfy it, first rearrange the equation.

$$-3y = -2x + 12$$
$$y = \tfrac{2}{3}x - 4$$

Next, give various values to x and solve for the corresponding values of y (for instance, when $x = 0$, $y = -4$),

$$x = -3, \ 0, \ 3, \ 6, \ 9$$
$$y = -6, \ -4, \ -2, \ 0, \ 2$$

Plotting these ordered pairs as coordinates of a point creates the graph of the equation $2x - 3y = 12$ shown in Fig. 12.1. When the points are connected, they lie in a straight line. An infinite number of values could be given to x and y.

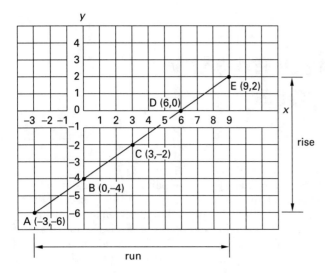

Figure 12.1 *Straight Line*

Example 12.1

Find three roots of each equation. Give your answers as (x, y).

(a) $2x + y = 10$ (b) $3x + 2y = 0$
(c) $x + y = 5$ (d) $x - y = 0$

Solution

(a) $(3,4); (2,6); (5,0)$ (b) $(2,-3); (4,-6); (0,0)$
(c) $(1,4); (2,3); (0,5)$ (d) $(-4,-4); (0,0); (2,2)$

Example 12.2

Find the coordinates of the points at which each of the following equations intersects the x- and y-axes.

(a) $x - y + 5 = 0$ (b) $3x - y + 6 = 0$
(c) $2x - 5y = 10$ (d) $2x + 3y + 6 = 0$

Solution

Let $y = 0$ and solve for x. Then let $x = 0$ and solve for y.

(a) $(0,5); (-5,0)$ (b) $(0,6); (-2,0)$
(c) $(0,-2); (5,0)$ (d) $(0,-2); (-3,0)$

3. SLOPE OF A LINE

Slope is the ratio of the change in the vertical distance to the change in the horizontal distance. If m represents the slope for the line connecting points (x_1, y_1) and (x_2, y_2), then

$$m = \frac{y_2 - y_1}{x_2 - x_1} \qquad 12.1$$

To describe the steepness or *slope* of a line (such as the graph of the equation $2x - 3y = 12$ shown in Fig. 12.1), choose two points on it (such as A $(-3, -6)$ and E $(9, 2)$).

$$\text{slope} = \frac{\text{rise}}{\text{run}} = \frac{\text{ordinate of E} - \text{ordinate of A}}{\text{abscissa of E} - \text{abscissa of A}}$$
$$= \frac{2 - (-6)}{9 - (-3)} = \frac{8}{12} = 2/3$$

Alternatively,

$$\text{slope} = \frac{\text{ordinate of A} - \text{ordinate of E}}{\text{abscissa of A} - \text{abscissa of E}}$$
$$= \frac{-6 - 2}{-3 - 9} = \frac{8}{12} = 2/3$$

Example 12.3

Determine the slope of the line through each pair of points.

(a) $(1,2); (3,6)$ (b) $(-4,3); (3,-4)$
(c) $(-2,-3); (2,5)$ (d) $(-4,3); (4,3)$

Solution

(a) $m = \dfrac{6 - 2}{3 - 1} + 2 = +4$

(b) $m = \dfrac{-4 - 3}{3 - (-4)} = -1$

(c) $m = \dfrac{5 - (-3)}{2 - (-2)} = +2$

(d) $m = \dfrac{3 - 3}{4 - (-4)} = 0$

In plotting the lines it can be seen that when a line rises from left to right, its slope is positive; and when a line falls from left to right, its slope is negative.

4. LINEAR EQUATIONS

The slope of the line represented by the linear equation $2x - 3y = 12$ is $+2/3$. The graph of the equation shown in Fig. 12.1 rises from left to right. Assume the equation is written in the following form.

$$y = \frac{2}{3}x - \frac{12}{3}$$

The coefficient of x is $+2/3$, which is the slope of the line. The numerator 2 is the coefficient of x, and the denominator 3 is the coefficient of y when the equation is written in the following form.

$$x - 3y = 12$$

Also notice that when the coefficient of x is $+2$ and the coefficient of y is -3, the slope is positive.

Equation 12.2 represents the *general form* of a linear equation, where A is the positive coefficient of x, B is the coefficient of y, and C is a constant.

$$Ax + By + C = 0 \qquad 12.2$$

The slope is then

$$m = -\frac{A}{B} \qquad 12.3$$

Example 12.4

Rearrange the following equations to the form $Ax + By + C = 0$ with A positive. Determine the slope m of each.

(a) $3x + 4y = 8$

(b) $2x = 6y$

(c) $2y + 3x - 6 = 0$

(d) $-3x + 4y + 10 = 0$

Solution

(a) $\quad 3x + 4y - 8 = 0$

$$m = -\frac{A}{B} = -\frac{+3}{+4} = -3/4$$

(b) $\quad 2x - 6y = 0$

$$m = -\frac{+2}{-6} = 1/3$$

(c) $\quad 3x + 2y - 6 = 0$

$$m = -\frac{+3}{+2} = -3/2$$

(d) $\quad 3x - 4y - 10 = 0$

$$m = -\frac{+3}{-4} = 3/4$$

5. EQUATIONS OF HORIZONTAL AND VERTICAL LINES

Consider the line that contains the points $(-4, 5)$ and $(4, 5)$. The slope of the line is

$$m = \frac{5 - 5}{4 + 4} = \frac{0}{8} = 0$$

This is a horizontal line that has the equation $y = 5$ (Fig. 12.2). In considering the general equation of a line, $Ax + By + C = 0$, the coefficient of x for a horizontal line is 0. So the equation of a horizontal line is

$$m = \frac{5 - (-5)}{4 - 4} = \frac{10}{0}$$

$$By = C$$

The line that contains the points $(4, 5)$ and $(4, -5)$ has the slope

$$m = \frac{5 - (-5)}{4 - 4} = \frac{10}{0}$$

This number is infinite, so the line has an infinite slope. The equation of the line is $x = 4$. Therefore, the general equation of the vertical line is

$$Ax + 0y + C = 0$$

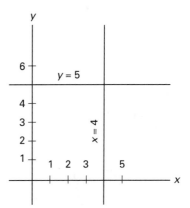

Figure 12.2 *Horizontal and Vertical Lines*

6. x- AND y-INTERCEPTS

If $x = 0$ in the equation $2x - 3y = 12$, then $y = -4$. The point $(0, -4)$ is the point where the graph of the equation crosses the y-axis (Fig. 12.1). The distance from the x-axis to this point is known as the *y-intercept* and is given the symbol b.

If $y = 0$ in the equation $2x - 3y = 12$, then $x = 6$. The point $(6, 0)$ is where the graph of the equation crosses the x-axis. The distance from the y-axis to this point is known as the *x-intercept* and is given the symbol a.

Therefore, the coordinates of the point where any line crosses the x-axis are $(a, 0)$ and the coordinates of the point where any line crosses the y-axis are $(b, 0)$. In Fig. 12.1, these points are $(0, -4)$ and $(6, 0)$.

If the equation $2x - 3y = 12$ is written as $y = {}^2\!/_3x - 4$, the constant term -4 is the y-coordinate of the point where the line crosses the y-axis. It is, in fact, the y-intercept. In finding this equivalent equation of $2x - 3y = 12$, both sides of the equation were divided by -3, which is the coefficient of y, so that the constant 12 was divided by -3 to obtain the quotient -4.

For any equation $Ax + By + C = 0$, the y-intercept is

$$b = -\frac{C}{B} \qquad 12.4$$

If the equation $2x - 3y = 12$ is written as $x = {}^3\!/_2y + 6$, the constant term 6 is the x-coordinate of the point where the line crosses the x-axis; it is the x-intercept. For any equation $Ax + By + C = 0$, the x-intercept is

$$a = -\frac{C}{A} \qquad\qquad 12.5$$

In summary, for any linear equation $Ax + By + C = 0$,

$$m = \text{slope} = -\frac{A}{B} \qquad\qquad 12.6$$

$$a = x\text{-intercept} = -\frac{C}{A} \qquad\qquad 12.7$$

$$b = y\text{-intercept} = -\frac{C}{B} \qquad\qquad 12.8$$

Example 12.5

Find the slope, x-intercept, and y-intercept of the line $3x + 2y - 6 = 0$.

Solution

$$m = -\frac{A}{B} = -\frac{3}{2} = -15$$

$$a = -\frac{C}{A} = -\frac{-6}{3} = 2$$

$$b = -\frac{C}{B} = -\frac{-6}{2} = 3$$

7. PARALLEL LINES

Two different lines having the same slope are called *parallel lines*.

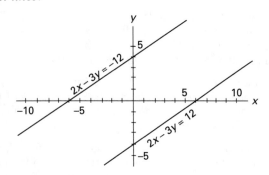

Figure 12.3 *Parallel Lines*

In Fig. 12.3, the lines $2x - 3y = -12$ and $2x - 3y = 12$ have the same slope, $+2/3$. Writing the equations in equivalent form shows that the slope for each line is the same; only the y-intercepts differ.

$$y = \frac{2}{3}x + 4$$

$$y = \frac{2}{3}x - 4$$

8. PERPENDICULAR LINES

Two lines intersecting at right angles are called *perpendicular lines*. For two lines to be perpendicular, the slope of one must be the negative reciprocal of the other, or

$$m_1 = -\frac{1}{m_2} \qquad\qquad 12.9$$

In Fig. 12.4, the lines $2x - 3y = 12$ and $3x + 2y = 8$ are perpendicular lines, as can be seen when written in the form $y = mx + b$.

$$y = \frac{2}{3}x - 4$$

$$y = -\frac{3}{2}x + 4$$

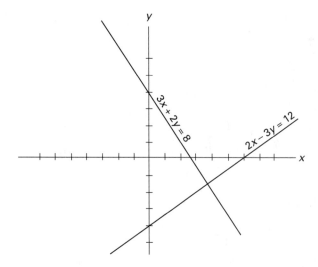

Figure 12.4 *Perpendicular Lines*

9. PERPENDICULAR DISTANCE FROM A POINT TO A LINE

A formula for finding the perpendicular distance from a point of known coordinates (x, y) to a line of known equation can be found from Eq. 12.10.

$$D = \frac{|Ax + By + C|}{\sqrt{A^2 + B^2}} \qquad\qquad 12.10$$

Example 12.6

Find the perpendicular distance D from the point $P(-2, 4)$ to the line $4x - 3y - 16 = 0$.

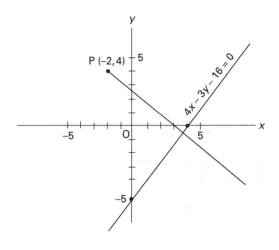

Solution

$$D = \frac{|(4)(-2) - (3)(4) - 16|}{\sqrt{(4)^2 + (-3)^2}} = \frac{|(-36)|}{5} = \frac{36}{5}$$

$$= 7.2$$

10. WRITING THE EQUATION OF A LINE

An equation may be written for a straight line if sufficient information is known, for example, if any of the following is true.

- one point on the line and the slope are known
- two points on the line are known
- the x-intercept and y-intercept are known
- the slope of the line and the y-intercept are known

11. POINT-SLOPE FORM OF THE EQUATION OF A LINE

If P (x, y) is any point on a line with slope m through point P$_1$ (x_1, y_1), then the *point-slope form* of the line is

$$\frac{y - y_1}{x - x_1} = m \qquad \textit{12.11}$$

$$y - y_1 = m(x - x_1) \qquad \textit{12.12}$$

Example 12.7

Write the equation of the line through the given point with the given slope m in the form $Ax + By + C = 0$.

(a) $(4, -2)$; $m = 2$

(b) $(3, -2)$; $m = -\dfrac{3}{2}$

(c) $(-4, 5)$; $m = 0$

Solution

(a)
$$y - y_1 = m(x - x_1)$$
$$y - (-2) = (2)(x - 4)$$
$$y + 2 = 2x - 8$$
$$-2x + y + 10 = 0$$
$$2x - y - 10 = 0$$

(b) The slope m can be written $-(3/2)$, $(-3)/2$, or $3/(-2)$. For ease in performing the algebraic operation, the numerator should carry the negative sign.

$$y - y_1 = m(x - x_1)$$
$$y + 2 = \frac{(-3)(x - 3)}{2}$$
$$2y + 4 = -3x + 9$$
$$3x + 2y - 5 = 0$$

(c)
$$y - y_1 = m(x - x_1)$$
$$y - 5 = (0)(x + 4)$$
$$y = 5$$

This is a linear equation. The graph is a line parallel to the x-axis, 5 units above.

12. TWO-POINT FORM OF THE EQUATION OF A LINE

If P (x, y) is any point on a line that passes through the points P$_1$ (x_1, y_1) and P$_2$ (x_2, y_2), then the *two-point form* of the line is

$$y - y_1 = m(x - x_1) \qquad \textit{12.13}$$

$$y - y_1 = \left(\frac{y_2 - y_1}{x_2 - x_1}\right)(x - x_1) \qquad \textit{12.14}$$

$$\frac{y - y_1}{x - x_1} = \frac{y_2 - y_1}{x_2 - x_1} \qquad \textit{12.15}$$

In writing the two-point form, either point may be designated as point 1.

Example 12.8

Write the equation of the line through the following two points.

(a) $(1, 4)$; $(3, -2)$ (b) $(-2, 2)$; $(1, -3)$

(c) $(1, -3)$; $(-2, 1)$ (d) $(3, 4)$; $(1, 4)$

Solution

(a)
$$\frac{y - y_1}{x - x_1} = \frac{y_2 - y_1}{x_2 - x_1}$$
$$\frac{y - 4}{x - 1} = \frac{-2 - 4}{3 - 1} = \frac{-6}{2} = -3$$
$$y - 4 = (-3)(x - 1)$$
$$y - 4 = -3x + 3$$
$$3x + y - 7 = 0$$

(b)
$$\frac{y - y_1}{x - x_1} = \frac{y_2 - y_1}{x_2 - x_1}$$
$$\frac{y - 2}{x + 2} = \frac{-3 - 2}{1 + 2} = -5/3$$
$$(-5)(x + 2) = (3)(y - 2)$$
$$-5x - 10 = 3y - 6$$
$$-5x - 3y - 4 = 0$$
$$5x + 3y + 4 = 0$$

(c)
$$\frac{y - y_1}{x - x_1} = \frac{y_2 - y_1}{x_2 - x_1}$$
$$\frac{y + 3}{x - 1} = \frac{1 + 3}{-2 - 1} = 4/-3$$
$$(4)(x - 1) = (-3)(y + 3)$$
$$4x - 4 = -3y - 9$$
$$4x + 3y + 5 = 0$$

(d)
$$\frac{y - y_1}{x - x_1} = \frac{y_2 - y_1}{x_2 - x_1}$$
$$\frac{y - 4}{x - 3} = \frac{4 - 4}{1 - 3} = \frac{0}{-2} = 0$$
$$(-2)(y - 4) = (0)(x - 3)$$
$$-2y + 8 = 0$$
$$2y - 8 = 0$$
$$y = 4$$

13. INTERCEPT FORM OF THE EQUATION OF A LINE

A line with x-intercept a and y-intercept b (where both a and b are not zero) has the equation

$$\frac{x}{a} + \frac{y}{b} = 1 \qquad 12.16$$

Example 12.9

Write the equation of the line with x-intercept 3 and y-intercept -4.

Solution

$$\frac{x}{3} + \frac{y}{-4} = 1$$
$$-4x + 3y = -12$$
$$4x - 3y = 12$$

14. SLOPE-INTERCEPT FORM OF THE EQUATION OF A LINE

If the slope of a line and its y-intercept are known, the *slope-intercept form* of the line is

$$y = mx + b \qquad 12.17$$

Example 12.10

Write the equation of the line of slope 2 and y-intercept -3.

Solution

$$y = 2x - 3$$
$$2x - y = 3$$

Example 12.11

Write the equation of the line through $(3, -1)$ perpendicular to the line $2x + 3y = 6$.

Solution

The slope of $2x + 3y$ is $m_1 = -2/3$.

The slope of the perpendicular line is $m_2 = 3/2$.

The equation of the perpendicular line is

$$y - y_1 = m_2(x - x_1)$$
$$y - (-1) = \left(\frac{3}{2}\right)(x - 3)$$
$$3x - 2y = 11$$

15. SYSTEMS OF LINEAR EQUATIONS

If the graphs of two linear equations lie in the same xy plane, then one of three conditions must be true:

- The two lines are parallel and will never intersect.
- The two lines coincide.
- The two lines will intersect at a point.

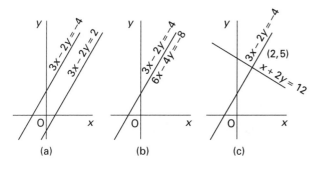

Figure 12.5 *System of Linear Equations*

When a point must be located on two straight lines, the line equations form a system of simultaneous equations. Any ordered pair that satisfies both equations is called a *solution* or a *root* of the system.

When two lines intersect at a point, there can be only one root, and this root can be found by solving the two equations simultaneously. The ordered pair found to be the root of a system will be the coordinates of the point of intersection of the two graphs of the equations.

In Fig. 12.5(a), the two lines are parallel and will not intersect. Therefore, the two equations cannot be solved simultaneously. The slopes of the two lines will indicate whether or not they are parallel.

In Fig. 12.5(b), the two lines intersect everywhere because they have the same solution set. It can be seen that the equation $6x - 4y = -8$ is equivalent to the equation $3x - 2y = -4$. If both sides of the equation are divided by 2, the result will be $3x - 2y = -4$. Thus, both equations have the same graph.

In Fig. 12.5(c), the two lines will intersect at a point. This point can be found by solving the equations simultaneously.

16. SOLVING SYSTEMS OF SIMULTANEOUS EQUATIONS

Several methods can be used to solve a system of equations. One method is known as the *method of reduction*.

Consider the equations $x + y = 9$ and $x - y = 3$. If the two equations are added,

$$\begin{aligned} x + y &= 9 \\ x - y &= 3 \\ \hline 2x \phantom{{}+y} &= 12 \\ x &= 6 \end{aligned}$$

The set of equations has been reduced to an equation of one variable. Substituting the value of x in either equation and solving for y,

$$y = 3$$

The same results can be obtained by subtracting one equation from the other.

$$\begin{aligned} x + y &= 9 \\ x - y &= 3 \\ \hline 2y &= 6 \\ y &= 3 \end{aligned}$$

Substituting,

$$x = 6$$

In this example, the coefficient of x and the coefficient of y are the same, but this will not always be the case. Consider the equations $3x + 2y = 4$ and $2x - 3y = 7$. Adding or subtracting the two equations will not eliminate one of the unknowns as it did in the first example. However, one or both of the two equations can be converted into an equivalent equation that will make it possible to do so.

$$\begin{aligned} 3x + 2y &= 4 \\ 2x - 3y &= 7 \end{aligned}$$

Multiplying the first equation by 3 and the second equation by 2 and adding will reduce the system of equations to a single variable equation.

$$\begin{aligned} 9x + 6y &= 12 \\ 4x - 6y &= 14 \\ \hline 13x \phantom{{}+6y} &= 26 \\ x &= 2 \\ y &= -1 \quad \text{[obtained by substitution]} \end{aligned}$$

If the graphs of the two equations are plotted, the two lines will intersect at $(2, -1)$.

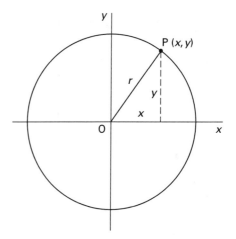

Figure 12.6 *Circle Centered at (0, 0)*

17. EQUATION OF A CIRCLE

A *circle* is a curve, all points on which are equidistant from a point called the *center*. The distance of all points from the center is known as the *radius*.

If the center of the circle is at the origin as in Fig. 12.6, the equation of the circle is

$$x^2 + y^2 = r^2 \qquad 12.18$$

If P is any point on the circle, its coordinates must satisfy the equation $x^2 + y^2 = r^2$.

If the center of the circle is at point Q (h, k), the equation becomes

$$(x - h)^2 + (y - k)^2 = r^2 \qquad 12.19$$

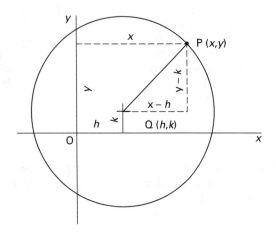

Figure 12.7 *Circle Centered at (h, k)*

The *general form* for this equation is

$$x^2 + y^2 + Dx + Ey + F = 0 \qquad 12.20$$

Example 12.12

Find the equation of the circle with center $(2, -1)$ and radius 3.

Solution

$$(x - 2)^2 + (y + 1)^2 = (3)^2$$
$$x^2 + y^2 - 4x + 2y - 4 = 0$$

Example 12.13

Write the equation $x^2 + 10x + y^2 - 6y + 18 = 0$ in the form $(x - h)^2 + (y - k)^2 = r^2$.

Solution

Complete the square.

$$x^2 + 10x + y^2 - 6y = -18$$
$$x^2 + 10x + 25 + y^2 - 6y + 9 = -18 + 25 + 9$$
$$(x + 5)^2 + (y - 3)^2 = 16$$

The center of the circle is at $(-5, 3)$, and the radius is 4.

18. LINEAR-QUADRATIC SYSTEMS

The intersections of a circle and a straight line can be found by solving the system of the linear equation and the quadratic equation. This is illustrated in Ex. 12.14.

Example 12.14

Find the intersections of the graphs of the following system.

$$x^2 + y^2 = 100$$
$$2x - y = 8$$

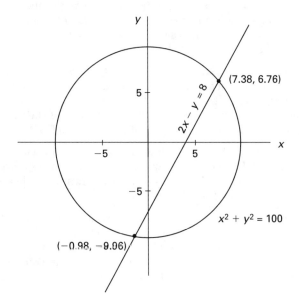

Solution

Transform the linear equation by isolating one of the variables.

$$y = 2x - 8$$

Substitute this value of y into the quadratic equation.

$$x^2 + (2x - 8)^2 = 100$$
$$x^2 + 4x^2 - 32x + 64 = 100$$
$$5x^2 - 32x - 36 = 0$$

Use the quadratic formulas to solve for x.

$$x = \frac{-B \pm \sqrt{B^2 - 4AC}}{2A}$$

$$x = \frac{-(-32) + \sqrt{-(32)^2 - (4)(5)(-36)}}{(2)(5)}$$

$$= \frac{-(-32) - \sqrt{-(32)^2 - (4)(5)(-36)}}{(2)(5)}$$

$$= 7.38 \text{ and } -0.98$$

Substituting the values of x in the equation $y = 2x - 8$ gives

$$y = (2)(7.38) - 8 = 6.76$$

and

$$y = (2)(-0.98) - 8 = -9.96$$

The intersections are $(7.38, 6.76)$ and $(-0.98, -9.96)$.

19. INCLINATION OF A LINE

The *inclination of a line* not parallel to the x-axis is the angle measured counterclockwise from the positive direction of the x-axis. (The inclination of a line parallel to the x-axis is zero.) The symbol α denotes inclination.

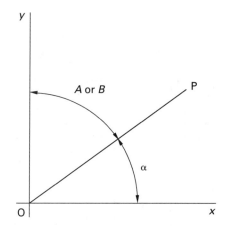

Figure 12.8 *Inclination of a Line*

In Fig. 12.8, the inclination of line OP is α. Considering the trigonometric ratios,

$$m = \frac{y}{x} = \tan \alpha \qquad 12.21$$

The *azimuth from north A* (*bearing angle B*) of a line calculated from the slope is the complement of inclination, and it can be calculated from the slope.

$$m = \cot A \qquad 12.22(a)$$
$$m = \cot B \qquad 12.22(b)$$

20. THE ACUTE ANGLE BETWEEN TWO LINES

If the equations of two intersecting lines are known, the acute angle between them can be found by using the *law of tangents*.

$$\tan \theta = \tan(\alpha_2 - \alpha_1) \qquad 12.23(a)$$
$$= \frac{\tan \alpha_2 - \tan \alpha_1}{1 + \tan \alpha_1 \tan \alpha_2} \qquad 12.23(b)$$
$$= \left| \frac{m_2 - m_1}{1 + m_1 m_2} \right| \quad [\text{for } m_1 m_2 \neq -1] \quad 12.23(c)$$

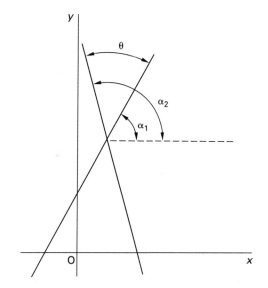

Figure 12.9 *Angle Between Two Lines*

Example 12.15

Find the acute angle θ between the two lines.

$$2x + 3y - 12 = 0$$
$$3x - 4y - 12 = 0$$

Solution

$$m_1 = -\frac{A}{B} = -\frac{2}{3}$$
$$m_2 = -\frac{A}{B} = +\frac{3}{4}$$
$$\tan \theta = \frac{\frac{3}{4} + \frac{2}{3}}{1 - \left(\frac{2}{3}\right)\left(\frac{3}{4}\right)} = 2.833$$
$$\theta = 7°30'$$

21. TRANSLATION OF AXES

Solving simultaneous equations in which the coefficients of x and y are large numbers can be simplified by reducing the value of the coefficients. This can be done without changing the values of the equations by translating the axes.

In Fig. 12.10, let P be any point with coordinates (x, y) with respect to the axes OX and OY. Establish the new axes, O'X' and O'Y', respectively, parallel to the old axes, so that the new origin O' has the coordinates (h, k) with respect to the old axes. The coordinates of the point P will then be $(x'y')$ with respect to the new axes.

$$x = x' + h \qquad 12.24$$
$$x' = x - h \qquad 12.25$$
$$y = y' + k \qquad 12.26$$
$$y' = y - k \qquad 12.27$$

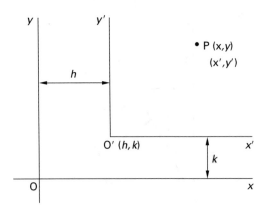

Figure 12.10 *Transformation of Axes*

Example 12.16

Point P has the coordinates (5,3). Find the coordinates of P from the origin $O'(3,1)$.

Solution

$$x' = x - h = 5 - 3 = 2$$
$$y' = y - k = 3 - 1 = 2$$

The coordinates of P are (2,2).

PRACTICE PROBLEMS

1. Find two sets of roots for each equation.

Example: $2x + y = 7$ $(2,3), (3,1)$

(a) $2x - 3y = 5$

(b) $x - 2y = 4$

(c) $3x + 2y = 6$

2. Find the coordinates of the points at which the graph of each equation intersects the two axes.

Example: $3x - 2y = 12$ $(0,-6), (4,0)$

(a) $x - y = 0$

(b) $2x + 3y = 18$

(c) $x + y = 4$

3. Determine the slope of the line through each pair of points.

Example: $(-2,4); (4,-3)$ $m = \dfrac{-3-4}{4+2} = -7/6$

(a) $(3,2); (6,8)$

(b) $(1,3); (4,5)$

(c) $(0,-2); (-3,5)$

(d) $(6,-4); (2,-3)$

(e) $(-3,4); (3,4)$

(f) $(1,-5); (-1,3)$

4. Rearrange the equation in the form $Ax + By + C = 0$ with A positive, and determine the slope m of each.

Example: $3y - 2x - 4 = 0$ $2x - 3y + 4 = 0$

$$m = -\frac{A}{B} = -\frac{+2}{-3} = 2/3$$

(a) $3x + 4y = 6$

(b) $y = -2x + 5$

(c) $-4x + 2y + 8 = 0$

(d) $y = -5x$

5. Find the slope of each line. Express as a common fraction showing the algebraic sign.

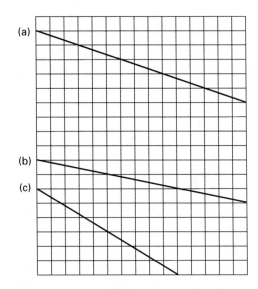

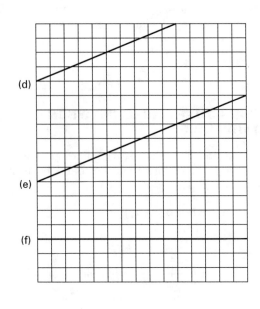

6. Graph each equation, plotting at least three points, and write the equation along the line.

(a) $3x - 2y = 0$

(b) $2x - 3y + 30 = 0$

(c) $3x + 2y + 6 = 0$

(d) $x + y = 0$

(e) $y + 11 = 0$

7. Write each equation in the form $y = mx + b$.

Example: $3x + 2y + 6 = 0$ $y = -\frac{3}{2}x - 3$

(a) $4x + 5y + 10 = 0$

(b) $2x - y - 10 = 0$

(c) $3x + 4y + 8 = 0$

(d) $x - y = 0$

(e) $2x - 3y - 12 = 0$

8. Write each equation in the form $Ax + By + C = 0$.

Example: $y = -\frac{2}{3}x - 2$ $2x + 3y + 6 = 0$

(a) $y = \frac{2}{5}x - 2$

(b) $y = 2x - 10$

(c) $y = \frac{3}{2}x + \frac{5}{2}$

(d) $y = -x + 5$

(e) $y = -\frac{2}{3}x + 4$

9. Write each equation in the form $y = mx + b$ and plot the graph. Write the equation in the form $Ax + By + C = 0$ along each graph.

Example: $2x - 3y + 30 = 0$ $y = \frac{2}{3}x + 10$

(a) $x + 5y - 60 = 0$

(b) $x - y = 0$

(c) $3x + 4y + 24 = 0$

(d) $x + y + 4 = 0$

(e) $2x - 5y - 50 = 0$

10. Write each equation in the form $Ax + By + C = 0$ and indicate the slope, the x-intercept, and the y-intercept of each. (Hint: $m = -A/B$, x-intercept $= -C/A$, and y-intercept $= -C/B$.)

Example: $4y - 3x = 10$ $3x - 4y + 10 = 0$

$m = 3/4$ $x_{y=0} = -10/3$ $y_{x=0} = 5/2$

(a) $x - y + 5 = 0$

(b) $-2x + 3y = 12$

(c) $x - y = 0$

(d) $6y = 4x + 2$

(e) $2x - 4y = 2$

11. Indicate which lines are parallel and which lines are perpendicular.

Example:
$$3x + 4y = 8$$
$$m = -\frac{3}{4}$$

(a) $2x - 3y - 3 = 0$

(b) $10x - 6y = 18$

(c) $y = -\frac{3}{2}x + 5$

(d) $5x - 3y = 9$

(e) $6x + 8y + 12$

(f) $4x - 3y = 7$

(g) $3x + 5y = -10$

(h) $y = -\frac{3}{4}x - 3$

12. Find the perpendicular distance from point P to the line indicated.

Example:
$$P\,(-8, -1);\ \ 2x - 3y - 6 = 0$$
$$D = \frac{|(2)(-8) + (-3)(-1) + (-6)|}{\sqrt{(2)^2 + (-3)^2}}$$
$$= \frac{|-16 + 3 - 6|}{\sqrt{13}} = 5.3$$

(a) $P\,(3, 3);\ \ 3x - 2y + 4 = 0$

(b) $P\,(-10, 8);\ \ x - y + 5 = 0$

(c) $P\,(9, 6);\ \ 5x - 2y + 10 = 0$

(d) $P\,(-8, -6);\ \ 3x + 2y = 0$

(e) $P\,(12, -6);\ \ 3x - 2y - 6 = 0$

13. Write the equation of the line through the given point with the given slope.

Example:
$$(1, 4);\ m = -\frac{1}{2}$$
$$y - 4 = \frac{(-1)(x - 1)}{2} \quad 2y - 8 = -x + 1 \quad x + 2y - 9 = 0$$

(a) $(3, 1);\ m = -2$

(b) $(-4, 3);\ m = \frac{2}{3}$

(c) $(-2, 5);\ m = 0$

(d) $(2, -3);\ m = -\frac{2}{3}$

Geometry II

14. Write the equation of the line through the two given points in the form $Ax + By + C = 0$.

Example:

$$(-4, 3); \quad (0, -2)$$

$$\frac{y - 3}{x + 4} = \frac{-2 - 3}{0 + 4}$$

$$\frac{y - 3}{x + 4} = \frac{-5}{4}$$

$$(-5)(x + 4) = (4)(y - 3)$$

$$-5x - 20 = 4y - 12$$

$$5x + 4y + 8 = 0$$

(a) $(-3, 2); \quad (1, 4)$

(b) $(2, -3); \quad (5, -2)$

(c) $(3, 4); \quad (-3, 4)$

(d) $(-2, -6); \quad (3, -4)$

15. Write the equation of the lines with the given x- and y-intercepts in the form $Ax + By + C = 0$.

Example:

$$a = -3; \quad b = 4$$

$$\frac{x}{-3} + \frac{y}{4} = 1$$

$$4x - 3y = -12$$

$$4x - 3y + 12 = 0$$

(a) $a = -4; \quad b = 3$

(b) $a = 1; \quad b = -4$

16. Write the equation of the line that has a slope of $^3/_4$ and a y-intercept of -3.

17. Write the equation of the line whose y-intercept is 4 and that is perpendicular to the line $4x + 3y + 9 = 0$.

18. Write the equation of the line through the point $(0, 8)$ and parallel to the line whose equation is $y = -3x + 4$.

19. Write the equations of two lines through the point $(5, 5)$, one parallel and one perpendicular to the line $2x + y - 4 = 0$.

20. Write the equation of the line through the point $(-2, 1)$ and parallel to the line through the points $(1, 4)$ and $(2, -3)$.

21. Solve the following systems of simultaneous equations by addition or subtraction.

(a) $x + y = 8$
    $x - y = 4$

(b) $x + 2y = 6$
    $x + 2y = 4$

(c) $2x + 5y = -8$
    $2x + 3y = 5$

(d) $5x - 4y = -15$
    $2x - 12y = 7$

(e) $7x - 2y = 3$
    $2x + 3y = 9$

(f) $7x - 2y = -11$
    $8x + 3y = -39$

(g) $5x - 7y = 3$
    $-3x + 6y = 4$

(h) $5x - 4y = -17$
    $2x - 12y = 14$

(i) $9x + 10y = 9$
    $6x - 25y = -13$

22. Graph the following systems of equations and find the intersection of the two lines in each system if they intersect.

(a) $4x + 3y = 24$
    $4x - 3y = -48$

(b) $3x + 2y = 24$
    $x - 2y = -11$

(c) $5x + 11y = -55$
    $5x + 11y = -11$

(d) $2x + 4y = -48$
    $x + 2y = -24$

23. Find the coordinates of the point of intersection of the diagonal lines CA and EB.

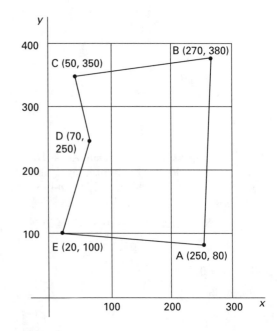

24. The figure ABCDEA represents a tract of land that is to be subdivided by a line from D parallel to EA. Find the coordinates of the point of intersecion of DH and AB. (Designate the point of intersection as H.)

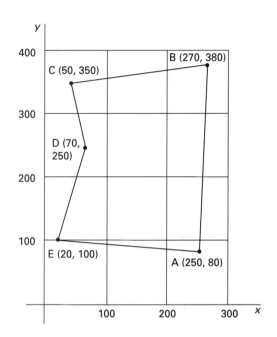

25. Find the acute angle θ between the two lines.

(a) $4x + 3y = 24$
$4x - 3y = -48$

(b) $3x + 2y = 24$
$x - 2y = -11$

(c) $2x - y = 5$
$4x + y = 2$

(d) $2x + y = -2$
$6x - 5y = 18$

(e) $x - y = 5$
$x + 2y = 2$

26. Find the new coordinates if the axes are translated to a new origin located at $(4, 3)$.

(a) $(4, 6)$

(b) $(-7, 3)$

(c) $(-3, -2)$

(d) $(0, 0)$

(e) $(8, -2)$

(f) $(7, 7)$

SOLUTIONS

1. (a) $\boxed{(7,3)}$ $\boxed{(4,1)}$ (b) $\boxed{(6,1)}$ $\boxed{(8,2)}$

(c) $\boxed{(4,-3)}$ $\boxed{(6,-6)}$

2. (a) $\boxed{(0,0)}$ (b) $\boxed{(9,0)}$ $\boxed{(0,6)}$

(c) $\boxed{(0,4)}$ $\boxed{(4,0)}$

3. (a) $m = \dfrac{8-2}{6-3} = \boxed{2}$

(b) $m = \dfrac{5-3}{4-1} = \boxed{2/3}$

(c) $m = \dfrac{5+2}{-3} = \boxed{-7/3}$

(d) $m = \dfrac{-3+4}{2-6} = \boxed{-1/4}$

(e) $m = \dfrac{4-4}{3+3} = \boxed{0}$

(f) $m = \dfrac{3+5}{-1-1} = \boxed{-4}$

4. (a) $3x + 4y = 6$

$\boxed{3x + 4y - 6 = 0}$

$m = -\dfrac{+3}{+4} = \boxed{-3/4}$

(b) $y = -2x + 5$

$\boxed{2x + y - 5 = 0}$

$m = -\dfrac{+2}{+1} = \boxed{-2}$

(c) $-4x + 2y + 8 = 0$

$\boxed{4x - 2y - 8 = 0}$

$m = -\dfrac{+4}{-2} = \boxed{2}$

(d) $y = -5x$

$\boxed{5x + y = 0}$

$m = -\dfrac{+5}{+1} = \boxed{-5}$

Geometry II

5. (a) $\boxed{-1/3}$ (b) $\boxed{-1/5}$ (c) $\boxed{-3/5}$

(d) $\boxed{2/5}$ (e) $\boxed{2/5}$ (f) $\boxed{0}$

(e) $2x - 3y - 12 = 0$

$-3y = -2x + 12$

$\boxed{y = \frac{2}{3}x - 4}$

6.

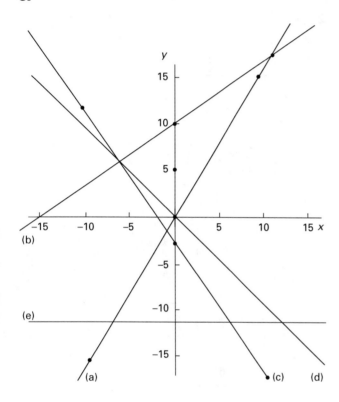

7. (a) $4x + 5y + 10 = 0$

$5y = -4x - 10$

$\boxed{y = -\dfrac{4}{5}x - 2}$

(b) $2x - y - 10 = 0$

$-y = -2x + 10$

$\boxed{y = 2x - 10}$

(c) $3x + 4y + 8 = 0$

$4y = -3x - 8$

$\boxed{y = -\frac{3}{4}x - 2}$

(d) $x - y = 0$

$-y = -x$

$\boxed{y = x}$

8. (a) $y = \frac{2}{5}x - 2$

$5y = 2x - 10$

$\boxed{2x - 5y - 10 = 0}$

(b) $y = 2x - 10$

$-2x + y + 10 = 0$

$\boxed{2x - y - 10 = 0}$

(c) $y = -\frac{3}{2}x + \dfrac{5}{2}$

$2y = -3x + 5$

$\boxed{3x + 2y - 5 = 0}$

(d) $y = -x + 5$

$\boxed{x + y - 5 = 0}$

(e) $y = -\frac{2}{3}x + 4$

$3y = -2x + 12$

$\boxed{2x + 3y - 12 = 0}$

9. (a) $x + 5y - 60 = 0$

$\boxed{y = -\frac{1}{5}x + 12}$

(b) $x - y = 0$

$\boxed{y = x}$

(c) $3x + 4y + 24 = 0$

$\boxed{y = -\frac{3}{4}x - 6}$

(d) $x + y + 4 = 0$

$\boxed{y = -x - 4}$

(e) $2x - 5y - 50 = 0$

$\boxed{y = \frac{2}{5}x - 10}$

(a) $x + 5y - 60 = 0$

(d) $x + y + 4 = 0$

(b) $x - y = 0$

(c) $3x + 4y + 24 = 0$

(e) $2x + 5y - 50 = 0$

	slope	parallel to problem no.	perpendicular to problem no.
(e) $6x + 8y = 12$	$-\dfrac{3}{4}$	h	f
(f) $4x - 3y = 7$	$\dfrac{4}{3}$	–	e & h
(g) $3x + 5y = -10$	$-\dfrac{3}{5}$	–	b & d
(h) $y = -\frac{3}{4}x - 3$	$-\dfrac{3}{4}$	e	f

12. (a) $D = \dfrac{(3)(3) - (2)(3) + 4}{\sqrt{(3)^2 + (2)^2}} = \boxed{1.9}$

(b) $D = \dfrac{(1)(-10) + (-1)(8) + 5}{\sqrt{(1)^2 + (1)^2}} = \boxed{9.2}$

(c) $D = \dfrac{(5)(9) - (2)(6) + 10}{\sqrt{(5)^2 + (-2)^2}} = \boxed{8.0}$

(d) $D = \dfrac{(3)(-8) + (2)(-6)}{\sqrt{(3)^2 + (2)^2}} = \boxed{10.0}$

(e) $D = \dfrac{(3)(12) + (-2)(-6) - 6}{\sqrt{(3)^2 + (-2)^2}} = \boxed{11.6}$

10.

	$Ax + By + C = 0$	m	x-int	y-int
(a) $x - y + 5 = 0$	$x - y + 5 = 0$	1	-5	5
(b) $-2x + 3y = 12$	$2x - 3y + 12 = 0$	$\dfrac{2}{3}$	-6	4
(c) $x - y = 0$	$x - y = 0$	1	0	0
(d) $6y = 4x + 2$	$4x - 6y + 2 = 0$	$\dfrac{2}{3}$	$-\dfrac{1}{2}$	$\dfrac{1}{3}$
(e) $2x - 4y = 2$	$2x - 4y - 2 = 0$	$\dfrac{1}{2}$	1	$-\dfrac{1}{2}$

13. (a) $y - 1 = (-2)(x - 3)$

$y - 1 = -2x + 6$

$\boxed{2x + y - 7 = 0}$

(b) $y - 3 = \dfrac{(2)(x + 4)}{3}$

$3y - 9 = 2x + 8$

$\boxed{2x - 3y + 17 = 0}$

(c) $y - 5 = 0$

$\boxed{y = 5}$

(d) $y + 3 = \dfrac{(-2)(x - 2)}{3}$

$3y + 9 = -2x + 4$

$2x + 3y + 5 = 0$

$\boxed{2x + 3y + 5 = 0}$

11.

	slope	parallel to problem no.	perpendicular to problem no.
(a) $2x - 3y - 3 = 0$	$\dfrac{2}{3}$	–	c
(b) $10x - 6y = 18$	$\dfrac{5}{3}$	d	g
(c) $y = -\frac{3}{2}x + 5$	$-\dfrac{3}{2}$	–	a
(d) $5x - 3y = 9$	$\dfrac{5}{3}$	b	g

Geometry II

14. (a) $\dfrac{y-2}{x+3} = \dfrac{4-2}{1+3} = \dfrac{1}{2}$

$x + 3 = (2)(y-2)$

$x + 3 = 2y - 4$

$$\boxed{x - 2y + 7 = 0}$$

(b) $\dfrac{y+3}{x-2} = \dfrac{-2+3}{5-2} = \dfrac{1}{3}$

$x - 2 = (3)(y+3)$

$x - 2 = 3y + 9$

$$\boxed{x - 3y - 11 = 0}$$

(c) $\dfrac{y-4}{x-3} = \dfrac{4-4}{-3-3} = 0$

$$\boxed{y = 4}$$

(d) $\dfrac{y+6}{x+2} = \dfrac{-4+6}{3+2} = \dfrac{2}{5}$

$(2)(x+2) = (5)(y+6)$

$2x + 4 = 5y + 30$

$$\boxed{2x - 5y - 26 = 0}$$

15. (a) $\dfrac{x}{-4} + \dfrac{y}{3} = 1$

$3x - 4y = -12$

$$\boxed{3x - 4y + 12 = 0}$$

(b) $\dfrac{x}{1} + \dfrac{y}{-4} = 1$

$-4x + y = -4$

$$\boxed{4x - y - 4 = 0}$$

16. $y = \dfrac{3x}{4} - 3$

$4y = 3x - 12$

$$\boxed{3x - 4y - 12 = 0}$$

17. For the perpendicular line,

$m = \dfrac{3}{4}$

$y = \dfrac{3x}{4} + 4$

$4y = 3x + 16$

$$\boxed{3x - 4y + 16 = 0}$$

18. For the parallel line,

$m = -3$

$y - 8 = (-3)(x - 0)$

$y - 8 = -3x$

$$\boxed{3x + y - 8 = 0}$$

19. $m = -2$

$y - 5 = (-2)(x - 5)$

$y - 5 = -2x + 10$

$$\boxed{2x + y - 15 = 0}$$

$y - 5 = \dfrac{(1)(x-5)}{2}$

$2y - 10 = x - 5$

$$\boxed{x - 2y + 5 = 0}$$

20. $m = \dfrac{4+3}{1-2} = -7$

$y - 1 = (-7)(x + 2)$

$y - 1 = -7x - 14$

$$\boxed{7x + y + 13 = 0}$$

21. (a)
$$
\begin{array}{r}
x + y = 8 \\
x - y = 4 \\
\hline
2x = 12
\end{array}
$$

$$\boxed{x = 6}$$

$$\boxed{y = 2}$$

(b)
$$
\begin{array}{r}
x + 2y = 6 \\
x + 2y = 4 \\
\hline
\end{array}
$$

$$\boxed{\text{parallel}}$$

(c)
$$
\begin{array}{r}
2x + 5y = -8 \\
2x + 3y = 5 \\
\hline
2y = -13
\end{array}
$$

$$\boxed{y = -6\tfrac{1}{2}}$$

$$\boxed{x = 12\tfrac{1}{4}}$$

(d)
$$
\begin{array}{r}
5x - 4y = -15 \\
2x - 12y = 7 \\
\hline
10x - 8y = -30 \\
10x - 60y = 35 \\
\hline
52y = -65
\end{array}
$$

$$\boxed{y = -1.25}$$

$$\boxed{x = -4}$$

(e)
$$7x - 2y = 3$$
$$2x + 3y = 9$$

$$21x - 6y = 9$$
$$4x + 6y = 18$$

$$25x = 27$$

$$\boxed{x = 1.08}$$

$$\boxed{y = 2.28}$$

(f)
$$7x - 2y = -11$$
$$8x + 3y = -39$$

$$21x - 6y = -33$$
$$16x + 6y = -78$$

$$37x = -111$$

$$\boxed{x = -3}$$

$$\boxed{y = -5}$$

(i)
$$9x + 10y = 9$$
$$6x - 25y = -13$$

$$54x + 60y = 54$$
$$54x - 225y = -117$$

$$285y = 171$$

$$\boxed{y = 0.6}$$

$$\boxed{x = 0.333}$$

22. (a) $4x + 3y = 24$
$\qquad 4x + 3y = -48$

(b) $3x + 2y = 24$
$\qquad x - 2y = -11$

(c) $5x + 11y = -55$
$\qquad 5x + 11y = -11$

(d) $2x + 4y = 48$
$\qquad x + 2y = -24$

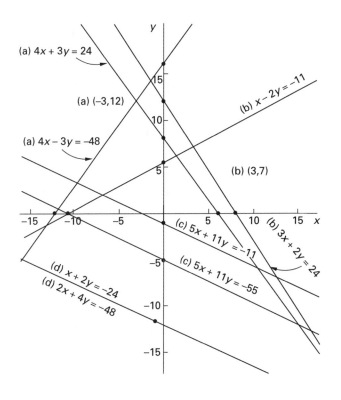

(a) 4x + 3y = 24

(a) (−3,12)

(a) 4x − 3y = −48

(b) x − 2y = −11

(b) (3,7)

(c) 5x + 11y = −11

(b) 3x + 2y = 24

(c) 5x + 11y = −55

(d) x + 2y = −24

(d) 2x + 4y = −48

23. The equation of line CA can be developed as follows.

$$\frac{y - 350}{x - 50} = \frac{80 - 350}{250 - 50} = \frac{-270}{200} = \frac{-27}{20}$$
$$(-27)(x - 50) = (20)(y - 350)$$
$$-27x + 1350 = 20y - 7000$$
$$-27x - 20y = -8350$$
$$27x + 20y = 8350$$

The equation of line EB can be developed as follows.

$$\frac{y - 100}{x - 20} = \frac{380 - 100}{270 - 20} = \frac{28}{25}$$
$$(28)(x - 20) = (25)(y - 100)$$
$$28x - 560 = 25y - 2500$$
$$28x - 25y = -1940$$

The solution of the system equations is

$$27x + 20y = 8350$$
$$28x - 25y = -1940$$

$$675x + 500y = 208{,}750$$
$$560x - 500h = -38{,}800$$

$$1235x = 169{,}950$$
$$x = 137.61$$
$$27x = 3715.47$$
$$3715.47 + 20y = 8350$$
$$20y = 4634.53$$
$$y = 231.73$$

$$\boxed{\text{The coordinates are } (138{,}232).}$$

24. The slope of EA is

$$m = \frac{80 - 100}{250 - 20} = \frac{-20}{230} = \frac{-2}{23}$$

The equation of line DH can be developed as follows.

$$y - 250 = \frac{(-2)(x - 70)}{23}$$
$$(-2)(x - 70) = (23)(y - 250)$$
$$-2x + 140 = 23y - 5750$$
$$-2x - 23y = -5890$$
$$2x + 23y = 5890$$

The equation of line AB can be developed as follows.

$$\frac{y - 80}{x - 250} = \frac{380 - 80}{270 - 250} = 15$$
$$15x - 3750 = y - 80$$
$$15x - y = 3670$$

The solution of the system of equations is

$$
\begin{aligned}
2x + 23y &= 5890 \\
15x - y &= 3670 \\
\hline
30x + 345y &= 88,350 \\
30x - 2y &= 7340 \\
\hline
347y &= 81,010 \\
y &= 233.46 \\
15x - 233.46 &= 3670 \\
x &= 260.21
\end{aligned}
$$

The coordinates are $(260, 233)$.

25. (a) $\quad m_1 = \dfrac{-4}{3}$

$\quad m_2 = \dfrac{4}{3}$

$$
\tan\theta = \dfrac{\dfrac{4}{3} + \dfrac{4}{3}}{1 + \left(\dfrac{-4}{3}\right)\left(\dfrac{4}{3}\right)} = \dfrac{\dfrac{8}{3}}{\dfrac{9}{9} - \dfrac{16}{9}}
$$

$$
= \dfrac{\dfrac{8}{3}}{\dfrac{-7}{9}} = \dfrac{24}{7}
$$

$\theta = \boxed{73°44'}$

(b) $\quad m_1 = \dfrac{-3}{2}$

$\quad m_2 = \dfrac{1}{2}$

$$
\tan\theta = \dfrac{\dfrac{1}{2} + \dfrac{3}{2}}{1 + \left(\dfrac{-3}{4}\right)} = \dfrac{\dfrac{4}{2}}{\dfrac{1}{4}} = 8
$$

$\theta = \boxed{82°52'}$

(c) $\quad m_1 = 2$

$\quad m_2 = -4$

$$
\tan\theta = \dfrac{-4 - 2}{1 + (-8)} = \dfrac{-6}{-7}
$$

$\theta = \boxed{40°36'}$

(d) $\quad m_1 = -2$

$\quad m_2 = \dfrac{6}{5}$

$$
\tan\theta = \dfrac{\dfrac{6}{5} + \dfrac{10}{5}}{1 + \left(\dfrac{-12}{5}\right)} = \dfrac{\dfrac{16}{5}}{\dfrac{-7}{5}} = \dfrac{16}{-7}
$$

$\theta = \boxed{66°22'}$

(e) $\quad m_1 = 1$

$\quad m_2 = -\dfrac{1}{2}$

$$
\tan\theta = \dfrac{\dfrac{-1}{2} - \dfrac{2}{2}}{1 + \left(-\dfrac{1}{2}\right)} = \dfrac{\dfrac{-3}{2}}{\dfrac{1}{2}} = -3
$$

$\theta = \boxed{71°34'}$

26. (a) $\boxed{(0, 3)}$ (b) $\boxed{(-11, 0)}$ (c) $\boxed{(-7, -5)}$

(d) $\boxed{(-4, -3)}$ (e) $\boxed{(4, 5)}$ (f) $\boxed{(3, 4)}$

13 Measurements and Field Practice

Nomenclature

C correction of a slope measurement
d length of smallest division on rod scale
F per 1000 ft
h error in feet caused by refraction
H horizontal distance
M per mile
n number of divisions on vernier
S slope distance
T temperature
v length of a vernier division
V difference in elevation between horizontal
     and slope distances

Part 1: Taping

1. LINEAR MEASUREMENT

The distance between two points can be determined by pacing, taping, electronic distance measurement (EDM), tacheometry (stadia), using an odometer, or scaling on a map. Of these methods, only taping will be discussed in this chapter.

Taping consists of aligning the tape, pulling the tape tight, using the plumb bob on unlevel ground, marking tape lengths, and reading the tape.

2. GUNTER'S CHAIN

The *Gunter's chain* was once used extensively in surveying the public lands of the United States, but that is no longer the case. However, the term *chaining* is still used to mean taping.

The Gunter's chain is 66 ft long and consists of 100 links, each link being 7.92 in in length. One chain = 1/80 mi, and 10 square chains = $(10)(66)^2 = 43{,}560 \text{ ft}^2 = 1$ ac.

Knowledge of the Gunter's chain is important to surveyors when retracing old surveys in which the Gunter's chain was used.

3. STEEL TAPES

Steel tapes are available in widths of from $3/8$ in to $5/16$ in with thicknesses varying from 0.016 in to 0.025 in. Lengths of steel tapes can be 50 ft, 100 ft, 200 ft, 300 ft, and 500 ft. The 100 ft tape is most common for surveying. Tapes are also made in 30 m, 50 m, and 100 m lengths.

4. INVAR TAPES

Invar tapes are made of a steel alloy containing 35% nickel. The expansion and contraction of the Invar tape because of changes in temperature is only about 3% of the change of a steel tape.

These tapes are used when measurements of extreme accuracy are required, such as measuring base lines or calibrating steel tapes. Invar tapes are, however, too fragile for normal use.

5. CHAINING PINS

Pins made of $3/16$ in steel are used to mark tape lengths. They are usually 12 in to 18 in long and are sharpened on one end and have a ring of about $2^1/2$ in in diameter at the other. A set of chaining pins contains 11 pins.

6. TYPES OF STEEL TAPES

Some 100 ft tapes measure 100 ft from the outer edges of the end loops. Most tapes, however, are in excess of 100 ft from end loop to end loop and have graduations for every foot from 0 to 100 ft.

An *add tape* has an extra graduated foot beyond the zero mark. The extra foot is usually graduated in tenths of a foot, but it is sometimes graduated in tenths and hundredths of a foot.

A *cut tape*, or a *subtract tape*, does not have the extra graduated foot, but the last foot at each end is graduated in tenths of a foot or in tenths and hundredths of a foot.

Figure 13.1 shows the distance between point A and point B measured by both an add tape and a cut tape. The distance using add tape is 26 ft + 0.18 ft = 26.18 ft. The distance using the cut tape is 27 ft − 0.82 ft = 26.18 ft.

7. HORIZONTAL TAPING

In surveying, the distance between two points is the horizontal distance, regardless of the slope.

8. TAPING WITH TAPE SUPPORTED THROUGHOUT ITS LENGTH

The rear chainperson wraps the leather thong at the end of the tape tightly around his right hand near the knuckles and faces at right angles to the line of measurement. He kneels with his left knee near the pin (or other mark) and braces his right arm against his right leg near the knee with the heel of his right hand firmly against the ground. To bring the end mark of the tape exactly on the pin, he shifts his weight to the left knee, or right foot, as desired, keeping the heel of his right hand firmly braced on the ground. In this position, he is off the line of sight, and his eyes are directly over the end mark of the tape and the pin. (Left-handed people will use the opposite positions.)

The forward chainperson wraps the leather thong around her left hand, faces at right angles to the line of measurement, and kneels on her right knee. She increases or decreases the pull on the tape by shifting her body weight. With her right hand, she sticks the pin at the zero mark on a call from the rear chainperson, which indicates the 100 ft mark is on the pin. In this position, she is also off the line of sight.

9. TAPING ON SLOPE WITH TAPE SUPPORTED AT ENDS ONLY

Taping downhill, the rear chainperson proceeds as in taping on level ground.

The forward chainperson wraps the leather thong around her left hand and takes a position facing at right angles to the line of sight as she did in taping on level ground, but in this procedure she remains standing. With her right hand, she makes one loop of the plumb

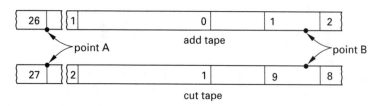

Figure 13.1 Add and Cut Tapes

bob string around the tape. With her forefinger under the tape and her thumb on the top of the tape, she can roll the string to the proper mark with her thumb. The last two fingers of her right hand grasp the loose end of the string. To get the proper length of string, she rests the plumb bob on the ground and feeds the string with her right hand.

She holds the tape as nearly horizontal as possible. Her feet should be placed well apart and her left elbow should be braced against her body. To apply tension, her left knee is bent so that the weight of her body pushes against the arm holding the tape. The plumb bob is steadied by lowering it to the ground. When the plumb bob is just slightly above ground and steady, the tape is horizontal, and when the chainperson feels the proper tension, she lets the plumb bob drop and then marks the point with a pin.

Taping uphill, the rear chainperson holds the plumb bob over the pin or point on the ground, and the forward chainperson proceeds as when taping on level ground. The rear chainperson holds the tape and the plumb bob as the foward chainperson does when taping downhill.

10. STATIONING WITH PINS AND RANGE POLE ON LEVEL GROUND

As stated earlier, a set of pins consists of 11 pins. A *station* is 100 ft (one tape) in length. In route surveying, stationing is carried along continuously from a starting point designated as sta 0+00.

The rear chainperson stations himself at the beginning point with one pin in hand or in the ground if the beginning point was not previously marked.

The forward chainperson takes 10 pins. With the zero end of the tape and the range pole, she advances in the direction of the stationing. She counts her paces from the beginning point so that, if she does not hear the rear chainsperson call, she knows when she has advanced approximately one station.

The rear chainperson watches the tape pass his beginning station. When the end is about 6 ft from his position, he calls, "chain," to the foward chainperson. He grabs the leather thong on the end of the tape as it nears him and proceeds as explained in Sec. 8.

On hearing "chain," the foward chainperson immediately turns and faces the rear. She observes the rear chainperson grab the leather thong. With the tape in her left hand and the range pole in her right, she puts tension on the tape, flips it to straighten it, and, holding the range pole vertically, places it near and slightly to the rear of the zero mark. She immediately drops the tape and, with legs spread fairly wide apart, takes the range between the forefinger and thumb of each hand and observes the transitperson (or rear chainperson) for alignment. She keeps the range pole vertical and her legs apart so that, on long sights, the transitperson will have a clear view of the range pole between her legs.

When she recieves an OK from the transitperson, she presses the point of the range pole in the ground, removes it, and places a chaining pin in the hole left by the range pole. She then wraps the thong around her left hand, flips the tape for alignment, and pulls the edge of the tape over to the pin.

On observing the forward chainperson reach this point, the rear chainperson checks the 100 ft mark to see that it is on the pin and calls out his station number, such as "eight." Besides keeping up with the station number, this call "eight" is also saying "I am on my mark" to the forward chainperson.

On hearing the rear chainperson call "eight," the forward chainperson quickly and carefully sticks her pin at the zero mark and calls her station, "nine." The call "nine," besides keeping up with the station number, says to the rear chainperson, "I have marked my point. Drop the tape and start walking forward."

The rear chainperson should never hang on the tape as he moves forward, but he should keep the end of the tape in view.

The system whereby both chainpersons call out the station numbers is a double-check on counting the pins. It is also a simple way to communicate when the forward pin should be set.

In taping long distances, both chainpersons should choose distance objects on their line to walk toward.

Chaining pins should be stuck at an angle of 45° with the ground and at right angles to the line of measurement.

11. STATIONING WHEN DISTANCE IS MORE THAN TEN LENGTHS

When the forward chainperson has set her last pin in the ground, she should have just heard the rear chainperson call "nine." She should have replied, "ten." Her last pin in the ground indicates that she has taped 10 stations, or 1000 ft. She waits at this last pin until the rear chainperson comes forward and hands her his pins. Both chainpersons count the pins to be certain there are ten in hand and one in the ground. As taping is resumed, the situation is the same as it was in the beginning. One pin is in the ground in front of the rear chainperson, and ten pins are in the hand of the forward chainperson.

12. STATIONING AT END OF LINE OR WHEN PLUS IS DESIRED AT POINT ON LINE

Using the add tape. The rear chainperson moves to the forward station and holds a foot mark on the pin.

If the forward chainperson needs more tape, she calls, "Give me a foot." The rear chainperson slides the next larger foot mark to the pin.

If the forward chainperson has too much tape, she calls, "Take a foot." The rear chainperson slides the next smaller foot mark to the pin.

The forward chainperson calls, "What are you holding?"

The rear chainperson calls, "Holding 46," for example.

The forward chainperson then calls "Reading 46.32," for example.

The forward chainperson then calls "Station?"

The rear chainperson counts the pins in his possession but does not count the pin in the ground at the last full station. The station number is the same as the number of pins in his hand if the station is less than ten. If it is more than ten, the station number is the same as the number of pins in his hand plus ten for each exchange of ten pins. He calls out the station number.

The forward chainperson checks the rear chainperson's count. The difference between ten and the number of pins in her hand, plus ten for each exchange of ten pins, is the station number.

The forward chainperson calls out the full station number and plus. Both chainpersons record it.

Using the cut tape. The procedure is the same for the rear chainperson in placing a foot mark on the pin.

The foreward chainperson calls, "What are you holding?"

The rear chainperson calls, "Holding 47."

The foreward chainperson then calls, "Cut 68."

The rear chainperson calls, "46.32."

The forward chainperson repeats, "46.32. Station?"

Both chainpersons then check the number of pins in hand as described in the procedure for using an add tape. They call the station number.

13. BREAKING TAPE

Where the slope is so great that a 100 ft length of the tape cannot be held horizontally without plumbing above the shoulders, a procedure known as *breaking tape* can be used as illustrated in Fig. 13.2.

The forward chainperson pulls the tape forward a full length as usual.

She puts the tape approximately on line and walks back along the tape to a point where the tape can be held horizontal below the shoulder level.

She then picks up a foot mark ending in 0 or 5 (70, for example) and measures a partial tape length (30 ft, for example), using the plumb bob as described in Sec. 9, and marking the point with a pin.

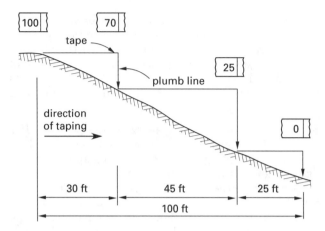

Figure 13.2 *Breaking Tape*

After the forward chainperson has placed the pin, she waits for the rear chainperson to come forward and then tells him what foot mark she was holding, such as, "Holding 70."

The rear chainperson repeats, "Holding 70." He hands the forward chainperson a pin to replace the one used to mark the intermediate point.

The chainpersons continue the procedure at as many intermediate points as necessary. The rear chainperson always picks up the intermediate pin. When he moves forward to an intermediate point, he hands the forward chainperson a pin. He does *not* hand a pin to the forward chainperson when he moves forward to the zero mark.

14. TAPING AT AN OCCUPIED STATION

When taping at a station that is occupied by an instrument, chainpersons must be extremely careful not to hit the leg of the instrument. If a plumb bob is needed at the point, the plumb bob string hanging from the instrument can be used. In some cases, it may be necessary to use the point on top of the instrument on the vertical axis as a measuring point.

15. CARE OF THE TAPE

The tape will not be broken by pulling on it unless there is a kink (loop) in it. Chainpersons should always be alert to "kinking." The tape is easily broken if it is pulled when there is a kink in it.

If the tape has been used in wet grass or mud, it should be cleaned and oiled lightly by pulling it through an oily rag.

16. SLOPE MEASUREMENTS

On fairly level ground where the slope is uniform, it is sometimes easier to determine the slope and make corrections for changing the slope measurement to horizontal measurement rather than to break tape every few feet. To determine horizontal distance, the correction will be subtracted from the slope distance.

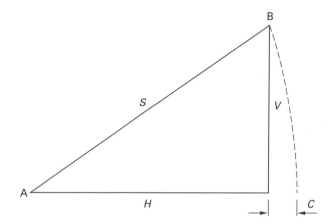

Figure 13.3 *Measurements on Slope*

In Fig. 13.3, H is the horizontal distance from A to B. S is the slope distance from A to B. V is the difference in elevation from A to B, and C is the correction.

Thus,

$$V^2 = S^2 - H^2 = (S - H)(S + H) \qquad 13.1$$

Where the slope difference is small, $S + H$ is approximately $2S$. Therefore,

$$V^2 \approx (2S)(S - H) \qquad 13.2$$

Because $S - H =$ correction C,

$$C \approx \frac{V^2}{2S} \qquad 13.3$$

Where $S = 100$ ft (one tape length),

$$C \approx \frac{V^2}{200 \text{ ft}} \qquad 13.4$$

The approximate value will be within 0.007 ft of the actual value when the difference in elevation per 100 ft of slope distance is not more than 15 ft. For steeper slopes, more exact formulas may be used.

Example 13.1

The difference in elevation between two points is 4.0 ft and the slope distance is 100.00 ft. What is the horizontal distance?

Solution

The correction is

$$C \approx \frac{(4 \text{ ft})^2}{200 \text{ ft}} = 0.08 \text{ ft}$$
$$= 100.00 \text{ ft} - 0.8 \text{ ft} = 99.92 \text{ ft}$$

17. TENSION

Tapes are not guaranteed by their manufacturers to be of exact length. The National Bureau of Standards will, for a fee, compare any tape with a standard tape or distance, and it will certify the exact length of the tape under certain conditions.

In the United States, steel tapes are standardized for use at 68°F. The standard pull for a 100 ft tape with the tape supported throughout its length is usually, depending on the cross-sectional area of the tape, 10 lbf. When tapes are standardized, they are usually standardized for use under two conditions: supported throughout, and supported only at the ends. When supported only at the ends, the pull is usually 30 lbf, but this can be varied on request. (See Fig. 13.4.)

When the tape is pulled with more or less than the standard amount of tension (10 lbf when supported), the actual distance is more or less than 100.00 ft. However, variations in pull when the tape is supported do not affect the distance greatly. (If a pull of 20 lbf is exerted on a tape that was standardized for a pull of 10 lbf, the increased length is 0.006 ft.)

By using spring-balance handles, chainpersons can get the feel of a 10 lbf pull so that, in ordinary taping with the tape supported, the error caused by variation in tension is neglible.

18. CORRECTION FOR SAG

When the tape is supported only at the ends, it sags and takes the form of a catenary. The correction for sag can be determined by formula or can be offset by increased tension. For a medium-weight tape standardized for a pull of 10 lbf, a pull of 30 lbf will offset the difference in length caused by sag. If the tape is supported at 25 ft intervals, the pull only need be 14 lbf.

Chainpersons should use the spring-balance handle to familiarize themselves with various "pulls."

19. EFFECT OF TEMPERATURE ON TAPING

Steel tapes are standardized for 68°F in the United States. For a change in temperature of 15°F, a steel tape will undergo a change in length of about 0.01 ft, introducing an error of about 0.5 ft per mile.

The coefficient of thermal expansion for steel is approximately 0.00000645 per unit length per degree Fahrenheit.

For a 100 ft tape where T is the temperature (in °F) at time of measurement, the correction in length C due to change in temperature is

$$C = (0.00000645)(T - 68°)(100) \qquad 13.5$$

UNITED STATES DEPARTMENT OF COMMERCE

WASHINGTON

National Bureau of Standards

Certificate
100-Foot
Steel Tape

Maker: Keuffel & Esser Co.   Submitted by   NBS No. 10565

This tape has been compared with the standards of the United States, and the intervals indicated have the following lengths at 68... Fahrenheit (20... centigrade) under the conditions given below:

Supported on a horizontal flat surface

Tension (pounds)	Interval (feet)	Length (feet)
10	0 to 100	100.002
5 1/2	0 to 100	100.000

Supported at the 0, 50, and 100-foot points

Tension (pounds)	Interval (feet)	Length (feet)
30	0 to 100	100.000

Supported at the 0 and 100-foot points

Tension (pounds)	Interval (feet)	Length (feet)
38 1/2	0 to 100	100.000

See Note 3(a) on the reverse side of this certificate.

the Director
National Bureau of Standards

Test No.   2.4/1426
Date: June 13, 1955

Lewis V. Judson

Lewis V. Judson
Chief Length Section
Optics and Metrology Division

Figure 13.4 *National Bureau of Standards Certificate*

For example, assume that a line was measured to be 675.48 ft at 30°F. The change in the recorded length due to temperature change is

$$C = (0.00000645)(30°F - 68°)(675.48 \text{ ft}) = -0.17 \text{ ft}$$

The corrected length is

$$L = 675.48 \text{ ft} - 0.17 \text{ ft} = 675.31 \text{ ft}$$

If the same line were measured when the temperature was 106°F, the change would be

$$C = (0.00000645)(106°F - 68°F)(675.48 \text{ ft}) = +0.17 \text{ ft}$$

The corrected length would be

$$L = 675.48 \text{ ft} + 0.17 \text{ ft} = 675.65 \text{ ft}$$

20. EFFECT OF IMPROPER ALIGNMENT

Improper alignment is probably the least important error in taping. Many transitpersons and chainpersons spend time aligning that is not justified by the effect of improper alignment. The linear error when one end of the tape is off line can be computed in the same way slope correction is computed. For example, for a 100 ft tape with one end off line 1.0 ft,

$$C = \frac{V^2}{200 \text{ ft}} = 0.005 \text{ ft}$$

When the error in alignment is 0.5 ft, the linear error is 0.001 ft per tape length, or about 0.05 ft per mile.

21. INCORRECT LENGTH OF TAPE

A standardized tape can be used to check other tapes. If a 100 ft tape is known to be of incorrect length, the correction factor to be used for measurements made with the correct tape 100 ft is

$$C = \text{actual length} - 100.00 \text{ ft}$$

For example, the correction for a tape found to be 100.02 ft long after comparison with a standardized tape is

$$C = 100.02 \text{ ft} - 100.00 \text{ ft} = +0.02 \text{ ft per 100 ft}$$

If a line is measured to be 662.35 ft with this tape, the corrected length would be

$$662.35 \text{ ft} + (6.6235 \text{ ft})(+0.02) = 662.48 \text{ ft}$$

For a tape found to be 99.98 ft long after comparison with the standardized tape, the correction would be

$$C = 99.98 \text{ ft} - 100.00 \text{ ft} = -0.02 \text{ ft per 100 ft}$$

If a line is measured to be 662.35 ft with this tape, the corrected length would be

$$662.35 \text{ ft} + (6.6235 \text{ ft})(-0.02) = 662.22 \text{ ft}$$

A line measured with a tape that is longer than 100 ft is actually longer than the measurement shown. A line measured with a tape that is shorter than 100 ft is acutally shorter than the measurement shown. A rule to remember is, "for a tape too long, add; for a tape too short, subtract." This rule can also be applied to temperature correction.

22. COMBINED CORRECTIONS

Corrections for incorrect length of tape, temperature, and slope can be combined.

Example 13.2

A tape that is 100.03 ft long was used to measure a line that was recorded as 1238.22 ft when the temperature was 18°F. The difference in elevation from beginning to end was 12.1 ft. What is the corrected length?

Solution

The tape correction is

$$(12.3822)(+0.03) = +0.37$$

The temporary correction is

$$(0.00000645)(18° - 68°)(1238.22) = -0.40$$

The slope correction is

$$\frac{(12.1)^2}{(2)(1238.22)} = -0.06$$

The total correction is −0.09.

The corrected length is

$$1238.22 - 0.09 = 138.13 \text{ ft}$$

Part 2: Leveling

23. DEFINITIONS

Understanding leveling requires a vocabulary of terms used in the study of the earth's surface. The following terms are important to know.

- *vertical line:* a line from any point on the earth to the center of the earth
- *plumb line:* a vertical line, usually established by a pointed metal bob hanging on a string or cord

- *level surface:* Because the earth is round, a level surface is actually a curved surface. Although a lake appears to have a flat surface, it follows the curvature of the earth. A level surface is a curved surface that, at any point, is perpendicular to a plumb line.

- *horizontal line:* a line perpendicular to the vertical

- *datum:* any level surface to which elevations are referred. Mean sea level is usually used for a datum.

- *elevation:* the vertical distance from a datum to a point on the earth

- *leveling:* the process of finding the difference in elevation of points on the earth

- *spirit level:* a device for establishing a horizontal line by centering a bubble in a slightly curved glass tube (vial) filled with alcohol or another liquid

- *bench mark:* a marked point of known elevation from which other elevations may be established

- *turning point:* a temporary point on which an elevation has been established and which is held while an engineer's level is moved to a new location

- *height of instrument:* the vertical distance from the datum to the line of sight of the level

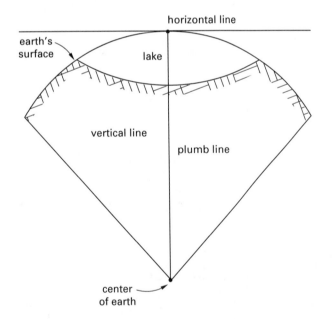

Figure 13.5 *Leveling Terms*

24. DIFFERENTIAL LEVELING

As the term implies, *differential leveling* is the process of finding the difference in elevation between two points. An engineer's level and level rod are used in differential leveling. A rod is merely a piece of wood, fiberglass, or aluminum that is marked off from one end to the other in meters or in feet, tenths of a foot, and hundredths of a foot. It is held in vertical position. The level is a telescope with crosshairs attached to a spirit level. By keeping the level bubble centered in the vial, the horizontal crosshair in the telescope can be kept on the same elevation while the telescope turns in any direction in the horizontal plane. It establishes a horizontal plane in space from which measurements can be made with the rod. As the levelperson focuses on the level rod, a measurement can be made from the horizontal plane to the point on which the rod rests merely by reading the measured markings on the rod where the horizontal crosshair is imposed on it.

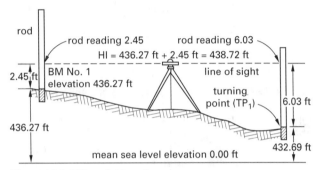

Figure 13.6 *Differential Leveling*

Figure 13.6 illustrates how the level and rod can be used to find the difference in elevation between two points.

Bench mark 1 (BM No. 1) is a semipermanent object, the elevation of which is 436.27 ft above mean sea level. Turning point 1 (TP_1) is a temporary object (in this case, the top of a stake), the elevation of which is to be determined.

The level is set up so that both objects can be seen through the telescope.

The rod is first placed on BM No. 1, and a reading of the rod is made and recorded with the bubble centered. This reading is known as a *backsight* (BS) and is added to the elevation of BM No. 1 to find the height of the instrument (HI). Backsight is commonly called a *plus sight* because it is always added to the known elevation.

The rod is then placed on the TP_1, and a reading of the rod is made and recorded with the bubble centered. This reading is known as a *foresight* (FS) and is subtracted from the HI to find the elevation of TP_1. It is commonly called a *minus sight* because it is always subtracted from a known HI.

The elevations of a continuing line of objects can be determined by moving the level along the line. While the level is being moved forward, the rodperson must hold the turning point so that the levelperson may make a backsight reading on it from the new location of the level.

Figure 13.7 shows a profile view of differential levels between BM No. 3 and BM No. 4 and field notes recorded at the time the levels were run. It can be seen that leveling is a series of vertical measurements that alternate in sequence from a plus sight to a minus sight.

Field notes show columns for plus readings (+), for minus readings (−), for HIs, and for elevations of bench marks and turning points.

Notice that a plus reading is shown on the same horizontal line as BM No. 3, but a minus reading is not shown on that line. Only one reading was taken on BM No. 3. Also notice that a plus reading and a minus reading are shown on the same horizontal line as each

turning point. The minus reading is subtracted from HI on the line above it to determine the elevation of the turning point, which is shown on the same horizontal line.

After the elevation of the turning point has been determined, the level is moved forward. Then a backsight (+) is read on the same turning point. This plus reading is added to the elevation just determined to new HI. Notice that a minus reading is shown on the same horizontal line as BM No. 4, but a plus reading is not shown on that line. BM No. 4 is the end of the level line; only one reading was made on it.

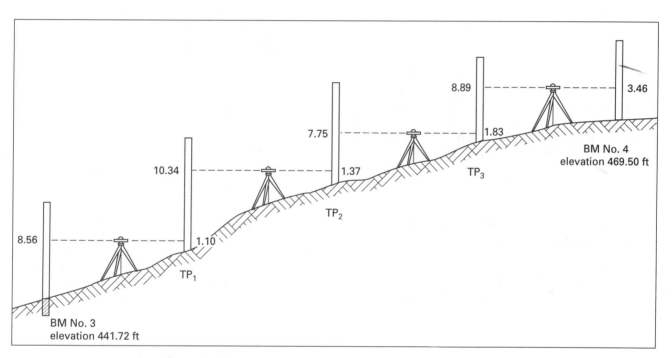

	differential levels				
sta	+	HI	−	rod	elev
BM No. 3	8.56	450.28			441.72
TP	10.34	459.52	1.10		449.18
TP	7.75	465.90	1.37		458.15
TP	8.89	472.96	1.83		464.07
BM No. 4			3.46		469.50
	35.54		7.76		469.50
	7.76				441.72
	27.78	– – – –	check	– – – –	27.78

hwy 345 — 3-28-72 — Jones ⅄ Smith ⅏

bolthead in concrete mon. – Lt. sta. 10+00

spike in 12 in elm 100 ft – Lt. sta. 25+50

Figure 13.7 Continuous Differential Leveling

At the bottom of the field notes, the plus column and the minus column are totalled and the smaller total is subtracted from the larger total. This difference should be the same as the difference between the beginning and ending elevations. If it is not, a mistake has been made in arithmetic.

The "rod" column is not used for differential leveling but it is used for other types of leveling.

25. THE PHILADELPHIA ROD

The *Philadelphia rod* is commonly used in leveling. It consists of two sliding parts and can be extended from 7 ft to 13 ft in length. When it is 7 ft long, it is known as a "low rod." When it is 13 ft long, it is known as a "high rod."

The graduations on the face of the rod are in hundredths of a foot. They measure continuously from zero at the bottom to 13 ft at the top. Each full foot is marked with a red number (white in Fig. 13.8); each tenth of a foot is marked with black numbers (1 to 9) between two red numbers.

The hundredth graduations alternate from black to white. At each tenth mark, the black graduation is extended and slashed so that the top edge is read for 0.10, 0.20, 0.30, and so on. The black graduation at the halfway distance between each tenth mark (0.05) is also extended and slashed. The bottom edge of the black graduation is read.

26. USING BLACK NUMBERS TO READ THE PHILADELPHIA ROD

In Fig. 13.8(a), the top edge of the black 1 is in line with the bottom edge of a black hundredth graduation; the reading, then, is 5.13. All the black numbers are

accurately and consistently marked on the rod so that whenever the top edge of any black number aligns with the crosshair, the reading ends in 0.03. Whenever the bottom edge of any black number aligns with the crosshair, the reading ends in 0.07.

In Fig. 13.8(b) the top edge of the base of the black 2 is in line with the top edge of a black hundredth graduation, so that the reading is 5.18. The bottom edge of the top part of the black 2 is in line with the top edge of a black hundredth graduation, so that, if the middle crosshair were on that line, the reading would be 5.22.

The black numbers are easier to read than the black hundredth graduations, so the numbers can be used from bottom to top to read 0.07, 0.08, 0.02, and 0.03 at any black number. The 0.05 reading can be identified by the extended black graduation, which is slashed. Therefore, almost any reading can be made by using the black numbers, except near the red numbers. The red numbers are not made so that they can be read in a useful way other than for the full foot.

27. TARGETS

For long sights or for readings to the thousandth of a foot, a target attached to a vernier is used.

For readings less than 7 ft, the target is moved up or down the rod at the direction of the levelperson until it coincides with the middle horizontal crosshair. The rodperson then clamps it at that position. The reading is made by the rodperson using the vernier.

For readings greater than 7 ft, the target is clamped at 7.000 ft, and the top section of the rod is moved up or down at the direction of the levelperson. The reading is made by use of the vernier on the back of the rod.

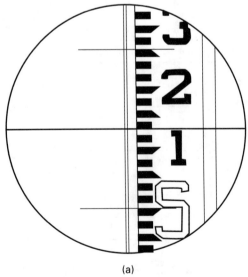

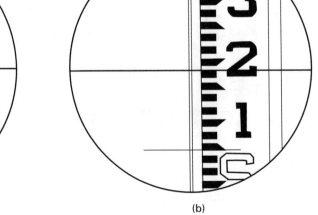

(a)   (b)

Figure 13.8 *Graduations on a Philadelphia Rod*

28. VERNIERS

A *vernier* can be used to find a fractional part of the smallest division of a scale. Using a vernier on a Philadelphia rod, readings can be made to the thousandth of a foot.

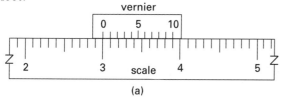

(a)

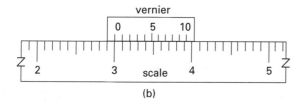

(b)

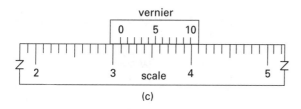

(c)

Figure 13.9 Verniers

It can be seen in Fig. 13.9(a) that ten spaces on the vernier cover nine of the smallest divisions on the scale, which on a Philadelphia rod are hundredths of a foot. For any vernier,

$$(n-1)d = nv \qquad 13.6$$

n is the number of divisions on the vernier, $n-1$ is the number of the smallest divisions on the scale, d is the length of the smallest scale division, and v is the length of a vernier division.

For a Philadelphia rod, $n = 10$, $d = 0.01$ ft, and $v = 0.09/10 = 0.009$ ft or $(10-1)(0.01) = (10)(0.009)$.

In Fig. 13.9(b), the vernier has moved so that the first division past the zero on the vernier is in line with the 0.31 mark on the scale. The vernier has moved a distance of 0.001, so the reading is 0.301.

In Fig. 13(c), the eighth division on the vernier is in line with a mark on the main scale, so the reading is 0.308.

In summary, to determine the rod reading, read the red number directly, determine the tenths reading by observing the zero mark on the vernier, and determine the thousandths reading by determining which mark on the vernier is in line with a hundredth graduation on the rod.

29. LEAST COUNT OF A VERNIER

The *least count of a vernier* is the smallest reading that can be determined without interpolation. For any vernier, the least count is d/n, where d is the length of smallest scale division and n is the number of divisions on the vernier. It can easily be remembered as

$$\text{least count} = \frac{\text{value of smallest division on the scale}}{\text{number of divisions on the vernier}}$$

$$13.7$$

30. EFFECT OF CURVATURE OF THE EARTH

By definition, a level surface is a curved surface and a horizontal line is a straight line (see Fig. 13.5).

If a level sight were made on a level rod 1 mi away from any point on the earth, the reading, if one could be made, would be greater by 0.667 ft because of the curvature of the earth. The departure of the earth from the horizontal line varies as the square of the distance from the level to the rod. Two formulas can be used to find this distance.

$$C = 0.667M^2 \qquad 13.8$$
$$C = 0.024F^2 \qquad 13.9$$

C is the departure in feet, M is the distance in miles, and F is the distance in thousands of feet.

At a distance of 100 ft, $C = 0.00024$ ft. At a distance of 300 ft, $C = 0.0022$ ft.

Where sights are held to 300 ft and read to hundredths of a foot, the effect on elevations that are expressed to hundredths of a foot is very small when it is considered that the effect of the curvature is offset somewhat by refraction.

31. REFRACTION

Light passing through the atmosphere is bent so that in reading a rod, the reading is less. This offsets the effect of the earth's curvature by about 14%. A formula for the combined effect of curvature and refraction is

$$h = 0.574M^2 = 0.0206F^2 \qquad 13.10$$

h is the error in feet, M is in miles, and F is in thousands of feet.

For a sight of 200 ft, $h = 0.008$ ft, and for a sight of 300 ft, $h = 0.0019$ ft.

32. WAVING THE ROD

It is extremely important that the rod be plumb when a reading is taken. The levelperson can bring the rod into plumb in one direction by observing the vertical

Measurements

crosshair and signaling the rodperson to plumb the rod, but the levelperson cannot tell whether the rod is leaning toward or away from the rodperson. For low rods, the rodperson can hold the rod lightly between finger tips just in front of his nose and balance the rod. For high rods, however, this is more difficult.

If the bench mark or turning point is not a flat surface (it should not be), the plumb position can be found by a method known as *waving the rod*. The rodperson moves the rod slowly toward and then away from the level while the levelperson observes the horizontal crosshair on the rod. The rod is plumb at the lowest reading. The error caused by the rod being out of plumb can be found by adapting Eq. 13.3 from Part 1, Sec. 16.

$$E = \frac{D^2}{2L} \qquad \textit{13.11}$$

E is the error caused by the rod being out of plumb, D is the distance in feet that the top of the rod is out of plumb, and L is the length of the rod.

Example 13.3

A rod reading of 12.000 was made on a 12 ft rod when it was 9 in out of plumb. What is the error, E?

Solution

$$D = \frac{9 \text{ in}}{12 \frac{\text{in}}{\text{ft}}} = 0.75 \text{ ft}$$

The error E is

$$E = \frac{(0.75 \text{ ft})^2}{(2)(12 \text{ ft})} = 0.0234 \text{ ft}$$

If the rod reading were 10.500 under the same conditions, the error E would be

$$E = \frac{(10.500)(0.75 \text{ ft})^2}{(12)(2)(12 \text{ ft}^2)} = 0.021 \text{ ft}$$

33. PARALLAX

Parallax is the apparent change in the position of the crosshair as viewed through the telescope. Because the reticle (the ring that holds the crosshairs in the telescope) is stationary, the distance between it and the eyepiece must be adjusted to suit the eye of each individual observer. The eyepiece is adjusted by turning it slowly until the crosshair is as black as possible. After the eyepiece is adjusted, the object viewed should be brought into sharp focus by means of the focusing knob for the objective lens. If the crosshairs seem to move across the object when the viewer moves his eye slightly, parallax exists. It is eliminated by carefully adjusting the eyepiece and the objective lens. If parallax is not eliminated, it can affect the accuracy of the rod readings.

34. BALANCING SIGHTS

The most common cause of errors in leveling is imperfect adjustment of the level. Centering the level bubble establishes a horizontal plane for the observer. If the level is not properly adjusted, however, the line of sight may not be parallel to the axis of the level vial, causing the rod reading to be greater or less than the true reading. The error can be offset by *balancing sights*—that is, by making the horizontal length of plus sights and minus sights approximately equal for each setup of the level. Leveling uphill or downhill makes this impossible, but if the total length of plus sights equals the total length of minus sights for a line of levels, the result will be the same. Distances can be determined by means of the stadia hairs in modern levels.

It is extremely important to make sure that the level bubble is exactly centered at the instant of a rod reading. The bubble should be centered, the telescope should be focused on the rod, a check on the bubble should be made, and a final reading should be made without touching the level. All this can be done in a few seconds.

35. RECIPROCAL LEVELING

Running a line of levels across a river or other obstacle where the horizontal distance is more than the desired maximum can be performed using *reciprocal leveling*.

The level is set up on the bank of the river, and turning point A is established nearby on the same side of the river. Turning point B is established on the other side of the river. A reading is taken on turning point A, and several readings are made on turning point B by unleveling, releveling, and then averaging the readings. The level is then set up on the side of the river opposite point A and near point B. Readings are made on B and A in the same manner as before. The difference in elevation between A and B is determined from the average readings.

36. DOUBLE-RODDED LEVELS

When saving time is important, using two rods and two sets of turning points on the same line of levels provides a good check on the difference in elevations. The notekeeper will, in effect, have two sets of notes.

37. THREE-WIRE LEVELING

Reading the two stadia hairs and the middle crosshair at each turning point and bench mark provides an excellent check for a line of levels. The difference in the middle crosshair reading and the upper stadia hair reading should be very near the difference in the middle crosshair reading and the lower stadia hair reading. If there is a discrepancy, one of the readings can be disregarded; otherwise, differences in elevations can be determined by averaging the three readings.

38. PROFILE LEVELING

In planning highways, canals, pipelines, and so on, a vertical section of the earth is needed to determine the vertical location of the centerline of the project. This vertical section is known as a *profile*. It is plotted on paper from field notes. *Profile leveling* is similar to differential leveling except that many minus readings are taken in addition to the usual plus and minus readings taken on bench marks and turning points.

At each setup of the level, readings are taken on the ground along the centerline at each full station and at each break on the ground. (A *break* on the ground is a point on the ground where the slope changes.) These readings are all minus readings. They are measurements made to determine the elevation at each point on the profile, and they are subtracted from the HI at each level setup. For clarity, these ground readings are recorded in the rod column. Bench mark and turning point readings are recorded as they are in differential leveling. To determine elevations of ground points, all readings taken at one level setup are subtracted from the HI at that level setup.

After elevations at each ground point are determined, they are plotted on specially ruled paper known as *profile paper* or *profile sheets*.

Fig. 13.11 shows the profile plotted from the field notes in Fig. 13.10.

	profile levels				
sta	B+S	HI	FS	rod	elev
BM no. 4	4.87	483.13			478.26
32+00			—	11.5	471.6
33+00				9.4	473.7
+75				10.1	473.0
34+00				8.2	474.9
35+00				3.0	480.1
+15				1.9	481.2
+70				2.3	480.8
36+00				5.2	477.0
+50				6.8	476.3
37+00				5.9	477.2
38+00				13.3	469.8
TP	4.54	476.95	10.72		472.41
38+60				13.2	463.8
39+00				12.0	465.0
40+00				3.9	473.1
41+00				1.2	475.8
42+00				0.8	476.2
+70				0.7	476.3
+80				1.5	475.5
43+00				0.4	476.6
BM no. 5			0.17		476.78
	9.41		10.89		
	10.89 − 9.41 = 1.48		478.26 − 476.78 = 1.48		

FM ROAD 123						
R.R. spike in 12 in oak 75 ft Lt. sta. 33+50						
		T/stake 38+00 Lt				
R.R. spike in 16 in elm 100 ft rt. sta. 42+50						

Figure 13.10 Profile Leveling Field Notes

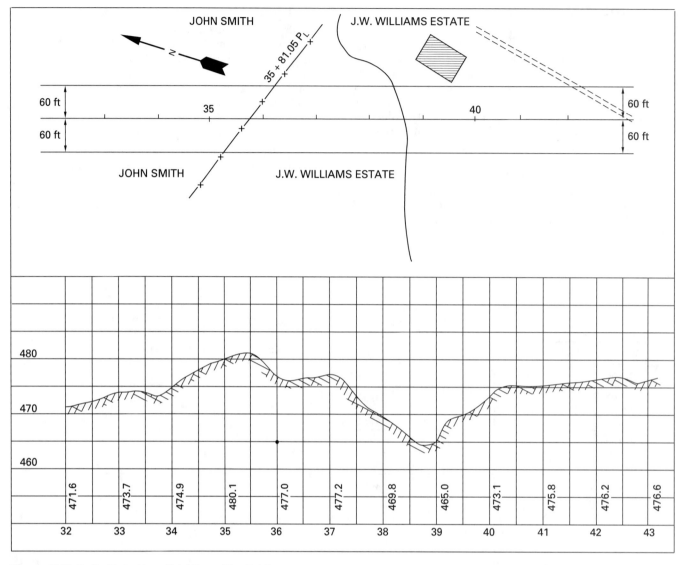

Figure 13.11 *Profile Plotted from Field Notes (Fig. 13.10)*

Part 3: Compass Surveying

39. MAGNETIC NEEDLE

A *magnetic needle* is a slender, magnetized steel rod that, when freely suspended at its center of gravity, points to magnetic north.

40. MAGNETIC DIP

In the northern hemisphere, the magnetic needle dips toward the north magnetic pole. In the southern hemisphere, the needle dips toward the south magnetic pole. To counteract the dip so that the needle will be horizontal, a counterweight is attached to the south end of the needle in the northern hemisphere and to the north end in the southern hemisphere. This weight is usually a short piece of fine brass wire.

41. THE MAGNETIC COMPASS

The *magnetic compass* consists of a magnetic needle mounted on a pivot at the center of a graduated circle in a metal box covered with a glass plate. It is constructed so that the angle between a line of sight and the magnetic meridian can be measured. The line of sight, with the horizontal circle, can be rotated in the horizontal plane while the needle continues to point to magnetic north. The point of the needle marks the angle made by the magnetic meridian and the line of sight.

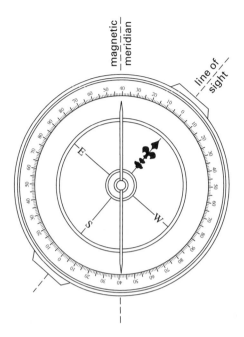

Figure 13.12 *Typical Compass*

42. THE SURVEYOR'S COMPASS

The horizontal circle in the *surveyor's compass* is usually graduated in half degrees. (Figure 13.12 does not show these graduations.) The letters E and W on the compass are reversed so that direct readings of bearings can be made. In the figure, the bearing of the line of sight is N 40° E. When the letters are reversed, the north end of the needle lies in the northeast quadrant of the horizontal circle.

43. MAGNETIC DECLINATION

The magnetic poles do not coincide with the axis of the earth. The horizontal angle between the magnetic meridian and the true (geodetic) meridian is known as *declination*. In some areas, the needle points east of true north, and in some areas it points west of true north. Zero declination is found along a line between the areas. This line is known as the *agonic line*. It passes, generally, through Florida and the Great Lakes, but it is constantly changing its location. East of this line, declination is west (−); west of this line, declination is east (+). In the United States, declination varies from 0° to 23°.

44. VARIATIONS IN DECLINATION

Declination in any one point varies daily, annually, and secularly (over a long period of time such as a century). Declination for a particular location for a particular year can be obtained from the United States Geological Survey.

45. IMPORTANCE OF COMPASS SURVEYING

Compass surveying is as obsolete as the Gunter's chain, but it is important for the modern surveyor to understand it when retracing old lines. The modern surveyor needs to be able to convert magnetic bearings to true bearings. In order to do so, the surveyor must know whether the declination is east or west and what the declination is, or was, on a certain day. Typical problems are given as follows.

Example 13.4

Convert the following magnetic bearings to true bearings.

(a) N 68°20′ E, decl 8°00′ W

(b) S 12°30′ W, decl 3°45′ E

(c) S 20°30′ E, decl 6°30′ W

(d) N 3°15′ W, decl 4°20′ E

Solution

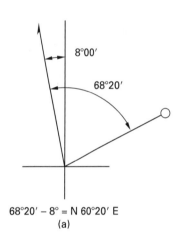

68°20′ − 8° = N 60°20′ E
(a)

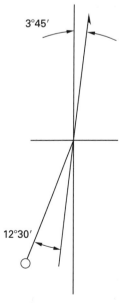

12°30′ + 3°45′ = S 16°15′ W
(b)

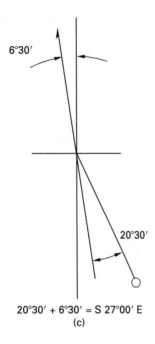

20°30′ + 6°30′ = S 27°00′ E
(c)

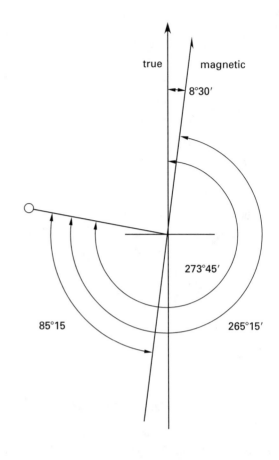

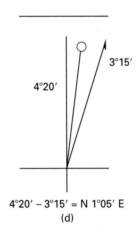

4°20′ − 3°15′ = N 1°05′ E
(d)

Magnetic bearings can also be converted to true bearings by use of azimuths.

Example 13.5

The magnetic bearing of a line was S 85°15′ W at a location where the declination was 8°30′ E. Find the true bearing.

Solution

$$\text{magnetic bearing} = \text{S } 85°15′ \text{ W}$$
$$\text{magnetic azimuth} = 265°15′$$
$$\text{declination} = +8°30′$$
$$\text{true azimuth} = 273°45′$$
$$\text{true bearing } \beta = \text{N } 86°15′ \text{ W}$$

PRACTICE PROBLEMS

1. Compared to steel tape, the expansion and contraction of an Invar tape caused by temperature change is
(A) slightly more
(B) slightly less
(C) much less
(D) about the same

2. How many pins are in a set of chaining pins?
(A) 10
(B) 11
(C) 12
(D) 20

3. An add tape has
(A) the last foot of each end graduated in tenths or hundredths of a foot.
(B) an extra graduated foot beyond the zero mark.
(C) both (A) and (B).
(D) neither (A) nor (B).

4. A Gunter's chain
  (A) is 66 ft long
  (B) has 66 links
  (C) is 100 ft long
  (D) both (A) an (B)

5. A square, one chain on a side, equals
  (A) 4356 ft^2
  (B) 43,560 ft^2
  (C) 1 ac
  (D) none of the above

6. The area of a rectangle 6 chains by 5 chains equals
  (A) 3 ac
  (B) 30 ac
  (C) 300 ac
  (D) none of the above

7. In marking 100 ft tape lengths, at what angle should the pin be set?
  (A) 30° to the ground
  (B) 45° to the ground
  (C) 60° to the ground
  (D) 90° to the ground

8. In taping from A to B, a distance of 1646.32 ft, how many pins should the rear chainperson have in his hand when he holds the tape on the last full station?
  (A) 4
  (B) 5
  (C) 6
  (D) 7

9. In taping as in Prob. 8, how many pins should the foreward chainperson have in her hand when she reaches point B?
  (A) 3
  (B) 4
  (C) 5
  (D) 6

10. A distance of 1000.00 ft was measured when the temperature was 68°F with a tape calibrated to be 100.000 ft in length. The difference in elevation from sta 0+00 to sta 10+00 was 11 ft. What is the corrected distance?
  (A) 999.82 ft
  (B) 999.94 ft
  (C) 1000.06 ft
  (D) 1000.18 ft

11. The coefficient of thermal expansion for steel is
  (A) 0.00000645 per unit length per degree Fahrenheit.
  (B) 0.00000465 per unit length per degree Fahrenheit.
  (C) 0.00000654 per unit length per degree Fahrenheit.
  (D) 0.0000645 per unit length per degree Fahrenheit.

12. A distance of 787.35 ft was measured on level ground with a tape that was 100.000 ft in length when the temperature was 98°F. What is the corrected length?
  (A) 787.20 ft
  (B) 787.50 ft
  (C) 787.55 ft
  (D) none of the above

13. A line was measured on level ground when the temperature was 68°F and was found to be 582.32 ft in length. Later, the tape was calibrated and found to be 100.03 ft in length. What is the corrected length of the line?
  (A) 582.15 ft
  (B) 582.49 ft
  (C) 582.65 ft
  (D) none of the above

14. For a change in temperature of 15°F, a 100 ft steel tape will undergo a change in length of about
  (A) 0.005 ft
  (B) 0.01 ft
  (C) 0.02 ft
  (D) 0.04 ft

15. At 28°F, a tape that measured 100.000 ft in length at 68°F will measure
  (A) 99.97 ft
  (B) 100.01 ft
  (C) 100.59 ft
  (D) none of the above

16. A tape that was 100.02 ft in length at 68°F was used to measure a line recorded as 1196.44 ft when the temperature was 28°F. The difference in elevation from the beginning to the end of the line was 10 ft. What is the corrected length?
  (A) 1196.27 ft
  (B) 1196.33 ft
  (C) 1196.55 ft
  (D) 1196.61 ft

Measurements

17. A tape that was 99.97 ft in length at 68°F was used to measure a line recorded as 713.19 ft when the temperature was 98°F. The difference in elevation from the beginning to the end of the line was 8 ft. What is the corrected length?

(A) 712.99 ft
(B) 713.05 ft
(C) 713.07 ft
(D) 713.14 ft

18. A level surface is

(A) a flat surface perpendicular to a horizontal line.
(B) a horizontal surface perpendicular to a plumb line.
(C) a curved surface perpendicular, at any point, to a plumb line.
(D) all of the above.

19. A plumb line is

(A) a horizontal line established by a spirit level.
(B) a line established by a plumb bob.
(C) a vertical line.
(D) both (B) and (C).

20. A datum is

(A) a horizontal plane.
(B) any level surface to which elevations are referred.
(C) the field notes kept by a survey party.
(D) none of the above.

21. The term *backsight* in leveling means

(A) a sight toward the beginning bench mark.
(B) a rod reading on a turning point.
(C) a rod reading on a point whose elevation is known.
(D) all of the above.

22. The term *foresight* in leveling means

(A) a sight on point whose elevation is to be determined.
(B) a minus sight.
(C) both (A) and (B).
(D) neither (A) nor (B).

23. A plus sight is taken on

(A) a point whose elevation is known.
(B) a point whose elevation is to be determined.
(C) the starting bench mark.
(D) both (A) and (C).

24. The height of instrument, as used in leveling, is the

(A) distance from the ground to the axis of the telescope.
(B) elevation of the line of sight above the datum plane.
(C) height of the line of sight above a bench mark.
(D) all of the above.

25. A point observed through a level telescope appears to be higher than it actually is because of

(A) parallax
(B) refraction
(C) curvature of the earth
(D) all of the above

26. Balancing distances to backsights and foresights eliminates

(A) errors caused by curvature of the earth.
(B) errors caused by refraction.
(C) parallax.
(D) both (A) and (B).

27. A rod reading of 10.00 was made on a 12 ft rod when it was 10 in out of plumb. What is the error caused by the rod's being out of plumb?

(A) 0.01 ft
(B) 0.02 ft
(C) 0.04 ft
(D) 0.20 ft

28. If a level sight were made on a level rod 1000 ft away, what would be the correction, C, for curvature of the earth? (Note: $C = 0.024F^2$.)

(A) -2.4 ft
(B) -0.024 ft
(C) $+0.0024$ ft
(D) $+0.24$ ft

29. What would be the combined error, h, for curvature and refraction for a 500 ft sight? (Note: $h = 0.0206$ ft^2.)

(A) 0.004 ft
(B) 0.005 ft
(C) 0.01 ft
(D) 0.02 ft

30. Record the rod reading for each of the following figures.

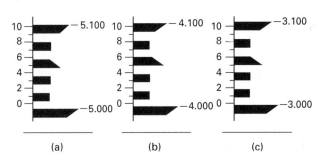

(a)   (b)   (c)

32. Complete the following level notes.

	differential levels				
sta	+	HI	–	rod	elev
BM No. 1		455.58			447.23
TP		465.14	0.88		
TP	9.56	473.67			464.11
TP	7.16	480.08			472.92
TP		488.76	1.65		478.43
BM No. 2	9.22		2.70		486.06
TP	4.04	491.41	7.91		484.37
TP	3.51	487.82	7.10		
TP		482.04			479.21
BM No. 3			5.17		476.87

31. Complete the following level notes and show the arithmetic check.

	differential levels				
sta	+	HI	–	rod	elev
BM No. 1	9.45				431.71
TP	11.88		1.08		
TP	10.99		0.62		
TP	8.33		1.30		
TP	11.21		2.47		
BM No. 2	10.54		3.90		
TP	5.60		6.77		
TP	4.18		8.32		
TP	3.90		9.74		
BM No. 3			5.12		

33. Complete the following field notes, show the arithmetic check, and plot the profile on the blank profile sheet on the following page.

profile levels					
sta	+	HI	–	rod	elev
BM No. 1	0.17				476.78
32+00				0.4	
33+00				0.8	
34+00				1.2	
35+00				3.9	
36+00				12.0	
+40				1.2	
TP	10.72		4.54		
37+00				13.3	
38+00				5.9	
+50				6.8	
39+00				5.2	
+30				2.3	
+85				1.9	
40+00				3.0	
41+00				8.2	
+25				10.1	
42+00				9.4	
43+00				11.5	
BM No. 2			4.87		

For Probs. 34 through 36, convert the magnetic bearings to true (geodetic) bearings. Draw sketches.

34. bearing: S 56°10′ W

declination: 7°30′ W

35. bearing: N 48°30′ W

declination: 4°20′ E

36. bearing: S 68°10′ E

declination: 5°40′ W

For Probs. 37 and 38, convert the bearings to azimuths and determine the true bearings. Draw sketches showing all angles.

37. magnetic bearing: S 89°50′ E

declination: 6°30′ W

38. magnetic bearing: N 88°15′ W

declination: 3°20′ E

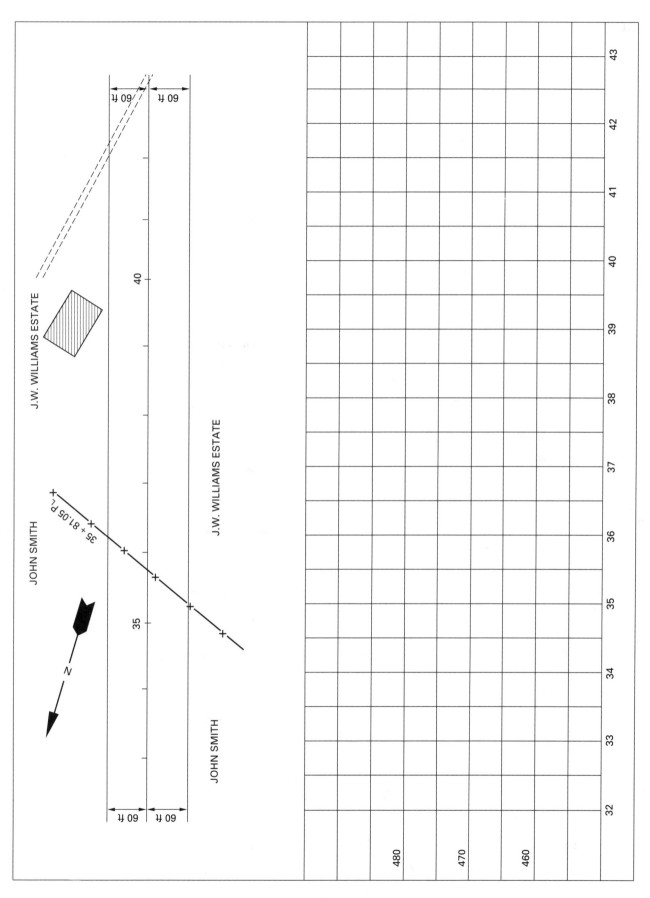

Problem 33: Blank Profile Sheet

SOLUTIONS

1. The answer is C.

2. The answer is B.

3. The answer is B.

4. The answer is A.

5. The answer is A.

6. The answer is A.

7. The answer is B.

8. The answer is C.

9. The answer is A.

10. The answer is B.

11. The answer is A.

12. The answer is B.

13. The answer is B.

14. The answer is B.

15. The answer is A.

16. The answer is B.

17. The answer is C.

18. The answer is C.

19. The answer is D.

20. The answer is B.

21. The answer is D.

22. The answer is C.

23. The answer is D.

24. The answer is B.

25. The answer is B.

26. The answer is D.

27. The answer is B.

28. The answer is B.

29. The answer is A.

30.

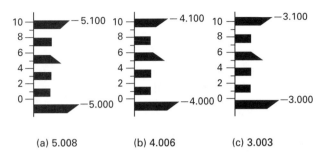

(a) 5.008   (b) 4.006   (c) 3.003

31.

		differential levels			
sta	+	HI	–	rod	elev
BM no. 1	9.45	**441.16**			431.71
TP	11.88	**451.96**	1.08		**440.08**
TP	10.99	**462.33**	0.62		**451.34**
TP	8.33	**469.36**	1.30		**461.03**
TP	11.21	**478.10**	2.47		**466.89**
BM no. 2	10.54	**484.74**	3.90		**474.20**
TP	5.60	**483.57**	6.77		**477.97**
TP	4.18	**479.43**	8.32		**475.25**
TP	3.90	**473.59**	9.74		**469.69**
BM no. 3			5.12		**468.47**
	76.08		39.32		
	39.32				468.47
	36.76				431.71
					36.76

32.

		differential levels			
sta	+	HI	–	rod	elev
BM no. 1	**8.35**	455.58			447.23
TP	**10.44**	465.14	0.88		**454.70**
TP	9.56	473.67	**1.03**		464.11
TP	7.16	480.08	0.75		472.92
TP	**10.33**	488.76	1.65		478.43
BM no. 2	9.22	**495.28**	2.70		486.06
TP	4.04	491.41	7.91		484.37
TP	3.51	487.82	7.10		**484.31**
TP	2.83	482.04	**8.61**		479.21
BM no. 3			5.17		476.87
	65.44		35.80		
	35.80				476.87
	29.64				447.23
					29.64

33. See the profile sheet on the following page.

profile levels					
sta	+	HI	−	rod	elev
BM no. 1	0.17	**476.95**			476.78
32+00				0.4	**476.6**
33+00				0.8	**476.2**
34+00				1.2	**475.8**
35+00				3.9	**473.1**
36+00				12.0	**465.0**
+40				1.2	**475.8**
TP	10.72	**483.13**	4.54		**472.41**
37+00				13.3	**469.8**
38+00				5.9	**477.2**
+50				6.8	**476.3**
39+00				5.2	**477.9**
+30				2.3	**480.8**
+85				1.9	**481.2**
40+00				3.0	**480.1**
41+00				8.2	**474.9**
+25				10.1	**473.0**
42+00				9.4	**473.7**
43+00				11.5	**471.6**
BM no. 2			4.87		**478.26**
	10.89		**9.41**		476.78
	9.41				**1.48**
	1.48				

35.

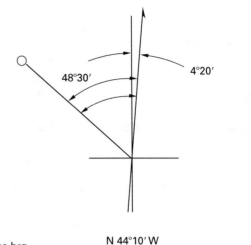

true brg _____ N 44°10′ W

36.

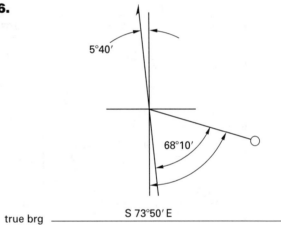

true brg _____ S 73°50′ E

37.

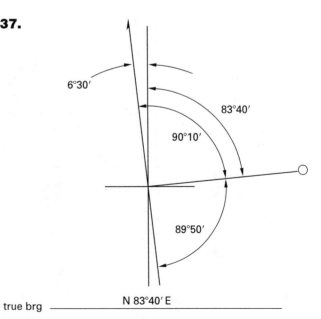

true brg _____ N 83°40′ E

34.

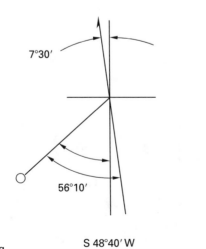

true brg _____ S 48°40′ W

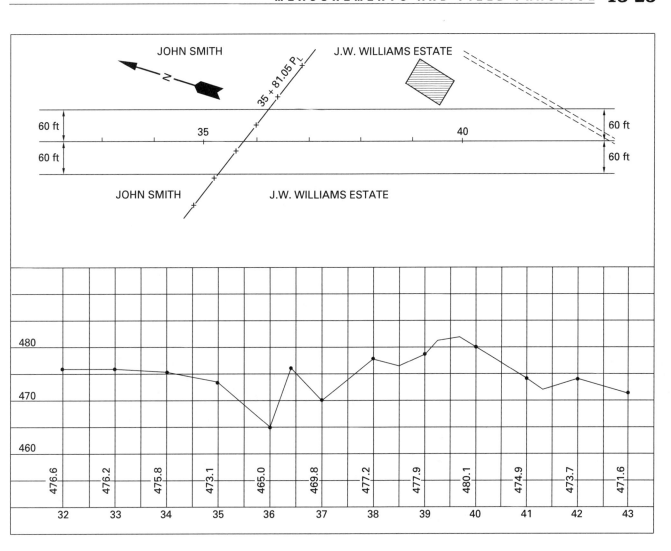

Solution 33: Profile Sheet

Measurements

38.

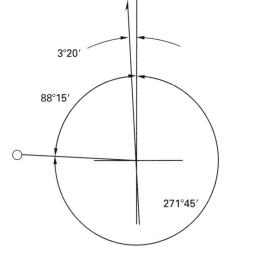

3°20'

88°15'

271°45'

true brg ——————— N 84°55' W ————————

14 The Traverse

1. INTRODUCTION

A *traverse* is a series of lines connecting successive instrument stations of a survey. The relative position of the stations is determined by the direction and length of the lines. In land surveys the lines are the boundaries of the land; in topographic surveys the lines are the control net to which physical features are tied. The traverse is also used as control for construction surveys.

2. OPEN TRAVERSE

An *open traverse* is a series of lines that do not return to the starting point. It is used in route surveying for the location of highways, pipelines, canals, and so on. To check the accuracy of an open traverse, the traverse must start and end at points of known position.

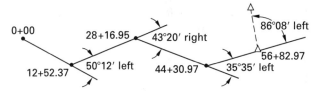

Figure 14.1 Open Traverse

3. CLOSED TRAVERSE

A *closed traverse*, also called a *loop traverse*, starts and ends at the same point. Because it is a closed polygon, the interior angles and the lengths of the sides may be checked for accuracy and mathematically adjusted.

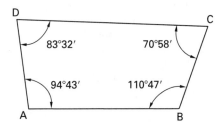

Figure 14.2 Closed Traverse

4. HORIZONTAL ANGLES

Angles measured for open traverses are usually deflection angles as shown in Fig. 14.1. Angles measured for closed traverses are usually interior angles as shown in

Traverse I

Fig. 14.2, but they can be deflection angles as shown in Fig. 14.9.

Interior angles can be turned clockwise (right) or counterclockwise (left), but usually are turned by the method known as *angles to the right*.

5. DEFLECTION ANGLES

A *deflection angle* is an angle between a line and the extension of the preceding line, as shown in Fig. 14.1. It may be turned either right or left from the extension, but the direction of turning must be recorded with the angular measurement. In an open traverse, the straight lines between the points of change in direction are known as *tangents*, and the points of change in direction are known as *points of intersection* (PI).

6. ANGLES TO THE RIGHT

The interior angles of the closed traverse in Fig. 14.2 are known as *angles to the right*. They are measured from the backsight station to the foresight station in a clockwise direction. With the instrument at station A, a backsight was made on station D, the telescope was turned in a clockwise direction to station B, and the angle 94°43′ was read and recorded. With the instrument at station B, the backsight was on station A and foresight was on station C. At each point, the angle was measured in a clockwise direction, or to the right.

The stations in the traverse ABCDA run in a counterclockwise direction, alphabetically, which is appropriate for the angles-to-the-right method. Most theodolites measure angles only to the right, which makes the counterclockwise lettering of the stations necessary. But it is not improper to letter traverse points in a clockwise direction.

7. DIRECTION OF SIDES

The *direction of the sides* of the traverse may be determined if the direction of one of the sides is known. If the direction of none of the sides is known, the direction of one side may be determined by measuring the angle at the intersection of this line and a line outside the traverse that has a known direction, or by making an observation on the sun or the stars. Otherwise, the direction of one of the lines must be assumed.

8. ANGLE CLOSURE

The sum of the interior angles of a polygon depends on the number of sides of the polygon. For a triangle, the sum is 180°; for a four-sided polygon, the sum is 360°. For any polygon with n sides, the sum is

$$\text{sum of angles} = (n-2)(180°) \qquad 14.1$$

After the interior angles of a traverse have been measured, they should be adjusted so that their sum agrees with Eq. 14.1. The error may be distributed evenly at each angle, or it may be distributed arbitrarily in accordance with the surveyor's knowledge of the conditions of the survey. The error should be within the limits allowed in specifications for the survey. In land surveying, these specifications are usually based on the value of the land. Surveys of metropolitan areas are performed at much more rigid standards than are surveys of arid ranch lands.

The interior angles of an arbitrary five-sided traverse are balanced in the following example.

angles as measured	
A	96°03′30″
B	95°19′30″
C	65°13′00″
D	216°19′30″
E	67°06′00″
	540°01′30″

balanced angles $(n-2)(180°) = 540°$	
A	96°03′
B	95°19′
C	65°13′
D	216°19′
E	67°06′
	540°00′

The angles in this traverse were measured with a one-minute vernier, and the balanced angles reflect this accuracy. However, the total discrepancy of 90 sec could be distributed evenly over the five angles by subtracting 18 sec from each recorded angle.

A	96°03′12″
B	95°19′12″
C	65°12′42″
D	216°19′12″
E	67°05′42″
	540°00′00″

9. METHODS OF DESIGNATING DIRECTION

The *direction of a line* is expressed as the angle between a meridian and the line. The *meridian* may be a *true meridian* (a great circle of the earth passing through the poles), a *magnetic meridian* (the direction of which is defined by a compass needle), or a *grid meridian* (established for a plane coordinate system).

The direction of a line may be expressed as its *bearing* or its *azimuth*. *True bearing* or *true azimuth* is measured from true north, referred to as *geodetic north*.

Old land surveys were usually referenced to the magnetic meridian. The direction of a line was determined by reading the angle between the line and the compass needle. This method has long since been discarded for most surveys; modern surveys refer to geodetic north or grid north. But since old surveys must be retraced, an understanding of the magnetic meridian is essential to the land surveyor.

10. BEARING

The *bearing of a line* is the horizontal acute angle between the meridian and the line. Because the bearing of a line cannot exceed 90°, the full horizontal circle is divided into four *quadrants*: northeast, southeast, southwest, and northwest. An angle of 40°, measured between the meridian and a line in each of the four quadrants, is shown in Fig. 14.3

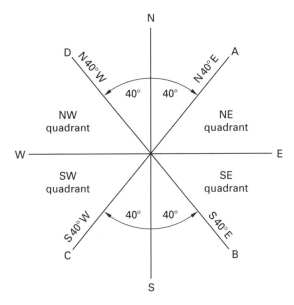

Figure 14.3 *Bearing Quadrants*

Angles are measured from either the north or south, but are never measured from the east or west. The quadrants are designated in the bearing of a line by preceding the angle with North (N) or South (S) and following the angle with East (E) or West (W).

11. BACK BEARING

If the bearing of a line AB is N 65° E, the back bearing of AB is S 65° W. In other words, someone standing at A looking at B is looking northeast; if the person stands at B and looks at A, the person will be looking southwest. The angle with the meridian is the same. The prefix is changed from N to S, and the suffix is changed from E to W. This is illustrated in Fig. 14.4.

When two parallel lines are cut by a transversal, the alternate interior angles are equal. In Fig. 14.4, the meridians through A and B are parallel lines. The line AB is the transversal, and the alternate interior angles

are the bearing angles at A and B. The interior angle at A is the bearing angle of AB, and the interior angle at B is the bearing angle of BA. This also demonstrates that the bearing angle of AB is equal to the back bearing angle. Only the prefix and the suffix differ.

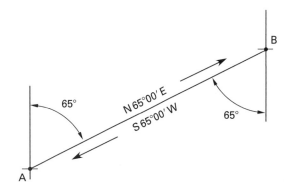

Figure 14.4 *Back Bearing*

12. COMPUTATION OF BEARINGS OF A CLOSED TRAVERSE

Before the directions of the sides of a closed traverse are computed, it is essential that the interior angles of the traverse be adjusted so that their sum agrees with Eq. 14.1.

In Fig. 14.5, the bearing AB is known and the interior angles have been measured and adjusted. The bearings of the other sides must be computed.

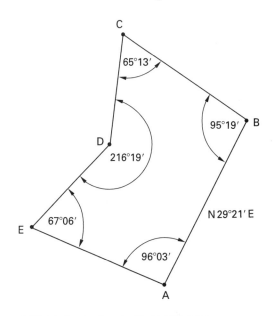

Figure 14.5 *Interior Angles of a Closed Traverse*

The first step in computing the bearing of BC is to draw a sketch around point B showing:

• the meridian through B

- the angle made by BA with the meridian. (The bearing of BA is the back bearing of AB, which is known.)

- the interior angle at B with its field measurement

- the bearing angle of BC, which is the angle to be computed and, by definition, is the angle between the meridian and the line BC

A straight angle is an angle that equals 180°. The meridian makes an angle of 180° at point B. Therefore, the bearing angle of BC is

$$180° - (29°21' + 95°19') = 55°20'$$

From Fig. 14.6, BC bears northwest. Therefore, the bearing of BC is

$$N\,55°20'\,W$$

Computations for bearing of CD, DE, and EA are shown in Fig. 14.7. In each case, a meridian is drawn through the next traverse point after a bearing has been computed. At each point, three angles are identified:

- the angle between the meridian and the preceding side

- the measured interior angle

- the angle between the meridian and the succeeding side

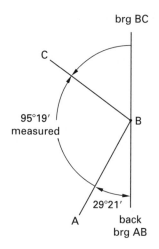

Figure 14.6 *Angles About a Point*

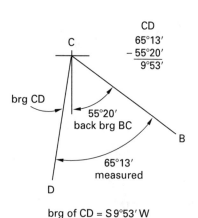

brg of CD = S 9°53' W

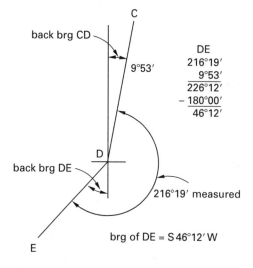

brg of DE = S 46°12' W

brg of EA = S 66°42' E

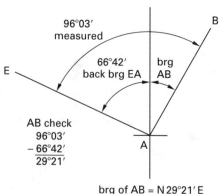

brg of AB = N 29°21' E

Figure 14.7 *Bearing Computations for Traverse in Figure 14.5*

Accuracy of the computations can be checked by calculating the bearing of AB (the known bearing) using the computed bearing of EA. The computed bearing of AB must be equal to the given bearing.

The importance of a sketch at each traverse point cannot be overemphasized. No set rule for computing bearings can be made, but a sketch that properly identifies the three angles mentioned makes a lengthy explanation unnecessary. The meridian through each traverse point always makes an angle of 180°, and the bearing angle is the angle between the meridian and the line (not between an east-west line and the line).

13. AZIMUTH

The *azimuth* of a line is the horizontal angle measured clockwise from the meridian. Azimuth is usually measured from the north.

Azimuths are not limited to 90°. Therefore, there is no need to divide the circle into quadrants. Azimuths vary from 0° to 360°. Fig. 14.8 shows azimuths of 40°, 140°, 220°, and 320°.

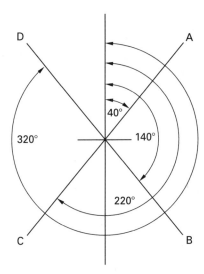

Figure 14.8 *Representative Azimuths*

14. BACK AZIMUTH

As with bearing, the *back azimuth* of a line AB is the azimuth of line BA. The back azimuth of a line may be found by either adding 180° to the azimuth of the line or by subtracting 180° from the azimuth of the line. If the azimuth is less than 180°, subtract 180°.

Referring to Fig. 14.8,

- for azimuth 40°, the back azimuth = 220°
- for azimuth 140°, the back azimuth = 320°
- for azimuth 220°, the back azimuth = 40°
- for azimuth 320°, the back azimuth = 140°

15. CONVERTING BEARING TO AZIMUTH

Bearing is used to give the direction of a course in most land surveys. Azimuth is used in topographic surveys and some route surveys. Bearing may be converted to azimuth, and azimuth may be converted to bearing.

In converting bearing to azimuth in the northeast quadrant, the azimuth angle equals the bearing angle.

$$N\,76°30'\,E = 76°30'$$

In the southeast quadrant, the azimuth angle equals 180° minus the bearing angle.

$$S\,42°28'\,E = 180° - 42°28' = 137°32'$$

In the southwest quadrant, 180° is added to the bearing angle.

$$S\,36°47'\,W = 180° + 36°47' = 216°47'$$

In the northwest quadrant, the bearing angle is subtracted from 360°.

$$N\,62°56'\,W = 360° - 62°56' = 297°04'$$

16. CONVERTING AZIMUTH TO BEARING

In the northeast quadrant, the bearing angle equals the azimuth. The prefix N and the suffix E must be added.

In the southeast quadrant, the azimuth is subtracted from 180° and the prefix S and suffix E are added.

$$168°40' = 180° - 168°40' = S\,11°20'\,E$$

In the southwest quadrant, 180° is subtracted from the azimuth.

$$195°22' = 195°22' - 180° = S\,15°22'\,W$$

In the northwest quadrant, the azimuth is subtracted from 360°.

$$314°35' = 360° - 314°35' = N\,45°25'\,W$$

17. CLOSED DEFLECTION ANGLE TRAVERSE

In a *closed deflection angle traverse* (Fig. 14.9), the difference between the sum of the right deflection angles and the sum of the left deflection angles is 360°. Before bearings are computed, deflection angles must be adjusted.

If, during the adjustment of angles, it is found that the sum of the right deflection angles is greater than the sum of the left deflection angles, the sum of the right deflection angles must be reduced and the sum of the left deflection angles must be increased. The correction may be distributed arbitrarily or evenly.

Bearings of the sides of the traverse are computed in much the same manner as with the interior angle traverse, with a sketch drawn at each traverse point showing the angles involved in the computation.

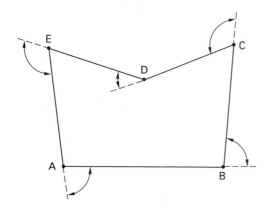

Figure 14.9 *Closed Deflection Angle Traverse*

18. ANGLE-TO-THE-RIGHT TRAVERSE

Open or closed traverses can be run by the angle-to-the-right method. All angles are measured from the backsight to the foresight in a clockwise direction, as shown in Fig. 14.10.

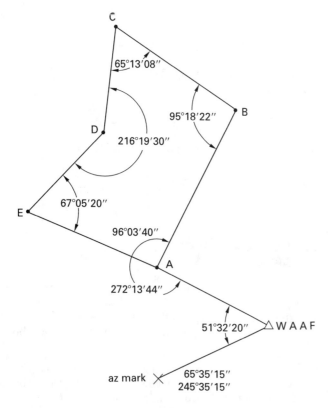

Figure 14.10 *Angle-to-the-Right Traverse*

To determine the forward azimuth from a traverse point, the angle to the right is added to the back azimuth of the preceding line. In other words, the angle to the right is added to the azimuth of the preceding line $\pm 180°$. (It is sometimes necessary to subtract 360°.)

Traverse ABCDEA in Fig. 14.10 is tied to triangulation station WAAF for direction. National Geodetic Survey data show the azimuth to the azimuth mark from station WAAF to be 65°35′15″ (from the south). Converted to azimuth from the north, this azimuth is $180° + 65°35′15″ = 245°35′15″$. Angles to the right, adjusted, are shown in Fig. 14.10. Computations for the azimuths of the traverse courses are shown in Table 14.1.

Table 14.1 *Computation of Azimuth*

245°35′15″	az WAAF-mark	124°39′41″	az CB
51°32′20″	angle right	65°13′08″	angle right
297°07′35″		189°52′49″	az CD
−180°		−180°	
117°07′35″	az A-WAAF	9°52′49″	az DC
272°13′44″	angle right	216°19′30″	angle right
389°21′19″		226°12′19″	az DE
−360°		−180°	
29°21′19″	az AB	46°12′19″	az ED
+180°		67°05′20″	angle right
209°21′19″	az BA	113°17′39″	az EA
95°18′22″	angle right	180°	
304°39′41″	az BC	293°17′39″	az AE
−180°		96°03′40″	angle right
124°39′41″	az CB	389°21′19″	
		−360°	
		check 29°21′19″	az AB

19. LATITUDES AND DEPARTURES

Latitudes and departures are similar in concept to the projections of a line. The projection of a line can be compared to the shadow of a building. When the sun is nearly overhead, the shadow is short; when the sun is sinking in the west, the shadow becomes long. The height has not changed, but the length of its shadow has changed.

In Fig. 14.11, the line AB is projected on the y-axis of a rectangular coordinate system by dropping perpendiculars from A to the y-axis and from B to the y-axis. The interval between these two perpendiculars along the y-axis is the projection of AB on the y-axis. As AB changes its position relative to the y-axis, as shown in Fig. 14.11, the length of the projection becomes longer or shorter. As the position of AB nears the vertical, the projection nears the length of the line. As the projection of AB nears the horizontal, the projection of AB becomes very short.

In Fig. 14.12, AB is projected on the x-axis. As in Fig. 14.11, the length of the projection of AB on the x-axis changes as the position of AB changes relative to the x-axis.

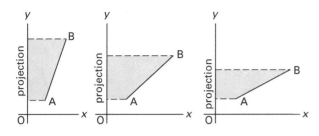

Figure 14.11 *Projections on the y-Axis*

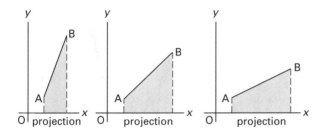

Figure 14.12 *Projections on the x-Axis*

In surveying, the projection of a side of a traverse on the north-south (y-) axis of a rectangular coordinate system is known as its *latitude*. The projection on the east-west (x-) axis is known as its *departure*.

Figure 14.13 shows the projection of a line AB on the y- and x-axes. The latitude of line AB is the length of the right angle projection of AB on a meridian. The departure of line AB is the length of the right angle projection of line AB on a line perpendicular to the meridian, an east-west line.

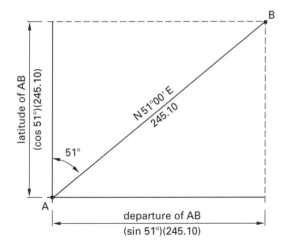

Figure 14.13 *Latitude and Departure*

The *latitude of a course* is equal to the cosine of its bearing angle multiplied by its length. The *departure of a course* is equal to the sine of its bearing angle multiplied by its length.

$$\text{latitude} = (\cos \text{bearing})(\text{length}) \qquad 14.2$$

$$\text{departure} = (\sin \text{bearing})(\text{length}) \qquad 14.3$$

The latitude of the course AB in Fig. 14.13 is

$$\text{latitude} = (\cos 51°00')(245.10 \text{ ft}) = 154.13$$

The departure of the course AB is

$$\text{departure} = (\sin 51°00')(245.10 \text{ ft}) = 190.48$$

Figure 14.14 shows the latitude and departure for each of the courses in the traverse ABCDEA. Course AB has a north latitude and an east departure, BC has a south latitude and an east departure, BC has a south latitude and a west departure, DE has a south latitude and a west departure, and EA has a north latitude and a west departure.

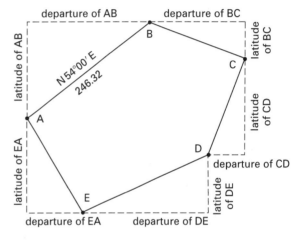

Figure 14.14 *Latitudes and Departures of Traverse Legs*

20. ERROR OF CLOSURE

If the traverse ABCDEA in Fig. 14.14 starts at point A and ends at point A, it is obvious that the distance traversed north equals the distance traversed south. Likewise, the distance traversed east equals the distance traversed west. In other words, the sum of the north latitudes must equal the sum of the south latitudes; and the sum of the east departures must equal the sum of the west departures.

However, because measurements are not exact due to human and instrument errors, these sums will not be equal and their differences may be used to determine the *linear error of closure* due to errors caused by angular and linear measurements. The actual error for each course cannot be determined, but the total error can be distributed over the entire length of the traverse so that north latitudes equal south latitudes and east departures equal west departures. This is called *balancing the traverse* or *closing the traverse*.

The *error of closure* of a traverse is a measure of the precision of a survey.

Traverse I

Figure 14.15 shows a traverse that does not close. Point A is the *point of beginning*. Point A' is the *ending point*, found by plotting the latitude and departure of each course before the traverse is balanced. The distance from A' to A is the error of closure. For the traverse to close, A' would move in the direction of A'A. Balancing the traverse does just that—it makes A' coincide with A.

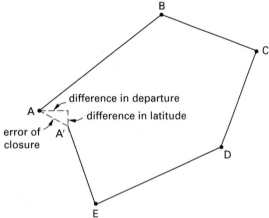

Figure 14.15 *Error of Closure*

The error of closure is found by the Pythagorean theorem.

$$\text{error of closure} = \sqrt{\begin{array}{c}(\text{difference in latitude})^2 \\ + (\text{difference in departure})^2\end{array}}$$

14.4

The direction of the line A'A is found by determining the angle it makes with the meridian. The tangent of this bearing angle is equal to the difference in departures divided by the difference in latitudes.

$$\text{tangent bearing } A'A = \frac{\text{difference in departure}}{\text{difference in latitude}} \quad 14.5$$

21. BALANCING THE TRAVERSE

The traverse can be balanced by using one of several methods: the least squares adjustment, the compass rule, the transit rule, or the Crandall method.

The *least squares method* is adaptable to any traverse, whether angular accuracy is higher, equal to, or less than linear accuracy. It is difficult and seldom used.

The *compass rule*, also known as the *Bowditch rule*, is adaptable to traverses in which angular accuracy and linear accuracy are about the same. In this type of traverse, the compass rule will give very nearly the same results as the least squares adjustment. It is used in a great majority of traverse closures.

The *transit rule* is adaptable to traverses in which angular accuracy is much higher than linear accuracy. It is seldom used.

The *Crandall method* is also used where angular accuracy is much higher than linear accuracy. It is more accurate than the transit rule but requires more time. It, also, is seldom used.

22. THE COMPASS RULE

With the compass rule, the difference in the sums of the north and south latitudes is distributed over the latitudes of the traverse. A correction is made in the latitude of each side to bring the north and south latitudes into balance. The difference in the sums of the east and west departures is distributed in the same way.

The correction to be applied to the latitude of each side is a fraction of the total difference in the north and south latitudes. The fraction is the ratio of the length of each side to the perimeter of the entire traverse.

For example, if the difference in north and south latitudes of a traverse is 0.86, the length of side AB is 356.73, and the perimeter of the traverse is 2156.78, the correction in latitude for the side AB is

$$\text{correction AB} = \left(\frac{356.73}{2156.78}\right)(0.86) = 0.14$$

This can be expressed as a proportion.

$$\frac{\text{correction for latitude of AB}}{\text{difference in N-S latitudes}} = \frac{\text{length of AB}}{\text{traverse perimeter}}$$

14.6

$$\text{correction for AB} = \left(\frac{\text{length AB}}{\text{perimeter}}\right)\left(\begin{array}{c}\text{difference} \\ \text{in latitude}\end{array}\right)$$

14.7

This can be written as Eq. 14.8.

$$\text{correction for AB} = \left(\frac{\begin{array}{c}\text{difference} \\ \text{in latitude}\end{array}}{\text{perimeter}}\right)(\text{length AB})$$

14.8

Using Eq. 14.8 in the example,

$$\text{correction for AB} = \left(\frac{0.86}{2156.78}\right)(356.73) = 0.14 \text{ ft}$$

Equation 14.8 is more efficient because 0.86/2156.78 is a constant that can be applied to the other sides.

After the corrections are made in the latitude and departure for each side, the traverse is balanced.

23. RATIO OF ERROR

The *ratio of error*, or *precision*, of a traverse is the ratio of the error of closure to the perimeter. It is expressed with the numerator as one (1) and the denominator in round numbers. It is a measure of the precision of a traverse.

If the error of closure of a traverse is 0.76 and the perimeter is 5214.75, the ratio of error is

$$\frac{0.76}{5214.75}$$

Dividing the numerator and denominator by 0.76 (in order to have 1 in the numerator), and rounding off,

$$\text{ratio of error} = \frac{1}{6900}$$

This indicates an error of 1 ft per 6900 ft in distance.

24. SUMMARY OF COMPUTATIONS FOR BALANCING A TRAVERSE

The steps in balancing a traverse are summarized as follows.

step 1: Compute the angular error and adjust to make the sum of the angles agree with Eq. 14.1.

step 2: Compute the bearings for each course.

step 3: Compute the latitudes and departures.

step 4: Compute the error of closure.

step 5: Compute the ratio of error.

step 6: Compute the latitude and departure corrections for each course.

step 7: Adjust the latitudes and departures.

Example 14.1

Given the traverse ABCDEA shown with angles adjusted, balance the traverse using the compass rule and compute error of closure and ratio of error.

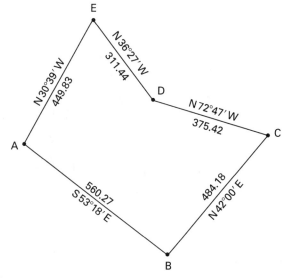

Solution

For each course, the cosine of the bearing angle times the distance has been recorded in one of the two latitude columns. Latitudes of courses with a north prefix

are recorded under N; latitudes of courses with a south prefix are recorded under S.

Likewise, the sine of the bearing angle times the distance has been recorded in one of the two departure columns. Departures of courses with an east suffix are recorded under E; departures of courses with a west suffix are recorded under W.

Next, the N, S, E, and W columns have been added and the totals recorded. The smaller of the two latitude totals has been subtracted from the larger total, and the smaller of the departure totals has been subtracted from the larger.

The error of closure is the square root of the sum of the squares of these two numbers.

The ratio of error, or precision, is the ratio of the error of closure to the perimeter. Ratio of error, in this case, indicates that for every 5000 ft measured, an error of 1 ft was made.

To balance the traverse, the difference in the north and south latitude totals has been prorated to the latitudes of each course. A correction is made for each latitude in proportion to the ratio of the length of the course to the perimeter. Computations for corrections to latitudes and departures are shown in the following table.

course	latitude corrections	departure corrections
AB	$\left(\dfrac{0.37}{2181}\right)(560.27) = -0.10$	$\left(\dfrac{0.24}{2181}\right)(560.27) = -0.06$
BC	$\left(\dfrac{0.37}{2181}\right)(484.18) = +0.08$	$\left(\dfrac{0.24}{2181}\right)(484.18) = -0.05$
CD	$\left(\dfrac{0.37}{2181}\right)(375.42) = +0.06$	$\left(\dfrac{0.24}{2181}\right)(375.42) = +0.04$
DE	$\left(\dfrac{0.37}{2181}\right)(311.44) = +0.05$	$\left(\dfrac{0.24}{2181}\right)(311.44) = +0.04$
EA	$\left(\dfrac{0.37}{2181}\right)(449.83) = -0.08$	$\left(\dfrac{0.24}{2181}\right)(449.83) = +0.05$

In the first table, the sum of the north latitudes is smaller than the sum of the south latitudes. Therefore, corrections for north latitudes are positive and are added to the computed latitudes. Corrections for south latitudes are negative and are subtracted from the computed latitudes.

The sum of the west departures is smaller than the sum of the east departures. Therefore, corrections for west departures were added, and corrections for east departures were subtracted. After corrections were made, north latitudes equal south latitudes, and east departures equal west departures. The traverse is balanced.

Note that the sum of the correction column equals the difference in the sums of the north and south latitudes. Likewise, the departure corrections equal the difference in east and west departures.

point	bearing	distance	cosine sine	latitude N	latitude S	correction	departure E	departure W	correction	latitude N	latitude S	departure E	departure W
A			0.597625										
	S 53°18′ E	560.27	0.801776		334.83	−0.10	449.21		−0.06		334.73	449.15	
B			0.743145										
	N 42°00′ E	484.18	0.669131	359.81		+0.08	323.98		−0.05	359.89		323.93	
C			0.295986										
	N 72°47′ W	375.42	0.955192	111.12		+0.06		358.60	+0.04	111.18			358.64
D			0.804376										
	N 36°27′ W	311.44	0.594121	250.52		+0.05		185.03	+0.04	250.57			185.07
E			0.860298										
	S 30°39′ W	449.83	0.509792		386.99	−0.08		229.32	+0.05		386.91		229.37
A		2181.14		721.45	721.82		773.19	772.95		721.64	721.64	773.08	773.08
					721.45		772.95						
					0.37		0.24						

$$\text{error of closure} = \sqrt{(0.37)^2 + (0.24)^2} = .44$$

$$\text{ratio of error} = 0.44/2181 = 1/5000$$

25. COORDINATES

After latitudes and departures have been computed, co-ordinates of the traverse points are easily computed. Coordinates for one of the points may be known, or they can be assumed. If coordinates are assumed, they should be large enough that no coordinates will be negative.

Coordinates of a point are computed by adding a north latitude to or subtracting a south latitude from the y-coordinate of the preceding point, and by adding an east departure to or by subtracting a west departure from the x-coordinate of the preceding point.

Example 14.2

Latitudes and departures for the traverse ABCDEA have been computed and balanced as shown in the following table. Assume the coordinates of point A to be $y = 1000.00$, $x = 1000.00$. Compute the coordinates of points B, C, D, and E. Check the arithmetic by computing coordinates of point A from coordinates of point E.

Solution

point	latitude north	latitude south	departure east	departure west	coordinates y	coordinates x
A					1000.00	1000.00
		334.73	449.15		−334.73	+449.15
B					665.27	1449.15
	359.89		323.93		+359.89	+323.93
C					1025.16	1773.08
	111.18			358.64	+111.18	−358.64
D					1136.34	1414.44
	250.57			185.07	+250.57	−185.07
E					1386.91	1229.37
		386.91		229.37	−386.91	−229.37
A					1000.00	1000.00
	721.64	721.64	773.08	773.08		

In the solution, north latitudes are given a positive sign and south latitudes are given a negative sign; east departures are given a positive sign and west departures are given a negative sign. y-coordinates are associated with latitude; x-coordinates are associated with departure. The cosine function is associated with latitude; the sine function is associated with departure.

26. FINDING BEARING AND LENGTH OF A LINE FROM COORDINATES

It is often necessary to find the bearing and length of a line between two points of known coordinates. Figure 14.16 illustrates that the tangent of the bearing angle can be determined from the latitude and departure.

$$\text{tangent of bearing angle} = \frac{\text{departure}}{\text{latitude}} \qquad 14.9$$

$$\text{tangent of bearing angle} = \frac{\text{difference in } x}{\text{difference in } y} \qquad 14.10$$

$$\begin{aligned} \text{length} &= \sqrt{(\text{departure})^2 + (\text{latitude})^2} \\ &= \sqrt{(\text{difference in } x)^2 + (\text{difference in } y)^2} \\ &= \frac{\text{latitude}}{\text{cosine bearing}} \\ &= \frac{\text{departure}}{\text{sine bearing}} \end{aligned} \qquad 14.11$$

It is also true that

$$\begin{aligned} \text{cotangent of bearing angle} &= \frac{\text{latitude}}{\text{departure}} \\ &= \frac{\text{difference in } y}{\text{difference in } x} \end{aligned} \qquad 14.12$$

For large angles, Eq. 14.12 is recommended.

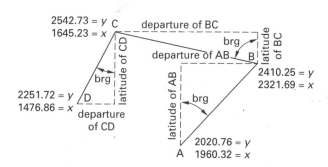

Figure 14.16 *Bearing and Length from Coordinates*

Example 14.3

Using the coordinates of points A, B, C, and D as shown in Fig. 14.16, find the bearing and lengths of AB, BC, and CD.

Solution

$$\tan \text{brg AB} = \frac{2321.69 - 1960.32}{2410.25 - 2020.76}$$

$$\text{brg AB} = \text{N}\,42°51'\,\text{E}$$

AB is northerly because the y-coordinate of B is greater than the y-coordinate of A. AB is easterly because the x-coordinate of B is greater than the x-coordinate of A.

$$\text{AB} = \sqrt{(2321.69 - 1960.32)^2 + (2410.25 - 2020.76)^2}$$
$$= 531.31 \text{ ft}$$

$$\tan \text{brg BC} = \frac{2321.69 - 1645.23}{2542.73 - 2410.25}$$
$$\text{brg BC} = \text{N}\,78°55'\,\text{W}$$

$$\text{BC} = \sqrt{(2321.69 - 1645.23)^2 + (2542.73 - 2410.25)^2}$$
$$= 689.31 \text{ ft}$$

$$\tan \text{brg CD} = \frac{1645.23 - 1476.86}{2542.73 - 2251.72}$$
$$\text{brg CD} = \text{S}\,30°03'\,\text{W}$$

$$\text{CD} = \sqrt{(1645.23 - 1476.86)^2 + (2542.73 - 2251.72)^2}$$
$$= 336.21 \text{ ft}$$

27. COMPUTING TRAVERSES WHERE TRAVERSE POINTS ARE OBSTRUCTED

In land surveying, it is common to find boundary corners occupied by fence posts or other obstructions, making it impractical to retrace courses as they were originally run. In such cases, a traverse can be run very near the original one, and ties can be made to the original corners. The adjacent traverse can be closed, and the original survey can be computed. By making the points on the adjacent traverse very close to the corresponding points on the original survey, error in measurement of the ties is lessened. However, error of closure cannot be computed for the original survey.

The direction and distance from an adjacent point to the corresponding original corner can be determined by measurement, and from this information latitude and departure for this tie can be computed. If the coordinates of each point in the adjacent traverse are known, the coordinates of original corners can be determined by using these coordinates and the latitudes and departures of each of the ties to the original corners.

If the coordinates of the original corners are known, the bearing and length of each course of the original survey can be computed.

$$\frac{\text{tangent}}{\text{bearing}} = \frac{\text{difference in } x}{\text{difference in } y} \qquad 14.13$$

$$\text{length} = \sqrt{(\text{difference in } x)^2 + (\text{difference in } y)^2}$$
$$14.14$$

Example 14.4

The traverse ABCDEA has been run inside the original survey represented by the traverse MNOPQM as shown. The traverse ABCDEA has been closed, and balanced latitudes and departures are shown. Ties to the original survey have been made from corresponding points on the adjacent traverse, and azimuth and distance for each are shown. Coordinates of the point A are assumed to be $x = 1000.00$, $y = 1000.00$.

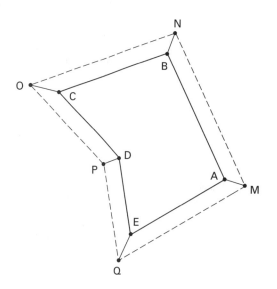

			balanced		balanced	
line	bearing	length	north	south	east	west
AB	N 25°45′ W	560.27	504.60			243.45
BC	S 69°33′ W	484.18		169.19		453.70
CD	S 45°14′ E	375.42		264.40	266.51	
DE	S 08°54′ E	311.44		307.70	48.16	
EA	N 58°11′ E	449.83	236.69		382.48	
			741.29	741.29	697.15	697.15

Corner Ties

line	azimuth	length
AM	106°20′	58.30
BN	33°52′	64.65
CO	271°59′	71.22
DP	256°11′	72.30
EQ	203°47′	77.35

Solution

The coordinates of points B, C, D, and E are computed as follows.

	latitude		departure		coordinates	
point	north	south	east	west	y	x
A					1000.00	1000.00
	504.60			243.45		
B					1504.60	756.55
	169.19			453.70		
C					1335.41	302.85
		264.40	266.51			
D					1071.01	569.36
		307.70	48.16			
E					763.31	617.52
	236.69		382.48			
A					1000.00	1000.00

After the coordinates of the traverse points of ABCDEA are computed, latitudes and departures of the ties to each survey corner are determined. Then the coordinates of each survey corner are computed. Directions of the ties have been recorded in azimuth, so these azimuths need to be converted to bearings. The tie from traverse point A to survey corner M is shown below.

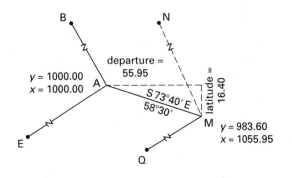

The coordinates of M are

$$\text{bearing AM} = 180° - 106°20' = \text{S } 73°40' \text{ E}$$

$$\text{latitude AM} = 58.30 \cos 73°40' = 16.40 \text{ ft}$$

$$\text{departure AM} = 58.30 \sin 73°40' = 55.95 \text{ ft}$$

$$y\text{-coordinate of M} = 1000.00 - 16.40 = 983.60$$

$$x\text{-coordinate of M} = 1000.00 + 55.95 = 1055.95$$

Computations of the coordinates of N, O, P, and Q are performed in a similar manner. Tabulations showing data used in these computations are shown in the table on the following page.

With the coordinates of M and N known, the bearing and length of MN are computed.

$$\tan \text{brg MN} = \frac{1055.95 - 792.58}{1558.28 - 983.60}$$

$$\text{brg MN} = \text{N } 24°37' \text{ W}$$

$$\text{length MN} = \sqrt{(263.37)^2 + (574.68)^2} = 632.16 \text{ ft}$$

Tabulations of bearings and lengths of NO, OP, PQ, and QM are shown in the table on the following page.

Example 14.5

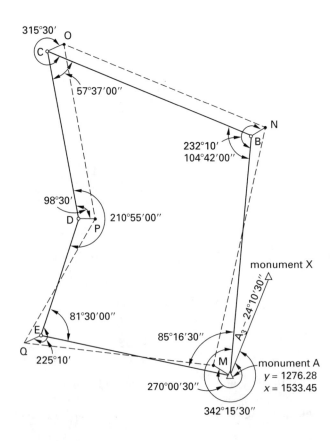

point	bearing	length	latitude north	latitude south	departure east	departure west	coordinates y	coordinates x
A							1000.00	1000.00
	S 73°40′ E	58.30		16.40	55.95			
M							983.60	1055.95
B							1504.60	756.55
	N 33°52′ E	64.65	53.68		36.03			
N							1558.28	792.58
C							1335.41	302.85
	N 88°01′ W	71.22	2.46			71.18		
O							1337.87	231.67
D							1071.01	569.36
	S 76°11′ W	72.30		17.27		70.21		
P							1053.74	499.15
E							763.31	617.52
	S 23°47′ W	77.35		70.78		31.19		
Q							692.53	586.33

computation of traverse MNOPQM

point	bearing	length	latitude north	latitude south	departure east	departure west	coordinates y	coordinates x
M							983.60	1055.95
	N 24°37′ W	632.16	574.68			263.37		
N							1558.28	792.58
	S 68°33′ W	602.66		220.41		560.91		
O							1337.87	231.67
	S 43°16′ E	390.22		284.13	267.48			
P							1053.74	499.15
	S 13°34′ E	371.58		361.21	87.18			
Q							692.53	586.33
	N 58°12′ E	552.51	291.07		469.62			
M							983.60	1055.95
			865.75	865.75	824.28	824.28		

The tract of land represented by the figure MNOPQM above has been resurveyed. It is completely enclosed by a fence, so that corners could not be occupied. The traverse ABCDEA was run as close to the survey as possible. Point A is a monument of known position, and the azimuth from A to monument X is known to be 24°10′30″. The traverse was tied to this line for direction. All angles were turned to the right, beginning with the instrument at A with backsight on X, using the known azimuth. The coordinates of A are known to be $y = 1276.28$, $x = 1533.45$, and coordinates of the traverse points and the original survey corners were referred to the coordinates of A. The interior angle at each point of the traverse and the angle to the survey corner were turned from the same instrument setup in the order shown in the field notes. Horizontal distances were also measured in the sequence shown in the notes.

Field Notes

⊼ station	angle right	measured angle	adjusted angle	distance from-to	distance
A	X-M	270°00′30″		A-M	85.20
A	X-B	342°15′30″		A-B	986.10
A	E-B	85°16′30″	85°16′00″		
B	A-C	104°42′00″	104°42′00″	B-N	72.67
B	A-N	232°10′00″		B-C	930.01
C	B-D	57°37′00″	57°37′00″	C-O	81.31
C	B-O	315°30′00″		C-D	690.70
D	C-P	98°30′00″		D-P	70.54
D	C-E	210°55′00″	210°55′00″	D-E	510.22
E	D-A	81°30′00″	81°30′00″	E-Q	77.80
E	D-Q	225°10′00″	_____	E-A	809.00
			540°00′00″		

Traverse I

Solution

The interior angles were adjusted and recorded in the field notes. Using the azimuth of A-X, azimuths of the traverse sides and corner ties were computed and converted to bearings. Computations are shown in the first two of the following tables.

Traverse closure and coordinate computations for the traverse ABCDEA are shown in the third table. Computations for coordinates of M, N, O, P, and Q and for bearings and lengths of MN, NO, OP, and PQ are shown in the last table.

Traverse Bearing Computations

point	angle right	azimuth	bearing
X			
		24°10′30″	
A	342°15′30″		
		6°26′00″	N 06°26′ E
B	104°42′00″		
		291°08′00″	N 68°52′ W
C	57°37′00″		
		168°45′00″	S 11°15′ E
D	210°55′00″		
		199°40′00″	S 19°40′ W
E	81°30′00″		
		101°10′00″	S 78°50′ E
A	85°16′00″		
		6°26′00″	check
B			

Corner Tie Bearing Computation

point	angle right	azimuth	bearing
X			
		24°10′30″	
A	270°00′30″		
		294°11′00″	N 65°49′ W
M			
A			
		6°26′00″	
B	232°10′00″		
		58°36′00″	N 58°36′ E
N			
B			
		291°08′00″	
C	315°30′00″		
		66°38′00″	N 66°38′ E
O			
C			
		168°45′00″	
D	98°30′00″		
		87°15′00″	N 87°15′ E
P			
D			
		199°40′00″	
E	225°10′00″		
		244°50′00″	S 64°50′ W
Q			

Traverse Computation

point	bearing	length	latitude N	latitude S	cor	departure E	departure W	cor	latitude N	latitude S	departure E	departure W	coordinates y	coordinates x
A													1276.28	1533.45
	N 06°26′ E	986.10	979.89		−0.16	110.49		−0.06	979.73		110.55			
B													2256.01	1644.00
	N 68°52′ W	930.01	335.31		−0.15		867.46	−0.06	335.16			867.40		
C													2591.17	776.60
	S 11°15′ E	690.70		677.43	+0.11	134.75		+0.04		677.54	134.79			
D													1913.63	911.39
	S 19°40′ W	510.22		480.46	+0.08		171.71	−0.03		480.54		171.68		
E													1433.09	739.71
	S 78°50′ E	809.00		156.67	+0.14	793.68		+0.06		156.81	793.74			
A													1276.28	1533.45
		3926.03	1315.20	1314.56		1038.92	1039.17		1314.89	1314.89	1039.08	1039.08		
			1314.56				1038.92							
			0.64				0.25							

$$\text{error of closure} = \sqrt{(0.64)^2 + (0.25)^2} = 0.69$$

$$\text{precision} = \frac{0.69}{3926.03} = \frac{1}{5700}$$

28. LATITUDES AND DEPARTURES USING AZIMUTH

In the preceding example, it was not necessary to convert azimuth to bearing; latitudes and departures can be found from azimuth with calculators that will give the algebraic sign of the latitudes and departures.

Example 14.6

Given the azimuths and lengths of the sides of the traverse ABCDEA as shown tabulated, compute latitudes and departures, balance the traverse, and compute the errors of closure and precision.

line	azimuth	length
AB	29°21′23″	560.06
BC	304°39′45″	484.14
CD	189°52′53″	375.48
DE	226°12′23″	311.53
EA	113°17′43″	449.79

Solution

line	azimuth	length	latitude	departure	balanced latitude	balanced departure
AB	29°21′23″	560.06	+488.14	+274.56	+488.11	+274.52
BC	304°39′45″	484.14	+275.35	−398.21	+275.33	−398.25
CD	189°52′53″	375.48	−369.91	−64.44	−369.93	−64.47
DE	226°12′23″	311.52	−215.59	−224.87	−215.61	−224.89
EA	113°17′43″	449.79	−177.88	+413.12	−177.90	+413.09
		2180.99	+0.11	+0.16	0.00	0.00

$$\text{error of closure} = \sqrt{(0.11)^2 + (0.16)^2} = 0.19$$

$$\text{precision} = \frac{0.19}{2180.99} = 1/11{,}500$$

29. ROUTE LOCATION BY DEFLECTION ANGLE TRAVERSE

Deflection angle traverses are suitable for highway locations because the deflection angle at the point of intersection of two tangents along the centerline of a

Summary of Ex. 14.5

point	bearing	length	latitude north	latitude south	departure east	departure west	coordinates y	coordinates x
				computations of coordinates				
A							1276.28	1533.45
	N 65°49′ W	85.20	34.90			77.72		
M							1311.18	1455.73
B							2256.01	1644.00
	N 58°36′ E	72.67	37.86		62.03			
N							2293.87	1706.03
C							2591.17	776.60
	N 66°38′ E	81.31	32.25		74.64			
O							2623.42	851.24
D							1913.63	911.39
	N 87°15′ E	70.54	3.38		70.46			
P							1917.01	981.85
E							1433.09	739.71
	S 64°50′ W	77.80		33.08		70.41		
Q							1400.01	669.30
				computation of MNOPQM				
M							1311.18	1455.73
	N 14°17′ E	1014.07	982.69		250.30			
N							2293.87	1706.03
	N 68°55′ W	916.12	329.55			854.79		
O							2623.42	851.24
	S 10°28′ E	718.38		706.41	130.61			
P							1917.01	981.85
	S 31°09′ W	604.13		517.00		312.55		
Q							1400.01	669.30
	S 83°33′ E	791.43		88.83	786.43			
M							1311.18	1455.73

highway is equal to the central angle of the circular arc that is inserted to connect two tangents. Straight sections along the centerline are known as *tangents*, and circular arcs are known as *simple curves*. Curves are not always circular arcs, but normally they are. Curves will be discussed in detail in Chap. 17.

The deflection angle traverse shown in Fig. 14.17 begins at point A (a point on the line XA, the azimuth of which is 346°06′) and ends at point E (on the line YE, the azimuth of which is 17°34′). The azimuths of these two lines are used to check the angular closure of the traverse. The azimuth of AB is found by adding the deflection angle at A to the azimuth XA. Azimuths of the other lines of the traverse are found by adding right deflection angles to the forward azimuth of the preceding line and subtracting left deflection angles from the forward azimuth of the preceding line.

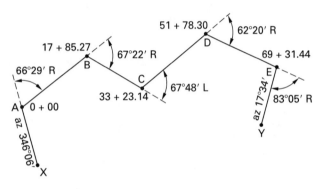

Figure 14.17 *Deflection Angle Traverse and Calculations*

Table 14.2 *Calculations for Fig. 14.17*

line	azimuth	correction	adjusted azimuth	adjusted bearing
XA	346°06′	fixed	346°06′00″	
	+66°29′			
	412°35′			
	−360°00′			
AB	52°35′	−0′30″	52°34′30″	N 52°34′30″ E
	+67°22′			
BC	119°57′	−1′00″	119°56′00″	S 60°04′00″ E
	−67°48′			
CD	52°09′	−1′30″	52°07′30″	N 52°07′30″ E
	+62°20′			
DE	114°29′	−2′00″	114°27′00″	S 65°33′00″ E
	+83°07′			
EY	197°36′	fixed	197°34′00″	S 17°34′00″ W
	−180°00′			
YE	+17°36′			
	−17°34′			
	+02′ = closure error			

30. CONNECTING TRAVERSE

Figure 14.18 shows a *connecting traverse* between triangulation station WAAF and triangulation station PRICE. Traverse stations A, B, and C are established as part of the connecting traverse. *x*- and *y*-coordinates for these stations are to be computed.

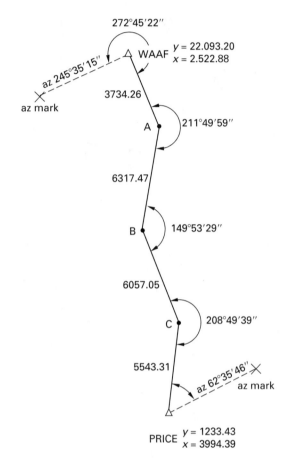

Figure 14.18 *Connecting Traverse*

Coordinates for triangulation stations WAAF and PRICE are known and shown in Fig. 14.18. Azimuth from each triangulation station to an azimuth mark is also known and shown in Fig. 14.18.

Computations for determining coordinates of stations A, B, and C are shown in Tables 14.3 through 14.5.

Table 14.3 shows computations for direction of traverse courses. The direction of the WAAF azimuth mark is fixed, as is the direction of the PRICE azimuth mark. Angular closure was found to be 0′35″, which was distributed evenly to angles at stations WAAF, A, B, C, and PRICE.

Table 14.4 shows traverse computations with the error of closure and precision.

Table 14.5 shows latitudes and departures balanced by the compass rule and coordinates of stations A, B, and C.

Table 14.3 *Computations of Directions for Fig. 14.18*

azimuth		correction	adjusted azimuth
245°35′15″	WAAF-mk		245°35′15″
+272°45′22″	angle right	−07″	272°45′15″
518°20′37″			518°20′30″
−360°00′00″			360°00′00″
158°20′37″	WAAF-A		158°20′30″
+211°49′59″	angle right	−07″	211°49′52″
370°10′36″			370°10′22″
−180°00′00″			180°00′00″
190°10′36″	A-B		190°10′22″
+149°53′29″	angle right	−07″	149°53′22″
340°04′05″			340°03′44″
−180°00′00″			180°00′00″
160°04′05″	B-C		160°03′44″
+208°49′39″	angle right	−07″	208°49′32″
368°53′44″			368°53′16″
−180°00′00″			180°00′00″
188°53′44″	C-PRICE		188°53′16″
+ 53°42′37″	angle right	−07″	53°42′30″
+242°36′21″			242°35′46″
−180°00′00″			180°00′00″
62°36′21″	PRICE-mk	check	62°35′46″
− 62°35′46″	PRICE-mk		
clos = +35″			

Table 14.4 *Traverse Computations for Fig. 14.18*

line	azimuth	length	latitude	departure
WAAF-A	158°20′30″	3734.26	−3470.63	+1378.21
A-B	190°10′22″	6317.47	−6218.16	−1115.77
B-C	160°03′44″	6057.05	−5694.01	+2065.45
C-PRICE	188°53′16″	5543.31	−5476.75	−856.44
		21652.09	−20859.55	+1471.45
			+20859.77	−1471.51
			+0.22	−0.06

$$\text{error of closure} = \sqrt{(0.22)^2 + (0.06)^2} = 0.23$$

$$\text{precision} = 1/94{,}000$$

Table 14.5 *Coordinate Computations for Fig. 14.18*

station	balanced latitude	balanced departure	coordinates	
			y	x
WAAF			22,093.20	2522.88
	−3470.67	+1378.22		
A			18,622.53	3901.10
	−6218.22	−1115.75		
B			12,404.31	2785.35
	−5694.07	+2065.47		
C			6710.24	4950.82
	−5476.81	−856.43		
PRICE			1233.43	3994.39
	−20,859.77	+1471.51		

31. ERRORS IN TRAVERSING

Errors made in linear and angular measurements in traversing are of two types: (a) *systematic errors* and (b) *accidental errors*, often called *random errors*.

Systematic errors are inherent in surveying instruments such as steel tapes, EDM instruments, theodolites, and engineer's levels. These errors can be computed and corrected. They can be prevented or minimized by using calibrated tapes, by repeating and averaging horizontal angular measurements, and by balancing backsight and foresight distances in leveling.

32. SYSTEMATIC ERRORS IN TAPING

A steel tape that is calibrated to be 100.01 ft long introduces a systematic positive error of 0.01 ft each time it is used; a tape calibrated to be 99.99 ft long introduces a systematic negative error of 0.01 ft each time it is used. Changes in length brought about by temperature changes can be computed and the algebraic sign of the correction determined, depending on whether the temperature was below or above the calibration temperature.

33. SYSTEMATIC ERRORS IN ANGULAR MEASUREMENT

Systematic errors in angular measurement occur because of the inherent inaccuracy of the transit or theodolite. The angular closure for the interior angles of a traverse is expressed by the formula

$$c = k\sqrt{n} \qquad \text{14.15}$$

n = number of angles, and k = the least count of the instrument circle.

For a five-sided closed traverse measured with a 20 in theodolite, the permissible closure error is

$$20 \text{ in}\sqrt{5} = 45 \text{ in} \qquad \text{14.16}$$

For a closure greater than 45 in, the surveyor of this traverse should examine the notes for an accidental error.

34. ACCIDENTAL (RANDOM) ERRORS

Accidental errors are not always easily found and cannot be corrected by known formulas. They may be large or small; they may be positive or negative. A 100 ft station may be inadvertently dropped or added in taping, numbers in angle measurements may be transposed in recording, and so on. Large accidental errors are sometimes referred to as *blunders*.

35. LOCATING ERRORS IN A TRAVERSE

After a traverse is plotted, large errors are often found by close inspection of the plot. If there is only one large

Traverse I

error in angle measurement, the perpendicular bisector of the line of closure will point to the traverse station where the error was probably made.

In Fig. 14.19, the perpendicular bisector of the closure line A'A points toward station B as the possible point of angular error. If the perpendicular bisector points toward two traverse stations, it is possible that there are two accidental errors.

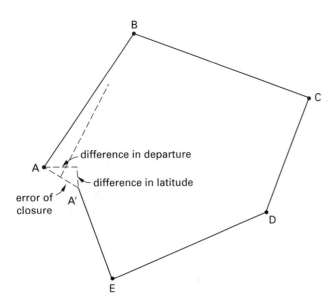

Figure 14.19 *Locating an Error in a Traverse*

If there are no accidental errors in angle measurement, the line of closure may nearly parallel the side that contains an accidental error in distance. The length of the closure line may indicate the magnitude of the accidental error. In Fig. 14.19, the side BC would be suspect.

36. INTERSECTIONS OF TRAVERSE LINES

The method for finding the point of intersection of two traverse lines using analytic geometry was explained in Chap. 12. It is more common, however, to determine the point of intersection by using trigonometry. There are three trigonometric methods for determining the point of intersection: (a) the bearing-bearing method, (b) the bearing-distance method, and (c) the distance-distance method. Selection of the method to use depends on the known information.

37. BEARING-BEARING METHOD OF DETERMINING INTERSECTIONS

If the coordinates of the end points of one side of a triangle are known and the bearings of the other two sides are known, the coordinates of the point of intersection of the other two sides can be found by using the *bearing-bearing* method.

Example 14.7

Find the coordinates of the point of intersection of the sides AC and BC of the triangle ABC, given the information shown.

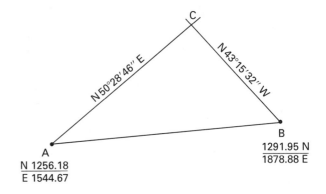

Solution

$$\text{tan brg AB} = \frac{1878.88 - 1544.67}{1291.95 - 1256.18}$$

$$\text{brg AB} = \text{N}\,83°53'28''\,\text{E}$$

$$\text{dist AB} = \sqrt{\begin{array}{l}(1878.88 - 1544.67)^2 \\ + (1291.95 - 1256.18)^2\end{array}}$$

$$= 336.12$$

$$\text{angle } A = 33°24'42''$$

$$\text{angle } B = 52°51'00''$$

$$\text{angle } C = \frac{93°44'18''}{180°00'00''}$$

$$AC = \frac{336.12 \sin 52°51'00''}{\sin 93°44'18''} = 268.48$$

$$BC = \frac{336.12 \sin 33°24'42''}{\sin 93°44'18''} = 185.48$$

$$\text{north coord. of C} = 1256.18 + 268.48 \cos 50°28'46''$$

$$= 1427.03$$

$$\text{east coord. of C} = 1544.67 + 268.48 \sin 50°28'46''$$

$$= 1751.77$$

38. BEARING-DISTANCE METHOD OF DETERMINING INTERSECTIONS

If the coordinates of the end points of one side of a triangle are known and the bearing of one of the other sides and the distance of the third are known the coordinates of the point of intersection of the latter two sides can be found using the *bearing-distance* method.

Example 14.8

Find the coordinates of the point of intersection of the sides AC and BC of the triangle ABC, given the information shown.

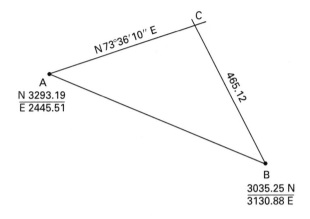

Solution

$$\tan \text{brg AB} = \frac{3130.88 - 2445.51}{3293.19 - 3035.25}$$

$$\text{brg AB} = \text{S}\,69°22'34''\,\text{E}$$

$$\text{dist AB} = \sqrt{\begin{array}{c}(3130.88 - 2445.51)^2 \\ + (3293.19 - 3035.25)^2\end{array}}$$

$$= 732.30$$

$$\text{angle } A = 37°01'16''$$

Note: When two sides of a triangle and the angle opposite one of them are known (the SSA case), it is possible to construct two triangles from the known data. In this case, if an arc with a radius of 465.12 ft is swung from point B, it will intersect a line from A with bearing N 73°36'10'' at two points. (See Ex. 11.19.)

$$\sin C' = \frac{732.30 \sin 37°01'16''}{465.12}$$

$$C' = 71°26'17''$$

$$\text{angle } C = 180° - 71°26'17'' = 108°33'43''$$

$$\text{angle } B = 34°25'01''$$

$$\text{north coord. of C} = 3293.19 + (436.62)(\cos 73°36'10'')$$

$$= 3416.45$$

$$\text{east coord. of C} = 2445.51 + (436.62)(\sin 73°36'10'')$$

$$= 2864.37$$

39. DISTANCE-DISTANCE METHOD OF DETERMINING INTERSECTIONS

If the coordinates of the end points of one side of a triangle are known and the distances of the other two sides are known, the coordinates of the point of intersection of the other two sides can be found by using the *distance-distance* method.

Example 14.9

Find the coordinates of the point of intersection of sides AC and BC given the information shown.

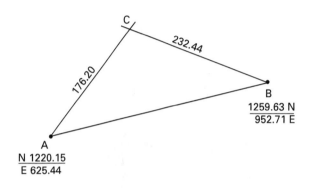

Solution

$$\tan \text{brg AB} = \frac{952.71 - 625.44}{1295.63 - 1220.15}$$

$$\text{brg AB} = \text{N}\,77°00'46''\,E$$

$$\text{dist AB} = \sqrt{\begin{array}{c}(952.71 - 625.44)^2 \\ + (1295.63 - 1220.15)^2\end{array}}$$

$$= 335.86$$

$$\cos A = \frac{\text{AC}^2 + \text{AB}^2 - \text{BC}^2}{(2)(\text{AC})(\text{AB})}$$

$$\qquad - \frac{(176.20)^2 + (335.86)^2 - (232.44)^2}{(2)(176.20)(335.86)}$$

$$\text{angle } A = 40°38'01''$$

$$\text{brg AC} = \text{N}\,36°22'45''\,E$$

$$\text{north coord. of C} = 1220.15 + 176.20 \cos 36°22'45''$$

$$= 1362.01$$

$$\text{east coord. of C} = 625.44 + 176.20 \sin 36°22'45''$$

$$= 729.95$$

Traverse I

PRACTICE PROBLEMS

1. Adjust the interior angles of the following closed traverses arbitrarily.

(a)

point	measured angle
A	67°06′30″
B	216°19′
C	65°12′30″
D	95°18′30″
E	96°02′

(b)

point	measured angle
A	92°38′
B	117°21′30″
C	129°13′
D	261°44′30″
E	55°28′30″
F	207°10′
G	70°52′30″
H	145°34′

2. Adjust the interior angles of the following closed traverse by applying the same correction to each angle.

point	measured angle
A	90°19′59″
B	106°19′55″
C	318°51′26″
D	48°29′12″
E	150°55′29″
F	195°44′47″
G	87°11′56″
H	193°08′23″
I	212°18′51″
J	47°27′37″
K	288°30′28″
L	60°43′21″

3. Compute the bearings of the sides of the traverse ABCDEA. Interior angles are as shown. Draw a sketch for each traverse station showing the two known angles and designate the unknown bearing angle with a question mark. Show computations at each station. Bearing of AB is N 15°22′ E.

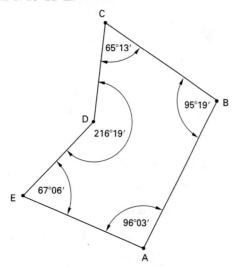

4. Compute the bearings of the sides of the traverse ABCDEA. Interior angles are as shown. Draw a sketch for each traverse station showing the two known angles and designate the unknown bearing angle with a question mark. Show computations at each station. Bearing of AB is N 65°04′ W.

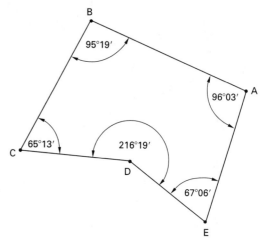

5. Use the figure showing the traverse ABCDEA to compute the interior angles of the traverse. Draw a meridian (north line) through each traverse station, identify the angles at each station that are used in the computations, and show the computations. Then record the interior angles on the figure.

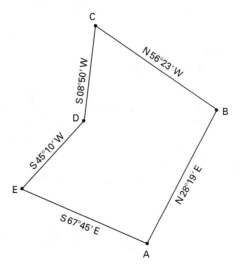

6. Using a protractor and a scale, plot the traverse ABCDEFGHA to a scale of 1 in = 100 ft.

AB:	N 21°12′ E, 289.07
BC:	S 78°41′ E, 366.63
CD:	S 25°24′ W, 228.60
DE:	S 35°23′ E, 232.44
EF:	S 47°51′ W, 281.24
FG:	N 78°30′ W, 316.68
GH:	N 29°29′ E, 192.13
HA:	N 25°57′ W, 174.36

7. On a Cartesian *x-y* plane, draw lines with azimuths: OA = 30°, OB = 60°, OC = 105°, OD = 135°, OE = 180°, OF = 210°, OG = 265°, and OH = 270°, and OI = 315°. Show the back azimuth of each line.

8. Traverse ABCDEA is tied to station WAAF for direction. Azimuth to the azimuth mark from station WAAF is 14°37′19″ (from the north). The angle to the right at station WAAF (az mk to A) is 21°14′46″. Angles to the right at other stations are shown. Compute the azimuth of each side of the travese.

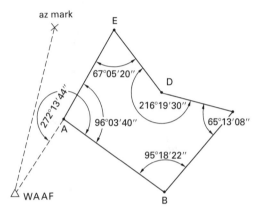

9. Convert the following bearings to azimuths.

(a) S 85°13′16″ W   (b) N 74°24′01″ W
(c) S 08°19′19″ E   (d) N 84°28′13″ E
(e) S 83°03′28″ E   (f) N 07°26′33″ W
(g) N 27°57′45″ E   (h) S 05°17′25″ E
(i) S 04°18′12″ W   (j) N 57°08′02″ W

10. Convert the following azimuths to bearings.

(a) 291°37′06″   (b) 12°13′47″
(c) 106°12′46″   (d) 232°31′18″
(e) 93°04′02″   (f) 65°11′37″
(g) 337°15′11″   (h) 267°25′51″
(i) 102°27′38″   (j) 317°40′53″

11. Determine the latitude and departure of each side of the traverse ABCDA and record in the proper column. Total the north and south latitudes and the east and west departures.

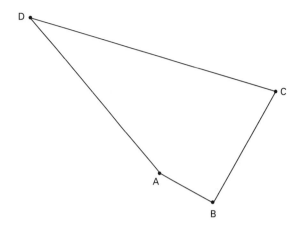

point	bearing	length
A		
	S 60°00′ E	50.0
B		
	N 30°00″ E	100.0
C		
	N 75°00′ W	200.0
D		
	S 41°30′ E	151.0
A		

12. Compute the error of closure and ratio of error, and balance the traverse by the compass rule. Show computations for corrections for each side of the traverse.

point	bearing	length
A		
	N 28°19′ E	560.27
B		
	N 56°23′ W	484.18
C		
	S 08°50′ W	375.42
D		
	S 45°10′ W	311.44
E		
	S 67°45′ E	449.83
A		

13. Balance the following traverse by the compass rule.

point	bearing	length
A		
	S 53°18′ E	560.27
B		
	N 42°00′ E	484.18
C		
	N 72°47′ W	375.42
D		
	N 36°27′ W	311.44
E		
	S 30°39′ W	449.83
A		

Traverse I

14. Balance the following traverse.

point	bearing	length
A		
	N 80°00′ W	1015.43
B		
	S 66°30′ W	545.52
C		
	S 12°00′ E	480.97
D		
	S 88°30′ E	750.26
E		
	N 69°00′ E	639.18
F		
	N 10°00′ E	306.28
A		

15. Given the balanced latitudes and departures of the traverse ABCDEA with the coordinates of station A being $y = 9012.34$ and $x = 1234.56$, determine the coordinates of stations B, C, D, and E.

pt	bearing	length	latitude N	latitude S	departure E	departure W	coordinates y	coordinates x
A							9012.34	1234.56
	N 28°19′ E	560.27	493.12		265.66			
B								
	N 56°23′ W	484.18	267.97			403.29		
C								
	S 08°50′ W	375.42		371.04		57.72		
D								
	S 45°10′ W	311.44		219.64		220.91		
E								
	S 67°45′ E	449.83		170.41	416.26			
A								

16. Given the open traverse ABCD with coordinates of A, B, C, and D as shown, compute the bearings and lengths of AB, BC, and CD.

point	coordinates y	coordinates x
A	9090.90	9090.90
B	9480.39	9452.27
C	9612.87	8890.70
D	9321.86	8775.81

17. The traverse ABCDEA has been run inside the tract MNOPQM because corners of the tract cannot be occupied. The traverse has been balanced, and coordinates of each traverse station have been computed. Ties to each corner of the tract have been made from the adjacent traverse station and are shown. Compute the bearing and length of each course of the trace MNOPQM.

pt	bearing	length	latitude N	latitude S	departure E	departure W	coordinates y	coordinates x
A							1000.00	1000.00
	N 28°19′ E	560.27	493.12		265.66			
B							1493.12	1265.66
	N 56°23′ W	484.18	267.97			403.29		
C							1761.09	862.37
	S 08°50′ W	375.42		371.04		57.72		
D							1390.05	804.65
	S 45°10′ W	311.44		219.64		220.91		
E							1170.41	583.74
	S 67°45′ E	449.83		170.41	416.26			
A							1000.00	1000.00

corner ties

line	azimuth	length
AM	167°58′	59.20
BN	88°13′	65.13
CO	346°08′	70.85
DP	303°51′	72.45
EQ	258°31′	76.51

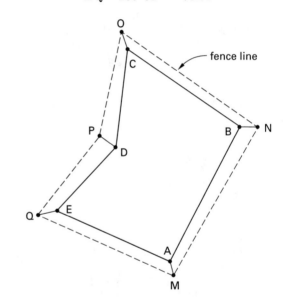

18. The traverse ABCDEA has been run and ties to the corners of the tract MNOPQM, which cannot be occupied, have been made. The traverse has been balanced and coordinates of each traverse station have been computed as shown. Compute the bearing and length for each course of the tract MNOPQM.

pt	bearing	length	latitude N	latitude S	departure E	departure W	coordinates y	coordinates x
A							1000.00	1000.00
	N 28°19′ E	560.27	493.12		265.66			
B							1493.12	1265.66
	N 56°23′ W	484.18	267.97			403.29		
C							1761.09	862.37
	S 08°50′ W	375.42		371.04		57.72		
D							1390.05	804.65
	S 45°10′ W	311.44		219.64		220.91		
E							1170.41	583.74
	S 67°45′ E	449.83		170.41	416.26			
A							1000.00	1000.00

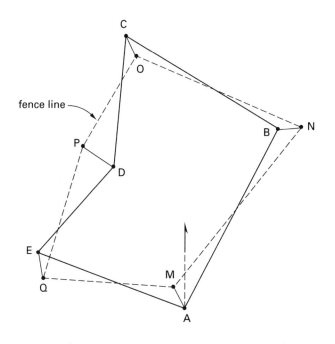

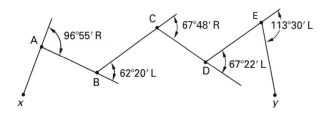

	corner ties	
line	azimuth	length
AM	331°12′	58.46
BN	87°33′	66.72
CO	152°13′	68.89
DP	304°31′	74.13
EQ	166°22′	66.52

21. Traverse FOX-A-B-C-DOG connects triangulation stations FOX and DOG, coordinates of which are known and shown. The direction FOX-az mk and DOG-az mk are fixed. The traverse stations A, B, and C were set. Find the angular closure, correct azimuths, and compute the coordinates of A, B, and C. Az FOX-az mk = 246°45′46″. Az DOG-az mk = 69°45′15″.

line	length
FOX	
−A	5543.31
AB	6057.05
BC	6317.47
C−	3734.26
DOG	

19. Balance the following traverse, and find coordinates of points B, C, D, E, F, G, and H. Enter the azimuth directly into a calculator with north and east plus (+) and south and west minus (−).

	corner ties	
line	azimuth	length
AB	21°12′	289.07
BC	101°19′	366.63
CD	205°24′	228.60
DE	144°37′	232.44
EF	227°51′	281.24
FG	281°30′	316.68
GH	29°29′	192.13
HA	334°03′	174.06

20. The open traverse ABCDE for a highway location begins at point A (which lies on line XA, the azimuth of which is known to be 21°44′) and ends at point E (which lies on line YE, the azimuth of which is known to be 350°14′). Using the azimuth of XA and YE as fixed, find the angular error in the traverse, correct arbitrarily, and convert azimuths to bearings.

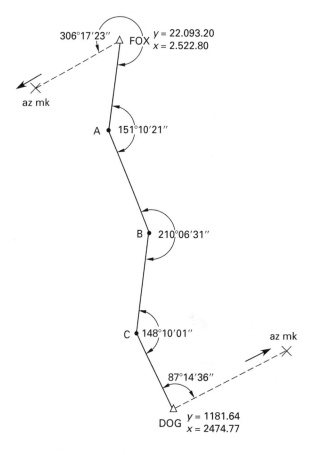

22. Corner C of the traverse ABCDA is the center of a 24 in tree. Traverse ABC'DA was run as detailed in the figure with C' being 18.6 ft off the center of the tree. Find bearing and length of BC and CD.

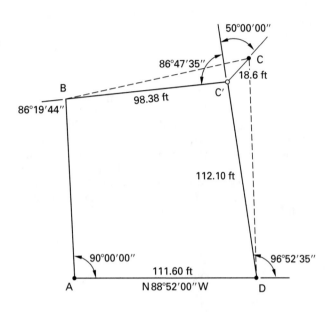

23. Given the figure shown with bearing of AB due south and length of lines as shown, find the bearing of each line and the area of the total figure.

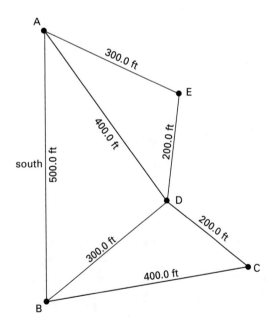

24. Given the tract ABCDA with sides CD and DA, and angles C and D as shown, and area of 43,560 ft², find sides AB and BC and angles A and B.

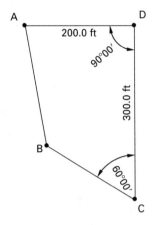

25. The trapezoidal tract ABCDA is to be divided into two equal areas by a line originating at the midpoint of the south line. Determine the necessary information to describe the property.

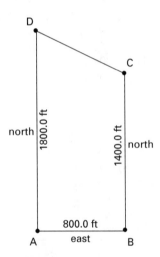

26. Find the coordinates of the point of intersection of the sides AC and BC of the triangle ABC given the information shown.

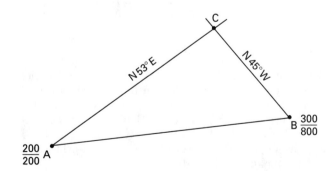

27. Find the coordinates of the point of intersection of the sides AC and BC of the triangle ABC given the information shown.

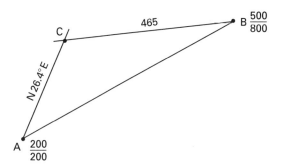

28. Find the coordinates of the point of intersection of the sides AC and BC of the triangle given the information shown.

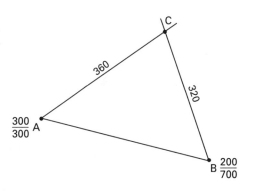

SOLUTIONS

1. (a)

point	measured angle	balanced angle
A	67°06′30″	67°07′
B	216°19′	216°19′
C	65°12′30″	65°13′
D	95°18′30″	95°19′
E	96°02′	96°02′
	539°58′30″	540°00′

(b)

point	measured angle	balanced angle
A	92°38′	92°38′
B	117°21′30″	117°21′
C	129°13′	129°13′
D	261°44′30″	261°44′
E	55°28′30″	55°28′
F	207°10′	207°10′
G	70°52′30″	70°52′
H	145°34′	145°34′
	1080°02′	1080°00′

2.

point	measured angle	balanced angle
A	90°19′59″	90°19′52″
B	106°19′55″	106°19′48″
C	318°51′26″	318°51′19″
D	48°29′12″	48°29′05″
E	150°55′29″	150°55′22″
F	195°44′47″	195°44′40″
G	87°11′56″	87°11′49″
H	193°08′23″	193°08′16″
I	212°18′51″	212°18′44″
J	47°27′37″	47°27′30″
K	288°30′28″	288°30′21″
L	60°43′21″	60°43′14″
	1800°01′24″	1800°00′00″

3.

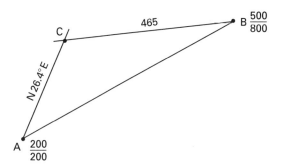

$$95°19'$$
$$\underline{+ \ 15°22'}$$
$$110°41'$$

$$179°60'$$
$$\underline{- \ 110°41'}$$
$$69°19'$$

bearing BC: N 69°19′ W

station C

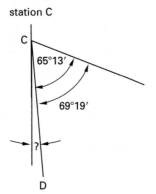

$$69°19'$$
$$\underline{-65°13'}$$
$$4°06'$$

bearing CD: S 04°06′ E

station D

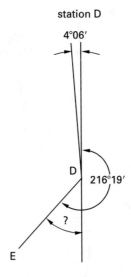

$$216°19'$$
$$\underline{- \ \ \ 4°06'}$$
$$212°13'$$
$$\underline{-180°00'}$$
$$32°13'$$

bearing DE: S 32°13′ W

station E

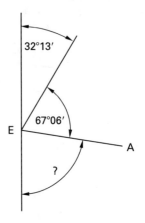

$$32°13'$$
$$\underline{+67°06'}$$
$$99°19'$$

$$179°60'$$
$$\underline{- \ 99°19'}$$
$$80°41'$$

bearing EA: S 80°41′ E

station A

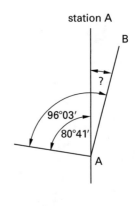

$$96°03'$$
$$\underline{-80°41'}$$
$$15°22'$$

bearing AB: N 15°22′ E

4.

station B

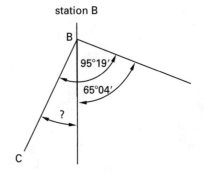

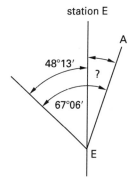
station E

$$95°19'$$
$$\underline{-65°04'}$$
$$30°15'$$

bearing BC: S 30°15′ W

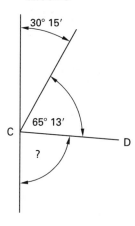
station C

$$67°06'$$
$$\underline{-48°13'}$$
$$18°53'$$

bearing EA: N 18°53′ E

$$65°13'$$
$$\underline{+30°15'}$$
$$95°28'$$

$$179°60'$$
$$\underline{-\ 95°28'}$$
$$84°32'$$

bearing CD: S 84′°32′ E

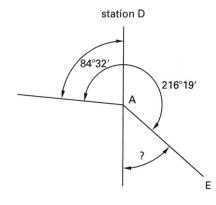
station D

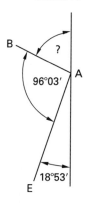
station A

$$216°19'$$
$$\underline{-\ 84°32'}$$
$$132°47'$$

$$179°60'$$
$$\underline{-132°47'}$$
$$48°13'$$

bearing DE: S 48°13′ E

$$96°03'$$
$$\underline{+\ 18°53'}$$
$$114°56'$$

$$179°60'$$
$$\underline{-114°56'}$$
$$65°04'$$

bearing AB: N 65°04′ W

Traverse I

5.

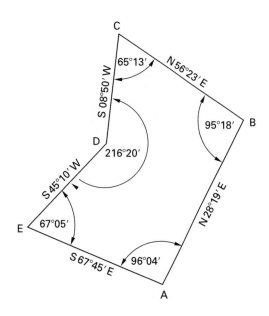

angle A	angle B	angle C	angle D	angle E
67°45′	179°60′	56°23′	179°60′	179°60′
+28°19′	−56°23′	+8°50′	−8°50′	−45°10′
96°04′	−28°19′	65°13′	+45°10′	67°45′
	95°18′		216°20′	67°05′

6.

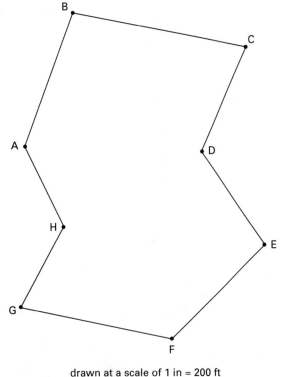

drawn at a scale of 1 in = 200 ft

7.

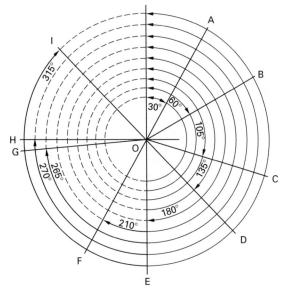

line	azimuth	back azimuth
OA	30°	210°
OB	60°	240°
OC	105°	285°
OD	135°	315°
OE	180°	0°
OF	210°	30°
OG	265°	85°
OH	270°	90°
OI	315°	135°

8.

$$
\begin{array}{rl}
14°37′19″ & \\
+\ 21°14′46″ & \\
\hline
35°52′05″ & \text{WAAF-A} \\
+180°00′00″ & \\
\hline
215°52′05″ & \\
+272°13′44″ & \\
\hline
488°05′49″ & \\
-360°00′00″ & \\
\hline
128°05′49″ & \text{AB} \\
+180°00′00″ & \\
\hline
308°05′49″ & \\
+\ 95°18′22″ & \\
\hline
403°24′11″ & \\
-360°00′00″ & \\
\hline
43°24′11″ & \text{BC} \\
+180°00′00″ & \\
\hline
223°24′11″ & \\
+\ 65°13′08″ & \\
\hline
288°37′19″ & \text{CD} \\
-180°00′00″ & \\
\hline
108°37′19″ & \\
+216°19′30″ & \\
\hline
324°56′49″ & \text{DE} \\
-180°00′00″ & \\
\hline
144°56′49″ & \\
+\ 67°05′20″ & \\
\hline
212°02′09″ & \text{EA} \\
-180°00′00″ & \\
\hline
32°02′09″ & \\
+\ 96°03′40″ & \\
\hline
128°05′49″ & \text{AB} \\
\end{array}
$$

9. (a) $\boxed{265°13'16''}$ (b) $\boxed{171°40'41''}$ **10.** (a) $\boxed{N\,68°22'54''\,W}$ (b) $\boxed{S\,73°47'14''\,E}$

(c) $\boxed{285°35'59''}$ (d) $\boxed{84°28'13''}$ (c) $\boxed{N\,12°13'47''\,E}$ (d) $\boxed{S\,52°31'18''\,W}$

(e) $\boxed{96°56'32''}$ (f) $\boxed{352°33'27''}$ (e) $\boxed{S\,86°55'58''\,E}$ (f) $\boxed{N\,65°11'37''\,E}$

(g) $\boxed{27°57'45''}$ (h) $\boxed{185°17'25''}$ (g) $\boxed{N\,22°44'49''\,W}$ (h) $\boxed{S\,87°25'51''\,W}$

(i) $\boxed{175°41'48''}$ (j) $\boxed{302°51'58''}$ (i) $\boxed{S\,77°32'22''\,E}$ (j) $\boxed{N\,42°19'07''\,W}$

11.

point	bearing	length	cos	sin	latitude N	latitude S	departure E	departure W
A								
	S 60°00′ E	50.0	0.5000	0.8660		25.0	43.3	
B								
	N 30°′00′ E	100.0	0.8660	0.5000	86.6		50.0	
C								
	N 75°00′ W	200.0	0.2588	0.9659	51.8			193.2
D								
	S 41°30′ E	151.0	0.7490	0.6626		113.1	100.1	
A								
					138.4	138.1	193.4	193.2

12.

point	bearing	length	latitude N	latitude S	cor	departure E	departure W	cor	latitude N	latitude S	departure E	departure W
A												
	N 28°19′ E	560.27	493.23		−0.11	265.76		−0.10	493.12		268.66	
B												
	N 56°23′ W	484.18	268.06		−0.09		403.21	+0.08	267.97			403.29
C												
	S 08°50′ W	375.42		370.97	+0.07		57.65	+0.07		371.04		57.72
D												
	S 45°10′ W	311.44		219.58	+0.06		220.86	+0.05		219.64		220.91
E												
	S 67°45′ E	449.83		170.33	+0.08	416.34		−0.08		170.41	416.26	
A		2181.14	761.29	760.88		682.10	681.72		761.09	761.09	681.92	681.92
				760.88		681.72						
				0.41		0.38						

error of closure $= \sqrt{(0.37)^2 + (0.24)^2} = 0.44$

ratio of error $= \dfrac{1}{5000}$

Traverse I

course	latitude corrections	course	latitude corrections
AB	$\left(\dfrac{0.41}{2181.14}\right)(560.27) = 0.11$	AB	$\left(\dfrac{0.38}{2181.14}\right)(560.27) = 0.10$
BC	$\left(\dfrac{0.41}{2181.14}\right)(484.18) = 0.09$	BC	$\left(\dfrac{0.38}{2181.14}\right)(484.18) = 0.08$
CD	$\left(\dfrac{0.41}{2181.14}\right)(375.42) = 0.07$	CD	$\left(\dfrac{0.38}{2181.14}\right)(375.42) = 0.07$
DE	$\left(\dfrac{0.41}{2181.14}\right)(311.44) = 0.06$	DE	$\left(\dfrac{0.38}{2181.14}\right)(311.44) = 0.05$
EA	$\left(\dfrac{0.41}{2181.14}\right)(449.83) = 0.08$	EA	$\left(\dfrac{0.38}{2181.14}\right)(449.83) = 0.08$

13.

point	bearing	length	latitude N	latitude S	cor	departure E	departure W	cor	latitude N	latitude S	departure E	departure W
A												
	S 53°18′ E	560.27		334.83	−0.10	449.21		−0.06		334.73	449.15	
B												
	N 42°00′ E	484.18	359.82		+0.08	323.98		−0.05	359.90		323.93	
C												
	N 72°47′ W	375.42	111.12		+0.06		358.60	+0.04	111.18			358.64
D												
	N 36°27′ W	311.44	250.51		+0.05		185.03	+0.04	250.56			185.07
E												
	S 30°39′ W	449.83		386.99	−0.08		229.32	+0.05		386.91		229.37
A			721.45	721.82		773.19	772.95		721.64	721.64	773.08	773.08
				721.45		772.95						
				0.37		0.24						

14.

point	bearing	length	latitude N	latitude S	cor	departure E	departure W	cor	latitude N	latitude S	departure E	departure W
A												
	M 80°00′ W	1015.43	176.33		+0.17		1000.00	−0.10	176.50			999.90
B												
	S 66°30′ W	545.52		217.53	−0.09		500.27	−0.05		217.44		500.22
C												
	S 12°00′ E	480.97		470.46	−0.08	100.00		+0.05		470.38	100.05	
D												
	S 88°30′ E	750.26		19.64	−0.12	750.00		+0.07		19.52	750.07	
E												
	N 69°00′ E	639.18	229.06		+0.10	596.73		+0.06	229.16		596.79	
F												
	N 10°00′ E	306.28	301.63		+0.05	53.18		+0.03	301.68		53.21	
A		3737.64	707.02	707.63		1499.91	1500.27		707.34	707.34	1500.12	1500.12

error of closure $= \sqrt{(0.61)^2 + (0.36)^2} = 0.71$

ratio of error $= \dfrac{1}{5300}$

15.

point	bearing	length	latitude N	latitude S	departure E	departure W	coordinates x	coordinates y
A							9012.34	1234.56
	N 28°19′ E	560.27	493.12		265.66			
B							9505.46	1500.22
	N 56°23′ W	484.18	267.97			403.29		
C							9773.43	1096.93
	S 08°50′ W	375.42		371.04		57.72		
D							9402.39	1039.21
	S 45°10′ W	311.44		219.64		220.91		
E							9182.75	818.30
	S 67°45′ E	449.83		170.41	416.26			
A							9012.34	1234.56

16.

point	coordinates y	coordinates x	latitude N	latitude S	departure E	departure W	bearing	length
A	9090.90	9090.90						
			389.49		361.37		N 42°51′ E	531.31
B	9480.39	9452.27						
			132.48			561.57	N 76°44′ W	576.99
C	9612.87	8890.70						
				291.01		114.89	S 21°33′ W	312.87
D	9321.86	8775.81						

17.

point	bearing	length	latitude N	latitude S	departure E	departure W	coordinates x	coordinates y
A							1000.00	1000.00
	S 12°02′ E	59.20		57.90	12.34			
M							942.10	1012.34
B							1493.12	1265.66
	N 88°13′ E	65.13	2.03		65.10			
N							1495.15	1330.76
C							1761.09	862.37
	N 13°52′ W	70.85	68.79			16.98		
O							1829.85	845.39
D							1390.05	804.65
	N 56°09′ W	72.45	40.36			60.17		
P							1430.41	744.48
E							1170.41	583.74
	S 78°31′ W	76.51		15.23		74.98		
Q							1155.18	508.76
M							942.10	1012.34
	N 29°56′ E	638.17	553.05		318.42			
N							1495.15	1330.76
	N 55°24′ W	589.60	334.73			485.37		
O							1829.88	845.39
	S 14°11′ W	412.02		399.49		100.91		
P							1430.41	744.48
	S 40°35′ W	362.37		275.23		235.72		
Q							1155.18	508.76
	S 67°04′ E	546.81		213.08	503.58			
M							942.10	1012.34

18.

point	bearing	length	latitude N	latitude S	departure E	departure W	coordinates y	coordinates x
A							1000.00	1000.00
	N 28°48′ W	58.46	51.23			28.16		
M							1051.23	971.84
B							1493.12	1265.66
	N 87°33′ E	66.72	2.85		66.66			
N							1495.97	1332.32
C							1761.09	862.37
	S 27°47′ E	68.89		60.75	32.11			
O							1700.14	894.48
D							1390.05	804.65
	N 55°29′ W	74.13	42.01			61.08		
P							1432.06	743.57
E							1170.41	583.74
	S 13°38′ E	66.52		64.65	15.68			
Q							1105.76	599.42
M							1051.23	971.84
	N 39°02′ E	572.49	444.74		360.48			
N							1495.97	1332.32
	N 65°00′ W	483.10	204.17			437.84		
O							1700.14	894.48
	S 29°23′ W	307.64		268.08		150.91		
P							1432.06	743.57
	S 23°50′ W	356.72		326.30		144.15		
Q							1105.76	599.42
	S 81°40′ E	376.39		54.53	372.42			
M							1051.23	971.84

19.

point	line	azimuth	length	latitude	departure	latitude	departure	coordinates y	coordinates x
A								1000.00	1000.00
	AB	21°12′	289.07	269.55	104.53	269.51	104.51		
B								1269.55	1104.51
	BC	101°19′	366.63	−71.94	359.50	−71.89	359.48		
C								1197.66	1463.99
	CD	205°24′	228.60	−206.50	−98.05	−206.47	−98.06		
D								991.19	1365.93
	DE	144°37′	232.44	−189.51	134.59	−189.48	134.57		
E								801.71	1500.50
	EF	227°51′	281.24	−188.73	−208.51	−188.69	−208.53		
F								613.02	1291.97
	FG	281°30′	316.68	63.14	−310.32	63.18	−310.34		
G								676.20	981.63
	GH	29′°29′	192.13	167.25	94.56	167.27	94.55		
H								843.47	1076.18
	HA	334′°03′	174.06	<u>156.51</u>	<u>−76.17</u>	156.53	−76.18		
A								1000.00	1000.00
				−0.27	+0.13				

error of closure $= \sqrt{(0.27)^2 + (0.13)^2} = 0.30$

precision $= \dfrac{1}{6900}$

20.

line	azimuth	correction	adjusted azimuth	adjusted bearing
XA	21°44'	fixed		N 21°44' E
	96°55'			
AB	118°39'	−0°30'	118°38'30"	S 61°21'30" E
	−62°20'			
BC	56°19'	−1°00'	56°18'00"	N 56°18' E
	67°48'			
CD	124°07'	−1°30'	124°05'30"	S 55°54'30" E
	−67°22'			
DE	56°45'	−2°00'	56°43'	N 56°43' E
	113°31'			
EY	170°16'			S 09°46' E
	180°			
YE	350°16'			
YE	350°14'	fixed		

21.

	azimuth	cor	adjusted azimuth
	246°45'46"		
	+306°17'23"	+07"	
	553°03'09"		
	−360°		
FOX-A	193°03'09"		193°03'16"
	−180°		
	13°03'09"		
	+151°10'21"	+07"	
AB	164°13'30"		164°13'44"
	−180°		
	344°13'30"		
	+210°06'31"	+07"	
BC	194°20'01"		194°20'22"
	−180°		
	14°20'01"		
	+148°10'01"	+07"	
C-DOG	162°30'02"		162°30'30"
	+180°		
	342°30'02"		
	+ 87°14'38"	+07"	
	429°44'40"		
	−360°		
	69°44'40"		
DOG-MK	−69°45'15"		69°45'15"
	35"		

Traverse I

| line | azimuth | length | latitude | departure | latitude | departure | coordinates | | point |
							y	x	
FOX							22,093.20	2522.88	FOX
$-$A	193°03′16″	5543.31	$-$5400.05	$-$1252.11	$-$5400.11	$-$1252.12			
							16,693.09	1270.76	A
AB	164°13′44″	6057.05	$-$5829.03	1646.28	$-$5829.10	$+$1646.26			
							10,863.99	2917.02	B
BC	194°20′22″	6317.47	$-$6120.65	$-$1564.62	$-$6120.72	$-$1564.64			
							4743.27	1352.38	C
$-$C	162°30′30″	3734.26	$-$3561.59	1122.40	$-$3561.63	$+$1122.39			
							1181.64	2474.77	DOG
DOG		21,652.09	$-$20,911.32	$-$48.05					
			$-$20,911.56	48.11					
			$+$0.24	$-$0.06					

$$\sqrt{(0.24)^2 + (0.06)^2} = 0.25$$

22. Solve triangle BCC′ and triangle CDC′.

ln $\triangle$BCC′:

$$BC = \sqrt{\begin{array}{l}(98.38)^2 + (18.6)^2 \\ - (2)(98.38)(18.6)(\cos 136°47′35″)\end{array}}$$

$$= 112.66$$

$$\sin B = \frac{18.6 \sin 196°47′35″}{112.66}$$

$$B = 6°29′25″$$

$$\begin{array}{l}\text{bearing} \\ \text{BC}\end{array} = \text{N}\,80°58′39″\,\text{E}$$

ln $\triangle$CDC′:

$$CD = \sqrt{\begin{array}{l}(18.6)^2 + (112.10)^2 \\ - (2)(18.6)(112.10)(\cos 130°00′00″)\end{array}}$$

$$= 124.87$$

$$\sin D = \frac{18.6 \sin 130°00′00″}{124.87}$$

$$D = 6°33′07″$$

$$\begin{array}{l}\text{bearing} \\ \text{CD}\end{array} = \text{S}\,0°48′32″\,\text{W}$$

$$\begin{array}{l}\text{bearing} \\ \text{BC}\end{array} = \text{N}\,80°58′39″\,\text{E} \qquad \text{distance BC} = 112.66$$

$$\begin{array}{l}\text{bearing} \\ \text{CD}\end{array} = \text{S}\,0°48′32″\,\text{W} \qquad \text{distance CD} = 124.87$$

23. Solve ABD:

$$\cos A = \frac{(500)^2 + (400)^2 - (300)^2}{(2)(500)(400)} \quad A = 36°52′$$

$$\cos B = \frac{(500)^2 + (300)^2 - (400)^2}{(2)(500)(400)} \quad B = 53°08′$$

$$D = \frac{90°00′}{180°00′}$$

$\triangle$ABD is a right triangle (3-4-5).

Solve BCD:

$$\cos B = \frac{(300)^2 + (400)^2 - (200)^2}{(2)(300)(400)} \quad B = 28°57′$$

$$\cos C = \frac{(400)^2 + (200)^2 - (300)^2}{(2)(400)(200)} \quad C = 46°34′$$

$$\cos D = \frac{(300)^2 + (200)^2 - (400)^2}{(2)(300)(200)} \quad D = \frac{104°29′}{180°00′}$$

Solve DEA:

$$\text{congruent to BCD} \qquad \begin{array}{l} D = 46°34′ \\ E = 104°29′ \\ A = \underline{28°57′} \\ 180°00′ \end{array}$$

Bearings:

AB = South
AD = S 36°52′ E
AE = S 65°50′ E
BD = N 53°08′ E
BC = N 82°05′ E
CD = N 51°21′ W
ED = S 09°42′ W

Area:

$$ABC = \sqrt{\begin{array}{c}(600)(600-500)\\ \times (600-400)(600-300)\end{array}} = 60{,}000 \text{ ft}^2$$

$$BCD = \sqrt{\begin{array}{c}(450)(450-400)\\ \times (450-300)(450-200)\end{array}} = 29{,}047 \text{ ft}^2$$

$$DEA = \text{congruent} \qquad\qquad = \underline{29{,}047} \text{ ft}^2$$
$$2.711 \text{ ac}$$

24. Construct triangles ABC, ACD, and BCE.

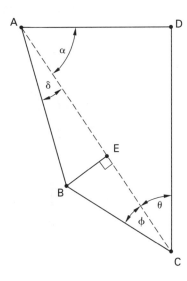

Solve ACD:

$$\theta = \tan^{-1}\left(\frac{200.0}{300.0}\right) = 33°41'$$

$$\alpha = 90° - 33°41' = 56°19'$$

$$AC = \sqrt{(200)^2 + (300)^2} = 360.6$$

Solve ABC:

$$\text{area ACD} = \left(\tfrac{1}{2}\right)(200.0)(300.0) = 30{,}000.0$$

$$ABC = 43{,}560.0 - 30{,}000 = 13{,}560.0$$

$$13{,}560.0 = \left(\tfrac{1}{2}\right)(BE)(360.0)$$

$$BE = (2)\left(\frac{13{,}560.0}{360.0}\right) = 75.2$$

Solve BCE:

$$\theta = 60° - 33°41' = 26°19'$$

$$BC = \frac{75.2}{\sin 26°19'} = 169.6$$

Solve ABC:

$$AB = \sqrt{\begin{array}{c}(169.6)^2 + (360.6)^2\\ - (2)(169.6)(360.6)(\cos 26°19')\end{array}}$$
$$= 221.7$$

$$\delta = \sin^{-1}\left(\frac{169.6 \sin 33°41'}{221.7}\right) = 19°50'$$

$$A = 19°50' + 56°19' = 76°09'$$

$$B = 180° - (19°50' + 26°19') = 133°51'$$

Summary:

$$AB = 221.7 \text{ ft}$$
$$BC = 169.6 \text{ ft}$$
$$\text{angle } A = 76°09'$$
$$\text{angle } B = 133°51'$$

25. Construct line MN parallel to BC and DA and originating at midpoint M. Construct MO to represent the boundary between the two tracts when they have been partitioned. Construct right triangle CPN with CP perpendicular to MN.

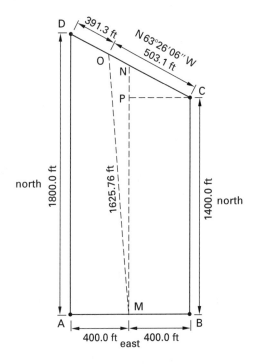

By proportion, MN = 1600.

$$\text{area ABCDA} = \left(\tfrac{1}{2}\right)(1400 + 1800)(800)$$
$$= 1{,}280{,}000$$

$$\text{area MBCN} = \left(\tfrac{1}{2}\right)(1400 + 1600)(400)$$
$$= 600{,}000$$

$$\text{area MNO} = 640{,}000 - 600{,}000 = 40{,}000$$

Solve triangle CPN:

$$C = \tan^{-1}\left(\frac{200}{400}\right) = 26°33'54''$$

$$CN = \sqrt{(200)^2 + (400)^2} = 447.21$$

ln ΔMNO:

$$N = 90° + 26°33'54'' = 116°33'54''$$

$$\text{area MNO} = \left(\tfrac{1}{2}\right)(NO)(MN)(\sin 116°33'54'')$$

$$40,000 = \left(\tfrac{1}{2}\right)(NO)(1600)(\sin 116°33'54'')$$

$$NO = \frac{(2)(40,000)}{1600 \sin 116°33'54''} = 55.90$$

$$MO = \sqrt{(1600)^2 + (55.90)^2}$$
$$+ \sqrt{(2)(1600)(55.90)(\cos 116°33'54'')}$$
$$= 1625.76$$

$$\sin M = \frac{55.90 \sin 116°33'54''}{1625.76}$$

$$M = 1°45'44''$$

$$CO = 447.21 + 55.90 = 503.11$$

$$OD = 447.21 - 55.90 = 391.31$$

$$\text{bearing MO} = N\,1°45'44''\,W$$

$$\text{bearing CD} = N\,63°26'06''\,W$$

26. *Bearing-Bearing Method:*

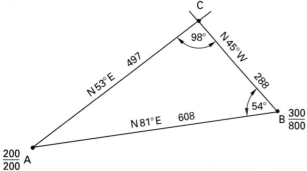

$$\tan\text{ bearing AB} = \frac{800 - 200}{300 - 200} = 6$$

$$\text{bearing AB} = N\,81°\,E$$

$$\text{distance AB} = \sqrt{(800 - 200)^2 + (300 - 200)^2}$$
$$= 608$$

$$\text{angle } A = 28°$$

$$\text{angle } B = 54°$$

$$\text{angle } C = 98°$$

$$AC = \frac{608 \sin 54°}{\sin 98°} = 497$$

$$BC = \frac{608 \sin 28°}{\sin 98°} = 288$$

$$\text{north coord. of C} = 200 + 497 \cos 53° = 499$$

$$\text{east coord. of C} = 200 + 497 \sin 53° = 597$$

27. *Bearing-Distance Method:*

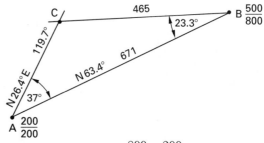

$$\text{tangent bearing AB} = \frac{800 - 200}{500 - 200} = 2$$

$$\text{bearing AB} = N\,63.4\,E$$

$$\text{distance AB} = \sqrt{600^2 + 300^2} = 671$$

$$\text{angle } A = 37.0°$$

$$\sin C' = \frac{671 \sin 37.0°}{465}$$

$$C' = 60.3°$$

$$C = 180° - 60.3° = 119.7°$$

$$B = 180° - A - C$$
$$= 180° - 37.0° - 119.7°$$
$$= 23.3°$$

$$b = \frac{\sin 23.3°(465)}{\sin 37°} = 306$$

$$\text{north coord.} = 200 + 306 \cos 26.4° = 474$$

$$\text{east coord.} = 200 + 306 \sin 26.4° = 336$$

28. *Distance-Distance Method:*

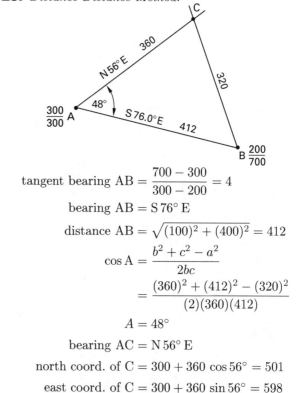

$$\text{tangent bearing AB} = \frac{700 - 300}{300 - 200} = 4$$

$$\text{bearing AB} = S\,76°\,E$$

$$\text{distance AB} = \sqrt{(100)^2 + (400)^2} = 412$$

$$\cos A = \frac{b^2 + c^2 - a^2}{2bc}$$
$$= \frac{(360)^2 + (412)^2 - (320)^2}{(2)(360)(412)}$$

$$A = 48°$$

$$\text{bearing AC} = N\,56°\,E$$

$$\text{north coord. of C} = 300 + 360 \cos 56° = 501$$

$$\text{east coord. of C} = 300 + 360 \sin 56° = 598$$

15 Area of a Traverse

1. METHODS FOR COMPUTATION OF AREA

The area within a traverse can be computed by the *double meridian distance* (DMD) *method*, by the coordinate method, and by use of geometric or trigonometric formulas. The most frequently used method for land area calculations is the DMD method.

2. DOUBLE MERIDIAN DISTANCES

The DMD method makes use of balanced latitudes and departures. It is widely used because it allows areas to be computed quickly.

The DMD method sets up a series of trapezoids and triangles, both inside and outside of the traverse. It calculates each of these areas and determines the area of the traverse from them.

The double meridian distance is simply twice the meridian distance (see Sec. 3). The DMD is used instead of the meridian distance (MD) to simplify the arithmetic. If the MD were used, division by two would be required several times. Using DMD, division by two is required only once.

The area of a trapezoid is one-half the sum of the bases times the altitude (i.e., the average of the bases times the altitude.) In the DMD method, the meridian distance for each course of the traverse serves as the average of the bases of a trapezoid. The DMDs of the courses are obtained from the departures of the courses. Thus, the only data needed are latitudes and departures.

3. MERIDIAN DISTANCES

The *meridian distance* (MD) of a course is the right angle distance from the midpoint of the course to a reference meridian. MDs are illustrated in Fig. 15.1. Since east and west departures are used, algebraic signs must be considered. To simplify the use of plus and minus values for departure, the entire traverse should be placed in the northeast quadrant. This can be done by taking the reference meridian through the most westerly point in the traverse.

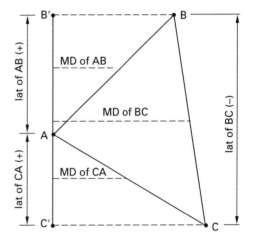

Figure 15.1 *Meridian Distance*

4. DETERMINING THE MOST WESTERLY POINT

The most westerly point can usually be determined by studying the east and west departures.

5. RULES FOR USING DMD CALCULATIONS

In Fig. 15.2, the MD of EA = ½ departure of EA. The DMD of EA = departure of EA.

The MD of AB = MD of EA + ½ departure EA + ½ departure AB. DMD of AB = DMD of EA + departure EA + departure AB.

Traverse II

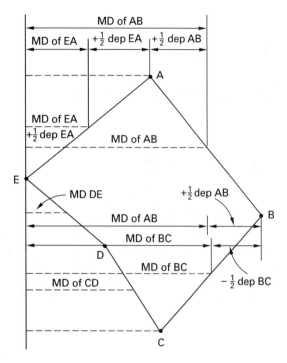

Figure 15.2 *DMD Rules*

The MD of BC = MD of AB + ½ departure AB − ½ departure BC. The DMD of BC = DMD of AB + departure AB − departure BC.

From these examples and from further examination, the following rules can be derived for computing DMDs.

step 1: The DMD of the first course is equal to the departure of the first course.

step 2: The DMD of any course is equal to the DMD of the preceding course plus the departure of the preceding course plus the departure of the course itself.

step 3: The DMD of the last course is equal to the departure of the last course with opposite sign. ("Opposite sign" means that for a westerly course, the DMD will be positive.)

6. AREAS BY DMD

The area of a traverse can be found by multiplying the DMD of each course by the latitude of that course, with north latitudes producing positive areas and south latitudes producing negative areas, adding the areas algebraically, and dividing by two. The algebraic sign of DMDs is always positive.

In Fig. 15.1,

$$\text{area ABC} = \text{area B}'\text{BCC}' - \text{area AB}'\text{B} - \text{area ACC}'$$

If MDs are positive, north latitudes are positive, and south latitudes are negative, then

$$\text{area of B}'\text{BCC}' = (\text{MD of BC})(\text{latitude of BC})$$
[negative]
$$\text{area of AB}'\text{B} = (\text{MD of AB})(\text{latitude of AB})$$
[positive]
$$\text{area of ACC}' = (\text{MD of CA})(\text{latitude of CA})$$
[positive]
$$\text{area of ABC} = \text{algebraic sum of B}'\text{BCC}', \text{AB}'\text{B}, \text{ACC}'$$

The sign of the sum can be either plus or minus.

If DMDs are used, the area will be double the area determined by using MDs, and the area of the traverse can be determined by dividing the double area by two.

Example 15.1

The following table shows the tabulation of computations for the area of a traverse ABCDEA by the DMD method. Latitudes and departures shown are balanced. Find the DMD.

Example 15.1

(All distances in feet.)

line	bearing	distance	latitude north	latitude south	departure east	departure west	DMD	double plus	area minus
AB	N 65°04′ W	560.27	236.11			507.97	995.41	235,026	
BC	S 30°14′ W	484.14		418.39		243.72	243.72		101,970
CD	S 84°33′ E	375.42		35.71	373.77		373.77		13,347
DE	S 48°13′ E	311.44		207.56	232.27		979.81		203,369
EA	N 18°53′ E	449.83	425.55		145.65		1357.73	577,782	
								812,808	318,686

Solution

The first step is to determine the most westerly traverse point. In lieu of a sketch showing the traverse, it is found as follows.

step 1: Because AB has a northwest direction, B is west of A. Looking at the departure column, it is 507.97 ft west of A.

step 2: C is 243.72 ft west of B, so B is not the most westerly point.

step 3: Courses, CD, DE, and EA have east departures; therefore, C is the most westerly point.

With C the most westerly point, the first DMD computed is for the course CD. Remembering that the DMD for the first course is the departure of the first course, and also remembering the definition of the DMD for any course,

departure of CD = +373.77 ft = DMD of CD

departure of CD = +373.77 ft

departure of DE = +232.27 ft

$\qquad$ + 979.81 ft = DMD of DE

departure of DE = +232.27 ft

departure of EA = +145.16 ft

$\qquad$ + 1357.73 ft = DMD of EA

departure of EA = +145.65 ft

departure of AB = −507.97 ft

$\qquad$ + 995.41 ft = DMD of AB

departure of AB = −507.97 ft

departure of BC = −243.72 ft

$\qquad$ + 243.72 ft = DMD of BC

Note that the DMD of BC is the same as its departure except that it has a positive sign.

The area of the traverse ABCDEA is found from the double area sums in the table.

area = $\left(\frac{1}{2}\right)$ (812,808 ft² − 318,686 ft²) = 247,061 ft²

If necessary, this area can be converted to acres by dividing by 43,560.

$$\text{area} = \frac{247{,}061 \text{ ft}^2}{43{,}560 \dfrac{\text{ft}^2}{\text{ac}}} = 5.672 \text{ ac}$$

7. AREA BY COORDINATES

After the coordinates of the corners of a tract of land are determined, the area of the tract can be computed by the *coordinate method*.

The coordinate formula is derived by forming trapezoids and determining their areas just as is done in the DMD method. Meridian distances are not used; trapezoids

are formed by the abscissas of the corners. Ordinates of the corners serve as the altitude of the trapezoids. Alternately, the trapezoids can be formed by the ordinates of the corners, and the abscissas serve as the altitudes. In Fig. 15.3, the abscissas are used to form the trapezoids.

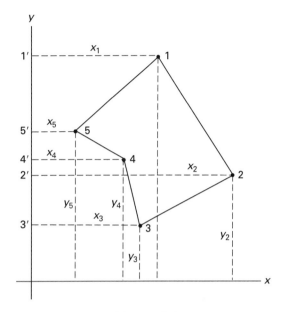

Figure 15.3 *Coordinates of Traverse Points*

$$\begin{aligned}
\text{area } 1\text{-}2\text{-}3\text{-}4\text{-}5\text{-}1 &= \text{area } 1'\text{-}1\text{-}2\text{-}2' + \text{area } 2'\text{-}2\text{-}3\text{-}3' \\
&\quad - \text{area } 1'\text{-}1\text{-}5\text{-}5' - \text{area } 5'\text{-}5\text{-}4\text{-}4' \\
&\quad - \text{area } 4'\text{-}4\text{-}3\text{-}3' \\
&= \left(\tfrac{1}{2}\right) \big[x_1(y_5 - y_2) + x_2(y_1 - y_3) \\
&\quad + x_3(y_2 - y_4) + x_4(y_3 - y_5) \\
&\quad + x_5(y_4 - y_1) \big]
\end{aligned}$$

The same results can be found by using the ordinates to form the trapezoids.

$$\begin{aligned}
\text{area } 1\text{-}2\text{-}3\text{-}4\text{-}5\text{-}1 &= \left(\tfrac{1}{2}\right) \big[y_1(x_5 - x_2) + y_2(x_1 - x_3) \\
&\quad + y_3(x_2 - x_4) + y_4(x_3 - x_5) \\
&\quad + y_5(x_4 - x_1) \big]
\end{aligned}$$

Example 15.2

Given the traverse 1-2-3-4-5-1 in Fig. 15.3 with the coordinates as shown, find the area inside the traverse by the coordinate method.

	coordinates	
point	y (ft)	x (ft)
1	1000.00	1000.00
2	1236.11	492.03
3	817.72	248.31
4	782.01	622.01
5	574.45	854.35

Traverse II

Solution

station	coordinates (ft)		$y_1 - y_2$	double area (ft^2)	
	y	x		plus	minus
1	1000.00	1000.00	$(1000.00)(574.45 - 1236.11)$		661,660
2	1236.11	492.03	$(492.03)(1000.00 - 817.72)$	89,687	
3	817.72	248.31	$(248.31)(1236.11 - 782.01)$	112,758	
4	782.01	622.08	$(622.08)(817.72 - 574.45)$	151,333	
5	574.45	854.35	$(854.35)(782.01 - 1000.00)$		186,240

$$353,778 \qquad 847,900$$
$$353,778$$
$$2)\ \overline{494,122}$$
$$247,061$$

$$\text{area} = \frac{847,900\ \text{ft}^2 - 353,778\ \text{ft}^2}{(2)\left(43,560\ \dfrac{\text{ft}^2}{\text{ac}}\right)} = 5.672\ \text{ac}$$

8. AREA BY TRIANGLES

When small traverses do not warrant computations of latitudes and departures, their areas can be determined by using formulas for the area of a triangle.

$$\text{area} = \tfrac{1}{2}ab \sin C \qquad \textit{15.1}$$

(a and b are any two sides, and C is the angle included between them.)

In Fig. 15.4(a), a tract of land has been divided into two triangles. Two sides and an included angle have been measured in each triangle. The areas of the triangles can be computed by using Eq. 15.1. Their sum is the area of the tract.

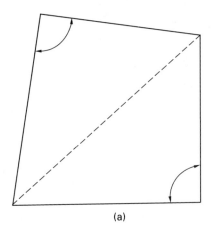

(a)

In Fig. 15.4(b), the property line has become covered with brush so that the four triangles have been formed from a central point. Angles at the central point have been measured for each triangle, and distances from the central point to each corner have also been measured. Areas can again be computed by using Eq. 15.1.

Also applicable in computing areas is Eq. 15.2.

$$A = \sqrt{(s)(s-a)(s-b)(s-c)} \qquad \textit{15.2}$$

s is one-half of the perimeter of the triangle, and a, b, and c are the sides of the triangle.

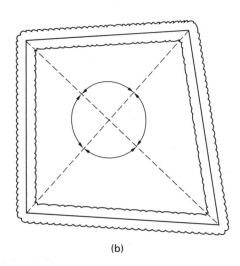

(b)

Figure 15.4 *Area by Triangles*

Example 15.3

Find the area of a triangle with sides 32 ft, 46 ft, and 68 ft.

Solution

$$s = \left(\tfrac{1}{2}\right)(32 \text{ ft} + 46 \text{ ft} + 68 \text{ ft}) = 73 \text{ ft}$$

$$\text{area} = \sqrt{\begin{array}{c}(73 \text{ ft})(73 \text{ ft} - 32 \text{ ft}) \\ \times\, (73 \text{ ft} - 46 \text{ ft})(73 \text{ ft} - 68 \text{ ft})\end{array}}$$

$$= 636 \text{ ft}^2$$

9. AREA ALONG AN IRREGULAR BOUNDARY

When a tract of land is bounded on one side by an irregular boundary such as a stream or lake, the traverse can be composed of straight lines so that closure can be computed. Points along the irregular side can be tied to one of the sides of the traverse by right angle offset measurements. The area between the irregular side and the traverse line is approximated by dividing the area into trapezoids and triangles formed by the ties to the breaks in the irregular side. This irregular area is then added to the traverse area. The irregular area can be computed by applying the trapezoidal rule or Simpson's one-third rule.

10. THE TRAPEZOIDAL RULE

In using the *trapezoidal rule*, it is assumed that the irregular boundary is made up of a series of straight lines. When the ties are taken close enough, a curved line connecting the ends of any two ties is very nearly a straight line, and no significant error is introduced.

The trapezoidal rule applies only to the part of the area where the ties are at regular intervals and form trapezoids. Triangles and trapezoids that do not have altitudes of the regular interval are computed separately and are added to the area found by applying the trapezoidal rule.

The rule is given by Eq. 15.3.

$$\text{area} = D\left(\frac{T_1}{2} + T_2 + T_3 + T_4 + \cdots + \frac{T_n}{2}\right) \quad 15.3$$

D is the regular interval, and T is the tie distance.

Example 15.4

Using the trapezoidal rule, find the area in the following illustration.

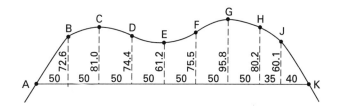

Solution

$$\text{area} = (50)\left(\frac{72.6}{2}\right)$$

$$+ (50)\left(\frac{72.6}{2} + 81.0 + 74.4 + 61.2\right.$$

$$\left. + 75.5 + 95.8 + \frac{80.2}{2}\right)$$

$$+ (35)\left(\frac{80.2 + 60.1}{2}\right)$$

$$+ (40)\left(\frac{60.1}{2}\right)$$

$$= 28{,}687$$

11. AREA OF A SEGMENT OF A CIRCLE

Land along highways, streets, and railroads often has a circular arc for a boundary. A traverse of straight lines can be run by using the *long chord* (LC) of the circular arc as one of the sides of the traverse. The area of the tract can be found by adding the area of the segment formed by the chord and the arc to the area within the traverse. It is usually practical to measure the chord length and the middle ordinate length. Using these two lengths and formulas derived for computing circular curves, the area of the segment can be found.

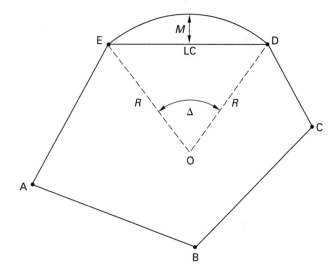

Figure 15.5 *Circular Segment Areas*

Δ is the *central angle*, M is the *middle ordinate*, LC is the *long chord*, and R is the *radius*.

$$\text{area of sector} = \frac{\Delta° \pi R^2}{360°} \qquad 15.4$$

$$\text{area of triangle} = \frac{R(R\sin\Delta)}{2}$$

$$= \frac{R^2 \sin\Delta}{2} \qquad 15.5$$

The radius R is not known, but it can be determined if the long chord DE and the middle ordinate M are known. These lengths can be measured. Two formulas used to compute R are

$$\frac{\tan\Delta}{4} = \frac{2M}{LC} \qquad 15.6$$

$$R = \frac{LC}{2\sin\dfrac{\Delta}{2}} \qquad 15.7$$

If A = the area of the segment, A_s = area of the sector, and A_t = area of the triangle, then

$$A = A_s - A_t = \frac{\Delta\pi R^2}{360°} - \frac{R^2 \sin\Delta}{2}$$

$$= R^2\left(\frac{\Delta\pi}{360°} - \frac{\sin\Delta}{2}\right) \qquad 15.8$$

Example 15.5

The tract of land shown in Fig. 15.5 consists of the area within the traverse ABCDEA plus the area in the segment bounded by side DE and arc DE. The area of the segment is equal to the area of the sector DOE minus the area of the triangle DOE.

Find the area in the segment bounded by side DE and arc DE in Fig. 15.5 if the long chord is 325.48 ft and the middle ordinate is 42.16 ft.

$$\frac{\tan\Delta}{4} = \frac{(2)(42.16 \text{ ft})}{325.48 \text{ ft}}$$

$$\Delta = 58.0958°$$

$$R = \frac{LC}{2\sin\dfrac{\Delta}{2}} = \frac{325.48 \text{ ft}}{2\sin\dfrac{58.0958°}{2}} = 335.17 \text{ ft}$$

$$A = (335.17 \text{ ft})^2\left(\frac{(58.0958°)\pi}{360°} - \frac{\sin 58.0958°}{2}\right)$$

$$= 9270 \text{ ft}^2$$

12. SPECIAL FORMULA

The formula for the area of a triangle given in Sec. 8, $A = \frac{1}{2}ab\sin C$, can be used in deriving another useful formula for the area of a triangle.

Use the law of sines and substitute $b = a\sin B/\sin A$ in the formula $A = \frac{1}{2}\sin C$ to get

$$\text{area} = \frac{a^2 \sin B \sin C}{2\sin A} \qquad 15.9$$

Also,

$$\text{area} = \frac{b^2 \sin C \sin A}{2\sin B} = \frac{c^2 \sin A \sin B}{2\sin C} \qquad 15.10$$

Example 15.6

A 3 ac triangular tract is to be cut off the northeast corner of a larger tract with bearings as shown. Find the lengths of the sides of the triangle.

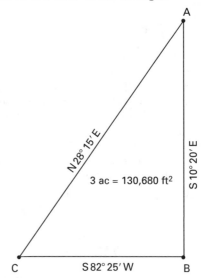

Solution

$A = 38°35'$, $B = 87°15'$, and $C = 54°10'$.

$$\text{area} = \frac{a^2 \sin B \sin C}{2\sin A}$$

$$a^2 = \frac{(\text{area})(2\sin A)}{\sin B \sin C}$$

Also,

$$\text{area} = \frac{b^2 \sin C \sin A}{2\sin B}$$

$$b^2 = \frac{(\text{area})(2\sin B)}{\sin C \sin A}$$

Also,

$$\text{area} = \frac{c^2 \sin A \sin B}{2\sin C}$$

$$c^2 = \frac{(\text{area})(2\sin C)}{\sin A \sin B}$$

Therefore,

$$a = \sqrt{\frac{(\text{area})(2\sin A)}{\sin B \sin C}} = \sqrt{\frac{(130{,}680 \text{ ft}^2)(2\sin 38°35')}{(\sin 87°15')(\sin 54°10')}}$$

$$= 448.6 \text{ ft}$$

$$b = \sqrt{\frac{(\text{area})(2\sin B)}{\sin C \sin A}} = \sqrt{\frac{(130{,}680 \text{ ft}^2)(2\sin 87°15')}{(2\sin 54°10')(\sin 38°35')}}$$

$$= 718.5 \text{ ft}$$

$$c = \sqrt{\frac{(\text{area})(2\sin C)}{\sin A \sin B}} = \sqrt{\frac{(130{,}680 \text{ ft}^2)(2\sin 54°10')}{(\sin 38°35')(\sin 87°13')}}$$

$$= 583.2 \text{ ft}$$

In summary,

$$AB = 583.2 \text{ ft}$$
$$BC = 448.6 \text{ ft}$$
$$CA = 718.5 \text{ ft}$$

Example 15.7

Compute the area of the tract 1-2-3-4-5-6-1.

1-2	N 02°27′50″ W	761.49
2-3	N 87°35′37″ E	1076.62
3-4	S 02°24′23″ E	290.00
4-5	S 09°49′21″ W	826.10
5-6	S 30°30′21″ W	68.00
6-1	N 67°56′54″ W	949.09

Solution to Example 15.7

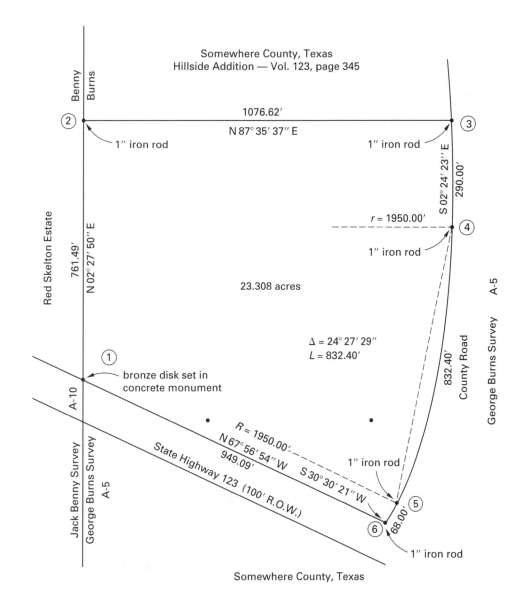

Area of Traverse 1-2-3-4-5-6-1

	bearing	distance	latitude	departure	DMD	area
1-2	N 02°27′50″ W	761.49	+760.79	−32.74	32.74	+24,908
2-3	N 87°35′37″ E	1076.62	+45.20	+1075.67	1075.67	+48,620
3-4	S 02°24′23″ E	290.00	−289.74	+12.18	2163.52	−626,858
4-5	S 09°49′21″ W	826.10	−813.99	−140.93	2034.77	−1,656,282
5-6	S 30°30′21″ W	68.00	−58.59	−34.52	7859.32	−108,938
6-1	N 67°56′54″ W	949.09	+356.33	−879.66	945.14	− 336,782
						−1,981,768

$$\frac{1{,}981{,}768 \text{ ft}^2}{2} = 990{,}884 \text{ ft}^2$$

$$\text{area } 1\text{-}2\text{-}3\text{-}4\text{-}5\text{-}6\text{-}1 = 990{,}884 \text{ ft}^2$$

$$\text{area of segment} = R^2 \left(\frac{\Delta \pi}{360} - \frac{\sin \Delta}{2} \right) = (1950 \text{ ft})^2 \left(\frac{(24°27′29″)\pi}{360} - \frac{\sin 24°27′29″}{2} \right)$$

$$= 24{,}425 \text{ ft}^2$$

$$\text{total area} = 990{,}884 \text{ ft}^2 + 24{,}425 \text{ ft}^2$$

$$= 1{,}015{,}309 \text{ ft}^2$$

PRACTICE PROBLEMS

(All dimensions and distances are in feet.)

1. Find the most westerly point in each traverse, and indicate which side would be the first for DMD computations.

(a)

line	departure east	departure west
AB	600	
BC	50	
CD		1000
DE		500
EF	100	
FA	750	

(b)

line	departure east	departure west
AB		500
BC	100	
CD	750	
DE	600	
EF	50	
FA		1000

(c)

line	departure east	departure west
AB	559.88	
BC	60.84	
CD		357.90
DE		294.87
EA	32.05	

(d)

line	departure east	departure west
AB		507.90
BC		243.75
CD	373.76	
DE	232.26	
EA	145.63	

(e)

line	departure
AB	+536.87
BC	+96.62
CD	+487.82
DE	−102.54
EF	−629.31
FG	−583.40
GA	+193.96

(f)

line	departure
AB	+629.31
BC	+102.54
CD	−487.82
DE	−96.62
EF	−536.87
FG	−193.94
GA	+583.40

2. Compute the area of the traverse ABCDEFA by the DMD method.

point	bearing	length	latitude N	latitude S	departure E	departure W
A						
	north	500.0	500		0	
B						
	N 45°00′ W	848.6	600			600
C						
	S 69°27′ W	854.4		300		800
D						
	S 11°19′ W	1019.8		1000		200
E						
	S 79°42′ E	1118.0		200	1100	
F						
	N 51°20′ E	640.3	400		500	
A						

3. Compute the area of the traverse ABCDEA by the DMD method.

point	bearing	length	latitude N	latitude S	departure E	departure W
A						
	N 28°19′ E	560.27	493.12		265.66	
B						
	N 56°23′ W	484.18	267.97			403.29
C						
	S 08°50′ W	375.42		371.04		57.72
D						
	S 45°10′ W	311.44		219.64		220.91
E						
	S 67°45′ E	449.83		170.41	416.26	
A						

4. Compute the area of the traverse ABCDEGHA by the DMD method.

line	bearing	length	latitude	departure
AB	S 25°57′ E	174.36	−156.78	+76.30
BC	S 29°29′ W	192.13	−167.25	−94.56
CD	S 78°30′ E	316.68	−63.14	+310.32
DE	N 47°51′ E	281.24	+188.73	+208.51
EF	N 35°23′ W	232.44	+189.51	−134.59
FG	N 25°24′ E	228.60	+206.50	+98.05
GH	N 78°41′ W	366.63	+71.94	−359.50
HA	S 21°12′ W	289.07	−269.51	−104.53

5. Compute the area of the traverse ABCDA by the coordinate method.

point	coordinates y	x
A	1000.00	1000.00
B	1493.12	1265.66
C	1761.09	862.37
D	1390.05	804.65
E	1170.41	583.74

6. Compute the area of the city lot shown using the formula $A = \frac{1}{2}ab \sin C$. AB = 218.5 ft, BC = 199.8 ft, CD = 231.2 ft, and DA = 231.2 ft.

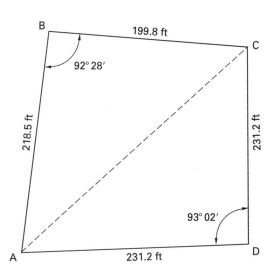

7. Compute the area of the city lot shown by using the formula $\frac{1}{2}ab \sin C$. OE = 140.4 ft, OF = 131.8 ft, OG = 144.8 ft, and OH = 172.0 ft.

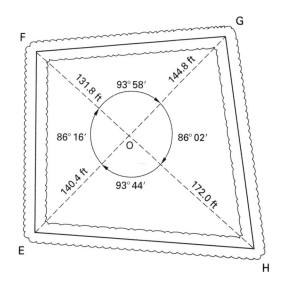

8. Find the area of the triangle with sides 12 ft, 14 ft, and 20 ft using the formula

$$A = \sqrt{s(s-a)(s-b)(s-c)}$$

9. Compute the area along the irregular boundary by using the trapezoidal rule.

Ties: B = 48.1, C = 52.6, D = 46.8, E = 39.9, F = 43.7, G = 58.0, H = 51.6, and J = 40.0.

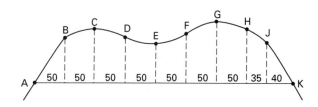

10. Compute the area of the segment with long chord = 491.67 ft and middle ordinate = 98.23 ft.

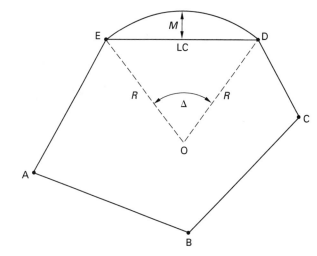

Traverse II

SOLUTIONS

1. (a)

most westerly point:	E

begin DMD:	EF

(b)

most westerly point:	B

begin DMD:	BC

(c)

most westerly point:	E

begin DMD:	EA

(d)

most westerly point:	C

begin DMD:	CD

(e)

most westerly point:	G

begin DMD:	GA

(f)

most westerly point:	G

begin DMD:	GA

2.

point	bearing	length	latitude N	latitude S	departure E	departure W	DMD	area plus	area minus
A									
	north	500.0	500		0		3200	1,600,000	
B									
	N 45°00′ W	848.6	600			600	2600	1,560,000	
C									
	S 69°27′ W	854.4		300		800	1200		360,000
D									
	S 11°19′ W	1019.8		1000		200	200		200,000
E									
	S 79°42′ E	1118.0		200	1100		1100		220,000
F									
	N 51°20′ E	640.3	400		500		2700	1,080,000	
A								4,240,000	780,000
								780,000	
								3,460,000	

$$\text{area} = \frac{3{,}460{,}000 \text{ ft}^2}{(2)\left(43{,}560 \dfrac{\text{ft}^2}{\text{ac}}\right)} = \frac{1{,}730{,}000 \text{ ft}^2}{43{,}560 \dfrac{\text{ft}^2}{\text{ac}}} = \boxed{39.715 \text{ ac}}$$

3.

point	bearing	length	latitude N	latitude S	departure E	departure W	DMD	area plus	area minus
A									
	N 28°19′ E	560.27	493.12		265.66		1098.18	541,535	
B									
	N 56°23′ W	484.18	267.97			403.29	960.55	257,399	
C									
	S 08°50′ W	375.42		371.04		57.72	499.54		185,349
D									
	S 45°10′ W	311.44		219.64		220.91	220.91		48,521
E									
	S 67°45′ E	449.83		170.41	416.26		416.26		70,935
A									
								798,934	304,805
								304,805	
								494,129	

$$\text{area} = \frac{494{,}129 \text{ ft}^2}{(2)\left(43{,}560 \ \dfrac{\text{ft}^2}{\text{ac}}\right)} = \frac{247{,}065 \text{ ft}^2}{43{,}560 \ \dfrac{\text{ft}^2}{\text{ac}}} = \boxed{5.672 \text{ ac}}$$

4.

line	bearing	length	latitude	departure	DMD	area
AB	S 25°57′ E	174.36	−156.78	+76.30	112.82	−17,688
BC	S 29°29′ W	192.13	−167.25	−94.56	94.56	−15,815
CD	S 78°30′ E	316.68	−63.14	+310.32	310.32	−19,594
DE	N 47°51′ E	281.24	+188.73	+208.51	829.15	+156,485
EF	N 35°23′ W	232.44	+189.51	−234.59	903.07	+171,141
FG	N 25°24′ E	228.60	+206.50	+98.05	866.53	+178,938
GH	N 78°41′ W	366.63	+71.94	−359.50	605.08	+43,529
HA	S 21°12′ W	289.07	−269.51	−104.53	141.05	−38,014
						458,982

$$\text{area} = \frac{458{,}982 \text{ ft}^2}{(2)\left(43{,}560 \ \dfrac{\text{ft}^2}{\text{ac}}\right)} = \frac{229{,}491 \text{ ft}^2}{43{,}560 \ \dfrac{\text{ft}^2}{\text{ac}}} = \boxed{5.268 \text{ ac}}$$

Traverse II

5.

	y	x	x_n	$(y_{n-1} - y_{n+1})$	area
A	1000.00	1000.00	$(1000.00)(1170.41 - 1493.12) =$		$-322{,}710$
B	1493.12	1265.66	$(1265.66)(1000.00 - 1761.09) =$		$-963{,}281$
C	1761.09	862.37	$(862.37)(1493.12 - 1390.05) =$		$+88{,}884$
D	1390.05	804.65	$(804.65)(1761.09 - 1170.41) =$		$+475{,}291$
E	1170.41	583.74	$(583.74)(1390.05 - 1000.00) =$		$+227{,}688$

$$\text{area} = \frac{494{,}128 \text{ ft}^2}{(2)\left(43{,}560 \, \dfrac{\text{ft}^2}{\text{ac}}\right)} = \boxed{5.67 \text{ ac}}$$

6. $A = \frac{1}{2}ab \sin C$

$= (\frac{1}{2})(218.5 \text{ ft})(199.8 \text{ ft}) \sin 92°28' = \quad 21{,}808 \text{ ft}^2$

$= (\frac{1}{2})(231.2 \text{ ft})(231.2 \text{ ft}) \sin 93°02' = \quad \underline{26{,}689 \text{ ft}^2}$

$$\boxed{48{,}497 \text{ ft}^2}$$

7. $A = (\frac{1}{2})(131.8 \text{ ft})(144.8 \text{ ft}) \sin 93°58' = \quad 9{,}519 \text{ ft}^2$

$= (\frac{1}{2})(144.8 \text{ ft})(172.0 \text{ ft}) \sin 86°02' = \quad 12{,}423 \text{ ft}^2$

$= (\frac{1}{2})(172.0 \text{ ft})(140.4 \text{ ft}) \sin 93°44' = \quad 12{,}049 \text{ ft}^2$

$= (\frac{1}{2})(140.4 \text{ ft})(131.8 \text{ ft}) \sin 86°16' = \quad \underline{9{,}233 \text{ ft}^2}$

$$\boxed{43{,}224 \text{ ft}^2}$$

8. $\text{area} = \sqrt{\begin{array}{c}(23 \text{ ft})(23 \text{ ft} - 12 \text{ ft})(23 \text{ ft} - 14 \text{ ft}) \\ \times (23 \text{ ft} - 20 \text{ ft})\end{array}}$

$= \boxed{83 \text{ ft}^2}$

9. $(\frac{1}{2})(50)(48.1)$

$+ (50)\left(\begin{array}{c}\dfrac{48.1}{2} + 52.6 + 46.8 + 39.9 \\ +43.7 + 58.0 + \dfrac{51.6}{2}\end{array}\right)$

$+ (35)\left(\dfrac{51.6}{2}\right) + (35)\left(\dfrac{51.6 + 40.0}{2}\right)$

$+ (\frac{1}{2})(40.0)(40.0)$

$= \boxed{18{,}148 \text{ ft}^2}$

10. $\tan \dfrac{\Delta}{4} = \dfrac{2M}{C} = \dfrac{(2)(98.23 \text{ ft})}{491.67 \text{ ft}}$

$\Delta = 87.122°$

$R = \dfrac{C}{2 \sin \dfrac{\Delta}{2}} = \dfrac{491.67 \text{ ft}}{2 \sin \left(\dfrac{87.122°}{2}\right)}$

$= 356.73 \text{ ft}$

$\text{area} = R^2 \left(\dfrac{\pi \Delta}{360°} - \dfrac{\sin \Delta}{2}\right)$

$= (356.73 \text{ ft})^2 \left(\dfrac{\pi(87.122°)}{360°} - \dfrac{\sin 87.122°}{2}\right)$

$= \boxed{33{,}203 \text{ ft}^2}$

16 Partitioning of Land

line	bearing	length (ft)
AB	N 28°19′ E	560.27
BC	N 56°23′ W	484.18
CD	S 08°50′ W	375.42
DE		
EA	S 67°45′ E	449.83

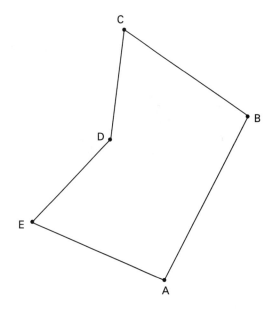

1. INTRODUCTION

An important part of the surveyor's work is the subdivision of tracts of land into two or more parts. Each partition of land is a separate problem, but there are basic techniques that can be used. These techniques will be illustrated in this chapter through the use of examples.

In solving the examples, latitudes, departures, and double meridian distances (DMDs) will be used where applicable. Unknown sides and angles of triangles will be solved by trigonometry. Formulas for area of triangles and trapezoids will be used to find lengths and bearings of sides.

2. LENGTH AND BEARING OF ONE SIDE UNKNOWN (THE CUTOFF LINE)

Example 16.1

The bearing and length of the side DE of the traverse ABCDEA are missing and are to be computed.

Solution

With the bearing and length of a side missing, the error of closure and precision cannot be computed. However, the traverse can be mathematically closed by giving a value to the latitude and departure of the side with missing bearing and length. These assigned values should be chosen to balance north and south latitudes and east and west departures of the traverse. This assumes that the latitudes and departures of the other sides are correct and any errors in them will be contained in the latitude and departure of the side with missing measurements.

line	bearing	length	latitude	departure
AB	N 28°19′ E	560.27	+493.23	+265.76
BC	N 56°23′ W	484.18	+268.06	−403.21
CD	S 08°50′ W	375.42	−370.97	−57.65
DE				
EA	S 67°45′ E	449.83	−170.33	+416.34
			+219.99	+221.24

tangent of bearing DE = $\dfrac{\text{dep}}{\text{lat}} = \dfrac{221.24 \text{ ft}}{219.99 \text{ ft}}$

bearing DE = S 45°09′ W

length DE = $\sqrt{\text{lat}^2 + \text{dep}^2}$

$\qquad = \sqrt{(219.99 \text{ ft})^2 + (221.24 \text{ ft})^2}$

$\qquad = 312.00 \text{ ft}$

In the solution, the sum of the latitude column is +219.99 ft, which represents the latitude of side DE with opposite sign. The sum of the departure column is +221.24 ft, which represents the departure of side DE with opposite sign.

3. LENGTHS OF TWO SIDES UNKNOWN

Example 16.2

The lengths for sides DE and EA of the traverse ABCDEA are unknown and are to be computed.

line	bearing	length (ft)
AB	N 65°04′ W	560.27
BC	S 30°14′ W	484.18
CD	S 84°33′ E	375.42
DE	S 48°13′ E	
EA	N 18°53′ E	

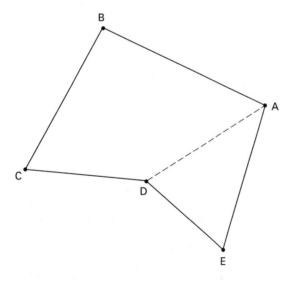

Solution

With measurements of two sides missing, the method of the *cutoff line* can be applied. The traverse ABCDA can be formed, excluding the sides with missing measurements. The side DA of this traverse will have an unknown bearing and length that can be computed by the method used in the preceding example. The side DA is the cutoff line. The cutoff line is usually used to isolate sides with missing measurements from sides with known measurements.

After the bearing and length of DA are computed, the side DA, together with sides DE and EA, can be considered to be the triangle ADE, which can be solved by the law of sines. When the triangle is solved, the solution is complete.

line	bearing	length	latitude	departure
AB	N 65°04′ W	560.27	+236.19	−508.05
BC	S 30°14′ W	484.18	−418.32	−243.80
CD	S 84°33′ E	375.42	−35.66	+373.72
DA				
			−217.79	−378.13

tangent of bearing DA = $\dfrac{378.13 \text{ ft}}{217.79 \text{ ft}}$

bearing DA = N 60°03′ E

length DA = $\sqrt{(378.13 \text{ ft})^2 + (217.79 \text{ ft})^2}$

$\qquad = 436.37 \text{ ft}$

In triangle DEA,

$$D = 180° - (60°03′ + 48°13′) = 71°44′$$

$$A = 60°03′ - 18°53′ = 41°10′$$

$$E = 180° - (71°44′ + 41°10′) = 67°06′$$

$$DE = \dfrac{(436.37 \text{ ft})(\sin 41°10′)}{\sin 67°06′}$$

$$= 311.82 \text{ ft}$$

$$EA = \dfrac{(436.37 \text{ ft})(\sin 71°44′)}{\sin 67°06′}$$

$$= 499.83 \text{ ft}$$

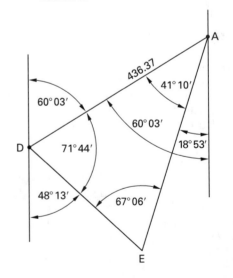

4. BEARING OF TWO SIDES UNKNOWN

Example 16.3

The bearing of the sides DE and EA of the traverse ABCDEA are unknown and are to be computed.

line	bearing	length (ft)
AB	N 65°04′ W	560.27
BC	S 30°14′ W	484.18
CD	S 84°33′ E	375.42
DE		311.82
EA		449.87

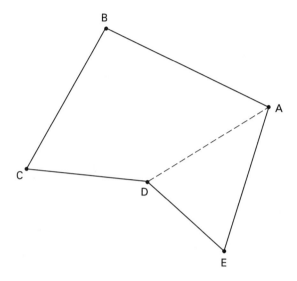

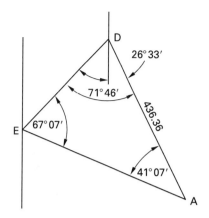

5. BEARING OF ONE SIDE AND LENGTH OF ANOTHER SIDE UNKNOWN

Example 16.4

The length of CD and the bearing of EA of the traverse ABCDEA are unknown and are to be computed.

line	bearing	length (ft)
AB	N 65°04′ W	560.27
BC	S 30°14′ W	484.18
CD	S 84°33′ E	
DE	S 48°13′ E	311.82
EA		449.87

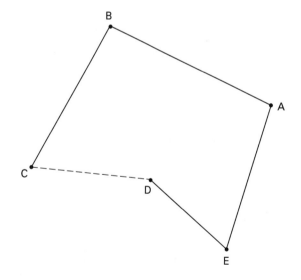

Solution

Bearings and lengths for sides AB, BC, and CD are the same as they are in the preceding example. Side DA is used as a cutoff line in traverse ABCDA again. The sides DA, DE, and EA form the triangle ADE again, but the known information is different from that in the preceding example. In this example, the lengths of all three sides of the triangle are known and the interior angles are to be determined in order to find bearings of the sides of the triangle. The law of cosines is appropriate for the solution.

$$\text{bearing DA} = \text{N } 60°03′ \text{ E}$$
$$\text{length DA} = 436.37 \text{ ft}$$

In triangle DEA,

$$\cos D = \frac{(436.37 \text{ ft})^2 + (311.82 \text{ ft})^2 - (449.87 \text{ ft})^2}{(2)(436.37 \text{ ft})(311.82 \text{ ft})}$$
$$D = 71°44′$$
$$\cos E = \frac{(311.82 \text{ ft})^2 + (449.87 \text{ ft})^2 - (436.37 \text{ ft})^2}{(2)(311.82 \text{ ft})(449.87 \text{ ft})}$$
$$E = 67°06′$$
$$A = 180° - (71°44′ + 67°06′)$$
$$= 41°10′$$
$$\text{bearing DE} = 180° - (60°03′ + 71°44′)$$
$$= \text{S } 48°13′ \text{ E}$$
$$\text{bearing EA} = 60°03′ - 41°10′ = \text{N } 18°53′ \text{ E}$$

Solution

The sides AB, BC, and DE, which have no missing measurements, can be connected in sequence to form a traverse with the missing side x in the following illustration. Arranging the sides in this order does not change the latitudes and departures of the sides. (Designating the sides in the manner shown is done to avoid confusion.) The bearing and length of the cutoff line x can be computed as in previous examples.

Partitioning

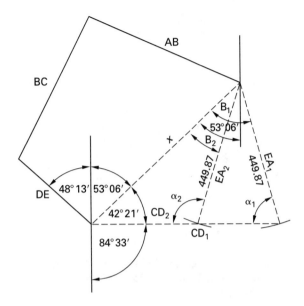

The line CD is given its correct bearing from the terminal of DE, but with an indefinite length. The length of EA is used as the radius of an arc, with center at the beginning point of AB, to intersect CD. In swinging the arc, it can be seen that it intersects CD at two points, giving two possible locations for EA and making two bearings. From the information given, the correct solution cannot be determined. Further information from the field would be necessary before determining the correct solution. (Not all problems of this nature will have two possible solutions.)

line	bearing	length	latitude	departure
AB	N 65°04′ W	560.27	+236.19	−508.05
BC	S 30°14′ W	484.18	−418.32	−243.80
DE	S 48°13′ E	311.82	−207.77	+232.51
			−389.90	−519.34

$$\text{tangent bearing } x = \frac{519.34 \text{ ft}}{389.90 \text{ ft}}$$

$$\text{bearing } x = \text{N } 53°06' \text{ E}$$

$$\text{length } x = \sqrt{(519.34 \text{ ft})^2 + (389.90 \text{ ft})^2}$$

$$= 649.41 \text{ ft}$$

In the triangle bounded by sides x, CD_1, and EA_1,

$$\sin \alpha_1 = \frac{(649.41 \text{ ft})(\sin 42°21')}{449.87}$$

$$\alpha_1 = 76°31'$$

$$\beta_1 = 180° - (76°31' + 42°21') = 61°08'$$

$$CD_1 = \frac{(449.87 \text{ ft})(\sin 61°08')}{\sin 42°21'} = 584.82 \text{ ft}$$

$$\text{bearing } EA_1 = 61°08' - 53°06' = \text{N } 08°02' \text{ W}$$

In the triangle bounded by sides x, CD_2, and EA_2,

$$\sin \alpha_2 = \frac{(649.41 \text{ ft})(\sin 42°21')}{449.87}$$

$$\alpha_2 = (103°29')(\text{related angle to } 76°31')$$

$$\beta_2 = 180° - (103°29' + 42°21') = 34°10'$$

$$CD_2 = \frac{(449.87 \text{ ft})(\sin 34°10')}{\sin 42°21'} = 375.04 \text{ ft}$$

$$\text{bearing } EA_2 = 53°06' - 34°10'$$

$$= \text{N } 18°56' \text{ E}$$

6. AREAS CUT OFF BY A LINE BETWEEN TWO POINTS ON THE PERIMETER

Example 16.5

The tract of land represented by the traverse ABCDEA is to be divided into two parts by a line from D to A.

line	bearing	length (ft)
AB	N 65°04′ W	560.27
BC	S 30°14′ W	484.18
CD	S 84°33′ E	375.42
DE	S 48°13′ E	311.44
EA	N 18°53′ E	449.83

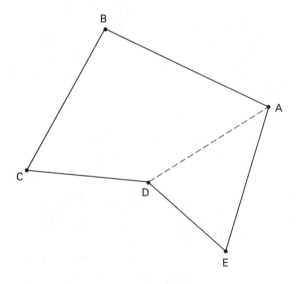

Solution

The area of the entire tract can be computed by DMD. The bearing and length for DA can be computed as in previous examples, and the areas of the two parts can be computed by DMD and their sum checked against the area of the entire tract.

line	bearing	length	total area		DMD	area
			latitude	departure		
AB	N 65°04′ W	560.27	+236.08	−507.90	995.40	+234,994
BC	S 30°14′ W	484.18	−418.38	−243.75	243.75	−101,980
CD	S 84°33′ E	375.42	−35.71	+373.76	373.76	−13,347
DE	S 48°13′ E	311.44	−207.56	+232.26	979.78	−203,363
EA	N 18°53′ E	449.83	+425.57	+145.63	1357.67	+577,784
			0.00	0.00		494,088

$$\text{area} = \frac{494{,}088 \text{ ft}^2}{2} = 247{,}044 \text{ ft}^2$$

line	bearing	length	cutoff line	
			latitude	departure
AB			+236.08	−507.90
BC			−418.38	−243.75
CD			−35.71	+373.76
DA				
			−218.01	−377.89

$$\text{tangent bearing DA} = \frac{377.80 \text{ ft}}{218.01 \text{ ft}}$$

$$\text{bearing DA} = \text{N } 60°01′ \text{ E}$$

$$\text{length DA} = \sqrt{(218.01 \text{ ft})^2 + (377.89 \text{ ft})^2}$$
$$= 436.27 \text{ ft}$$

line	bearing	length	area ABCDA		DMD	area
			latitude	departure		
AB			+236.08	−507.90	995.40	+234,994
BC			−418.38	−243.75	243.75	−101,980
CD			−35.71	+373.76	373.76	−13,347
DA	N 60°01′ E	436.27	+218.01	+377.89	1125.41	+245,351
			0.00	0.00		365,018

$$\text{area ABCDA} = \frac{365{,}018 \text{ ft}^2}{2} = 182{,}509 \text{ ft}^2$$

line	latitude	departure	area ADEA DMD	area
AD	−218.01	−377.89	377.89	−82,384
DE	−207.56	+232.26	232.26	−48,208
EA	+425.57	+145.63	610.15	+259,666
	0.00	0.00		129,074

$$\text{area ADEA} = \frac{129{,}074 \text{ ft}^2}{2} = 64{,}537 \text{ ft}^2$$

$$182{,}509 \text{ ft}^2 + 64{,}537 \text{ ft}^2 = 247{,}046 \text{ ft}^2$$

7. AREAS CUT OFF BY A LINE IN A GIVEN DIRECTION FROM A POINT ON THE PERIMETER

Example 16.6

The tract of land represented by the traverse ABCDEA is to be divided into two parts by a line from point D parallel to BC.

line	bearing	length (ft)
AB	N 65°04′ W	560.27
BC	S 30°14′ W	484.18
CD	S 84°33′ E	375.42
DE	S 48°13′ E	311.44
EA	N 18°53′ E	449.83

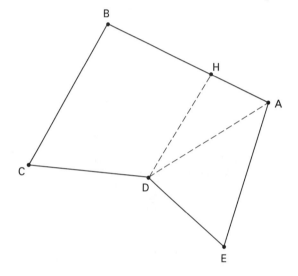

Solution

The line DH is drawn parallel to BC to represent the dividing line. The line DA is used as a cutoff line for the traverse ABCDA. The bearing and length for DA are computed as in previous examples. The triangle AHD is solved by the law of sines for sides AH and DH. HB is found by subtracting length AH from AB. Areas of HBCDH and AHDEA are computed by DMD and checked against the total area of ABCDEA.

From the preceding example,

$$\text{bearing DA} = \text{N } 60°01′ \text{ E}$$
$$\text{length DA} = 436.27 \text{ ft}$$
$$\text{bearing DH} = \text{bearing CB} = \text{N } 30°14′ \text{ E}$$
$$\text{bearing AH} = \text{bearing AB} = \text{N } 65°04′ \text{ W}$$

In triangle AHD,

$$\text{AH} = \frac{(436.27)(\sin 29°47′)}{\sin 95°18′}$$
$$= 217.64 \text{ ft}$$

$$DH = \frac{(436.27 \text{ ft})(\sin 54°55')}{\sin 95°18'}$$

$$= 358.53 \text{ ft}$$

$$HB = AB - AH = 560.27 \text{ ft} - 217.64 \text{ ft}$$

$$= 342.63 \text{ ft}$$

line	bearing	length (ft)
AB	N 65°04′ W	560.27
BC	S 30°14′ W	484.18
CD	S 84°33′ E	375.42
DE	S 48°13′ E	311.44
EA	N 18°53′ E	449.83

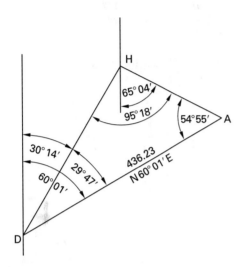

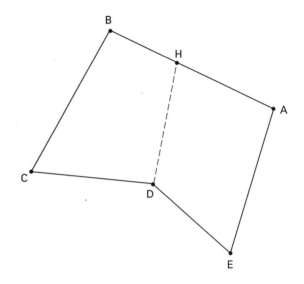

area HBCDH

line	bearing	length	latitude	departure	DMD	area
HB	N 65°04′ W	342.62	+144.44	−310.69	798.19	+115,291
BC	S 30°14′ W	484.18	−418.38	−243.75	243.75	−101,980
CD	S 84°33′ E	375.42	−35.71	+373.76	373.76	−13,347
DH*	S 30°14′ E	358.54	+309.65	+180.68	928.20	+287,417
			0.00	0.00		287,381

*closure forced

$$HBCDH = \frac{287,381 \text{ ft}^2}{2} = 143,690 \text{ ft}^2$$

area HBCDH

line	bearing	length	latitude	departure	DMD	area
AH	N 65°04′ W	217.65	+91.75	−197.36	558.42	+51,235
HD	S 30°14′ W	358.53	−309.75	−180.53	180.53	−55,921
DE	S 48°13′ E	311.44	−207.56	+232.26	232.26	−48,208
EA	N 18°43′ E	449.83	+425.57	+145.63	610.15	+259,662
			0.00	0.00		+206,768

$$\text{area AHDEA} = \frac{206,768 \text{ ft}^2}{2} = 103,384 \text{ ft}^2$$

$$143,690 \text{ ft}^2 + 103,384 \text{ ft}^2 = 247,074 \text{ ft}^2$$

8. DIVIDING TRACTS INTO TWO EQUAL PARTS BY A LINE FROM A POINT ON THE PERIMETER

Example 16.7

The tract represented by traverse ABCDEA is to be divided into two equal parts by a line from point D. The traverse ABCDEA is the same as that in the preceding example.

Solution

The line DH is drawn by inspection to divide the tract into two equal parts. In making computations for the solution of the problem, the line is considered to be in the exact location. The area of the entire tract is computed by DMD, and the area of the traverse HBCDH must be exactly one-half of the total area.

The area of the traverse ABCDA can be computed by DMD after the bearing and length of DA have been computed as in previous examples. Then the area of the triangle AHD can be found by subtracting area HBCDH from area ABCDA. Angle A of the triangle AHD can be found from bearings. Using the equation for the area of a triangle, the formula for triangle AHD is written $A = \frac{1}{2}(AH)(DA)\sin A$.

The triangle AHD is solved for DH by the law of cosines. Angle D is found by the law of sines. Bearing DH can now be found, as can length HB. Areas can be checked by DMD.

From Ex. 16.5,

$$\text{bearing DA} = N 69°01' E$$

$$\text{length DA} = 436.27 \text{ ft}$$

$$\text{area ABCDEA} = 247,044 \text{ ft}^2$$

$$\text{area ABCDA} = 182,509 \text{ ft}^2$$

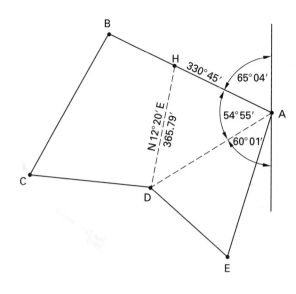

$$\text{area HBCDH} = \left(\tfrac{1}{2}\right)(\text{area ABCDEA}) = 123{,}522 \text{ ft}^2$$

$$\text{area AHD} = \text{area ABCDA} - \text{area HBCDH}$$

$$= 182{,}509 \text{ ft}^2 - 123{,}522 \text{ ft}^2$$

$$= 58{,}987 \text{ ft}^2$$

In triangle AHD,

$$\text{angle } A = 180° - (65°04' + 60°01')$$

$$= 54°55'$$

$$\text{area AHD} = \frac{(\text{AH})(\text{AD})(\sin 54°55')}{2}$$

$$\text{AH} = \frac{2A}{\text{AD} \sin 54°55'}$$

$$= \frac{(2)(58{,}987 \text{ ft}^2)}{(436.27 \text{ ft})(\sin 54°55')}$$

$$= 330.45 \text{ ft}$$

In triangle AHD, using the law of cosines,

$$\text{DH} = \sqrt{\begin{array}{l}(330.45 \text{ ft})^2 + (436.27 \text{ ft})^2 \\ \quad - (2)(330.45 \text{ ft})(436.27 \text{ ft})(\cos 54°55')\end{array}}$$

$$= 365.79 \text{ ft}$$

In triangle AHD, using the law of sines,

$$\sin D = \frac{(330.45 \text{ ft})(\sin 54°55')}{365.79}$$

$$D = 47°40'$$

$$\text{bearing DH} = 60°01' - 47°40' = \text{N} 12°21' \text{E}$$

Also,

$$\text{HB} = \text{AB} - \text{AH} = 560.27 \text{ ft} - 330.45 \text{ ft}$$

$$= 229.82 \text{ ft}$$

			area AHDEA			
line	bearing	length	latitude	departure	DMD	area
AH	N 65°04′ W	330.45	+139.31	−299.65	456.13	+63,543
HD	S 12°21′ W	365.79	−357.32	−78.24	78.24	−27,957
DE	S 48°13′ E	311.44	−207.56	+232.26	232.26	−48,106
EA	N 18°53′ E	449.83	+425.57	+145.63	610.15	+259,662
						247,142

$$\text{area AHDEA} = \frac{247{,}142 \text{ ft}^2}{2} = 123{,}571 \text{ ft}^2$$

9. DIVIDING AN IRREGULAR TRACT INTO TWO EQUAL PARTS

Example 16.8

The tract of land represented by the traverse ABCDEFA is to be divided into two equal parts by a line parallel to CD.

line	bearing	length (ft)
AB	N 80°00′ W	1015.43
BC	S 66°30′ W	545.22
CD	S 12°00′ E	480.97
DE	S 88°30′ E	750.26
EF	N 69°00′ E	639.18
FA	N 10°00′ E	306.78

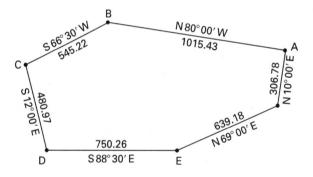

Solution

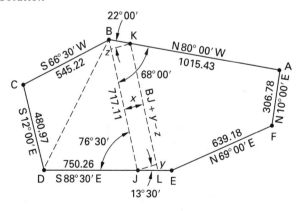

The problem can be solved in a manner similar to that used in the preceding example. A line KL can be drawn parallel to CD at the approximate location of the dividing line, using a scale drawing of the traverse. It can be seen by inspection that K will fall on AB and L will fall on DE.

From the point nearest to K, point B, the line BJ is drawn parallel to CD. (Point E could be used in place of point B.)

There are now four traverses formed:

(1) the original traverse ABCDEFA, for which the total area can be found by the DMD method

(2) the traverse KBCDLK, for which the area can be found by taking one-half of the total area

(3) the traverse BCDJB, for which the area can be found after first determining the bearing and length of BJ by use of the cutoff line DB as in previous examples

(4) the traverse KBJLK, a trapezoid, for which the area can be found by subtracting the area of BCDJB from the area of KBCDLK

The altitude x of a trapezoid can be found as the unknown quantity in the formula for the area of a trapezoid in which the area and the lengths of bases BJ and KL are known. KL can be expressed in terms of BJ and x to give a quadratic equation of the form $Ax^2 + Bx + C = 0$, which can be solved by the quadratic formula

$$x = \frac{-B \pm \sqrt{B^2 - 4AC}}{2A}$$

total area

line	bearing	length	latitude	departure	DMD	area
AB	N 80°00′ W	1015.43	+176.33	−1000.00	2000.00	+352,660
BC	S 66°30′ W	545.22	−217.41	−500.00	500.00	−108,705
CD	S 12°00′ E	480.97	−470.46	+100.00	100.00	−47,046
DE	S 88°30′ E	750.26	−19.64	+750.00	950.00	−18,658
EF	N 69°00′ E	639.18	+229.06	+596.73	2296.73	+526,089
FA	N 10°00′ E	306.78	+302.12	+53.27	2946.73	+890,266
			0.00	0.00		1,594,606

$$\text{total area} = \frac{1,594,606 \text{ ft}^2}{2} = 797,303 \text{ ft}^2$$

$$\text{area KBCDLK} = \frac{797,303 \text{ ft}^2}{2} = 398,652 \text{ ft}^2$$

To find the bearing and length of BJ, the cutoff line DB is found as the missing side of BCDB and the triangle BDJ is solved.

line	latitude	departure
BC	−217.41	−500.00
CD	−470.46	+100.00
DB	−687.87	−400.00

$$\text{tangent bearing DB} = \frac{400.00 \text{ ft}}{687.87 \text{ ft}}$$

$$\text{bearing DB} = \text{N } 30°11′ \text{ E}$$

$$\text{length DB} = \sqrt{(687.87 \text{ ft})^2 + (400.00 \text{ ft})^2}$$

$$= 795.72 \text{ ft}$$

In triangle DJB,

$$DJ = \frac{(795.72 \text{ ft})(\sin 42°11′)}{\sin 76°30′} = 549.46 \text{ ft}$$

$$JB = \frac{(795.72 \text{ ft})(\sin 61°19′)}{\sin 76°30′} = 717.94 \text{ ft}$$

$$\text{bearing JB} = \text{bearing DC} = \text{N } 12°00′ \text{ W}$$

total area

line	bearing	length	latitude	departure	DMD	area
BC	S 66°30′ W	545.22	−217.41	−500.00	500.00	−108,705
CD	S 12°00′ E	480.97	−470.46	+100.00	100.00	−47,046
DJ	S 88°30′ E	549.46	−14.38	+549.27	749.27	−10,774
JB	N 12°00′ W	717.94	+702.25	−149.27	1149.27	+807,075
			0.00	0.00		640,550

$$\text{area BCDJB} = \frac{640,550 \text{ ft}^2}{2} = 320,275 \text{ ft}^2$$

In the trapezoid BJLK,

$$y = x \tan 13°30′$$

$$z = x \tan 22°00′$$

$$LK = JB + y - z$$

$$= JB + x \tan 13°30′ - x \tan 22°00′$$

$$\text{area KBCDKL} = 398,652 \text{ ft}^2$$

$$\text{area BCDJB} = 320,275 \text{ ft}^2$$

$$\text{area BJLKB} = 78,377 \text{ ft}^2$$

$$\text{area BJLKB} = \frac{(JB + LK)x}{2}$$

$$78,377 = \frac{(JB + JB + x \tan 13°30′ - x \tan 22°00′)x}{2}$$

$$= \frac{(2JB + x \tan 13°30′ - x \tan 22°00′)x}{2}$$

$$= (JB)x - \frac{(\tan 22°00′ - \tan 13°30′)x^2}{2}$$

$$= (717.94)x - (0.0820)x^2 (0.0820)x^2$$

$$- 717.94x + 78,377$$

$$= 0$$

Substituting in the quadratic formula,

$$x = \frac{717.94 \pm \sqrt{(-717.94)^2 - (4)(0.0820)(78,377)}}{(2)(0.0820)}$$

$$= 110.57 \text{ ft}$$

Then,

$$JL = \frac{110.57 \text{ ft}}{\cos 13°30′} = 113.71 \text{ ft}$$

$$KB = \frac{110.57 \text{ ft}}{\cos 22°00′} = 119.25 \text{ ft}$$

$$LE = DE - DJ - JL$$
$$= 750.26 \text{ ft} - 549.46 \text{ ft} - 113.71 \text{ ft}$$
$$= 87.09 \text{ ft}$$
$$AK = AB - KB = 1015.43 \text{ ft} - 119.25 \text{ ft}$$
$$= 896.18 \text{ ft}$$
$$LK = JB + y - z$$
$$= 717.94 \text{ ft} + (110.57 \text{ ft})(\tan 13°30')$$
$$\quad - (110.57 \text{ ft})(\tan 22°00')$$
$$= 699.81 \text{ ft}$$
$$DL = DE - LE = 750.26 \text{ ft} - 87.09 \text{ ft}$$
$$= 663.17 \text{ ft}$$

line	bearing	length	latitude	departure	DMD	area
			\multicolumn{2}{c}{total area}		
AK	N 80°00' W	896.14	+155.61	−882.53	882.53	+137,330
KL	S 12°00' E	699.81	−684.51	+145.50	145.50	−99,596
LE	S 88°30' E	87.09	−2.28	+87.06	378.06	−862
EF	N 69°00' E	639.18	+229.06	+596.73	1061.85	+243,214
FA	N 10°00' E	306.78	+302.12	+53.27	1711.85	+517,166
			0.00	0.00		797,252

$$\text{area AKLEFA} = \frac{797{,}252 \text{ ft}^2}{2} = 398{,}626 \text{ ft}^2$$

$$\text{error in area} = 398{,}656 \text{ ft}^2 - 398{,}626 \text{ ft}^2 = 30 \text{ ft}^2$$

10. CUTTING A GIVEN AREA FROM AN IRREGULAR TRACT

Example 16.9

Two acres are to be cut off of the westerly end of the tract represented by the traverse ABCDA. At what easterly distance from point D along line DC will a boundary line parallel to AD intersect DC?

line	bearing	length
AB	N 51°00' E	647.81
BC	S 30°22' E	449.76
CD	S 64°14' W	596.15
DA	N 39°00' W	308.17

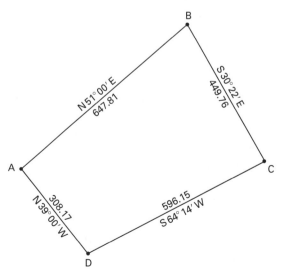

Solution

After plotting the traverse, it can be seen that the interior angle at A is a right angle.

Cutting two acres off the westerly end of the tract implies that exactly two acres are to be cut off and that the line cutting off the tract will be parallel to the westerly side DA.

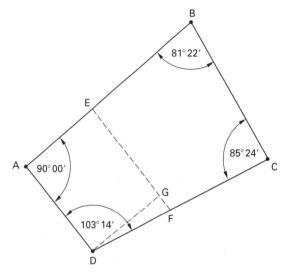

Computing the area of the entire tract by DMD gives an area of 223,440 ft^2 = 5.359 ac. The area to be cut off is 2 ac = 87,120 ft^2.

First draw EF parallel to DA at the approximate location of the dividing line, forming the trapezoid AEFD. Then drop a perpendicular to EF from D, which intersects EF at G. DG is parallel to AE. Using bearings,

$$\text{angle } FDG = 39°00' + 64°14' - 90°00'$$
$$= 13°14'$$

Also,

$$GF = EF - 308.17$$

In triangle DGF,

$$DG = \frac{GF}{\tan 13°14'} = \frac{EF - 308.17 \text{ ft}}{\tan 13°14'}$$

In trapezoid AEFD,

$$\text{area} = \frac{(\text{base} + \text{base})(\text{altitude})}{2}$$

$$87{,}120 = \frac{(EF + 308.17 \text{ ft})DG}{2}$$

$$87{,}120 = \frac{(EF + 308.17 \text{ ft})(EF - 308.17 \text{ ft})}{2 \tan 13°14'}$$

$$87{,}120 = \frac{(EF^2 - 308.17 \text{ ft})^2}{2 \tan 13°14'}$$

$$EF^2 = (2 \tan 13°14')(87{,}120 \text{ ft}^2) + (308.17 \text{ ft})^2$$

$$EF = 368.70 \text{ ft}$$

Partitioning

16-10 **LAND SURVEYOR REFERENCE MANUAL**

Solving for DG,

$$87{,}120 \text{ ft}^2 = \frac{(\text{EF} + 308.17 \text{ ft})\text{DG}}{2}$$

$$87{,}120 \text{ ft}^2 = \frac{(368.70 \text{ ft} + 308.17 \text{ ft})\text{DG}}{2}$$

$$\text{DG} = \frac{(2)(87{,}120 \text{ ft}^2)}{368.70 \text{ ft} + 308.17 \text{ ft}} = 257.42 \text{ ft}$$

$$\text{DF} = \frac{\text{DG}}{\cos 13°14'} = \frac{257.42 \text{ ft}}{\cos 13°14'} = 264.44 \text{ ft}$$

Check:

$$\frac{(368.70 \text{ ft} + 308.17 \text{ ft})(257.42 \text{ ft})}{2} = 87{,}120 \text{ ft}^2$$

11. ANALYTIC GEOMETRY IN PARTING LAND

In many situations, analytic geometry is more applicable to the solution of problems encountered in surveying than is trigonometry. Both analytic geometry and surveying deal with the coordinates of points. The equations of analytic geometry can be used in surveying problems as written or with slight modification. (Analytic geometry for surveyors has been discussed in Chap. 12.)

Example 16.10

For the traverse ABCDEA shown in Ex. 16.5, find the intersection of the line connecting points A and C and the line connecting points B and D.

point	coordinates	
	y	x
A	1000.00	1000.00
B	1236.08	492.10
C	817.70	248.35
D	782.01	622.11
E	574.45	854.37

Solution

Using the two-point form of the equation of a line and designating A as point 1 and C as point 2,

$$y - y_1 = \left(\frac{y_2 - y_1}{x_2 - x_1}\right)(x - x_1)$$

$$y - 1000.00 = \left(\frac{817.70 - 1000.00}{248.35 - 1000.00}\right)(x - 1000.00)$$

$$y - 1000.00 = (0.24253)(x - 1000.00)$$

$$y - 1000.00 = 0.24253x - 2242.53$$

$$0.24253x - y = -757.47$$

Designating B as point 1 and D as point 2,

$$y - y_1 = \left(\frac{y_2 - y_1}{x_2 - x_1}\right)(x - x_1)$$

$$y - y_1 = \left(\frac{782.01 - 1236.08}{622.08 - 492.03}\right)(x - x_1)$$

$$y - 1236.08 = (-3.49150)(x - 492.03)$$

$$y - 1236.08 = -3.49150x + 1717.92$$

$$3.49150x + y = 2954.00$$

Solving simultaneously,

$$\begin{aligned} 0.24253x - y &= -757.47 \\ 3.49150x + y &= 2954.00 \\ \hline 3.73403x &= 2196.53 \\ x &= 588.25 \end{aligned}$$

Substituting, $y = 900.14$.

The two lines intersect at $(588.25, 900.14)$.

12. AREAS CUT OFF BY A LINE IN A GIVEN DIRECTION FROM A POINT ON THE PERIMETER USING ANALYTIC GEOMETRY

Example 16.11

The tract of land represented by the traverse ABCDEA, Ex. 16.6, is to be divided into two parts by a line from point D parallel to side BC. Coordinates of traverse points are shown.

point	coordinates	
	y	x
A	1000.00	1000.00
B	1236.08	492.03
C	817.72	248.31
D	782.01	622.08
E	574.45	854.35

Solution

$$\text{slope DH} = \text{slope CB} = \frac{1236.08 - 817.72}{492.03 - 248.31} = 1.716560$$

(The slope of line DH can also be found as the cotangent of the bearing angle.)

The equation of line DH is

$$y - y_1 = m(x - x_1)$$

$$y - 782.01 = (1.716560)(x - 622.08)$$

$$y - 782.01 = 1.716560x - 1067.84$$

$$1.7167x - y = 285.83$$

PROFESSIONAL PUBLICATIONS, INC.

The equation of line AB is

$$y - y_1 = \left(\frac{y_2 - y_1}{x_2 - x_1}\right)(x - x_1)$$

$$y - 1000.00 = \left(\frac{1236.08 - 1000.00}{492.03 - 1000.00}\right)(x - 1000.00)$$

$$y - 1000.00 = (-0.464752)(x - 1000.00)$$

$$y - 1000.00 = -0.4648x + 464.75$$

$$0.4648x + y = 1464.75$$

Solving simultaneously,

$$
\begin{aligned}
1.7167x - y &= 285.83 \\
0.4648x + y &= 1464.75 \\
\hline
2.1815x &= 1750.58 \\
x &= 802.47 \\
y &= 1091.76
\end{aligned}
$$

The coordinates of point H are $(802.47, 1091.76)$.

$$\text{length DH} = \sqrt{\begin{array}{l}(802.47 - 622.08)^2 \\ + (1091.76 - 782.01)^2\end{array}}$$

$$= 358.45$$

$$\text{length AH} = \sqrt{\begin{array}{l}(1000.00 - 802.47)^2 \\ + (1000.00 - 1091.76)^2\end{array}}$$

$$= 217.80$$

$$\text{length HB} = 560.27 - 217.80 = 342.47$$

PRACTICE PROBLEMS

(All distances and dimensions are in feet.)

1. Compute the bearing and length of the side DE of the traverse ABCDEA.

line	bearing	length	latitude	departure
AB	N 28°19′ E	560.27	+493.23	+265.76
BC	N 56°23′ W	484.18	+268.06	−403.21
CD	S 08°50′ W	375.42	−370.97	−57.65
DE				
EA	S 67°45′ E	449.83	−170.33	+416.34

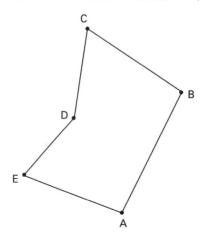

2. Compute the length of the sides DE and EA of the traverse ABCDEA. Use DA as a cutoff line. Draw a sketch of triangle DEA to a scale of 1 in = 200 ft. Show the size of angles in sketch.

line	bearing	length
AB	N 28°19′ E	560.27
BC	N 56°23′ W	484.18
CD	S 08°50′ W	375.42
DE	S 45°10′ W	
EA	S 67°45′ E	

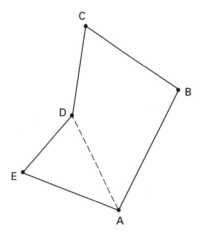

3. Compute the bearing of the sides DE and EA of the traverse ABCDEA. Use DA as a cutoff line. Draw a sketch of triangle DEA to a scale of 1 in = 200 ft. Show the size of angles in sketch.

line	bearing	length
AB	N 28°19′ E	560.27
BC	N 56°23′ W	484.18
CD	S 08°50′ W	375.42
DE		311.44
EA		449.83

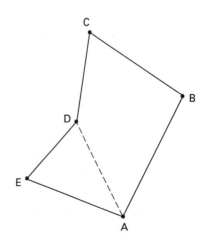

4. Compute the length of side CD and the bearing of side EA of the traverse ABCDEA. Use the following procedure.

step 1: Designate the cutoff line as XA.

step 2: From X draw CD in its given direction.

step 3: Using A as a compass point and length of side EA as the radius, draw an arc to intersect CD. A line from this intersection to A gives the direction of EA.

There are two intersections so there will be two solutions. AB: N 28°19′ E, 560.27; BC: N 56°23′ W, 484.18; CD: S 08°50′ W; DE: S 45°10′ W, 311.44; and EA: 449.83.

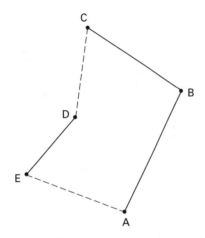

5. The tract of land represented by the traverse ABCDEA has been divided into two parts by a line from D to A. Balanced latitudes and departures are shown. Compute the area of the entire tract, the area of the tract ABCDA, and the area of the tract DEAD by the DMD method.

line	bearing	length	latitude	departure
AB	N 28°19′ E	560.27	+493.12	+265.66
BC	N 56°23′ W	484.18	+267.97	−403.29
CD	S 08°50′ W	375.42	−371.04	−57.72
DE	S 45°10′ W	311.44	−219.64	−220.91
EA	S 67°45′ E	449.83	−170.41	+416.26

6. The tract of land represented by the traverse ABCDEA is to be divided into two parts by a line from D parallel to the side BC, which intersects the side AB at point H. The length AH must be computed so that the point H can be located by measuring from point A. To find AH, the triangle AHD is formed by using the

cutoff line AD. The bearing of AD has been computed to be S 26°36′ E and the length of AD has been computed to be 436.23′. The area of the tract has been computed as shown. Find the lengths AH, DH, and HB, and compute the area of the two subdivided tracts by the DMD method, and check the sum of the two areas against the total area.

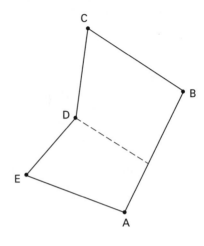

7. The tract of land represented by the traverse ABCDEA in Prob. 6 is to be divided into two equal parts by a line from point D. The dividing line intersects side AB at the point H. The tract has been surveyed and the following information has been found from computations.

(1) area ABCDEA = 247,065 ft^2

(2) area ABCDA = 182,300 ft^2

(3) bearing DA = S 26°36′ E

(4) length DA = 436.23 ft

Compute the length of AH, HB, and DH, and the bearing DH.

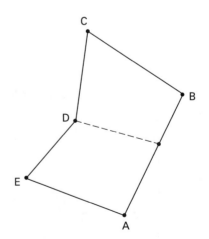

8. The tract of land represented by the traverse ABCDEFA is to be divided into two equal parts by a line parallel to AB. A survey has been made and computations for the area of the tract have been made as shown. Designate dividing line as KL and the line parallel to AB as FJ. Find the length of LC, BL, FK, JL, and KL, and compute the area of ABLKFA by the DMD method.

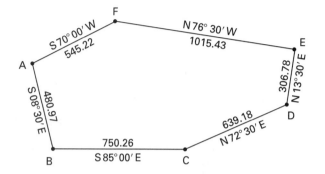

SOLUTIONS

1.

line	bearing	length	latitude	departure
AB	N 28°19′ E	560.27	+493.23	+265.76
BC	N 56°23′ W	484.18	+268.06	−403.21
CD	S 08°50′ W	375.42	−370.97	−57.65
DE				
EA	S 67°45′ E	449.83	−170.33	+416.34
			219.99	221.24

$$\text{bearing DE} = \tan^{-1} \frac{221.24 \text{ ft}}{219.99 \text{ ft}} = 45°10′$$

$$= \boxed{\text{S } 45°10′ \text{ W}}$$

$$\text{length DE} = \sqrt{(221.24 \text{ ft})^2 + (219.99 \text{ ft})^2}$$

$$= \boxed{312.00 \text{ ft}}$$

2.

line	bearing	length	latitude	departure
AB	N 28°19′ E	560.27	+493.23	+265.76
BC	N 56°23′ W	484.18	+268.06	−403.21
CD	S 08°50′ W	375.42	−370.97	−57.65
DA				
			390.32	−195.10

$$\text{bearing DA} = \tan^{-1} \frac{195.10 \text{ ft}}{390.32 \text{ ft}} = 26°33′ = \text{S } 26°33′ \text{ E}$$

$$\text{length DA} = \sqrt{(195.10 \text{ ft})^2 + (390.32 \text{ ft})^2}$$

$$= 436.36 \text{ ft}$$

Solution of triangle DEA:

$$\text{length DE} = \frac{(436.36 \text{ ft})(\sin 41°12′)}{\sin 67°05′} = \boxed{311.98 \text{ ft}}$$

$$\text{length EA} = \frac{(436.36 \text{ ft})(\sin 71°43′)}{\sin 67°05′} = \boxed{449.84 \text{ ft}}$$

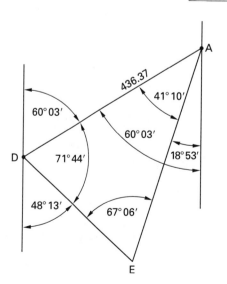

Partitioning

3.

line	bearing	length	latitude	departure
AB	N 28°19' E	560.27	+493.23	+265.76
BC	N 56°23' W	484.18	+268.06	−403.21
CD	S 08°50' W	375.42	−370.97	−57.65
DA			390.32	195.10

$$\text{bearing DA} = \tan^{-1}\left(\frac{195.10 \text{ ft}}{390.32 \text{ ft}}\right) = 26°33'$$

$$= \text{S } 26°33' \text{ E}$$

$$\text{length DA} = \sqrt{(195.10 \text{ ft})^2 + (390.32 \text{ ft})^2}$$

$$= 436.36 \text{ ft}$$

Solution of triangle DEA:

$$\cos E = \frac{(311.44 \text{ ft})^2 + (449.83 \text{ ft})^2 - (436.36 \text{ ft})^2}{(2)(311.44 \text{ ft})(449.83 \text{ ft})}$$

$$E = 67°07'$$

$$\cos D = \frac{(311.44 \text{ ft})^2 + (436.36 \text{ ft})^2 - (449.83 \text{ ft})^2}{(2)(311.44 \text{ ft})(436.36 \text{ ft})}$$

$$D = 71°46'$$

$$\cos A = \frac{(436.36 \text{ ft})^2 + (449.83 \text{ ft})^2 - (311.44 \text{ ft})^2}{(2)(436.36 \text{ ft})(449.83 \text{ ft})}$$

$$A = 41°07'$$

$$\text{bearing DE} = \boxed{\text{S } 45°13' \text{ W}}$$

$$\text{bearing EA} = \boxed{\text{S } 67°40' \text{ E}}$$

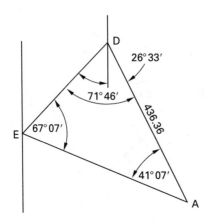

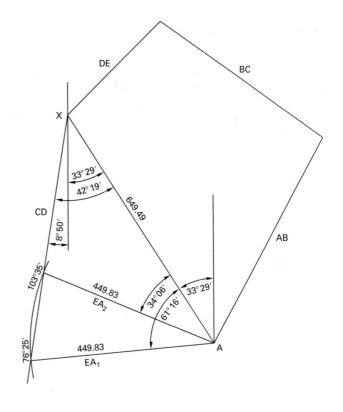

4. The computations for cutoff line are as follows.

line	bearing	length	latitude	departure
			493.23	265.76
			268.06	403.21
			219.58	220.86
			541.71	358.31

$$\text{bearing XA} = \tan^{-1}\frac{358.31 \text{ ft}}{541.71 \text{ ft}} = \text{S } 33°29' \text{ E}$$

$$\text{length XA} = 649.49 \text{ ft}$$

Solution of triangle no. 1:

$$\sin\theta_1 = \frac{(649.49 \text{ ft})(\sin 42°19')}{449.83 \text{ ft}}$$

$$\theta_1 = 76°25'$$

$$\text{CD}_1 = \frac{(449.83 \text{ ft})(\sin 61°16')}{\sin 42°19'} = 585.90 \text{ ft}$$

$$\text{bearing EA} = \boxed{\text{N } 85°15' \text{ E}}$$

$$\text{length CD} = \boxed{585.90 \text{ ft}}$$

Solution of triangle no. 2:

$$\sin\theta_2 = \frac{(649.49\ \text{ft})(\sin 42°19')}{449.83}$$

$$\theta_2 = 103°35'$$

$$CD_2 = \frac{(449.83\ \text{ft})(\sin 34°06')}{\sin 42°19'} = 374.60\ \text{ft}$$

bearing EA = S 67°35' E

length CD = 374.60 ft

line	latitude	departure	DMD	area
DE	−219.64	−220.91	220.91	−48,521
EA	−170.41	+416.26	416.26	−70,935
AD	+390.05	−195.35	637.17	−248,528
				129,072

$$\text{area} = \frac{129{,}072\ \text{ft}^2}{(2)\left(43{,}560\ \dfrac{\text{ft}^2}{\text{ac}}\right)} = \frac{64{,}536\ \text{ft}^2}{43{,}560\ \dfrac{\text{ft}^2}{\text{ac}}} = \boxed{1.482\ \text{ac}}$$

total area check: $182{,}528\ \text{ft}^2 + 64{,}536\ \text{ft}^2 = 247{,}064\ \text{ft}^2$

5.

line	bearing	length	latitude	departure	DMD	area
AB	N 28°19' E	560.27	+493.12	+265.66	1098.18	+541,535
BC	N 56°23' W	484.18	+267.97	−403.29	960.55	+257,399
CD	S 08°50' W	375.42	−371.04	−57.72	499.54	−185,349
DE	S 45°10' W	311.44	−219.64	−220.91	220.91	−48,521
EA	S 67°45' E	449.83	−170.41	+416.26	416.26	−70,935
						494,129

$$\text{area} = \frac{494{,}129\ \text{ft}^2}{(2)\left(43{,}560\ \dfrac{\text{ft}^2}{\text{ac}}\right)} = \frac{247{,}065\ \text{ft}^2}{43{,}560\ \dfrac{\text{ft}^2}{\text{ac}}} = 5.672\ \text{ac}$$

The computations for DA are as follows.

line	bearing	length	latitude	departure
DE			−219.64	−220.91
EA			−170.41	+416.26
			390.05	195.35

$$\text{bearing DA} = \tan^{-1}\frac{195.35\ \text{ft}}{390.05\ \text{ft}} = 26°36'$$

$$= \text{S } 26°36'\text{ E}$$

$$\text{length DA} = \sqrt{(390.05\ \text{ft})^2 + (195.35\ \text{ft})^2}$$

$$= 436.23\ \text{ft}$$

line	latitude	departure	DMD	area
AB	+493.12	265.66	656.36	323,664
BC	+267.97	−403.29	518.73	139,004
CD	−371.04	−57.72	57.72	−21,416
DA	−390.05	+195.35	195.35	−76,196
				365,056

$$\text{area} = \frac{365{,}056\ \text{ft}^2}{(2)\left(43{,}560\ \dfrac{\text{ft}^2}{\text{ac}}\right)} = \frac{182{,}528\ \text{ft}^2}{43{,}560\ \dfrac{\text{ft}^2}{\text{ac}}} = 4.109\ \text{ac}$$

6.

line	bearing	length	latitude	departure	DMD	area
AB	N 28°19' E	560.27	+493.12	+265.66	1098.18	+541,535
BC	N 56°23' W	484.18	+267.97	−403.29	960.55	+257,399
CD	S 08°50' W	375.42	−371.04	−57.72	499.54	−185,349
DE	S 45°10' W	311.44	−219.64	−220.91	220.91	−48,521
EA	S 67°45' E	449.83	−170.41	+416.26	416.26	−70,935
						494,129

$$\text{area} = \frac{494{,}129}{(2)\left(43{,}560\ \dfrac{\text{ft}^2}{\text{ac}}\right)} = \frac{247{,}065\ \text{ft}^2}{43{,}560\ \dfrac{\text{ft}^2}{\text{ac}}} = 5.672\ \text{ac}$$

Solution of triangle AHD:

$$DH = \frac{(436.23\ \text{ft})(\sin 54°55')}{\sin 95°18'} = \boxed{358.51\ \text{ft}}$$

$$AH = \frac{(436.23\ \text{ft})(\sin 29°47')}{\sin 95°18'} = \boxed{217.62\ \text{ft}}$$

$$HB = 560.27 - 217.62 = \boxed{342.65\ \text{ft}}$$

length DH = 358.51 ft

length AH = 217.62 ft

length HB = 342.65 ft

bearing DH = S 56°23' E

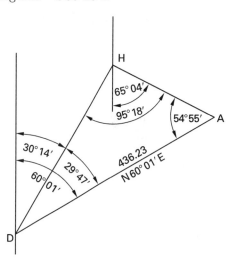

line	bearing	length	latitude	departure	DMD	area
HB	N 28°19′ E	342.65	+301.59	+162.48	759.48	+229,051
BC	N 56°23′ W	484.18	+267.98	−403.28	518.68	+138,996
CD	S 08°50′ W	375.42	371.03	−57.70	57.70	−21,408
DH	S 56°23′ E	358.51	−198.54	+298.50	298.50	−59,264
						287,375

$$\text{area} = \frac{287,375 \text{ ft}^2}{(2)\left(43,560 \dfrac{\text{ft}^2}{\text{ac}}\right)} = \frac{143,687 \text{ ft}^2}{43,560 \dfrac{\text{ft}^2}{\text{ac}}} = 3.299 \text{ ac}$$

line	bearing	length	latitude	departure	DMD	area
AH	N 28°19′ E	217.62	191.56	103.20	935.78	+179,258
HD	N 56°23′ W	358.51	198.44	298.59	740.39	+146,923
DE	S 45°10′ W	311.44	219.62	220.90	220.90	−48,514
EA	S 67°45′ E	449.83	170.38	416.29	416.29	−70,927
						206,740

$$\text{area} = \frac{206,740 \text{ ft}^2}{(2)\left(43,560 \dfrac{\text{ft}^2}{\text{ac}}\right)} = \frac{103,370 \text{ ft}^2}{43,560 \dfrac{\text{ft}^2}{\text{ac}}} = \boxed{2.373 \text{ ac}}$$

7. Solution of triangle AHD:

$$\text{area AHD} = 183,300 \text{ ft}^2 - \left(\tfrac{1}{2}\right)(247,065 \text{ ft}^2)$$
$$= 58,768 \text{ ft}^2$$

$$AH = \frac{(2)(58,768 \text{ ft}^2)}{(436.23 \text{ ft})(\sin 54°55')} = 329.56$$

$$DH = \sqrt{\begin{array}{c}(436.23 \text{ ft})^2 + (329.56 \text{ ft})^2 \\ - (2)(436.23 \text{ ft})(329.56 \text{ ft})(\cos 54°55')\end{array}}$$
$$= 365.57 \text{ ft}$$

$$\sin D = \frac{(3292.56 \text{ ft})(\sin 54°55')}{365.57 \text{ ft}}$$

$$D = 47°32'$$

$$HB = 560.27 \text{ ft} - 329.56 \text{ ft} = 230.71 \text{ ft}$$

$$AH = \boxed{329.56 \text{ ft}}$$

$$DH = \boxed{365.57 \text{ ft}}$$

bearing DH = S 74°08′ E

$$HB = \boxed{230.71 \text{ ft}}$$

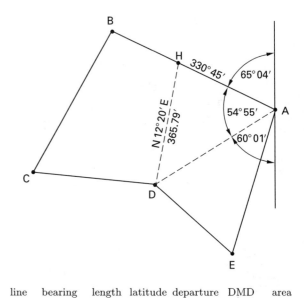

line	bearing	length	latitude	departure	DMD	area
AH	N 28°19′ E	329.56	+290.11	+156.31	988.83	+286,869
HD	N 74°08′ W	365.57	+99.94	−351.66	793.48	+79,300
DE	S 45°10′ W	311.44	−219.64	−220.91	220.91	−48,521
EA	S 67°45′ E	449.83	−170.41	+416.26	416.26	−70,935
						246,713

$$\text{area} = \frac{246,713 \text{ ft}^2}{(2)\left(43,560 \dfrac{\text{ft}^2}{\text{ac}}\right)} = \frac{123,356 \text{ ft}^2}{43,560 \dfrac{\text{ft}^2}{\text{ac}}} = 2.832 \text{ ac}$$

8.

line	bearing	length	lat	dep	DMD	area
AB	S 08°30′ E	480.97	−475.69	+71.09	71.09	−33,817
BC	S 85°00′ E	750.26	−65.39	+747.41	889.59	−58,170
CD	N 72°30′ E	639.18	+192.21	+609.60	2246.60	+431,819
DE	N 13°30′ E	306.78	+298.30	+71.62	2927.82	+873,369
EF	N 76°30′ W	1015.43	+237.05	−987.37	2012.07	+476,961
FA	S 70°00′ W	545.22	−186.48	−512.35	512.35	−95,543
						1,594,619

$$\text{area} = \frac{1,594,619 \text{ ft}^2}{(2)\left(43,560 \dfrac{\text{ft}^2}{\text{ac}}\right)} = \frac{797,310 \text{ ft}^2}{43,560 \dfrac{\text{ft}^2}{\text{ac}}} = 18.304 \text{ ac}$$

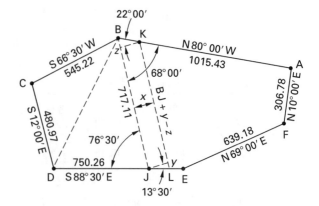

$$\text{area ABLKFA} = \frac{797,310 \text{ ft}^2}{2} = 398,655 \text{ ft}^2$$

The bearing and length of BF can be computed as follows.

line	bearing	length	latitude	departure
AB			-475.69	$+71.09$
BF			$(+662.17)$	$(+441.25)$
FA			-186.48	-512.34

$$\text{tangent bearing BF} = \frac{441.25 \text{ ft}}{662.17 \text{ ft}}$$

$$\text{bearing BF} = \text{N}\,33°41'\,\text{E}$$

$$\text{length} = \sqrt{(441.25 \text{ ft})^2 + (662.17 \text{ ft})^2}$$
$$= 795.72 \text{ ft}$$

Solution of triangle BJF:

$$\text{BJ} = \frac{(795.72 \text{ ft})(\sin 42°11')}{\sin 76°30'} = 549.51 \text{ ft}$$

$$\text{JF} = \frac{(795.72 \text{ ft})(\sin 61°19')}{\sin 76°30'} = 717.91 \text{ ft}$$

line	bearing	length	latitude	departure	DMD	area
AB	$\text{S}\,08°30'\,\text{E}$	480.97	-475.69	$+71.09$	71.09	$-33,817$
BJ	$\text{S}\,85°00'\,\text{E}$	549.51	-47.89	689.60	689.60	$-33,025$
JF	$\text{N}\,08°30'\,\text{W}$	717.91	$+710.06$	1130.85	1130.85	$+802,971$
FA	$\text{S}\,70°00'\,\text{W}$	545.22	-186.48	512.34	512.34	$-95,541$
						$\overline{640,588}$

$$\text{area} = \frac{640,588 \text{ ft}^2}{2} = 320,294 \text{ ft}^2$$

$$y = x \tan 13°30'$$
$$z = x \tan 22°00'$$

$$\text{LK} = \text{FJ} + y - z$$
$$= \text{FJ} + x \tan 13°30' - x \tan 22°00'$$

$$\text{area ABLKFA} = 398,655 \text{ ft}^2$$
$$\text{area ABJFA} = \underline{320,294 \text{ ft}^2}$$
$$\text{area FJLK} = 78,361 \text{ ft}^2$$

$$78,361 \text{ ft}^2 = \frac{(\text{FJ} + \text{FJ} + x \tan 13°30' - x \tan 22°00')x}{2}$$
$$= \frac{(2\text{FJ} + x \tan 13°30' - x \tan 22°00')x}{2}$$
$$= \frac{(\text{FJ})x - (\tan 22°00' - \tan 13°30')x^2}{2}$$
$$= 717.91x - 0.0819737x^2$$

$$0.819737x^2 - 717.91x + 78,361 = 0$$

$$x = \frac{717.91 \pm \sqrt{(-717.91)^2 - (4)(0.0819737)(78,361)}}{(2)(0.0819737)}$$
$$= 110.55 \text{ ft}$$

$$\text{JL} = \frac{110.55 \text{ ft}}{\cos 13°30'} = 113.69 \text{ ft}$$

$$\text{FK} = \frac{110.55 \text{ ft}}{\cos 22°00'} = 119.23 \text{ ft}$$

$$\text{LC} = 750.26 \text{ ft} - 549.51 \text{ ft} - 113.69 \text{ ft} = \boxed{87.06 \text{ ft}}$$

$$\text{BL} = 549.51 \text{ ft} + 113.69 \text{ ft} = \boxed{663.20 \text{ ft}}$$

$$\text{KL} = 717.91 \text{ ft} + (110.55 \text{ ft})(\tan 13°30')$$
$$- (1101.55 \text{ ft})(\tan 22°00')$$
$$= \boxed{699.79 \text{ ft}}$$

line	bearing	length	latitude	departure	DMD	area
AB	$\text{S}\,08°30'\,\text{E}$	480.97	-475.69	$+71.09$	71.09	35,817
BL	$\text{S}\,85°00'\,\text{E}$	663.20	-57.80	$+660.68$	802.86	$-46,405$
LK	$\text{N}\,08°30'\,\text{W}$	699.79	-692.12	-103.66	1360.08	$+941,335$
KF	$\text{N}\,76°30'\,\text{W}$	119.23	$+27.85$	-115.96	1140.66	$+31,767$
FA	$\text{S}\,70°00'\,\text{W}$	545.22	-186.48	-512.35	512.35	$-95,543$
						$\overline{797,341}$

$$\text{area} = \frac{797,341 \text{ ft}^2}{(2)\left(43,560 \dfrac{\text{ft}^2}{\text{ac}}\right)} = \frac{398,670 \text{ ft}^2}{43,560 \dfrac{\text{ft}^2}{\text{ac}}} = \boxed{9.152 \text{ ac}}$$

check: $398,670 \text{ ft}^2 - 398.655 \text{ ft}^2 = 15 \text{ ft}^2$
$$= 0.0003 \text{ ac} \quad [\text{error}]$$

17 Horizontal Curves

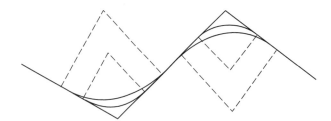

Figure 17.1 *Curves Connecting Tangent Lines*

Figure 17.1 shows that a choice of circular arcs, or curves, can be made after the tangent locations have been made. The curves are usually classified as to their *degree of curve*, the angle subtended by a portion of the curve 100 ft long. In selecting the degree of curve, consideration is given to design speed, topographic features, economy, and other variables.

2. GEOMETRY

Formulas for computations involving circular curves depend on certain principles of geometry and trigonometry.

3. INSCRIBED ANGLE

An *inscribed angle* is an angle that has its vertex on a circle and that has chords for its sides as shown in Fig. 17.2(a).

4. MEASURE OF AN INSCRIBED ANGLE

An inscribed angle is measured by one-half its intercepted arc as shown in Fig. 17.2(a).

5. MEASURE OF AN ANGLE FORMED BY A TANGENT AND A CHORD

An angle formed by a tangent and a chord is measured by one-half its intercepted arc as shown in Fig. 17.2(b).

1. SIMPLE CURVES

Highways consist of a series of straight sections joined by curved sections. The straight sections are known as *tangents*. The curves are most often circular arcs, known as *simple curves*, but may be spiral curves. Spiral curves are encountered more often on railroads.

Initial locations of highways usually consist of straight lines. Curves are later inserted to connect two intersecting tangents. Many curves of different radii, or degrees of curve, may be selected for any given intersection of tangents.

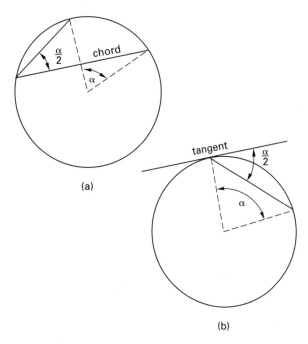

(a)

(b)

Figure 17.2 *Tangent and Arc Geometry*

6. RADIUS IS PERPENDICULAR TO TANGENT

The radius of a circle is perpendicular to a tangent at the point of tangency as shown in Fig. 17.3(a).

7. RADIUS IS PERPENDICULAR TO BISECTOR OF A CHORD

The perpendicular bisector of a chord passes through the center of the circle as shown in Fig. 17.3(b).

8. DEFINITIONS AND SYMBOLS

The following definitions of symbols are used in curve computations.

- C *(length of chord):* chord length

- Δ *(deflection angle):* central angle: angle at PI, or angle at center.

- D *(degree of curve):* central angle that subtends a 100 ft arc (arc basis)

- E *(external distance):* distance from PI to middle of curve

- L *(length of curve):* distance from PC to PT along the arc

- LC *(length of long chord):* distance from PC to PT; chord length for angle Δ

- M *(middle ordinate):* length of ordinate from middle of long chord to middle of curve

- PC *(point of curvature):* beginning of curve

- PI *(point of intersection):* point where two tangents intersect

- PT *(point of tangency):* end of curve

- R *(radius):* a straight line from the center of a circle to the circumference

- T *(tangent distance):* distance from PI to PC, or distance from PI to PT

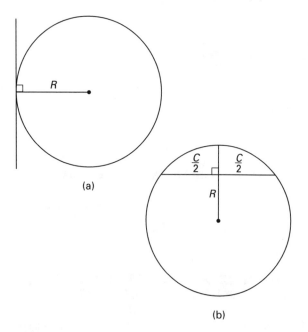

(a)

(b)

Figure 17.3 *Chord Geometry*

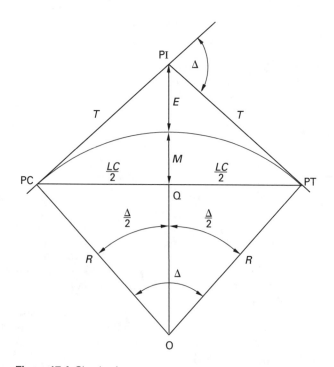

Figure 17.4 *Circular Arc*

9. DEFLECTION ANGLE EQUALS CENTRAL ANGLE

In Fig. 17.4, the sum of the interior angles of the polygon O-PC-PI-PT is 360°. The angles at the PC and PT each equal 90°. The sum of the interior angle at the PI and the deflection angle is 180°. The sum of the interior angle at the PI and the central angle is 180°. Therefore, the deflection angle equals the central angle.

10. HORIZONTAL CURVE FORMULAS

In computing various components of a circular curve, certain formulas derived from trigonometry are useful and necessary. In addition to the usual six trigonometric functions, two others are used in curve computations: *versed sine* (vers) and *external secant* (exsec).

Most surveying textbooks include tables for these functions.

$$\operatorname{vers}\phi = 1 - \cos\phi \qquad 17.1$$

$$\operatorname{exsec}\phi = \sec\phi - 1 \qquad 17.2$$

Other formulas commonly encountered are

$$T = R\tan\frac{\Delta}{2} \qquad 17.3$$

$$\mathrm{LC} = \frac{2\pi R\Delta}{360°} \qquad 17.4$$

$$C = 2R\sin\frac{D}{2} \qquad 17.5$$

$$M = R\left(1 - \cos\frac{\Delta}{2}\right) = R\operatorname{vers}\frac{\Delta}{2} \qquad 17.6$$

$$E = R\left(\sec\frac{\Delta}{2} - 1\right) = R\operatorname{exsec}\frac{\Delta}{2} \qquad 17.7$$

11. DEGREE OF CURVE

There are two definitions of *degree of curve*. The *arc definition* is used for highways and streets. The *chord definition* is used for railroads. By the arc definition, degree of curve, D, is the central angle that subtends a 100 ft arc. By the chord definition, degree of curve, D, is the central angle that subtends a 100 ft chord.

Using the arc definition,

$$R = \frac{(360°)(100\text{ ft})}{2\pi D} = \frac{5729.58}{D} \quad \left[\begin{array}{l}\text{for } D = 1°, \\ R = 5729.58\text{ ft}\end{array}\right] \quad 17.8$$

Using the chord definition,

$$R = \frac{50}{\sin\dfrac{D}{2}} \quad \left[\begin{array}{l}\text{for } D = 1°, \\ R = 5729.65\text{ ft}\end{array}\right] \quad 17.9$$

Values of R for various values of D are given in App. B.

When using the arc definition for curve computations with a 100 ft tape to lay out the curve in the field, measurements are actually chord lengths of 100 ft, and the arc length is somewhat greater. For curves up to 4°, the difference in arc length and chord length is negligible. For instance, the chord length for a 100 ft arc on a 4° curve is 99.980 ft. For curves of greater degree of curve, the actual chord length can be found in the tables in App. C. Chord lengths can be measured accordingly.

12. CURVE LAYOUT

Because of their long radii, most curves cannot be laid out by swinging an arc from the center of the circle. They must be laid out by a series of straight lines (chords). This is done by use of transit and tape.

13. DEFLECTION ANGLE METHOD

The *deflection angle method* is based on the fact that the angle between a tangent and a chord, or between two chords that form an inscribed angle, is one-half the intercepted arc (see Fig. 17.2). In Fig. 17.6, the angle formed by the tangent at the PC and a chord from the PC to a point 100 ft along the arc is equal to one-half the degree of curve. Likewise, the angle formed by this chord and a chord from the PC to a point 100 ft farther along the arc is also equal to one-half the degree of curve. These angles are known as *deflection angles*. The deflection angle from the PC to the PT is one-half the central angle Δ, which provides an important check in computing deflection angles.

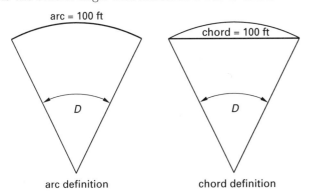

Figure 17.5 Arc and Chord Definitions

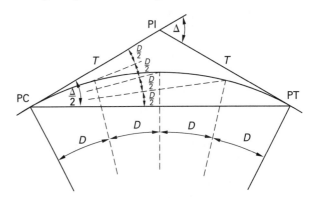

Figure 17.6 Laying Out a Curve

In laying out the curve, the transit is set up at the PC, the PT, or some other point on the curve, and deflection angles are turned for 100 ft arcs along the curve as 100 ft arcs are marked off by a 100 ft tape. For degree of curve up to 4°, the difference in length between chord and arc is slight. For sharper curves, this discrepancy can be corrected by laying out a chord slightly less than 100 ft. This length can be found in App. C or from Eq. 17.5.

On route surveys, stationing is carried continuously along tangent and curve. Thus, the PC and PT will seldom fall on a full station. The first full station will not likely be 100 ft from the PC, and the deflection angle to the first full station will not be $D/2$ but will be a fraction of $D/2$.

When stakes are required on closer intervals than full stations, such as 50 ft, the true length of the 50 ft chord (known as a *subchord*) will be less than 50 ft and this length can be found in App. C or from Eq. 17.5. (The central angle that subtends a 50 ft arc is $D/2$).

14. LENGTH OF CURVE

The *length of curve*, L (arc definition), is the distance along the arc from the PC to PT. As any two arcs are proportional to their central angles,

$$L = \frac{100\Delta}{D} \qquad 17.10$$

15. FIELD PROCEDURE IN STAKING A SIMPLE CURVE

The following steps can be used when staking out a curve from the PC or PT.

step 1: Measure the deflection angle.

step 2: Select D by considering the design criteria.

step 3: Compute the tangent distance T from the PI to the PC.

Tangent distances for a 1° curve for various values of Δ can be found in App. A. Tangent distance for any degree of curve can be found by dividing the tangent distance for a 1° curve by D.

step 4: Measure the tangent distance T from the PI to the PT and set the tacked hub. Measure T from the PI to the PC and set the tacked hub.

step 5: Compute PC station by subtracting T from PI station.

step 6: Compute the length of curve L. Compute the PT station by adding L to the PC station.

step 7: Compute the deflection angles at the PC for each station, checking to see that the deflection angle to the PT is $\Delta/2$.

step 8: (a) Set up a transit at PC and take a foresight on PI with the telescope normal and with the A vernier set on 0°00′. Turn the deflection angle for each station as chainpeople mark off corresponding arc length. Make the check at the PT for angle and distance.

(b) Set up the transit at the PT and take a backsight reading on the PC with the telescope normal and with the A vernier set on 0°00′. Turn the deflection angle for each station as chainpeople mark off corresponding arc lengths, with chainpeople starting at the PC. Deflection angles are the same as those computed for the staking curve from the PC. Make the check by sighting on the PI with the deflection angle to the PT.

16. CIRCULAR CURVE COMPUTATIONS

Example 17.1

Compute the parameters of a simple curve.

$$PI = \text{sta } 25{+}05$$
$$\Delta = 20°$$
$$D = 2°$$

Solution

$$L = (100)\left(\frac{\Delta}{D}\right) = \frac{(100)(20°)}{2°} = 1000 \text{ ft}$$

From App. B,

$$R = 2864.79 \text{ ft}$$

From Eq. 17.8,

$$R = \frac{5729.58}{2} = 2864.79 \text{ ft}$$

$$T = \frac{T \text{ for } 1° \text{curve}}{D} = \frac{1010.28 \text{ ft}}{2°}$$

$$= 505 \text{ ft}$$

From Eq. 17.3,

$$T = R\tan\frac{\Delta}{2} = (2864.79 \text{ ft})\tan\left(\frac{20°}{2}\right)$$

$$= 505 \text{ ft (Eq. 17.3)}$$

From App. A,

$$E = \frac{88.39 \text{ ft}}{2} = 44.19 \text{ ft}$$

From Eq. 17.7,

$$E = R\left(\sec\frac{\Delta}{2} - 1\right) = (2864.79 \text{ ft})\left(\sec\left(\frac{20°}{2}\right) - 1\right)$$
$$= 44.19 \text{ ft}$$

From Eq. 17.6,

$$M = R\left(1 - \cos\frac{\Delta}{2}\right) = (2864.79 \text{ ft})\left(1 - \cos\frac{20°}{2}\right)$$
$$= 43.52 \text{ ft}$$

$$
\begin{array}{rll}
\text{PI} = \text{sta} & 25+05 \\
T = & -5+05 \\
\text{PC} = \text{sta} & 20+00 \\
L = & +10+00 \\
\text{PT} = \text{sta} & 30+00
\end{array}
$$

The deflection angles are

point	station		deflection angle
PC	20+00		0°00′
	21+00 =	$D/2$	1°00′
	22+00 = deflection angle of station	21+00 + $D/2$ =	2°00′
	23+00 =	22+00 + $D/2$ =	3°00′
	24+00 =	23+00 + $D/2$ =	4°00′
	25+00 =	24+00 + $D/2$ =	5°00′
	26+00 =	25+00 + $D/2$ =	6°00′
	27+00 =	26+00 + $D/2$ =	7°00′
	28+00 =	27+00 + $D/2$ =	8°00′
	29+00 =	28+00 + $D/2$ =	9°00′
PT	30+00 =	29+00 + $D/2$ =	10°00′

(Check: deflection angle to the PT = $\Delta/2 = 10°00'$)

Example 17.2

Compute the parameters of a simple curve.

$$\text{PI} = \text{sta } 45+78.39$$
$$\Delta = 43°39'$$
$$D = 4°15'$$

Solution

$$L = \left(\frac{43°39'}{4°15'}\right)(100) = \left(\frac{43.65°}{4.25°}\right)(100) = 1027.06 \text{ ft}$$

From App. B,

$$R = 1348.14 \text{ ft}$$

From App. A,

$$T = \frac{T \text{ for } 1° \text{curve}}{D}$$
$$= \frac{2294.57 \text{ ft}}{4.25°} = 539.90 \text{ ft}$$

From App. B,

$$R = 1348.14 \text{ ft}$$

$$
\begin{array}{rll}
\text{PI} = \text{sta} & 45+78.39 \\
T = & -5+39.90 \\
\text{PC} = \text{sta} & 40+38.49 \\
L = & +10+27.06 \\
\text{PT} = \text{sta} & 50+65.55
\end{array}
$$

The deflection angles are

point	station		deflection angle
PC	40+38.49		0°00′
	41+00 =	$\left(\frac{100-38.49}{100}\right)\left(\frac{D}{2}\right)$ =	1°18′
	42+00 =	1°18′ + $D/2$ =	3°26′
	43+00 =	3°26′ + $D/2$ =	5°33′
	44+00 =	5°33′ + $D/2$ =	7°41′
	45+00 =	7°41′ + $D/2$ =	9°48′
	46+00 =	9°48′ + $D/2$ =	11°56′
	47+00 =	11°56′ + $D/2$ =	14°03′
	48+00 =	14°03′ + $D/2$ =	16°11′
	49+00 =	16°11′ + $D/2$ =	18°18′
	50+00 =	18°18′ + $D/2$ =	20°26′
PT	50+65.55 =	20°26′ + (0.6555)(2.125) =	21°49.5′

Field notes for this curve are shown in Table 17.1.

17. TRANSIT AT POINT ON CURVE

On long highway curves, it is often impossible to locate the entire curve from one point. Obstructions along the arc or along the line of sight from the transit to a station may prevent location of the curve from one point. In these situations, part of the curve can be located from the PC and then the transit can be moved to a point on the curve that has been located from the PC. The balance of the curve can then be located. On extremely long curves, part of the curve can be located from the PC and part from the PT. Any error will be located where the two parts meet.

In all cases, when it is necessary to move up on a curve, the deflection angles are computed as if the curve were to be located from the PC. When it is decided to move the transit, tack points are set at the station that is to be occupied and at the station that is be used as a backsight, unless the PC can be used as a backsight.

In either case, to orient the transit at the new station, the deflection angle of the station that is to be used for the backsight is set on the horizontal circle with the upper clamp before setting on the backsight with the lower clamp. If the PC is to be used for the backsight, the vernier would be set on 0°00′. After orientation, deflection angles as originally computed are turned for the balance of the curve.

Table 17.1 *Field Notes for Alignment of Reisel-Mart Highway*

		deflection calculated		
point	station	angle	bearing	curve data
⊙PT	50+65.55	21°49.5′	N 71°54′ E	
	50+00.00	20°26′		
	49+00.00	18°18′		
	48+00.00	16°11′		
	47+00.00	14°03′		
				$\Delta = 43°39′$ right
	46+00.00	11°56′		$D = 4°15′$
⊙PI	45+78.39			$T = 539.90$ ft
	45+00.00	9°48′		$L = 1027.06$ ft
				$R = 1348.14$ ft
	44+00.00	7°41′		
	43+00.00	5°33′		
	42+00.00	3°26′		
	41+00.00	1°18′		
⊙PC	40+38.49	0°00′	N 28°15′ E	

This procedure can be used in locating *culverts* on curves. The culvert station is occupied, the line of sight is made tangent at this point, and 90° is turned off this tangent. This puts the centerline of the culvert on a radial line.

Example 17.3

Sta 41+00, 42+00, and 43+00 for the curve tabulated in Table 17.1 have been set from the PC, but because of an obstruction in the line of sight, sta 44+00 cannot be located. The transit is to be moved to sta 43+00 for continuation of the curve location; the PC is to be used for the backsight. Locate sta 44+00.

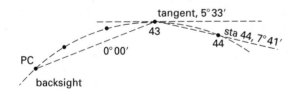

Solution

step 1: Set the transit on station 43+00.

step 2: Set 0°00′ on the *A* vernier.

step 3: Backsight on the PC with the telescope inverted.

step 4: Plunge the telescope and continue the location using the deflection angles as originally computed. (Note: When the vernier reads 5°33′, the line of sight is tangent to the curve at the point occupied.)

Example 17.4

Sta 41 through 46 of the curve tabulated in Table 17.1 have been set from the PC. The balance of the curve is to be located from sta 46+00, from which the PC is not visible. Sta 43+00 is to be used as a backsight. Outline the procedure.

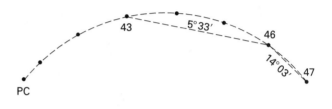

Solution

step 1: Set the transit on sta 46+00.

step 2: Set 5°33′ on the *A* vernier that is the deflection angle for sta 43+00.

step 3: Backsight on sta 43+00 with the telescope inverted.

step 4: Plunge the telescope and continue the location using deflection angles as originally computed. (The vernier is set on the deflection angle of the station sighted.)

18. COMPUTING TRANSIT STATIONS FOR HIGHWAY LOCATION

As has been mentioned, initial locations of highways usually consist of a series of tangents. Curves are inserted later to connect two intersecting tangents. The original tangents are chained from PI to PI, and deflection angles are measured. When curves are inserted, the original stations at PIs must be corrected. Stations of PCs and PTs must be computed by considering distances along the curves and not along the tangents. Obviously, the total length of the project will be less than the lengths of the tangents.

Example 17.5

A preliminary highway location has been made. PI stations, deflection angles, and the end of the line are as shown in Table 17.1. (The beginning station is 0+00.) Degree of curve for each curve has been selected. Stations for each PC and PT and the station for the end of the line are to be computed.

PI number	original station	Δ	D
1	12+24.31	21°28′R	1°30′
2	31+12.48	40°56′L	2°30′
3	51+90.13	22°12′R	2°00′
end	65+56.03	end of line	

Solution

A sketch is drawn showing curves connecting tangents. Distances from PI to PI, tangent distances and lengths of curves are then computed and tabulated. Using the tabulated distances and the sketch, stations are computed in sequence.

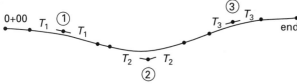

$$\text{distance PI}_1 - \text{PI}_2 = 3112.48 - 1224.31 = 1888.17$$

$$\text{distance PI}_2 - \text{PI}_3 = 5190.13 - 3112.48 = 2077.65$$

$$\text{distance PI}_3 - \text{end} = 6556.03 - 5190.13 = 1365.90$$

$$\text{PC}_1 = \text{PI}_1 - T_1 = 12{+}24.31 - 724.05 = 5{+}00.26$$

$$\begin{aligned}\text{PT}_1 &= \text{PC}_1 + L_1 = 5{+}00.26 + 1431.11 \\ &= 19{+}31.37\end{aligned}$$

$$\begin{aligned}\text{PC}_2 &= \text{PT}_1 + (\text{PI}_2 - \text{PI}_1 - T_1 - T_2) \\ &= 19{+}31.37 + (1888.17 - 724.05 - 855.36) \\ &= 22{+}40.13\end{aligned}$$

$$\begin{aligned}\text{PT}_2 &= \text{PC}_2 + L_2 = 22{+}40.13 + 1637.32 \\ &= 38{+}77.46\end{aligned}$$

$$\begin{aligned}\text{PC}_3 &= \text{PT}_2 + (\text{PI}_3 - \text{PI}_2 - T_2 - T_3) \\ &= 38{+}77.46 + (2077.65 - 855.36 - 562.05) \\ &= 45{+}37.70\end{aligned}$$

$$\begin{aligned}\text{PT}_3 &= \text{PC}_3 + L_3 = 45{+}37.70 + 1110.00 \\ &= 56{+}47.70\end{aligned}$$

$$\begin{aligned}\text{end} &= \text{PT}_3 + \text{end} - \text{PI}_3 - T_3) \\ &= 56{+}47.70 + (1365.90 - 562.05) = 64{+}51.55\end{aligned}$$

The stations can be computed by successive additions.

$$\begin{aligned}1224.31 &- 724.05 + 1431.11 + 3112.48 - 1224.31 \\ &- 724.05 - 855.36 + 1637.33 + 5190.13 - 3112.48 \\ &- 855.36 - 562.05 + 1110.00 + 6556.03 - 5190.13 \\ &- 562.05 = 6451.11 \text{ (end)}\end{aligned}$$

The data can now be summarized.

PI	original station	T	L	PC station	PT station
1	12+24.31	724.05	1431.11	5+00.26	19+31.37
	1888.17				
2	31+12.48	855.36	1637.33	22+40.13	38+77.46
	2077.65				
3	51+90.13	562.05	1110.00	45+37.70	56+47.70
	1365.90				
4	65+56.03				end = 64+51.55

19. LOCATING CURVE WHEN PI IS INACCESSIBLE

It is often necessary to find the deflection angle Δ and locate the PC and PT of a curve when the PI is inaccessible as shown in Fig. 17.7.

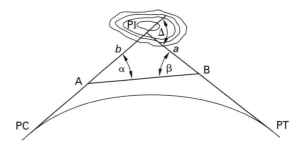

Figure 17.7 *Inaccessible PI*

To solve the problem, a *point on tangent* (POT) is established on the back tangent at any point A that is visible from a point B on the forward tangent. The station number of point A is established by chaining, and the angles α and β (the sum of which equals Δ) and the distance AB are measured.

Distances a and b in triangle PI-A-B are computed from the law of sines. The station number of the PI is found by adding distance b to the station number of point A. The tangent distance T is computed, and the PC and PT are then located by measuring from points A and B.

Example 17.6

Two tangents of a highway location intersect in a lake, which makes the PI inaccessible. A POT has been established at sta 23+45.67 and designated as point A. Locate a 4° curve on the ground.

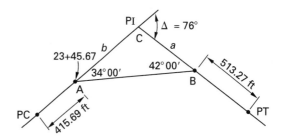

Solution

step 1: Point B is located on the forward tangent so that it is visible from point A on the back tangent.

step 2: Angle A is measured and found to be $334°00'$.

step 3: Angle B is measured and found to be $42°00'$.

step 4: Distance AB is measured and found to be 1020.00 ft.

step 5: The deflection angle Δ is the sum of angles A and B, which equals $76°00'$.

step 6: From the law of sines, in triangle PI-A-B,

$$a = \frac{(1020.00 \text{ ft})(\sin 34°)}{\sin 104°} = 587.84 \text{ ft}$$

$$b = \frac{(1020.00 \text{ ft})(\sin 42°)}{\sin 104°} = 703.41 \text{ ft}$$

step 7: Using $\Delta = 76°00'$ and $D = 4°$, $T = 1119.10$ ft.

step 8: Adding distance b to the station of point A, the PI is at sta 30+49.08. The PC is at sta 19+29.98.

step 9: The PC is located by measuring 415.69 ft from point A.

step 10: The PT is located by measuring 531.26 ft from point B.

20. SHIFTING FORWARD TANGENT

Route locations often require changes in curve locations. One such case involves shifting the forward tangent to a new location parallel to the original tangent and keeping the back tangent in its original location. This produces a change in both the PC and PT stations.

Example 17.7

The forward tangent of the highway curve shown is to be shifted outward so that it will be parallel to and 100 ft from the original tangent. Curve data for the original curve are shown in the figure. The degree of curve is to remain unchanged. Compute the PC and PT stations for the new curve.

Solution

$$PI_1 - PI_2 = \frac{100 \text{ ft}}{\sin 60°} = 115.47 \text{ ft}$$

$$PI_2 = 28+97.00 + 115.47 = 30+12.47$$

$$PC_2 = 30+12.47 - 551.33 = 24+61.14$$

$$PT_2 = 24+61.14 + 1000.00 = 34+61.14$$

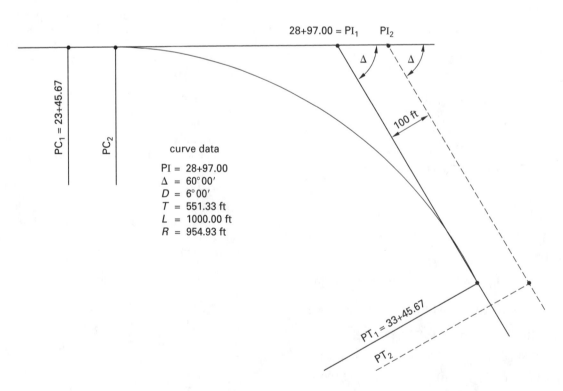

curve data

PI = 28+97.00
Δ = 60°00'
D = 6°00'
T = 551.33 ft
L = 1000.00 ft
R = 954.93 ft

Example 17.7

21. EASEMENT CURVES

Where simple curves are used, the tangent changes to a curved line at the PC. This means that a vehicle on a tangent arriving at the PC changes direction to a curved path instantaneously. At high speed this is impossible. What actually happens in an automobile is that the driver adjusts the steering wheel to make a gradual transition from a straight line to a curved line.

For railroad cars traveling at high speed, the problem is more acute than for automobiles. The rigidity and length of a railroad car cause a sharp thrust on the rails by the wheel flanges. To alleviate this situation, curves that provide a gradual transition from tangent to circular curve (and back again) are inserted between the tangent and the circular curve. These transition curves are known as *easement curves*. These curves also provide a place to increase superelevation from zero to the maximum required for the circular curve. The spiral is a curve that fulfills the requirements for the transition.

22. SPIRALS

The simple *spiral* has certain features that make it useful as an easement curve.

- The degree of curve of the spiral increases from zero at the beginning to the degree of curve of the circular arc where they meet. Likewise, the radius decreases from infinity at the beginning to the radius of the circular arc.

- Because the degree of curve changes uniformly, the central angle of the spiral equals the length of the spiral in stations times the average degree of curve.

$$\theta_s = \left(\frac{L_s}{100}\right)\left(\frac{D}{2}\right) = \frac{L_sD}{200} \qquad 17.11$$

- Spiral central angles are directly proportional to the squares of the lengths from the beginning of the spiral, as are the deflection angles.

23. LENGTH OF SPIRAL

The spiral provides a transition for superelevation. The length required to attain maximum superelevation is a function of the speed of the vehicle. Experiments have provided a formula for the length of the spiral (where L_s= length of spiral (in feet), v = design speed (in miles per hour), and R = radius of circular arc (in feet)).

$$L_s = \frac{1.6v^3}{R_c} \qquad 17.12$$

24. COMPUTATIONS AND PROCEDURE FOR STAKING

Spirals can be computed by using spiral tables. Symbols used on spirals should be memorized to facilitate the use of such tables in other texts. These symbols are illustrated in Fig. 17.8.

Figure 17.8 shows that the simple curve has been shifted inward in order to insert the easement curve, while the radius of the simple curve has been maintained.

Example 17.8

A circular curve with spiral transitions is to be computed and staked. Curve data are as follows.

$$PI = sta\ 72+58.00$$
$$\Delta = 42°00'$$
$$D_c = 5°$$
$$v = 60\ mph$$

Solution

(This solution requires the use of spiral curve tables).

$$R_c = 1145.92\ ft$$

$$L_s = 1.6\left(\frac{V^3}{R_c}\right) = 1.6\left(\frac{\left(60\ \frac{mi}{hr}\right)^3}{1145.92\ ft}\right)$$

$$= 301\ ft\ (300\ ft)$$

$$\theta_s = \frac{L_sD_c}{200} = \frac{(300\ ft)(5°)}{200} = 7°30'$$

$$\Delta_c = \Delta - 2\theta_s = 42° - (2)(7°30') = 27°00'$$

$$L_c = \left(\frac{\Delta_c}{D_c}\right)(100) = \left(\frac{27°}{5°}\right)(100\ ft) = 540\ ft$$

$$p = 3.27\ ft;\ k = 149.91\ ft$$

$$T_s = (R_c + p)\tan\frac{\Delta}{2} + k$$

$$= (1145.92\ ft + 3.27\ ft)(0.38386) + 149.91\ ft$$

$$= 591.04\ ft$$

The transit stations are as follows.

$$PI = 72+58.00$$
$$T_s = -5+91.04$$
$$TS = \overline{66+66.96}$$
$$L_s = +3+00.00$$
$$SC = \overline{69+66.96}$$
$$L_c = +5+40.00$$
$$SC = \overline{75+06.96}$$
$$L_s = +3+00.00$$
$$ST = \overline{78+06.06}$$

$$LT = 200.18\ ft$$
$$ST = 100.16\ ft$$
$$LC = 299.77\ ft$$

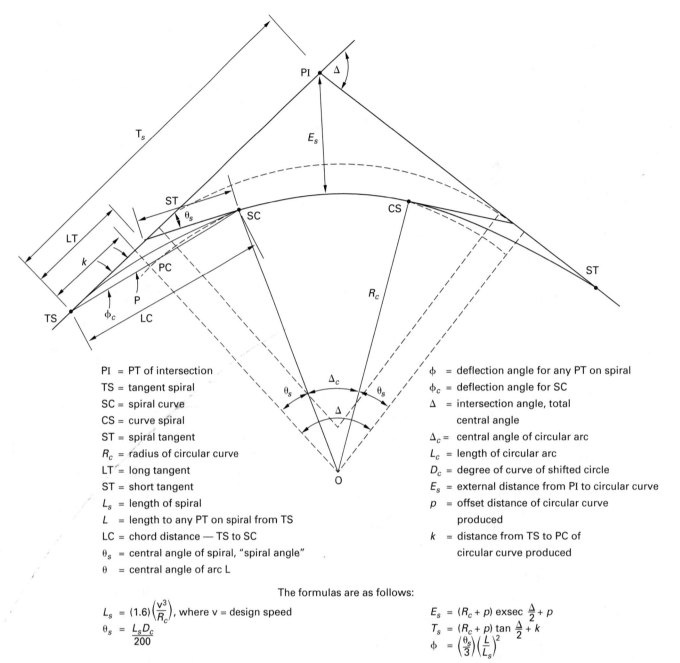

PI = PT of intersection
TS = tangent spiral
SC = spiral curve
CS = curve spiral
ST = spiral tangent
R_c = radius of circular curve
LT = long tangent
ST = short tangent
L_s = length of spiral
L = length to any PT on spiral from TS
LC = chord distance — TS to SC
θ_s = central angle of spiral, "spiral angle"
θ = central angle of arc L

ϕ = deflection angle for any PT on spiral
ϕ_c = deflection angle for SC
Δ = intersection angle, total central angle
Δ_c = central angle of circular arc
L_c = length of circular arc
D_c = degree of curve of shifted circle
E_s = external distance from PI to circular curve
p = offset distance of circular curve produced
k = distance from TS to PC of circular curve produced

The formulas are as follows:

$$L_s = (1.6)\left(\frac{v^3}{R_c}\right), \text{ where v = design speed}$$
$$\theta_s = \frac{L_s D_c}{200}$$

$$E_s = (R_c + p)\, \text{exsec} \, \frac{\Delta}{2} + p$$
$$T_s = (R_c + p)\, \tan \frac{\Delta}{2} + k$$
$$\phi = \left(\frac{\theta_s}{3}\right)\left(\frac{L}{L_s}\right)^2$$

Figure 17.8 *Spiral Curve Nomenclature*

For the deflection angles for the spiral,

$$\phi = \left(\frac{\theta_s}{3}\right)\left(\frac{L^2}{L_s^2}\right) = \left(\frac{1°}{3}\right)\left(\frac{7.5L^2}{(300 \text{ ft})^2}\right) = 0.0000278L^2$$

point	station	L	L^2	ϕ	ϕ
TS	66+66.96	0.00	0	0°	= 0°
	67+00	33.04	1092	0.0303°	= 0°02′
	67+50	83.04	6896	0.1915°	= 0°11.5′
	68+00	133.04	17,700	0.4916°	= 0°29.5′
	68+50	183.04	33,503	0.9306°	= 0°56′
	69+00	233.04	54,307	1.5082°	= 1°30.5′
	69+50	283.04	80,111	2.2253°	= 2°13.5′
SC	69+66.96	300.00	90,000	2.5000°	= 2°30′

The field procedure is:

step 1:   Locate the TS by measuring from the PI.

step 2:   Locate the intersection of the LT and ST (200.18 ft from TS).

step 3:   Set up at the intersection of LT and ST; sight on PI. Locate the SC by turning 7°30′ (θ_s) and measuring 100.16 (ST).

step 4:   Locate the CS as in step 2 and step 3.

step 5:   Set up on TS and sight on PI. Locate points on the spiral using deflection angles.

step 6: Locate points on the circular curve: (a) Set up on the SC. (b) Sight on the intersection of LT and ST with the telescope inverted and 0°00′ on the *A* vernier. (c) Set points on circular curve in normal manner.

step 7: Set up on the ST and locate spiral as in step 5.

25. STREET CURVES

Because *street curves* are usually short in radius, field procedures in staking them may differ from those used for highway curves. All formulas used for highway curves are valid, but the choice of formulas may vary. Street curves are computed by using the arc definition, just like highway curves, but stakes are usually set at 25 ft or 50 ft stations. The difference in the arc length and the chord length is much more pronounced on curves of short radius. Highway curves are usually designated by their degree of curve. Street curves are usually designated by a round-number radius, and the degree of curve concept is not used.

From Eq. 17.4, the long chord for a highway curve is equal to $2R\sin(\Delta/2)$, Δ being the central angle. Accordingly, $C = 2R\sin(D/2)$ for an arc of 100 ft. For street curves, the formula (where δ is the central angle for any arc) is expressed as

$$C = 2R\sin\frac{\delta}{2} \qquad 17.13$$

Some street curves are divided into 3, 4, or more equal arcs, but usually they are staked on the half- or quarter-stations. Chord lengths from quarter-station to quarter-station (or half-station to half-station) are usually measured, as is done on highway curves. For short curves, chords from the PC to each station can also be measured.

26. CURVE COMPUTATIONS

The degree of curve concept is not used. Therefore, the deflection angle will be expressed in terms of the central angle, δ. This is illustrated in Ex. 17.9.

Example 17.9

Computations are to be made for a street curve with a deflection angle Δ of 50°00′ and a centerline radius of 120 ft. The PI is at station 8+72.43.

Solution

$$T = R\tan\frac{\Delta}{2} = (120 \text{ ft})\left(\tan\frac{50°}{2}\right)$$
$$= 55.96 \text{ ft}$$
$$L = \left(\frac{50°}{360°}\right)2\pi R = \left(\frac{50°}{360°}\right)2\pi(120 \text{ ft})$$
$$= 104.72 \text{ ft}$$

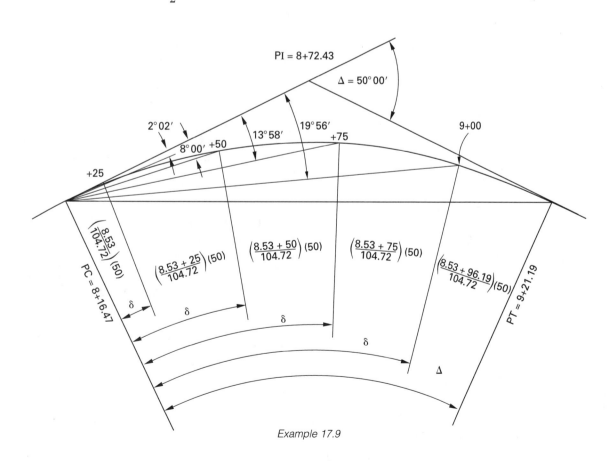

Example 17.9

$$PI = 8+72.43$$
$$T = -55.96$$
$$PC = 8+16.47$$
$$L = +1+04.72$$
$$PT = 9+21.19$$

The deflection angles are

PC $8+16.47 = 0°00'$

$$8+25 = \left(\frac{25.00 - 16.47}{104.72}\right)\left(\frac{50°}{2}\right)$$

$$= \left(\frac{8.53}{104.72}\right)(25°)$$

$$= 2.0364°$$

$$= 2°02'$$

$$8+50 = \left(\frac{8.53 + 25.00}{104.72}\right)\left(\frac{50°}{2}\right)$$

$$= \left(\frac{33.53}{104.72}\right)(25°)$$

$$= 8.0047°$$

$$= 8°00'$$

$$8+75 = \left(\frac{8.53 + 50.00}{104.72}\right)\left(\frac{50°}{2}\right)$$

$$= \left(\frac{58.53}{104.72}\right)(25°)$$

$$= 13.9730°$$

$$= 13°58'$$

$$9+00 = \left(\frac{8.53 + 75.00}{104.72}\right)\left(\frac{50°}{2}\right)$$

$$= \left(\frac{83.53}{104.72}\right)(25°)$$

$$= 19.9413°$$

$$= 19°56'$$

$$PT\ 9+21.19 = \left(\frac{8.53 + 96.19}{104.72}\right)\left(\frac{50°}{2}\right)$$

$$= \left(\frac{104.72}{104.72}\right)(25°)$$

$$= 25.0000°$$

$$= 25°00'$$

The chord lengths $\left(C = 2R\sin(\delta/2)\right)$ are

$8+16.47$ to $8+25.00$:

$$C = (240\text{ ft})(\sin 2.0364°) = 8.53\text{ ft}$$

$8+25.00$ to $8+50.00$:

$$C = (240\text{ ft})\sin(8.0047° - 2.0364°) = 24.95\text{ ft}$$

$9+00.00$ to $9+21.19$:

$$C = (240\text{ ft})\sin(25.0000° - 19.9413°) = 21.16\text{ ft}$$

The field notes are as follows.

point	station	$\delta/2$	deflection angle	C	curve data
PT	9+21.19		25°00'	21.16	
	9+00	19.9412°	19°56'	24.95	$\Delta = 50°00'$
	8+75	13.9730°	13°58'	24.95	$R = 120$ ft
	8+50	8.0047°	8°00'	24.95	$T = 55.96$ ft
	8+25	2.0364°	2°02'	8.53	$L = 104.72$ ft
PC	8+16.7	0.0000°	0°00'		

The length of the curve is only 104.72 ft. Therefore, the curve should be staked by measuring all chords from the PC. These chord lengths are computed as follows.

PC to $8+25.00$: $C = (240\text{ ft})(\sin 2.0364°) = 8.53$ ft
PC to $8+50.00$: $C = (240\text{ ft})(\sin 8.0047°) = 33.42$ ft
PC to $8+75.00$: $C = (240\text{ ft})(\sin 13.9730°) = 57.95$ ft
PC to $9+00.00$: $C = (240\text{ ft})(\sin 19.9412°) = 81.85$ ft
PC to $9+21.19$: $C = (240\text{ ft})(\sin 25.0000°) = 101.43$ ft

27. PARALLEL CIRCULAR ARCS

The design radius for a curve is usually to the centerline. However, it is often necessary to locate a parallel curve such as a right-of-way line, the edge of pavement, or a curb line. Because the central angle is the same for *parallel arcs*, deflection angles are the same.

The chord lengths are a function of the radius of the arc, and they will be different for arcs of different radius. Length of the curve is also a function of the radius. The PC stations for all parallel arcs will be the same. Arc lengths for inside and outside curves will be different, but the PT stations will be the same.

Example 17.10

Construction stakes are to be set for the curve in the preceding example. The street width is 34 ft. Stakes are to be set on 3 ft offsets, inside and outside of the edge of pavement.

Solution

Deflection angles are given in the field notes. Chord lengths are computed from the formula $C = 2R\sin(\delta/2)$ and are also shown in the field notes.

point	station	deflection angle	$R = 100$ ft C inside	$R = 140$ ft C outside	curve data
PT	9+21.19	25°00'			
			17.64	24.69	$\Delta = 50°00'$
	9+00	19°56'			
			20.80	29.11	$R = 120$ ft
	8+75	13°58'			
			20.80	29.11	$T = 55.96$ ft
	8+50	8°00'			
			20.80	29.11	$L = 104.72$ ft
	8+25	2°02'			
			7.11	9.95	
PC	8+16.47	0°00'			

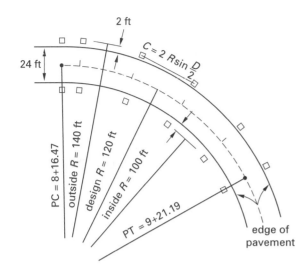

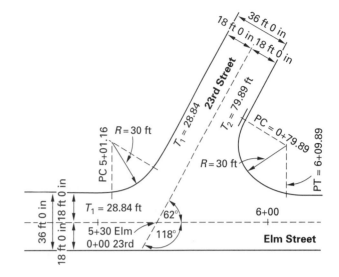

28. CURB RETURNS AT STREET INTERSECTIONS

Curb returns are the arcs made by the curbs at street intersections. The radius of the arc is selected by the designer with consideration given to the speed and volume of traffic. A radius of 30 ft to the back of the curb is common. Streets that intersect at right angles have curb returns of one quarter circle.

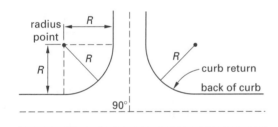

Figure 17.9 *Curb Returns*

The arcs can be swung from a *radius point*, the center of the circle. The radius point can be located by finding the intersection of two lines, each of which is parallel to one of the curb lines and at a distance equal to the radius. One stake at the radius is sufficient for the curb return.

Curb returns for streets that do not intersect at right angles are shown in Ex. 17.11. Computations for the location of PCs, PTs, and radius points are shown in Ex. 17.11.

Example 17.11

Elm Street and 23rd Street intersect as shown. The pavement width for each street is 36 ft. The radius to the edge of the pavement is 30 ft. Compute the location of the radius points.

Solution

The PC stations along Elm Street are

$$\Delta_1 = 62°00'$$

$$T_1 = R \tan \frac{\Delta}{2} = (48) \left(\tan \frac{62°}{2} \right)$$

$$= 28.84 \text{ ft}$$

$$\begin{aligned} \text{PI} &= \quad 5+30.00 \\ T_1 &= \quad -\,28.84 \\ \text{PC} &= \quad 5+01.16 \end{aligned}$$

$$\Delta_2 = 118°00'$$

$$T_2 = R \tan \frac{\Delta}{2} = (48) \left(\tan \frac{118°}{2} \right)$$

$$= 79.89 \text{ ft}$$

$$\begin{aligned} \text{PI} &= \quad 5+30.00 \\ T_2 &= \quad +\,79.89 \\ \text{PC} &= \quad 6+09.89 \end{aligned}$$

The PT stations along 23rd Street are

$$\begin{aligned} \text{PI} &= \quad 0+00.00 \\ T_1 &= \quad +\,28.84 \\ \text{PT} &= \quad 0+28.84 \end{aligned}$$

$$\begin{aligned} \text{PI} &= \quad 0+00.00 \\ T_2 &= \quad +\,79.89 \\ \text{PT} &= \quad 0+79.89 \end{aligned}$$

29. COMPOUND CURVES

A *compound curve* consists of two or more simple curves with different radii joined together at a common tangent point. Their centers are on the same side of the curve.

Compound curves are not generally used for highways except in mountainous country, because an abrupt change in degree of curve causes a serious hazard even at moderate speeds. They are sometimes used for curvilinear streets in residential subdivisions, however.

In Fig. 17.10, the subscript 1 is used for the curve of longer radius, and the subscript 2 is used for the curve of shorter radius. The point of common tangency is called the *point of common curvature*, PCC. The short tangents for the two curves are designated as t_1 and t_2.

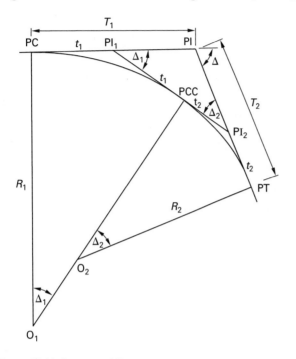

Figure 17.10 *Compound Curve*

There are seven major parameters of a compound curve: Δ, Δ_1, Δ_2, R_1, R_2, T_1, and T_2. Four of these must be known before computations can be made. Usually, Δ is measured. R_1, R_2, and either Δ_1 or Δ_2 are given. If Δ, Δ_1, R_1, and R_2 are given, Eqs. 17.14 through 17.17 can be used.

$$\Delta = \Delta_1 + \Delta_2 \qquad 17.14$$
$$\Delta_2 = \Delta - \Delta_1 \qquad 17.15$$
$$t_1 = R_1 \tan \frac{\Delta_1}{2} \qquad 17.16$$
$$t_2 = R_2 \tan \frac{\Delta_2}{2} \qquad 17.17$$

The triangle PI-PI$_1$-PI$_2$ can be solved by the law of sines. The sine of the angle at the PI equals the sine of Δ because they are related angles.

$$\text{PI} - \text{PI}_1 = \frac{\sin \Delta_2 (t_1 + t_2)}{\sin \Delta} \qquad 17.18$$
$$\text{PI} - \text{PI}_2 = \frac{\sin \Delta_1 (t_1 + t_2)}{\sin \Delta} \qquad 17.19$$
$$T_1 = t_1 + \text{PI} - \text{PI}_1 \qquad 17.20$$
$$T_2 = t_2 + \text{PI} - \text{PI}_2 \qquad 17.21$$

To stake out the curve, the PC and PT are located as is done for a simple curve. The PCC is located by establishing the common tangent from either point PI$_1$

or PI$_2$. Deflection angles for the two curves are computed separately. The first curve is staked from the PC, and the second curve is staked from the PCC using the common tangent for orientation with the vernier set on $0°00'$.

Example 17.12

PC, PCC, and PT stations, deflection angles, and chord lengths are to be computed from the following information.

$$\text{PI} = \text{sta } 15+56.32$$
$$\Delta = 68°00'$$
$$\Delta_1 = 35°00'$$
$$R_1 = 600 \text{ ft}$$
$$R_2 = 400 \text{ ft}$$

Solution

$$\Delta_2 = \Delta - \Delta_1 = 68°00' - 35°00'$$
$$= 33°00'$$
$$t_1 = R_1 \tan \frac{\Delta_1}{2} = (600 \text{ ft})(\tan 17°30')$$
$$= 189.18 \text{ ft}$$
$$t_2 = R_2 \tan \frac{\Delta_2}{2} = (400 \text{ ft})(\tan 16°30')$$
$$= 118.49 \text{ ft}$$
$$\text{PI} - \text{PI}_1 = \frac{(t_1 + t_2) \sin \Delta_2}{\sin \Delta} = \frac{(307.67 \text{ ft})(\sin 33°)}{\sin 68°}$$
$$= 180.73 \text{ ft}$$
$$\text{PI} - \text{PI}_2 = \frac{(t_1 + t_2) \sin \Delta_1}{\sin \Delta} = \frac{(307.67 \text{ ft})(\sin 35°)}{\sin 68°}$$
$$= 190.33 \text{ ft}$$
$$T_1 = t_1 + \text{PI} - \text{PI}_1 = 189.18 + 180.73$$
$$= 369.91 \text{ ft}$$
$$T_2 = t_2 + \text{PI} - \text{PI}_2 = 118.49 + 190.33$$
$$= 308.82 \text{ ft}$$
$$L_1 = \left(\frac{\Delta_1}{360°}\right) 2\pi R = \left(\frac{35°}{360°}\right) 2\pi(600)$$
$$= 366.52 \text{ ft}$$
$$L_2 = \left(\frac{\Delta_2}{360°}\right) 2\pi R = \left(\frac{33°}{360°}\right) 2\pi(400)$$
$$= 230.38 \text{ ft}$$

$$
\begin{array}{rl}
\text{PI} = & 15+56.32 \\
T_1 = & -3+69.91 \\
\hline
\text{PC} = & 11+86.41 \\
L_1 = & +3+66.52 \\
\hline
\text{PCC} = & 15+52.93 \\
L_2 = & +2+30.38 \\
\hline
\text{PT} = & 17+83.31
\end{array}
$$

The deflection angles are as follows.

point	station		deflection angles
PC	11+86.41		
	12+00	$\left(\dfrac{13.59}{366.52}\right)\left(\dfrac{35}{2}\right) =$	$0.6489° = 0°39'$
	13+00	$\left(\dfrac{113.59}{366.59}\right)\left(\dfrac{35}{2}\right) =$	$5.4235° = 5°25'$
	14+00	$\left(\dfrac{213.59}{366.52}\right)\left(\dfrac{35}{2}\right) =$	$10.1981° = 10°12'$
	15+00	$\left(\dfrac{313.59}{366.52}\right)\left(\dfrac{35}{2}\right) =$	$14.9728° = 14°59'$
PCC	15+52.93	$\left(\dfrac{366.52}{366.52}\right)\left(\dfrac{35}{2}\right) =$	$17.5000° = 17°30'$
	16+00	$\left(\dfrac{47.07}{230.38}\right)\left(\dfrac{33}{2}\right) =$	$3.3712° = 3°22'$
	17+00	$\left(\dfrac{147.07}{230.38}\right)\left(\dfrac{33}{2}\right) =$	$10.5332° = 10°32'$
PT	17+83.31	$\left(\dfrac{230.38}{230.38}\right)\left(\dfrac{33}{2}\right) =$	$16.5000° = 16°30'$

The chord lengths are

$$C = (1200 \text{ ft})(\sin 0.6489°) = 13.59 \text{ ft}$$
$$C = (1200 \text{ ft})(\sin 4.7746°) = 99.88 \text{ ft}$$
$$C = (1200 \text{ ft})(\sin 2.5272°) = 52.91 \text{ ft}$$
$$C = (800 \text{ ft})(\sin 3.3712°) = 47.04 \text{ ft}$$
$$C = (800 \text{ ft})(\sin 7.1620°) = 99.74 \text{ ft}$$
$$C = (800 \text{ ft})(\sin 5.9668°) = 83.16 \text{ ft}$$

The field notes showing the results of these computations are shown as follows.

Field Notes for Elm Street

point	station	deflection angle	chord	calculated bearing	curve data
PT	17+83.31	16°30'			
			83.16'		
	17+00.00	10°32'			
			99.74'		
	16+00.00	3°22.3'			$\Delta = 68°00'$
PI	15+56.32		47.04'		$R_1 = 600$ ft
PCC	15+52.93	17°30'			$\Delta_1 = 35°00'$
			52.91'		$R_2 = 400$ ft
	15+00.00	14°58.7'			$\Delta_2 = 33°00'$
			99.88'		$T_1 = 369.91$ ft
	14+00.00	10°12.1'			$T_2 = 308.81$ ft
			99.88'		$L_1 = 366.52$ ft
	13+00.00	5°25.5'			$L_2 = 230.38$ ft
			99.88'		
	12+00.00	0°38.9'			
			13.59'		
PC	11+86.41	0°00'			

PRACTICE PROBLEMS

1. Provide the missing word or words in each sentence.

(a) Highway curves are most often _____ arcs known as simple curves.

(b) An inscribed angle is an angle that has its vertex on a _____ and that has _____ for its sides.

(c) An inscribed angle is measured by _____ its intercepted arc.

(d) An angle formed by a tangent and a chord is measured by _____ its intercepted arc.

(e) The radius of a circle is _____ to a tangent at the point of tangency.

(f) A perpendicular bisector of a chord passes through the _____ of the circle.

(g) By the arc definition, degree of curve, D, is the central angle that subtends a 100 ft _____ .

(h) By the chord definition, degree of curve, D, is the central angle that subtends a 100 ft _____ .

(i) By the arc definition, the radius R of a 1° curve is _____ ft.

(j) The deflection angle for a full station for a 1° curve is _____ .

2. Place all symbols pertinent to a circular curve on the following figure.

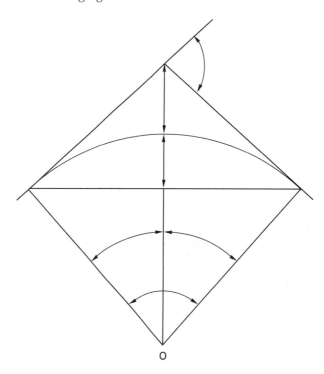

3. Using the information given, compute the PC and PT stations and the deflection angles for each full station of the simple highway curve. Round off T to the nearest foot.

$$PI = \text{sta } 25{+}01$$
$$\Delta = 10°$$
$$D = 1°$$

4. Using the information given, compute the PC and PT stations and the deflection angles for each full station of the simple highway curve. Express length to two decimal places.

$$PI = \text{sta } 45{+}11.75$$
$$\Delta = 30°$$
$$D = 3°$$

5. Using the following information, prepare field notes to be used in staking the centerline of a simple horizontal curve for a highway.

$$PI = 29 + 62.78$$
$$\Delta = 40°21'L$$
$$D = 5°15'$$
$$\text{back tangent bearing} = N\,56°12'\,W$$

6. A preliminary highway location has been made by locating tangents. Deflection angles have been measured at each PI, and the station number of each PI has been established by measuring along the tangents. Circular curves have not been located, but the degree of curve has been established for each curve. The beginning point is at sta 0+00. Using these data, make necessary computations to establish stations for PCs and PTs of the curves and for the end of the line.

PI no.	original station	Δ	D
1	10+35.27	13°24'R	1°30'
2	36+15.44	15°18'L	2°30'
3	52+98.40	18°05'R	3°00'
end	61+32.77	end of line	

7. In locating a highway, the PI of two tangents falls in a lake and is inaccessible. Point A on the back tangent has been established at sta 26+52.61. Point B has been established on the forward tangent and is visible from point A. Angle A has been measured and found to be $23°13'$; angle B has been measured and found to be $19°55'$. The length of AB has been found to be 434.87 ft. Find the deflection angle Δ and the PC and PT stations for a 3° curve.

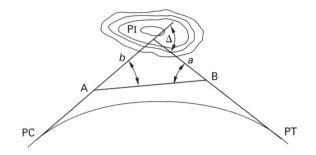

8. The forward tangent of the highway curve shown is to be shifted outward so that it will be parallel to and 50 ft from the original tangent. Data for the original curve are shown in the figure. The degree of curve is to be unchanged. Find the PC and PT stations for the new curve.

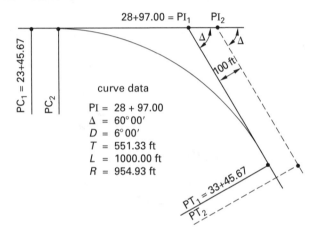

curve data
PI	= 28 + 97.00
Δ	= 60° 00'
D	= 6° 00'
T	= 551.33 ft
L	= 1000.00 ft
R	= 954.93 ft

9. Find L for Δ and D indicated.

(a) $\Delta = 32°18'$
    $D = 2°30'$

(b) $\Delta = 41°27'$
    $D = 3°15'$

10. Find T for Δ and D indicated.

(a) $\Delta = 41°51'$
    $D = 1°45'$

(b) $\Delta = 39°14'$
    $D = 2°15'$

11. Find E for Δ and D indicated.

(a) $\Delta = 31°30'$
    $D = 1°30'$

(b) $\Delta = 42°21'$
    $D = 2°45'$

12. Find D for the nearest full degree for Δ indicated and T approximately as indicated.

(a) $\Delta = 32°56'$

  $T = 600$ ft

(b) $\Delta = 40°10'$

  $T = 1000$ ft

13. Use trigonometric equations to solve the following problems.

(a) Find R for $D = 2°$.

(b) Find D for $R = 1909.86$ ft.

(c) Find T for $\Delta = 34°44'$ and $R = 800$ ft.

(d) Find E for $\Delta = 37°20'$ and $R = 650$ ft.

(e) Find M for $\Delta = 42°51'$ and $R = 800$ ft.

(f) Find LC for $\Delta = 32°55'$ and $R = 850$ ft.

(g) Find the chord length for $D = 8°$, $R = 716.20$ ft, and arc $= 50$ ft.

14. Compute deflection angles and chord lengths for quarter stations (25 ft) for a street curve. Chords are to be measured from quarter-station to quarter-station.

$$PI = \text{sta } 8+78.22$$
$$\Delta = 28°$$
$$R = 250 \text{ ft}$$

15. Prepare field notes to be used in staking the centerline of a horizontal street curve on the quarter-stations.

$$PI = \text{sta } 8+47.52$$
$$\Delta = 36°00' \text{ Right}$$
$$R = 400 \text{ ft}$$

16. Compute PC, PCC, and PT stations and deflection angles for full stations for the compound curve with the information given.

$$PI = \text{sta } 14+78.32$$
$$\Delta = 68°00'$$
$$\Delta_1 = 36°00'$$
$$R_1 = 400 \text{ ft}$$
$$R_2 = 300 \text{ ft}$$

17. Prepare field notes to be used in staking the centerline of the compound curve on full stations.

$$PI = \text{sta } 12+65.35$$
$$\Delta = 70°00'$$
$$\Delta_1 = 36°00'$$
$$R_1 = 900 \text{ ft}$$
$$R_2 = 600 \text{ ft}$$

18. Compute the area of the traverse to the nearest tenth of an acre.

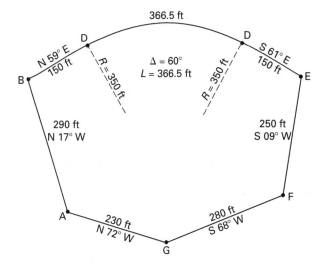

SOLUTIONS

1. (a) circular   (b) circle, chords

(c) one-half   (d) one-half

(e) perpendicular   (f) center

(g) arc   (h) chord

(i) 5729.58   (j) 30 ft

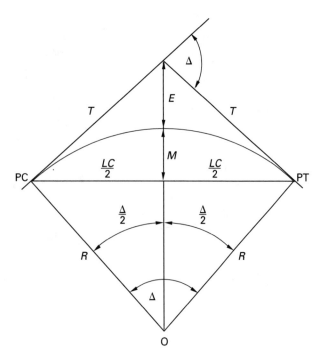

symbol	name	definition
PI	point of intersection	where two tangents intersect
Δ	deflection angle	central angle at PI; angle at center
D	degree of curve	central angle that subtends 100 ft arc
L	length of curve	distance from PC to PT along the arc
T	tangent distance	distance from PI to PI; from PI to PT
R	radius	
LC	length of long chord	
C	length of chord	chord length for angle D
PC	point of curvature	beginning of curve
PT	point of tangency	end of curve
M	middle ordinate	length from middle of LC to middle of curve
E	external distance	distance from PI to middle of curve

3.

$$L = (100 \text{ ft}) \left(\frac{100°}{1°} \right) = 1000 \text{ ft}$$

$$\boxed{T = 501 \text{ ft}}$$

$$R = 5729.58 \text{ ft}$$

The PC and PT stations are

$$PI = 25+01$$

$$T = 5+01$$

$$\boxed{PC = 20+00}$$

$$L = 10+00$$

$$\boxed{PT = 30+00}$$

The deflection angles are as follows.

point	station	deflection angle
PC	10+00	0°00′
	11+00	0°30′
	12+00	1°00′
	13+00	1°30′
	14+00	2°00′
	15+00	2°30′
	16+00	3°00′
	17+00	3°30′
	18+00	4°00′
	19+00	4°30′
PT	20+00	5°00′

4.

$$L = (100 \text{ ft}) \left(\frac{30°}{3°} \right) = 1000 \text{ ft}$$

$$T = 511.75 \text{ ft}$$

$$R = 1909.86 \text{ ft}$$

The PC and PT stations are

$$PI = 45+11.75$$

$$T = 5+11.75$$

$$\boxed{PC = 40+00}$$

$$L = 10+00$$

$$\boxed{PT = 50+00}$$

The deflection angles are as follows.

point	station	deflection angle
PC	40+00	0°00′
	41+00	1°30′
	42+00	3°00′
	43+00	4°30′
	44+00	6°00′
	45+00	7°30′
	46+00	9°00′
	47+00	10°30′
	48+00	12°00′
	49+00	13°30′
PT	50+00	15°00′

5.

point	station	deflection angle	chord	calculated bearing	curve data
PT	33+30.35	20°10.5′		S 83°27′ W	
	33+00	19°22.5′			
	32+00	16°45′			
	31+00	14°07.5′			$\Delta = 40°21'$L $D = 5°15'$ $T = 401.00$ ft $L = 768.57$ ft
	30+00	11°30′			
	29+00	8°52.5′			$R = 1091.35$ ft
	28+00	6°15′			
	27+00	3°37.5′			
	26+00	1°00′			
PC	25+61.78	0°00′		N 56°12′ W	

6.

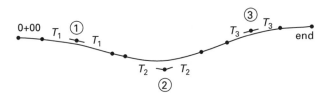

PI no.	original station	T	L	PC station	PT station	end
1	10+35.27	454.35	904.44	5+80.92	14+85.36	
2	36+15.44	307.83	612.00	33+03.35	39+15.35	
3	52+98.40	303.92	602.78	49+86.56	55+89.34	
end						61+19.79

7. $\Delta = 23°13' + 19°55' = \boxed{43°08'}$

$L = \left(\dfrac{43°08'}{3°}\right)(100 \text{ ft}) = 1437.78 \text{ ft}$

$T = \left(\dfrac{5729.58}{3}\right)\left(\tan\dfrac{43°08'}{2}\right) = 754.88 \text{ ft}$

$a = \dfrac{(434.87 \text{ ft})(\sin 23°13')}{\sin 43°08'} = 250.74 \text{ ft}$

$b = \dfrac{(434.87 \text{ ft})(\sin 19°55')}{\sin 43°08'} = 216.67 \text{ ft}$

$PC = 26{+}52.61 + 216.67 - 754.88 = \boxed{21{+}14.40}$

$PT = 21{+}14.40 + 1437.78 = \boxed{35{+}52.18}$

8. $PI_1 - PI_2 = \dfrac{50 \text{ ft}}{\sin 60°} = 57.74 \text{ ft}$

$PC = 23{+}45.67 + 57.74 = \boxed{24{+}03.41}$

$PT = 24{+}03.41 + 1000.00 = \boxed{34{+}03.41}$

9. (a) $L = \left(\dfrac{32°18'}{2°30'}\right)(100 \text{ ft}) = \boxed{1292.00 \text{ ft}}$

(b) $L = \left(\dfrac{41°27'}{3°15'}\right)(100 \text{ ft}) = \boxed{1275.38 \text{ ft}}$

10. (a) $T = \left(\dfrac{5729.58}{1°45'}\right)\left(\tan\dfrac{41°51'}{2}\right)$

$= \boxed{1251.87 \text{ ft}}$

(b) $T = \left(\dfrac{5729.58}{2°15'}\right)\left(\tan\dfrac{39°14'}{2}\right)$

$= \boxed{907.60 \text{ ft}}$

11. (a) $E = \dfrac{223.51 \text{ ft}}{1°30'} = \boxed{149.01 \text{ ft}}$

(b) $E = \dfrac{414.86 \text{ ft}}{2°45'} = \boxed{150.86 \text{ ft}}$

12. (a) $D = \left(\dfrac{5729.58}{600 \text{ ft}}\right)\left(\tan\dfrac{32°56'}{2}\right) = \boxed{3°}$

(b) $D = \left(\dfrac{5729.58}{1000 \text{ ft}}\right)\left(\tan\dfrac{40°10'}{2}\right) = \boxed{2°}$

13. (a) $R = \dfrac{5729.58}{2} = \boxed{2864.79 \text{ ft}}$

(b) $D = \dfrac{5729.58}{1909.86 \text{ ft}} = \boxed{3^\circ}$

(c) $T = (800 \text{ ft})\left(\tan \dfrac{34^\circ 44'}{2}\right) = \boxed{250.19 \text{ ft}}$

(d) $E = (650 \text{ ft})\left(\dfrac{1}{\cos \dfrac{37^\circ 20'}{2}}\right) = \boxed{36.09 \text{ ft}}$

(e) $M = (800 \text{ ft})\left(1 - \cos \dfrac{42^\circ 51'}{2}\right)$

$= \boxed{55.28 \text{ ft}}$

(f) $LC = (2)(850 \text{ ft})\left(\sin \dfrac{32^\circ 55'}{2}\right)$

$= \boxed{481.64 \text{ ft}}$

(g) $\text{chord} = (2)(716.20 \text{ ft})\left(\sin \dfrac{4^\circ}{2}\right)$

$= \boxed{49.99 \text{ ft}}$

14. $T = (250 \text{ ft})\left(\tan \dfrac{28^\circ}{2}\right) = \boxed{62.33 \text{ ft}}$

$L = \left(\dfrac{28^\circ}{360^\circ}\right) 2\pi (250 \text{ ft}) = \boxed{122.17 \text{ ft}}$

The PC and PT stations are

$$PI = 8+78.22$$
$$T = 62.33 \text{ ft}$$
$$PC = 8+15.89$$
$$L = 1+22.17$$
$$PT = 9+38.06$$

The deflection angles are as follows.

point	station			deflection angle
PC	8+15.89			0°00'
	8+25	$\left(\dfrac{9.11}{122.17}\right)\left(\dfrac{28}{2}\right)$	1.0440°	1°03'
	8+50	$\left(\dfrac{34.11}{122.17}\right)\left(\dfrac{28}{2}\right)$	3.9088°	3°55'
	8+75	$\left(\dfrac{59.11}{122.17}\right)\left(\dfrac{28}{2}\right)$	6.7737°	6°46'
	9+00	$\left(\dfrac{84.11}{122.17}\right)\left(\dfrac{28}{2}\right)$	9.6385°	9°38'
	9+25	$\left(\dfrac{109.11}{122.17}\right)\left(\dfrac{28}{2}\right)$	12.5034°	12°30'
	9+38.06	$\left(\dfrac{122.17}{122.17}\right)\left(\dfrac{28}{2}\right)$	14.0000°	14°00'

15.

Elm Street

point	station	deflection angle	chord	calculated bearing	curve data
PT	9+68.88	18°00'	18.88		
	9+50	16°39'			
			24.99		
	9+25	14°51'			
			24.99		
	9+00	13°04'			
			24.99		$\Delta = 36^\circ 00'$
	8+75	11°17'			$T = 129.97 \text{ ft}$
			24.99		$L = 251.33 \text{ ft}$
	8+50	9°29'			$R = 400 \text{ ft}$
	8+25	7°42'			
			24.99		
	8+00	5°54'			
			24.99		
	7+75	4°07'			
			24.99		
	7+50	2°19'			
			24.99		
	7+25	0°32'			
PC	7+17.55	0°00'	7.45		

16. $\Delta_2 = 68^\circ - 36^\circ = 32^\circ 00'$

$t_1 = (400 \text{ ft})(\tan 18^\circ) = 129.97 \text{ ft}$

$t_2 = (300 \text{ ft})(\tan 16^\circ) = 86.02 \text{ ft}$

$PI - PI_1 = \dfrac{(215.99 \text{ ft})(\sin 32^\circ)}{\sin 68^\circ} = 123.45 \text{ ft}$

$PI - PI_2 = \dfrac{(215.99 \text{ ft})(\sin 36^\circ)}{\sin 68^\circ} = 136.93 \text{ ft}$

$T_1 = 129.97 + 123.45 = 253.42 \text{ ft}$

$T_2 = 86.02 + 136.93 = 222.95 \text{ ft}$

$L_1 = \left(\dfrac{36}{360}\right) 2\pi (400) = 251.33 \text{ ft}$

$L_2 = \left(\dfrac{32}{360}\right) 2\pi (300) = 167.55 \text{ ft}$

$PI = 14+78.32$

$T_1 = 2+53.42$

$\boxed{PC = 12+24.90}$

$L_1 = 2+51.33$

$\boxed{PCC = 14+76.23}$

$L_2 = 1+67.55$

$\boxed{PT = 16+43.78}$

point	station			deflection angle
PC	12+24.90			$0°00'$
	13+00	$\left(\dfrac{75.10}{251.33}\right)$	$\left(\dfrac{36°}{2}\right) = 5.3786°$	$5°23'$
	14+00	$\left(\dfrac{175.10}{251.33}\right)$	$\left(\dfrac{36°}{2}\right) = 12.5405°$	$12°32'$
PCC	14+76.23	$\left(\dfrac{251.33}{251.33}\right)$	$\left(\dfrac{36°}{2}\right) = 18.0000°$	$18°00'$
	15+00	$\left(\dfrac{23.77}{167.55}\right)$	$\left(\dfrac{32°}{2}\right) = 2.2699°$	$2°16'$
	16+00	$\left(\dfrac{123.77}{167.55}\right)$	$\left(\dfrac{32°}{2}\right) = 11.8193°$	$11°49'$
PT	16+43.78	$\left(\dfrac{167.55}{167.55}\right)$	$\left(\dfrac{32°}{2}\right) = 16.0000°$	$16°00'$

17. Elm Street

point	station	deflection angle	curve data
PT	16+11.28	$17°00'$	
	16+00	$16°28'$	
	15+00	$11°41'$	
	14+00	$6°55'$	
	13+00	$2°08'$	$\Delta = 70°$
			$R_1 = 900$ ft
			$R_2 = 600$ ft
PCC	12+55.23	$18°00'$	$\Delta_1 = 36°$
			$\Delta_2 = 34°$
	12+00	$16°15'$	$T_1 = 575.61$ ft
			$T_2 = 481.10$ ft
	11+00	$13°04'$	$L_1 = 565.49$ ft
			$L_2 = 356.05$ ft
	10+00	$9°53'$	
	9+00	$6°42'$	
	8+00	$3°31'$	
	7+00	$0°20'$	
PC	6+89.74		

18.

line	bearing	distance	latitude	departure	DMD	area $(\text{ft})^2$
AB	N 17° W	290	+277.3	−84.8	84.8	+23,515
BC	N 59° E	150	+77.3	+128.6	128.6	+9941
CD	N 89° E	350	+6.1	+349.9	607.1	+3703
DE	S 61° E	150	−72.7	+131.2	1088.2	−79,112
EF	S 09° W	250	−246.9	−39.1	1180.3	−291,416
FG	S 68° W	280	−104.9	−259.6	881.6	−92,480
GA	N 74°15′ W	230	+63.8	−226.2	395.8	+ 25,252
			0.0	0.0		200,298

$$2\,\overline{)\,400{,}597}$$

area of segment: $\frac{1}{2}R^2(\theta - \sin\theta) = \left(\dfrac{1}{2}\right)(350\text{ ft})^2\left(\dfrac{\pi}{3} - 0.8660\right) = \underline{\ 11{,}098\ }$

$$211{,}396\text{ ft}^2$$

$$\text{total area} = \dfrac{211{,}396\text{ ft}^2}{43{,}560\ \dfrac{\text{ft}^2}{\text{ac}}} = \boxed{4.9\text{ ac}}$$

18 Topographic Surveying and Mapping

1. CARTOGRAPHY

Cartography is the profession of making maps. Topographic maps provide a plan view of a portion of the earth's surface showing natural and constructed features such as rivers, lakes, roads, buildings, and canals. The shape, or relief, of the area is shown by contour lines, hachures, or shading.

2. USES OF TOPOGRAPHIC MAPS

The planning of most construction begins with the *topographic map*, sometimes referred to as the *contour map*. A study of a topographic map should precede the planning of highways, canals, subdivisions, shopping centers, airports, golf courses, and other improvements.

3. TOPOGRAPHIC SURVEYS

Topographic surveys are made to determine the relative positions of points and objects so that the map maker can accurately represent their positions on the map.

4. TYPES OF MAPS

There are two basic types of maps: the strip map and the area map. The *strip map* is used in the development of highways, railroads, pipelines, powerlines, canals, and other projects that are narrow in width and long in length. The *area map* is used in the development of subdivisions, shopping centers, airports, and other localized projects.

5. CONTROL FOR TOPOGRAPHIC SURVEYS

Of great importance in topographic surveys is horizontal and vertical control. *Control* is the means of transferring the relative positions of points and objects on the surface of the earth to the surface of the map.

Topo

6. HORIZONTAL CONTROL

Relative position in the horizontal plane is maintained by *horizontal control*. Horizontal control consists of a series of points accurately fixed in position by distance and direction in the horizontal plane. For most topographic surveying, traverses furnish satisfactory control. For strip maps, the open traverse is used. For area maps, the closed traverse is used. The open traverse can be tied to fixed points at each end. The closed traverse can be closed to form a net that is accurate to the degree required.

For large areas, such as states, triangulation or trilateration furnish the most economical control.

7. VERTICAL CONTROL

Relative position in the vertical plane can be maintained by *vertical control*, a series of bench marks in the map area. These bench marks are referred to a known *datum*, usually mean sea level.

8. HORIZONTAL TIES

After the traverse is closed to the required specifications, objects that are to be included on the map are *tied* to the traverse. These *horizontal ties* are sometimes called the *detailing*. For large surveys, ties are made by photogrammetry. For smaller surveys, ties are made on the ground.

At least two measurements are required to tie one point to the traverse.

9. METHODS OF LOCATING POINTS IN THE FIELD

The two measurements required to tie one point to the traverse may consist of two horizontal distances, an angle and a horizontal distance, or two angles. There are several methods used to locate a point in the field. Only the four most common will be discussed.

10. RIGHT-ANGLE OFFSET METHOD OF TIES

The *right-angle offset method* of ties is the most common method used in route surveying for preparing strip maps. The ties are made after the centerline (or traverse line) has been established. Usually, stakes are driven at each station on the centerline, a 100 ft steel tape is stretched between successive stations with the 100 ft mark on the tape forward, and points on either side of the tape are tied to the traverse before the tape is moved forward to the next two stakes. To tie in the

corners of the house shown in Fig. 18.1, surveyors move along the tape to a point on the line where they estimate a perpendicular line from the traverse line would strike a corner of the building. They observe the plus at this point by glancing at the tape on the ground and then measure the distance from the traverse line to the corner with another (usually cloth) tape, and record both measurements in the field book. With the 100 ft mark of the steel tape forward, pluses are read directly. This procedure is repeated for the next corner. All sides of the house, including the side between the tied corners, are then measured with the cloth tape. A sketch of the house showing the dimensions of all sides of the house is placed in the field book.

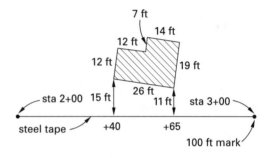

Figure 18.1 *Right Angle Offsets*

Unless the scale of the map is very large, measurements for ties are recorded to the nearest foot. It is usually impossible to scale a distance on the map for a tenth of a foot.

A right-angle mirror prism is convenient in establishing right angles. As a less accurate method, the surveyor may stand on the transit line facing the point to be tied with arms outstretched on each side pointing along the traverse line, and then bring both arms to the front of the body. If they do not point to the object to be tied, the surveyor should move along the traverse line until they do.

11. ANGLE AND DISTANCE METHOD OF TIES

The *angle and distance method* of ties, also called the *azimuth-stadia method*, is the most common of those used in preparation of area maps. The azimuth-stadia method allows direction and distance measurements to be made almost simultaneously. If the object is a house or building, two corners must be tied to the traverse, and all sides must be measured and recorded in the field book. The transit does not have to be confined to the traverse stations. Intermediate stations can be set from the traverse stations and ties made to the intermediate station. This method, using stadia for horizontal distance, is usually the most efficient.

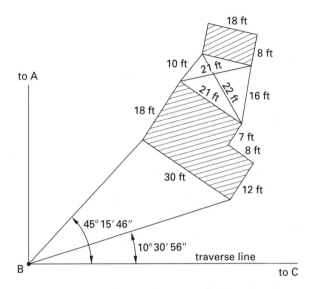

Figure 18.2 *Angle/Distance Ties and Two Distance Ties*

12. TWO DISTANCES METHOD OF TIES

The *two distances method* of ties can be used in conjunction with the angle and distance method. Where barns or out-buildings lie behind a house and are obscured from view, the house can be tied to the traverse by the angle and distance method. The out-building can be tied to the house by the two distances method. Two horizontal distances are required to locate one corner of the out-building. Two corners must be tied to the house as shown in Fig. 18.2. All sides of the out-building are measured and recorded in the field book.

13. TWO-ANGLE METHOD OF TIES

The *two-angle method* of ties is used in special cases where the object to be tied is inaccessible because it lies across a river, lake, or busy highway. The object is tied to the traverse by turning an angle to the object from two different points on the traverse.

14. STRENGTH OF TIES

When making horizontal ties, certain practices can be followed to reduce the error in locating points on the map.

- When the two distances method is used, the two distances should be as nearly at right angles as possible.

- When the two angles method is used, the two lines of sight to the object should be as nearly at right angles as possible.

15. VERTICAL TIES

Vertical ties can be made simultaneously with horizontal ties by stadia or by leveling. Leveling is used for strip maps with *cross-sectioning* as the usual method. For area maps, the *grid system* is common. This consists of laying out the area into a grid with 50 ft or 100 ft intervals and determining elevations at the grid intersections.

16. SUMMARY OF HORIZONTAL AND VERTICAL TIES

No one method of making ties excludes the possible use of others. The azimuth-stadia method is very efficient, does not require a large party, and is accurate enough for most work. Combinations of methods may be used. Where ditches or streams run through the map area, a combination of azimuth-stadia and cross-sectioning may by economical. The size of the area, slope of the terrain, and amount and size of vegetation also influence the selection of a method.

17. NOTEKEEPING

Examples of field notes for a right-angle offset survey are shown in Fig. 18.3. Most measurements are shown on the right. Transit stations and full stations are shown on the left. It is not necessary that the right half be drawn to scale, but it is often very convenient to do so. Each line space on the left represents 20 ft and the smallest line space on the right represents 10 ft. This scale makes for rapid plotting, but different topographic details require different scales. It is accepted practice to vary the scale from page to page if the amount of necessary detailing varies.

18. STADIA METHOD

The Greek word *stadia* denotes a unit of measure for horizontal distance. In surveying, the term is used to denote a system for measuring horizontal distances based on the optics of the transit telescope, theodolite, or level. This system eliminates the need for horizontal taping, and while not as accurate, it is satisfactory for making the horizontal measurements for topographic maps. The system is also known as *tacheometry*.

When employing the stadia method to obtain horizontal distances, the horizontal circle of the transit or theodolite indicates direction, and the vertical circle and a level rod determine elevation. Horizontal and vertical ties to the traverse can therefore be made simultaneously.

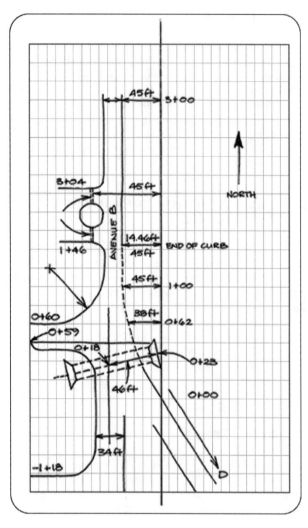

Figure 18.3 *Typical Field Notes for Right-Angle Offset Survey*

19. STADIA

Stadia sighting depends on two horizontal crosshairs, known as *stadia hairs*, within the telescope. These hairs are parallel to the horizontal crosshair and are equally spaced above and below it. The stadia hairs are shortened so that they will not be confused with the middle horizontal crosshair, although this may not be true in older transits (see Fig. 18.4).

The instrument operator sees the stadia hairs imposed on the stadia rod, as shown in Fig. 18.4. The distance on the rod is known as the *intercept*. If the rod is vertical and the telescope is horizontal, the distance from the center of the instrument to the rod is 100 times the intercept. Actually, the diverging lines of sight to the stadia hairs are from the vertex, which may be 1 in or 2 in from the center of the instrument. This discrepancy is ignored in topographic surveying. In older transits, a constant of 1 ft is added to the stadia distance because the vertex is about 1 ft forward of the center of the instrument.

The stadia principle is illustrated in Fig. 18.5. V_1, V_2, and V_3 represent rod intercepts at varying distances from the vertex of the transit. H_1, H_2, and H_3 represent corresponding horizontal distances. As the lines of sight from the vertex and the intercepts form similiar triangles, it can be seen that $H_1/V_1 = H_2/V_2 = H_3/V_3$. The stadia hairs are so constructed that when the stadia rod is 100 ft from the vertex, the intercept is 1.00 ft. Thus, $H_1 = 100V_1$ and $H_2 = 100V_2$. This means that the horizontal distance from the vertex to the rod is 100 times the intercept.

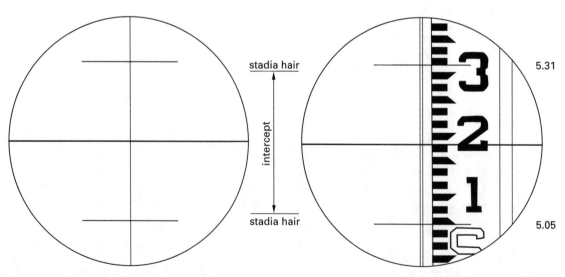

Figure 18.4 *Stadia Hairs and Use*

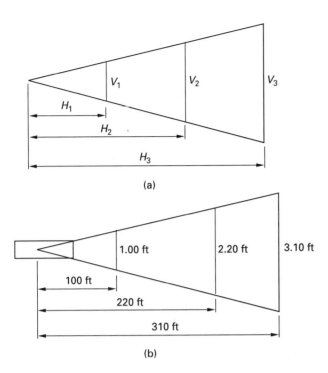

(a)

(b)

Figure 18.5 *Stadia Principles*

20. READING THE INTERCEPT

The intercept in Fig. 18.6(a) is 5.31 ft − 5.05 ft = 0.26 ft. The intercept in Fig. 18.6(b) is also 0.26 ft. In Fig. 18.6(b), the lower stadia hair has been placed on the nearest full foot by lowering the line of sight with the vertical tangent screw. The intercept can be determined easily by subtracting the full foot value from the reading at the upper stadia hair. This slight adjustment does not cause an appreciable error in the horizontal distance.

21. HORIZONTAL DISTANCE FROM INCLINED SIGHTS

In terrain steep enough to prevent all readings being made with the telescope horizontal, the horizontal distance is computed by trigonometry, using the rod intercept and the vertical angle of the line of sight.

Figure 18.7 indicates that the rod intercept is greater on inclined sights than it would be if the line of sight were perpendicular to the rod. As it would be impractical to hold the rod perpendicular to the line of sight, the rod is held plumb and the normal intercept is computed by trigonometry.

In Fig. 18.7, the horizontal crosshair is placed on a rod reading equivalent to the *height of the instrument, h.i.* (h.i. is the distance from the hub to the center of the instrument. It is not to be confused with HI, which is elevation above a datum.) This makes the vertical angle α at the instrument equal to the vertical angle α at the hub. In reading the intercept, the bottom stadia hair

is set on a full foot mark such that the middle hair is approximately on the h.i. After the intercept is read, the middle hair is set on the h.i. with the tangent screw and the angle α is read on the vertical circle.

If S' is the intercept actually read, S is the intercept normal to the line of sight, D is the slope distance, and H is horizontal distance, then

$$S = S' \cos \alpha \qquad 18.1$$

$$D = 100S = 100S' \cos \alpha \qquad 18.2$$

$$H = D \cos \alpha = 100S' \cos^2 \alpha \qquad 18.3$$

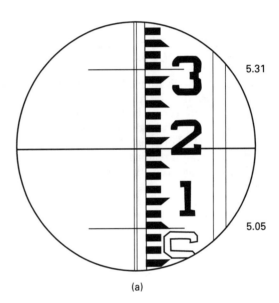

(a)

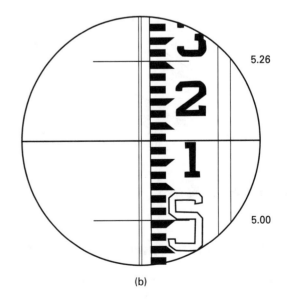

(b)

Figure 18.6 *Reading Intercepts*

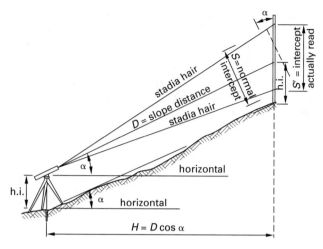

Figure 18.7 *Using Stadia Rod to Measure Horizontal Distance*

22. VERTICAL DISTANCE TO DETERMINE ELEVATION

The difference in elevation of the hub and any point can also be determined by trigonometry, using the angle α and the rod intercept. It can be seen in Fig. 18.8 that if the middle crosshair is on the h.i., the vertical distance V from the horizontal through the center of the instrument to this h.i. reading on the rod is the same as the vertical distance from the hub to the point on the ground where the rod rests.

$$V = D \sin \alpha = 100 S' \cos \alpha \sin \alpha \qquad 18.4$$

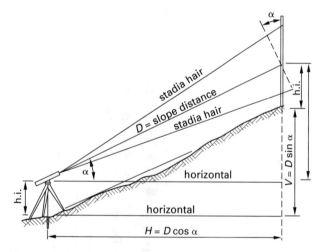

Figure 18.8 *Using Stadia Rod to Measure Vertical Distance*

23. USE OF STADIA REDUCTION TABLES

To avoid calculating $\cos^2 \alpha$ and $\cos \alpha \sin \alpha$, the horizontal distance H and the vertical distance V can be found in tables similar to Table 18.1.

Table 18.1 *Typical Stadia Reductions*

minutes	4° horizontal distance	4° vertical distance	5° horizontal distance	5° vertical distance
0	99.51	6.96	99.24	8.68
2	99.51	7.02	99.23	8.74
4	99.50	7.07	99.22	8.80
6	99.49	7.13	99.21	8.85
8	99.48	7.19	99.20	8.91
10	99.47	7.25	99.19	8.97
12	99.46	7.30	99.18	9.03
14	99.46	7.36	99.17	9.08

Table 18.1 lists horizontal and vertical distances for a rod intercept of 1.00 ft for 4° and 5° vertical angles. Vertical angles to the nearest full degree are shown at the top of the columns, and minutes are shown in the left column. Interpolation is required for angles of an odd number of minutes. To find horizontal and vertical distance for a particular vertical angle and intercept not equal to 1.00 ft, the intercept is multiplied by the distance found in the table.

Example 18.1

Find the horizontal distance H and the vertical distance V for a rod intercept of 3.68 ft and vertical angle α of 4°12′.

Solution

$$H = (3.68 \text{ ft})(99.46) = 366 \text{ ft}$$
$$V = (3.68 \text{ ft})(7.30) = 26.9 \text{ ft}$$

It can be seen from the table that for vertical angles up to 4°, the slope distance is very nearly the horizontal distance. For topographic detailing, no correction is needed. Of course, the vertical distance must still be computed.

24. AZIMUTH

Azimuth is the most efficient method of determining direction where a number of shots are taken from one station. By measuring all angles from the same reference line, plotting points on the map is simplified. A full circle protractor is oriented to north on each transit station. Azimuth readings from each transit station are plotted with one setting of the full circle protractor.

After the control traverse has been closed, the direction of all legs of the traverse should be recorded in azimuth in the field book. These directions are used in orienting the horizontal circle of the transit at each transit station.

To orient the horizontal circle, first set the vernier on the azimuth of the line along which the transit is to be sighted.

Set the vertical crosshair on that line by using the lower clamp, and then release the upper clamp. The circle will be oriented, but, as a check, the alidade is turned until the vernier reads 0° and the direction of the telescope is observed to see that it is pointing to the north. In Fig. 18.9, the transit is set up on point B. The azimuth of AB is now 150°00′. The horizontal circle is to be oriented for backsighting on point A.

To orient the circle, the vernier must be set on the azimuth of the line from the transit station B to the backsight A. The azimuth of BA is the back-azimuth of AB, so the vernier must be set on 330°00′ before the vertical crosshair is set on point A with the lower clamp.

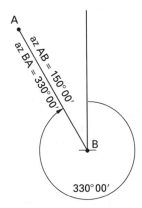

Figure 18.9 *Azimuth Measurments*

In locating points to be tied to the control net by azimuth, it is not necessary to read the vernier to determine the azimuth. In plotting with a protractor, it is impossible to plot to one-minute accuracy. Plotting to the nearest quarter of a degree is all that is practical. The horizontal circle can be read to 10′ without the use of the vernier. In establishing intermediate transit points not previously located, more care should be exercised in reading the circle.

After moving the transit to a new station, the circle will be oriented by setting on the back-azimuth of the line just established. The intercept from the new station to the previous station should be read as a check against the previous intercept. It is also good practice to take a shot on a known point as a check.

Where a number of shots are taken from one station, it is good practice to observe the backsight again before moving to a new station to see that the transit has not been disturbed.

25. ALGEBRAIC SIGN OF VERTICAL ANGLE

An *angle of elevation* is given a plus sign, and an *angle of depression* is given a minus sign. For small vertical angles, care should be taken that the wrong sign is not recorded. As a check, if the telescope bubble is forward, the sign is plus.

26. ELEVATION

As an example of determining vertical distance from the transit station to a point, assume in Fig. 18.8 that the elevation of the hub is 465.8 ft, the h.i. is 5.2 ft, the rod intercept is 4.22 ft, and the vertical angle is +4°12′ with the middle crosshair on the h.i. Using the tables,

$$V = (7.30)(4.22 \text{ ft}) = 30.8 \text{ ft}$$
$$\text{elevation} = 465.8 \text{ ft} + 5.2 \text{ ft} + 30.8 \text{ ft} - 5.2 \text{ ft}$$
$$= 465.8 \text{ ft} + 30.8 \text{ ft}$$
$$= 496.6 \text{ ft}$$

It can be seen that when the middle horizontal crosshair is on the h.i. when the vertical angle is read, the two h.i.'s will cancel. V can be added (or subtracted when α is minus) to the elevation of the transit station to obtain the elevation of a point. However, the reading of the h.i. on the rod may be obscured. In this case, the middle hair can be placed on the next higher (or lower) full foot mark and the vertical angle read. The h.i. and V will be added to the hub elevation and the rod reading will be subtracted.

Where possible, the rod readings should be taken with the telescope horizontal to eliminate work in reducing field notes. In this case, the h.i. is added to the elevation of the station and the rod reading is subtracted from this elevation just as in leveling. This eliminates the need to read the vertical angle.

27. FINDING THE h.i.

The h.i. can be found by placing the rod near the transit and moving a target along the rod until the horizontal line of the target lines up with the horizontal axis of the telescope.

28. SELECTING POINTS TO BE USED IN LOCATING CONTOURS

Contours are located on the map by assuming that there is a uniform slope between any two points that have been recorded in the notes. Elevations of points are written on the map and contour lines are interpolated between any two points. To ensure that the slope between any two points is uniform, shots must be taken in the field at certain key points.

29. KEY POINTS FOR CONTOURS

Key points are any points that will show breaks in the slope of the ground, just as in cross-sectioning. The most important of these are

- summits or peaks
- stream beds or valleys

- saddles (between two summits)
- depressions
- ridge lines
- ditch bottoms and tops of cuts
- tops of embankments and toes of slope

30. SPECIAL SHOTS

It is sometimes impossible for the rodperson to place the rod exactly on the point to be located, such as a corner of a house with a wide roof overhang. In this case, the rod is held near the house in such a position that the rod is the same distance from the transit as the house corner, the intercept is read at this point, and the house corner is used to determine horizontal direction.

For long shots where the intercept does not fall entirely on the rod or portion of the rod observed, the intercept between the upper crosshair and the middle crosshair can be observed and doubled for the intercept. On long shots, a stadia rod with bold markings and different colors is more suitable than a level rod.

31. EFFICIENCY OF THE SURVEY PARTY

The *efficiency* of the field party depends on the number of points located during a period of time. Time can be saved by following the order of taking readings on a point as follows.

step 1: Set the vertical crosshair on the rod.

step 2: Set the lower stadia crosshair on a full foot mark such that the middle crosshair is approximately on the h.i.

step 3: Read the upper crosshair, subtract the reading of the lower crosshair, and record the intercept.

step 4: With the tangent screw, place the middle crosshair on the h.i.

step 5: Wave the rodperson to the next point.

step 6: Read and record the horizontal angle.

step 7: Read and record the vertical angle.

It is important that the rodperson be waved on (step 5) before the two angle readings are made and recorded so that he can be moving to a new location while the instrument person is reading the angles. Many shots in a day's work will be lost by forgetting to do this.

32. COMPUTATIONS FROM FIELD NOTES

Field notes for an azimuth-stadia survey are shown in Table 18.2. In the notes, the transit is at station B at the beginning of the day. A backsight is made on station A (which was set at the close of the preceding day). Referring to the field notes of the preceding day, the azimuth of AB was 316°22′. Therefore, when the sight on station A is made, the vernier is set on 135°22′, which is the azimuth of BA and the back-azimuth of AB.

The intercept at station A is read as a check against what was read at B from A on the previous day. The vertical angle, −1°18′, is also read as a check.

The rod reading for station 1 was made with the telescope level. The elevation of station 1 is

$$467.2 \text{ ft} + 5.0 \text{ ft} - 8.0 \text{ ft} = 464.2 \text{ ft}$$

The rod reading on station 4 was made with a vertical angle of +4°34′. This is large enough to require a correction for horizontal distance. The correction factor for 4°34′, found in stadia reduction tables, is 99.37. The horizontal distance B to 4 is

$$H = (99.37)(3.40 \text{ ft}) = 338 \text{ ft}$$

The vertical angle was read when the horizontal crosshair was set on the h.i. If it had not been, the notes would indicate this fact. The factor for the vertical distance for an angle of 4°34′ is found to be 7.94. The vertical distance is

$$V = (7.94)(3.40 \text{ ft}) = 27.0 \text{ ft}$$

The elevation of station 4 is

$$467.2 \text{ ft} + 27.0 \text{ ft} = 494.2 \text{ ft}$$

Table 18.2 *Field Notes for an Azimuth-Stadia Survey*

sta	rod int	az	∠ or rod	*H*-dist.	*V*-dist.	elev.
		⊼ @ B elev. 467.2 h.i. 5.0				
A	6.82	136°22′	−1°18′			
1	1.15	88°10′	8.0			
2	1.84	96°30′	7.8			
3	2.28	124°45′	+3°21′			
4	3.40	206°20′	+4°34′			
5	3.25	318°00′	+4°20′			
6	4.36	345°30′	+2°47′			
C	6.25	318°52′	+1°12′			
		⊼ @ C elev. 480.3 h.i. 5.2				
B	6.24	138°52′	−1°13′			
1	1.78	96°10′	+2°41′			
2	2.49	128°45′	9.2			
3	3.12	186°00′	−4°56′			
4	3.40	232°40′	9.5			
D	7.18	285°30′	−1°16′			
		⊼ @ D elev. 464.4 h.i. 5.1				
C	7.18	105°30′	+1°16′			
1	2.42	145°15′	7.7			
2	3.18	180°20′	10.6			

The horizontal crosshair could not be placed on 5.0 (the h.i.) at station 4 because a limb of a tree obstructed the view. The horizontal crosshair was placed on 6.0, and the vertical angle was +4°45′. Then,

$$V = (8.25)(3.40 \text{ ft}) = 28.0 \text{ ft}$$

$$\text{elevation} = 467.2 \text{ ft} + 5.0 \text{ ft} + 28.0 \text{ ft} - 6.0 \text{ ft}$$

$$= 494.2 \text{ ft}$$

33. CONTOURS AND CONTOUR LINES

A *contour* is an imaginary line on the surface of the earth that connects points of equal elevation. A *contour line* is a line on a map that represents a contour on the ground.

34. CONTOUR INTERVAL

The *contour interval* of a map is the vertical distance between contour lines. The contour interval is selected by the mapmaker. In flat country, it may be 1 ft and in mountainous country it may be 100 ft, depending on the scale of the map and the character of the terrain. The contour interval can be too small, making the map a maze of lines that are not legible; the contour interval can be too large, not showing the true relief. The more accurate the contours, the more costly the map. The intended used of the map is a basic consideration in the selection of the contour interval.

Fig. 18.10 shows that the vertical distance between contour lines is constant, but the horizontal distance varies with the steepness of the ground.

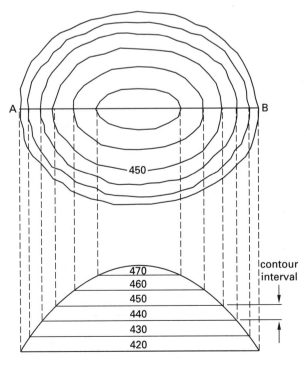

Figure 18.10 Contour Line Intervals

35. INDEX CONTOURS

To facilitate reading a topographic map, every fifth contour line may be darker. The elevation of that contour line is written in a break in the line as shown in Fig. 18.11. There are five spaces between any two heavier lines, called *index contours*, so that the contour interval can be computed by dividing the difference in elevation between two index contours by five. In Fig. 18.11, there are five spaces between the 700 ft contour and the 750 ft contour. The contour interval is

$$750 \text{ ft} - \frac{700 \text{ ft}}{5} = 10 \text{ ft}$$

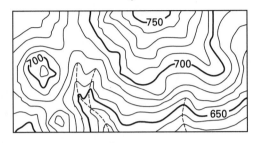

Figure 18.11 Typical Contour Map

36. CLOSED CONTOUR LINES

Contour lines that are closed represent either a hill or depression. Whether the closed contour lines represent a hill or depression can be determined by reading the elevation of the index contours. For a hill, the elevations increase as the contour lines become shorter. Depressions are often indicated by short hachures on the down slope side of the contour line.

37. SADDLE

The name *saddle* is given to the shape of contours that define two summits in the same vicinity. A profile view of the two summits looks somewhat like the profile view of a horse saddle, as shown in Fig. 18.12.

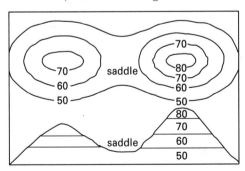

Figure 18.12 Saddle

38. CHARACTERISTICS OF CONTOURS

Certain fundamental characteristics of contours should be kept in mind when plotting contour lines and reading a contour map. The characteristics are illustrated in Fig. 18.13.

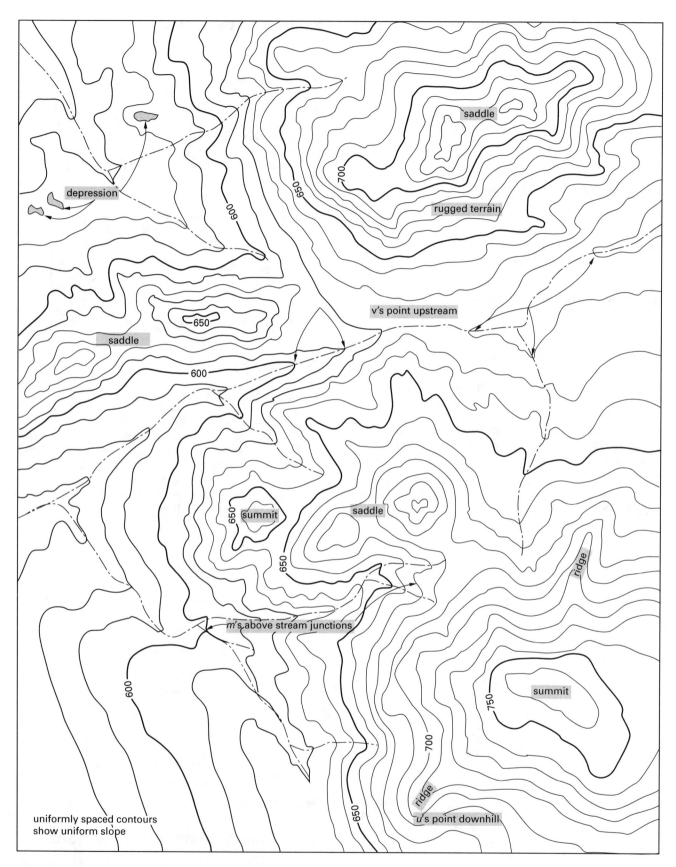

Figure 18.13 *Typical Contour Map Features*

- Each contour line must close upon itself either within or outside the borders of the map. Since all land areas on the surface of the earth rise above the sea, it can be seen that each contour will, no matter how long, finally close. This means that a contour line cannot end abruptly on a map.

- Contour lines cannot cross or meet, except in unusual cases of waterfalls or cliff overhangs. (If it were possible for two contour lines to cross, the intersection would represent two elevations for the same point.)

- A series of closed contour lines represents either a hill or depression. The elevations of the index contour lines will indicate which series represents a hill and which represents a depression.

- Contour lines crossing a stream form Vs that point upstream.

- Contour lines crossing a ridge form Us that point down the ridge.

- Contour lines tend to parallel streams. Rivers usually have a flatter gradient than do intermittent streams. Therefore, contour lines along rivers will be more nearly parallel than contours along intermittent streams, and they will run parallel for a longer distance.

- Contour lines form Ms just above stream junctions.

- Contour lines are uniformly spaced on uniformly sloping ground.

- Irregularly spaced contour lines represent rough, rugged ground.

- The horizontal distance between contour lines indicates the slope of the ground. Closely spaced contour lines represent steeper ground than widely spaced contour lines.

- Contours are perpendicular to the direction of maximum slope. The direction of rainfall run-off in a map area can be determined from this characteristic.

39. METHODS OF LOCATING CONTOURS

Several methods of locating contours are used in topographic surveying and mapping. They are the grid method, controlling points method, cross-section method, and tracing contours method. All methods depend on the assumption that there is a uniform slope between any two ground points located in the field.

40. GRID METHOD

The *grid method* is very effective in locating contours in a relatively small area of fairly uniform slope. The area is divided into squares or rectangles of 25 ft to 100 ft (depending on the scale of the map and the contour interval desired). Stakes are set at each intersection and at any points of slope change, such as at ridge lines or valleys.

The location of the point where the contour line crosses each side of each square is determined by interpolation (either by estimation or mathematical proportion).

In Fig. 18.14(a), contours are to be plotted on a 2 ft interval. Starting at A-1, Fig. 18.14(b), it can be seen that the 440 ft contour will cross between A-1 and B-1. The vertical distance between A-1 and B-1 is 4 ft, so the 440 ft contour will cross halfway between A-1 and B-1. The 442 ft contour will cross at B-1, so a mark is placed at each of these points.

The 444 ft contour will cross between B-1 and C-1. The vertical distance between B-1 and C-1 is 444.3 ft − 444.1 ft = 2.3 ft. The vertical distance between B-1 and the 444 ft contour is 2.0 ft. Therefore, the horizontal distance will be 2.0/2.3 of the way, or about 0.9 of the way. A mark is made at this point.

The 440 ft, the 442 ft, and the 444 ft contours will cross between A-2 and B-2. The crossing points are found in a similar manner and marked.

After the crossing points are located by interpolation, the crossing points for each contour are connected as shown in Fig. 18.14(c).

After all crossing points are connected, the contour lines are smoothed. Small irregularities are taken out so that the contour lines are more like the contours on the ground.

In following a particular contour line using the grid method, an inspection must be made of each grid line between each intersection to see if the contour line can cross. If it cannot, another line must be inspected. For example, in Fig. 18.14(c), the 444 ft contour line can cross between B-1 and C-1, between B-1 and B-2, between A-2 and B-2, or between B-2 and B-3, but not between B-2 and C-2. Each contour line must close or reach the border of the map at two points.

After contour lines are smoothed, index contour lines must be made heavier than the other lines, and the elevation of the index contour must be written in a break in the line.

41. CONTROLLING POINTS METHOD

The *controlling points method* is suitable for maps of large area and small scale. The selection of ground points is very important. The accuracy of the contours depends on the knowledge and experience of the survey party. Shots should be taken at stream junctions, at intermediate points in stream beds between junctions, and along ridge lines. Field notes should indicate these points so that ridges and streams can be plotted before interpolations are made. Interpolations are made in much the same way as they are made using the grid method.

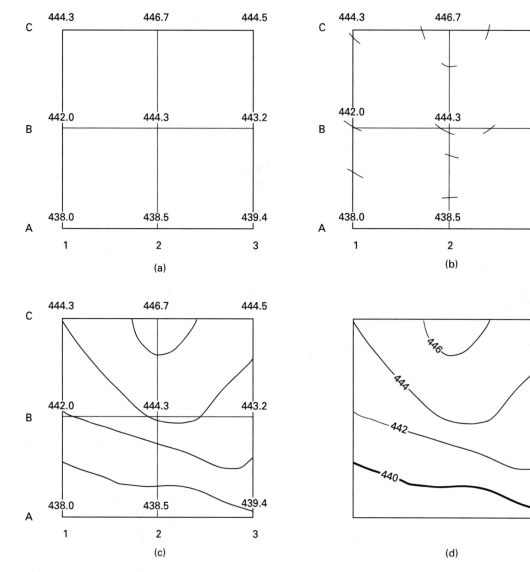

Figure 18.14 *Grid Method of Contour Location*

Figs. 18.15(a), 18.15(b), 18.15(c), and 18.15(d) show the progressive steps in plotting contour lines by the controlling points method.

42. CROSS-SECTION METHOD

The *cross-section method* is satisfactory for the preparation of strip maps. It can be accomplished by using level and tape or azimuth-stadia. Cross sections are taken at right angles to a centerline or baseline. Elevations of each cross-section shot are written on the strip map, and interpolation of contour lines is performed as in the grid and controlling points methods.

43. TRACING CONTOURS METHOD

The *tracing contours method* is used when the exact location of a particular contour line is needed. It is effectively performed by use of the plane table, but can be done by the azimuth-stadia survey method.

44. MAPPING

The first step in preparing a map is the selection of a scale and a contour interval. This selection is influenced by the size of the sheet to be used, the purpose of the map, and the required accuracy.

The traverse can be plotted by the coordinate method, the tangent method, or the protractor method.

45. COORDINATE METHOD

The *coordinate method* is the most accurate for plotting a traverse. Any error in plotting one point does not affect the location of the other points. Each point is plotted independently of the others.

Coordinates of each traverse station are computed prior to plotting. Each point is plotted on the grid system using the coordinates.

Before a traverse is plotted, the sheet is laid out with perpendicular grid lines. The distance between grid lines can be 50 ft, 100 ft, 500 ft, 1000 ft, or any other multiple that suits the scale of the map.

As an example, Fig. 18.16 shows a grid system laid out to scale of 1 in = 500 ft with grid lines 1000 ft apart.

The point P has the coordinates $x = 1660, y = 4705$. In plotting the point, the 50 scale is laid on the paper horizontally so that the 10 mark on the scale lines up with the vertical grid line marked 1000. A pencil dot is then made at 1660 on the scale. With a straight edge, a temporary vertical line is drawn through the pencil dot. The scale is then laid vertically along this vertical

Topo

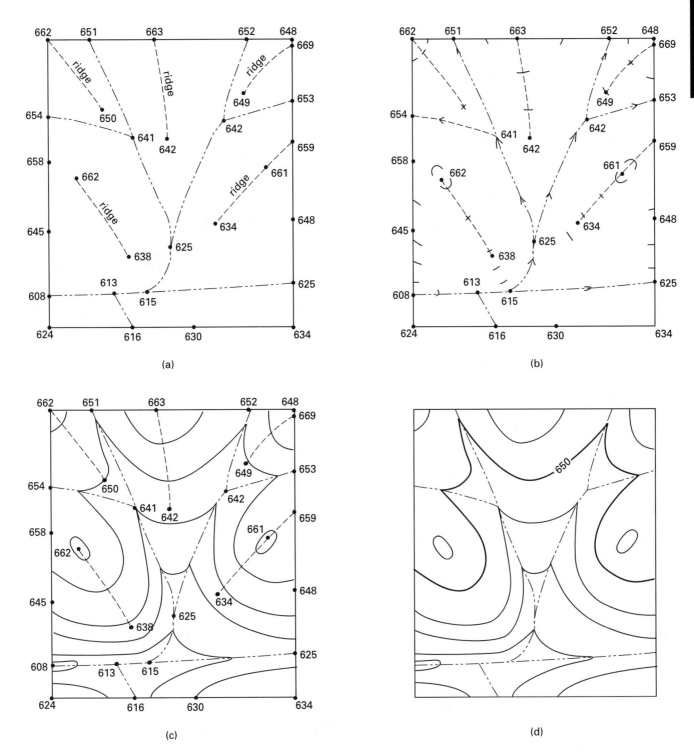

Figure 18.15 *Controlling Points for Contour Location*

line with the 40 mark on the scale lined up with the horizontal line on the paper marked 4000. A pencil dot is carefully made on the vertical line at 4705 on the scale (as close as can be read).

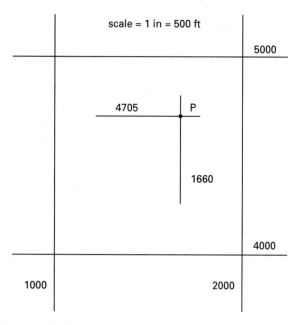

Figure 18.16 *Coordinates of a Traverse Point*

Figure 18.17 *Typical Map Symbols*

After all points of the traverse are plotted, they are connected with lines. Distances between points are scaled and checked against distances that were recorded in the field.

Detail points can be plotted with a protractor using grid lines to orient the protractor.

46. TANGENT METHOD

The *tangent method* is very convenient and accurate where deflection angles have been turned for a route survey. At any traverse station, the back line is produced past the points. A convenient distance (such as 10 in) is laid off from the traverse point along this prolongation. A perpendicular line is determined at this point, and the tangent distance for the deflection angle is marked on this line. A line from the traverse point defines the next leg of the traverse. Any error made in this plotting will be carried on to the next plotting.

47. PROTRACTOR METHOD

The *protractor method* is the fastest but least accurate method of plotting a traverse. Any error in plotting an angle or a distance will be carried on throughout the traverse. The protractor is commonly used for detailing and, for this, it is sufficiently accurate.

PRACTICE PROBLEMS

1. Complete the topography field notes using the right-angle offset method for ties to the road, stream, and buildings. Consider the enclosed area to be the right half of the page in the field notes. The vertical line is the baseline of the survey. The scale is 1/2 in = 100 ft.

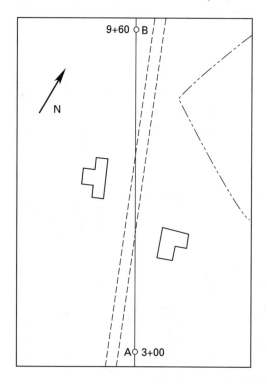

2. Complete the topography field notes using the angle and distance method and the two distances method. Use the two distances method to tie the small building to the adjacent building; use the angle and distance method for other ties. Consider the transit to be set up at station A on the baseline with the foresight on station B on the baseline. The scale is 1/2 in = 100 ft.

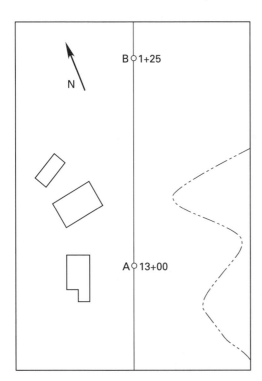

3. Tie the streets to the baseline by the right angle offset method. The scale is 1/2 in = 100 ft.

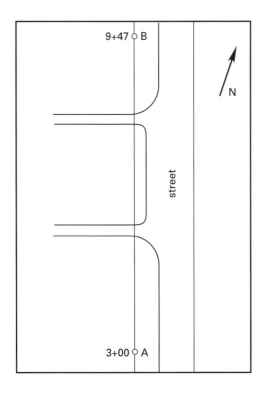

4. Find the horizontal distance H and the vertical distance V between two points when the stadia rod intercept and vertical angle α are as shown.

	rod intercept	α
(a)	3.48	4°00′
(b)	4.68	5°09′
(c)	5.75	5°12′
(d)	3.91	4°11′

5. Indicate the angle to be set on the vernier to orient the transit for an azimuth-stadia survey for each station of the traverse ABCDEA. AB: S 53°18′ E; BC: N 42°00′ E; CD: N 72°47′ W; DE: N 36°27′ W; and EA: S 30°39′ W.

6. Considering the transit to be set up at station A, elev 463.2 ft, with h.i. = 5.1 ft, compute the horizontal distance from A to points 1, 2, 3, and 4 and the elevation of points 1, 2, 3, and 4. (Consider middle crosshair on h.i. when vertical angle is read.)

point	intercept	V angle
1	3.85	+4°05′
2	2.98	+5°08′
3	2.56	−5°09′
4	5.44	−4°03′

7. Complete the following azimuth-stadia survey field notes.

			azimuth-stadia survey		
			V ang		
sta	int	az	or rod	H dist	elev
		TRAN @ B ELEV 467.2 h.i. 5.0			
A	6.74	148°04′	−4°14′		
1	0.91	90°45′	8.1		
2	1.66	120°20′	−5°12′		
3	2.10	135°15′	−5°07′		
4	3.25	143°00′	+4°11′		
C	4.00	60°10′	+5°08′		
		TRAN @ C ELEV 502.8 h.i. 5.2			
B	4.01	240°10′	−5°08′		
5	3.15	286°00′	−4°00′		
6	2.21	36°20′	9.2		

8. Plot 1 ft contours. Indicate index contours.

451.7	452.5	453.1	453.7	453.7	453.0	451.9	451.2
452.8	453.8	454.5	455.0	454.8	453.9	452.6	451.6
453.9	455.0	456.0	456.6	455.4	454.0	452.5	450.9
454.4	455.9	457.5	456.2	455.0	453.7	452.2	451.9
454.1	455.7	456.1	455.0	454.2	453.2	453.1	452.9
453.3	454.2	454.1	453.9	453.9	454.1	454.5	453.9
452.4	453.1	452.9	453.7	454.6	455.9	456.1	454.8
451.4	451.6	452.8	454.1	455.2	456.1	455.3	454.8
450.2	451.7	453.1	453.8	454.5	454.8	454.3	453.7

9. Plot 1 ft contours. Indicate index contours.

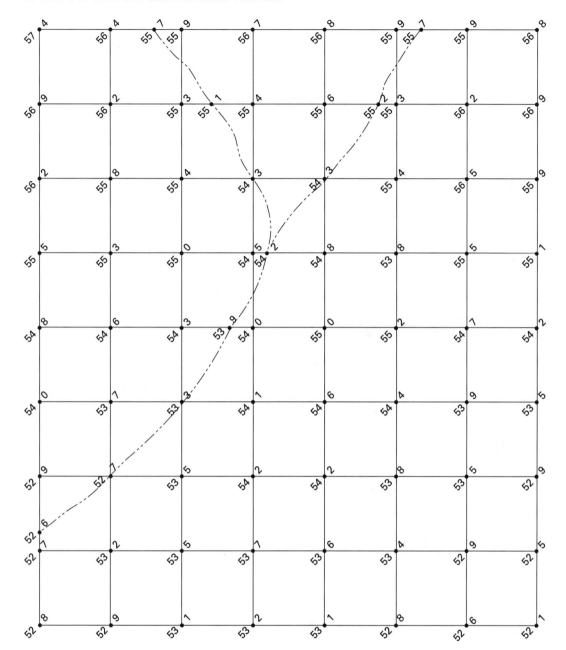

10. Plot 5 ft contours.

435.0 443.2 449.9 454.1 458.6 462.5 468.3 474.1

423.3 434.1 444.9 450.2 455.1 460.1 464.8 466.7

434.2 430.5 438.1 444.8 454.6 455.2 459.9 463.0

445.4 446.1 446.2 442.4 444.1 448.9 452.2 458.1

450.2 451.7 448.1 439.9 436.2 442.1 450.0 455.8

445.3 441.2 436.8 431.1 436.7 443.1 449.2 454.2

434.9 432.1 428.2 427.3 434.1 441.8 449.2 454.3

427.2 424.3 422.2 433.3 441.4 446.2 450.9 456.8

421.9 417.1 428.2 440.1 447.3 452.2 452.3 461.2

11. Plot 5 ft contours.

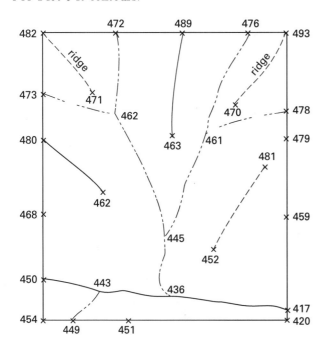

SOLUTIONS

1. Use an engineer's scale to determine distances. (See Fig. 18.17 in the text for map symbols.)

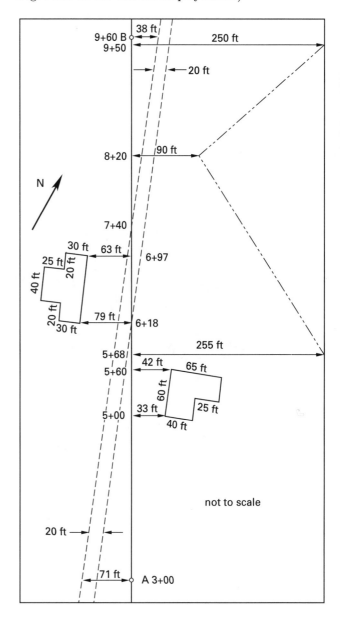

2. Measure angles with a protractor; measure distances with an engineer's scale.

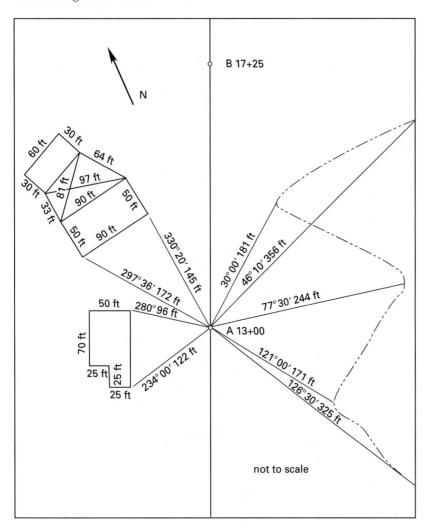

Topo

3. Use an engineer's scale for distances.

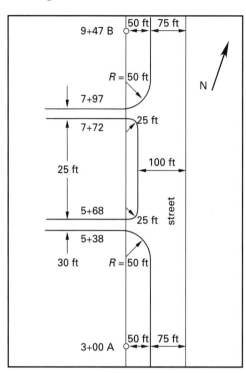

6.

pt	int	V ang	H dist	elev
1	3.85	+4°05′	383	490.5
2	2.98	+5°08′	296	489.8
3	2.56	−5°09′	254	440.3
4	5.44	−4°03′	541	424.9

7.

azimuth-stadia survey

sta	int	az	V ang or rod	H dist	elev
		TRAN @ B ELEV 467.2 h.i. 5.0			
A	6.74	148°04′	−4°14′	670	417.6
1	0.91	90°45′	8.1	91	464.1
2	1.66	120°20′	−5°12′	164	452.2
3	2.10	135°15′	−5°07′	208	448.6
4	3.25	143°00′	+4°11′	323	490.8
C	4.00	60°10′	+5°08′	397	502.8
		TRAN @ C ELEV 502.8 h.i. 5.2			
B	4.01	240°10′	−5°08′	398	467.2
5	3.15	286°00′	−4°00′	313	480.9
6	2.21	36°20′	9.2	221	498.8

8.

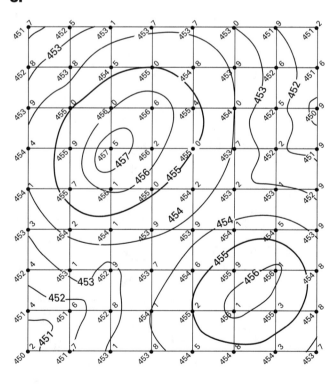

4. (a) $H = (3.48 \text{ ft})(99.51) = \boxed{346 \text{ ft}}$

$V = (3.48 \text{ ft})(6.96) = \boxed{24.2 \text{ ft}}$

(b) $H = (4.68 \text{ ft})(99.20) = \boxed{464 \text{ ft}}$

$V = (4.68 \text{ ft})(8.94) = \boxed{41.8 \text{ ft}}$

(c) $H = (5.75 \text{ ft})(99.08) = \boxed{570 \text{ ft}}$

$V = (5.75 \text{ ft})(9.03) = \boxed{51.9 \text{ ft}}$

(d) $H = (3.91 \text{ ft})(99.46) = \boxed{389 \text{ ft}}$

$V = (3.91 \text{ ft})(7.28) = \boxed{28.5 \text{ ft}}$

5.

transit at	sight on	vernier on
A	B	126°42′
B	A	306°42′
C	D	287°13′
D	E	323°33′
E	D	143°33′

9.

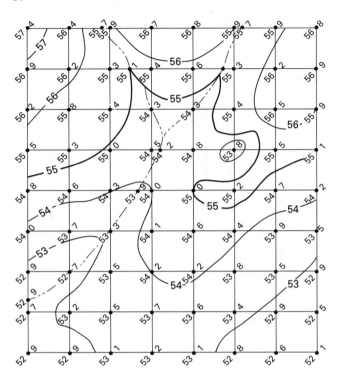

11.

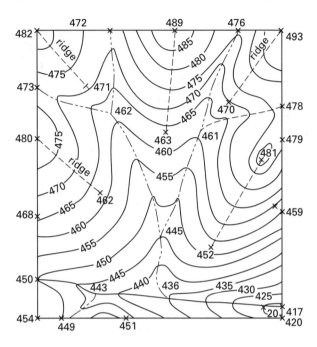

10.

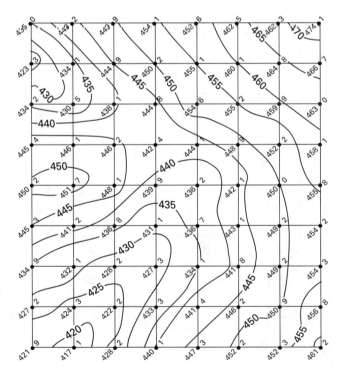

19 Astronomical Observations

1. ANCIENT ASTRONOMERS

People of ancient times were aware of their dependence on the cycles of earth and sky. The sun's life-giving warmth and radiance, the mysterious waxing and waning of the moon, and the regular movements of the stars and planets inspired reverence and gave rise to the belief that the heavenly bodies were gods who controlled the universe.

Temple priests were assigned the task of observing the movements of these deities. They were also responsible for noting other celestial events and interpreting them. Colorful stories about these gods and their interactions with humankind evolved.

A great deal of astronomical knowledge was accumulated and passed on orally through these myths. Only the Babylonians and Egyptians recorded their findings for future civilizations.

The Greeks, from 700 B.C. to A.D. 200 first gave serious thought to the size and shape of the earth, and they were the first to conclude that the earth is a sphere. Aristotle noticed that the positions of the stars change, and that during an eclipse the shadow of the earth on the moon is curved. Eratosthenes calculated the circumference of the earth and found it to be 25,000 miles. Ptolemy, the great cartographer, devised a system for locating a point on the earth using a system similar to present-day latitude and longitude.

After Ptolemy, civilization slipped into the Dark Ages. For more than a thousand years the theory that the earth is a sphere was forgotten.

In the 15th century, the beginning of the European Renaissance, there was renewed interest in astronomy. Copernicus concluded that the earth revolved around the sun, directly contradicting the beliefs of the Catholic Church. In the 17th century, Kepler discovered that the orbits of the planets around the sun are elliptical, with the sun at one focus of the ellipse. He also discovered that a line joining a planet and the sun sweeps out equal areas of space in equal time, proving that a planet moves fastest when nearest the sun and slowest when farthest away.

Galileo, who was the first to appreciate the significance of the telescope, accepted Copernicus' theories and was tried by the Inquisition for his heretical ideas.

2. THE EARTH

The earth rotates from west to east on its polar axis and revolves about the sun in an elliptical orbit with the sun at one focus of the ellipse. It completes one revolution in a period of 365.2564 days. The inclination of the earth ($23\frac{1}{2}°$ with the perpendicular to the orbital plane), combined with its revolution around the sun, causes the lengths of day and night to change and also causes the seasons (see Fig. 19.1).

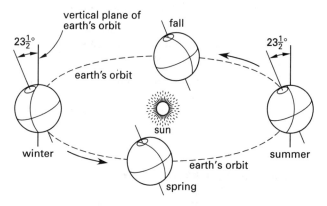

Figure 19.1 *Revolution of the Earth about the Sun*

On March 21 and September 23, the light from the sun reaches from one pole to the other. On these dates, shown as spring and fall in Fig. 19.1, the sun is directly overhead at the equator. On June 21, shown as summer in the figure, the sun is directly overhead at points $23\frac{1}{2}°$ above (north of) the equator. On December 22, shown as winter, the sun is overhead at points $23\frac{1}{2}°$ below (south of) the equator.

A line around the earth, parallel to and 23½° north of the equator, where the sun is directly overhead at its northernmost position, is known as the *Tropic of Cancer*. A line around the earth, parallel to and 23½° south of the equator, where the sun is directly overhead at its southernmost position, is known as the *Tropic of Capricorn*.

March 21 and September 23, when the sun crosses the equator and day and night are everywhere of equal length, are known as *equinoxes* (equal nights). The vernal equinox (March 21) marks the beginning of spring, and the autumnal equinox (September 23) marks the beginning of fall. June 21, the longest day of the year in the northern hemisphere, is known as the *summer solstice*. December 22, the shortest day of the year in the northern hemisphere, is known as the *winter solstice*.

In the southern hemisphere, the seasons are opposite those in the northern hemisphere.

3. GEODETIC NORTH OR GEODETIC AZIMUTH

The direction of a line is determined by the horizontal angle between the line and a reference line, usually true (geodetic) north. Determining the true azimuth of a line involves observations on a celestial body such as the sun or another star. The star usually used in the northern hemisphere is Polaris, the North Star. It is selected because it is very near true north from any point north of the equator. In making observations on celestial bodies, a surveyor is not interested in the distance to these bodies from the earth, but merely in their angular positions from the surveyor's observation point.

In addition to determining true azimuth, a surveyor can determine the latitude and longitude of a position by making observations on celestial bodies.

Observations on celestial bodies include making angular measurements, both horizontal and vertical. Determining true azimuth requires using an ephemeris, which is discussed in following sections.

4. PRACTICAL ASTRONOMY

Since ancient times, it has been convenient in practical astronomy to regard all celestial bodies as being fixed onto a sphere of infinite radius whose center is the earth's center. (However, stars are not the same distance from the earth, and they are much farther away than they appear to be.) This sphere of infinite radius is called the *celestial sphere* (see Fig. 19.2). To the observer, this sphere appears to be rotating about an axis, but it is not; the rotation of the earth causes the stars to be in rotation. In practical astronomy, it is assumed that the earth is stationary and that the celestial sphere revolves about the earth from west to east.

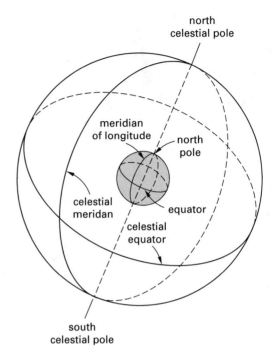

Figure 19.2 *Celestial Sphere*

The celestial sphere is assumed to rotate about the celestial axis, which is a prolongation of the earth's polar axis. The north and south poles become the *north celestial pole* and the *south celestial pole*.

The plane of the earth's equator extended to the celestial sphere becomes the *celestial equator*.

Comparable with the parallels of latitude of the earth are the *parallels of declination* of the celestial sphere. They measure the angular distance from the celestial equator to the north and south celestial poles (see Fig. 19.5).

Comparable with meridians of longitude of the earth are the *hour circles* of the celestial sphere, all of which converge at the celestial poles. They are also known as *celestial meridians*, as shown in Figs. 19.2 and 19.3.

The observer's position on the earth is located by latitude and longitude. If the observer's plumb line is extended upward to the celestial sphere, the point of intersection with the celestial sphere is the observer's *zenith*. If the plumb line is extended downward to the celestial sphere, the intersection is the observer's *nadir*.

The relative positions of an observer on the earth and on the celestial sphere are shown in Fig. 19.3. It can be seen that the angle at the center of the earth that measures latitude is the same for the earth and for the celestial sphere. Likewise, the angle that measures longitude is the same for the earth and for the celestial sphere.

In Fig. 19.3, the observer's position on earth is latitude 35° N and longitude 98° W. The arc distance (in degrees) between the zenith of the observer and the celestial equator is also 35° N. The arc distance (in degrees) along the celestial equator between the planes of the Greenwich meridian and the meridian of the observer, extended to the celestial sphere, is 98° W. This arc distance is also the angle at the celestial north pole between the planes of the two meridians.

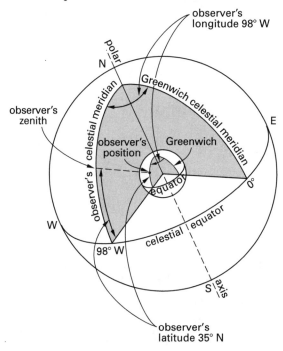

Figure 19.3 *Relation Between Earth and Celestial Sphere*

By using latitude and longitude and projecting it to the celestial sphere, the position of the observer's instrument is fixed at a point on the celestial sphere.

To identify the location of a celestial body on the celestial sphere, a coordinate system similar to latitude and longitude was devised (see Fig. 19.4). The celestial coordinates are called *declination* and *right ascension*. Using declination and right ascension as coordinates, the location of any celestial body on the celestial sphere can be determined with respect to the observer's zenith on the celestial sphere and to the north celestial pole.

Declination is the star's angular distance north or south of its celestial equator measured along the hour circle of the star (see Fig. 19.4). North declination is plus (+); south declination is minus (−). Declination corresponds to latitude on the earth.

Right ascension is the arc distance measured eastward along the celestial equator from the vernal equinox to the hour circle of the star. It may be measured in degrees, minutes, and seconds of arc or in hours, minutes, and seconds of time (see Fig. 19.4). Right ascension corresponds with longitude on the earth.

As with latitude and longitude, a celestial coordinate system requires points of origin, such as the equator for latitude and the Greenwich meridian for longitude. The celestial coordinate system uses the *celestial equator*, the *ecliptic*, and the *vernal equinox* in locating points of origin.

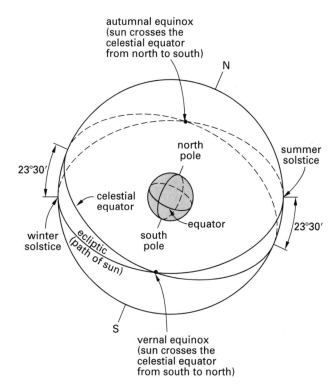

Figure 19.4 *Ecliptic and Vernal Equinox*

Because of the tilt of the earth as it follows its orbit around the sun, the sun traces a path called the *ecliptic* on the celestial sphere. The path of the sun moves from the southern hemisphere of the celestial sphere to the northern hemisphere and back (see Fig. 19.5).

The point where the sun crosses the celestial equator on its movement each year from south to north along the ecliptic is known as the *vernal equinox* (see Fig. 19.4). Astronomers designated the vernal equinox as the point of reference for right ascension.

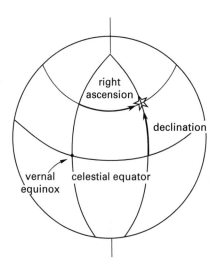

Figure 19.5 *Celestial Coordinates*

The vernal equinox is a point on the celestial sphere of infinite distance from the earth. Its location in time, relative to the Greenwich meridian, is known.

The point where the sky and earth meet, as seen by an observer on earth, is known as the *horizon*. The horizon for any place on the earth's surface is the great circle formed on the celestial sphere by the extension of the plane of the observer's horizon. In practical astronomy, the horizon is the plane tangent to the earth at the observer's position, perpendicular to the plumb line and extended to the celestial sphere. It is used as a reference for determining the altitude of a celestial body.

The *altitude* (h) of a celestial body is the angular distance measured from the horizon to a celestial body. It is the vertical angle measured by the observer from the horizon to the body (see Fig. 19.9).

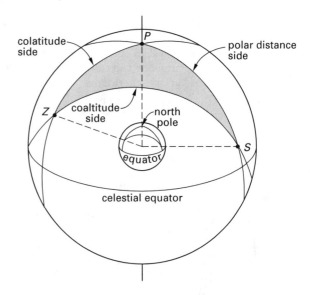

Figure 19.7 *Three Sides of the PZS Triangle*

5. THE ASTRONOMICAL TRIANGLE

To determine azimuth or latitude and longitude, the surveyor needs to be able to solve a spherical triangle on the celestial sphere known as the *astronomical triangle*, or the celestial triangle *PZS*.

The vertices of the *PZS* triangle are the north celestial pole (*P*), the observer's zenith (*Z*), and the position of the star or the sun (*S*), as shown in Fig. 19.6.

The sides of the triangle are arcs of great circles on the celestial sphere that pass through any two of the vertices, measured in degrees or hours. The angular value of each side is determined by the angle that the side subtends on the earth (see Fig. 19.7).

The three sides of the *PZS* triangle are known as (1) the polar distance, (2) the coaltitude, (3) the colatitude.

(1) The *polar distance* is the side *PS*. It is determined from the declination of the star or the sun. (Recall that declination is defined as the angular distance from a celestial body to the celestial equator.) When the celestial body lies north of the celestial equator, the declination has a positive sign; when it lies south of the celestial equator, it has a negative sign. The polar distance is determined algebraically by subtracting the declination of the celestial body from 90°. In Fig. 19.8, the polar distance is $90° - (-20) = 100°$. Observations on stars south of the equator are seldom made from the northern hemisphere.

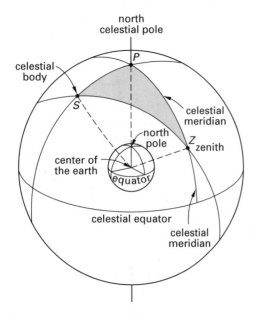

Figure 19.6 *PZS Triangle*

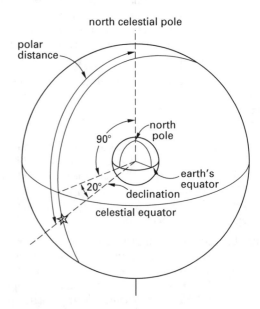

Figure 19.8 *Polar Distance Side of the PZS Triangle*

(2) The *coaltitude* is the side SZ, the arc distance from the celestial body to the observer's zenith. It is determined by subtracting the observed altitude of the celestial body (corrected for refraction and parallax) from 90°. Altitude is the vertical angle measured from the observer's horizon to the celestial body.

(3) The *colatitude* is the side PZ. It is the distance from the celestial north pole to the zenith. It is determined by subtracting the latitude of the observer from 90° (see Fig. 19.10).

The three interior angles of the PZS triangle are known as (1) the parallactic angle, (2) the azimuth or zenith angle, and (3) the angle t (see Fig. 19.11).

(1) The *parallactic angle* (angle S) is formed by the polar distance side and the coaltitude side.

(2) The *azimuth angle* (angle Z) is formed by the coaltitude side and the colatitude side. It is used to find the azimuth from the observer to the celestial body. When the celestial body is in the east, the azimuth angle is equal to the true azimuth. When the celestial body is in the west, true azimuth equals 360° minus the azimuth angle.

(3) The angle at the pole P formed by the colatitude side and polar distance side is known as the angle t, or the angle P.

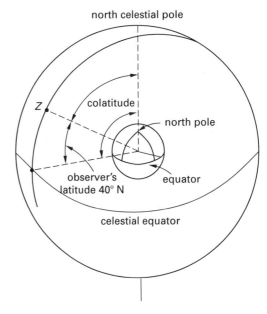

Figure 19.10 *Colatitude side of the PZS Triangle*

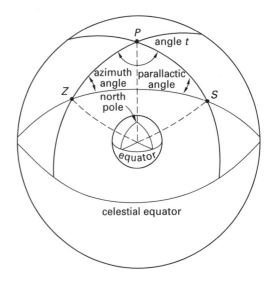

Figure 19.11 *Interior Angles of the PZS Triangle*

If any three elements of the PZS triangle are known, the other elements can be found by spherical trigonometry. However, each astronomical triangle is changing constantly because of the apparent rotation of the celestial sphere. The observer, then, must know the position of the PZS triangle at the time of the observation. Information concerning the position of the celestial bodies can be found in an ephemeris. (An *ephemeris* is similar to an almanac. It contains tables showing the positions of celestial bodies on certain dates in sequence. Typically, ephemerides are published by manufacturers of surveying instruments.)

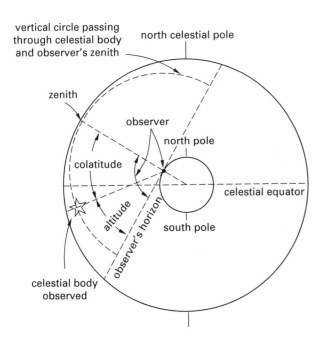

Figure 19.9 *Coaltitude Side of the PZS Triangle*

6. TIME

Because all celestial bodies are in constant apparent motion with respect to the observer, it is extremely important to know the precise time of an observation on a celestial body.

In practical astronomy, there are two categories of time: *sun time* and *sidereal time*. Both categories of time are based on the rotation of the earth with respect to a standard reference line.

Because the earth revolves around the sun in the plane of its orbit once each year, the reference line to the sun is changing constantly and the length of one solar day is not the true time of one rotation of the earth.

In practical astronomy, the true time of one rotation of the earth, which is known as the *sidereal day*, is based on one rotation with respect to the vernal equinox. One solar day is 3 minutes 56 seconds longer than one sidereal day.

Explaining the difference between the solar day and the sidereal day requires temporarily abandoning the theory that the earth is stationary and that the celestial sphere is revolving and returning to the true condition that the earth revolves around the sun.

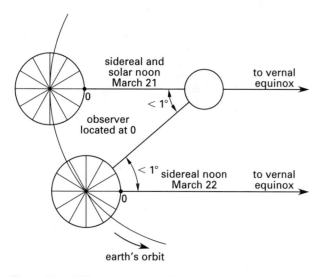

Figure 19.12 *Difference Between Solar Day and Sidereal Day*

The earth completes one revolution around the sun in 365.2564 days, although 1 year is 365 calendar days. In Fig. 19.12, an observer at zero on the earth on March 21 (the vernal equinox) would find the sun directly overhead at noon. From March 21 until the summer solstice, the observer would find that the sun is not directly overhead at noon, but advances about 1° north per day. This motion of the sun makes the intervals between the sun's transits (stated above) of the observer's meridian greater by about 3 minutes 56 seconds than the interval between transits of the vernal equinox of the observer's meridian. Therefore, the solar day is about 3 minutes 56 seconds longer than the sidereal day.

Because one apparent rotation of the celestial sphere is completed in one sidereal day, a star rises at nearly the same sidereal time throughout the year. On solar time, it rises about 4 minutes earlier from night to night, or 2 hours earlier from month to month. Thus, observed at the same hour night by night, the stars seem to move slowly westward across the sky as the year lengthens.

An apparent solar day is the interval between two successive transits of the sun over the same meridian. Because of the earth's tilt and the variation in the earth's velocity about the sun, the interval between two transits of the sun over the same meridian varies from day to day. This makes it impossible to use the variable *day* as a basis for accurate time. Therefore, a fictitious, or mean, sun was devised that is imagined to move at a uniform rate in its apparent path around the earth. It makes one apparent revolution around the earth in 1 year, the same as the actual sun. The average apparent solar day was used as a basis for the mean day. The time indicated by the position of the actual sun is called *apparent solar time*.

The difference between mean solar time and apparent solar time is called the *equation of time* (EOT). It varies from minus 14 minutes to plus 6 minutes (see Fig. 19.13). The value of EOT for any day can be found in an ephemeris.

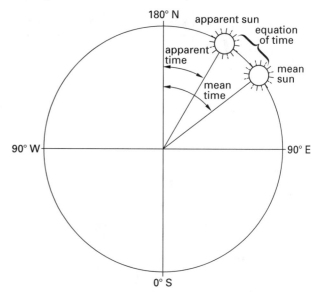

Figure 19.13 *Equation of Time*

In mean solar time, the length of the year is divided into 365.2422 mean solar days. Because the mean sun appears to revolve around the earth every 24 hours of the mean time, the apparent rate of movement of the mean sun is 15° of arc, or of longitude, per hour ($360 \div 24 = 15$).

In the system of latitude and longitude on the earth, the zero reference for latitude is the equator; the zero reference for longitude is the meridian that passes through Greenwich, England (the prime meridian, or 0° longitude). Using the Greenwich meridian as a basis for reference, time at a point 15° west of the Greenwich meridian is 1 hour earlier than the time at the Greenwich meridian because the sun passes the Greenwich meridian 1 hour before it crosses the meridian lying 15° to the west. The opposite is true along the meridian lying 15° to the east, where time is 1 hour later, because the sun crosses this meridian 1 hour before it arrives at the Greenwich meridian. Therefore, the difference in local time between the two places equals their difference in longitude (see Fig. 19.14).

Because the mean solar day has been divided into 24 equal units of time (hours), there are 24 time zones, each 15° wide, around the earth. Using the Greenwich meridian as the central meridian of a time zone and as the zero reference for the computation of time zones, each 15° zone extends 7½° east and west of the zone's central meridian (see Fig. 19.15).

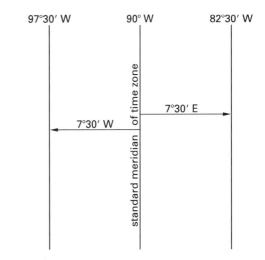

Figure 19.15 *Time Zone Boundaries*

The central meridian of each time zone, east or west of Greenwich, is a multiple of 15°. For example, the time zone of the 90° meridian extends from 82°30′ to 97°30′. Each 15° meridian or multiple thereof east or west of the Greenwich meridian is called a *standard time meridian*. Four of these meridians (75°, 90°, 105°, 120°) cross the continental United States (see Fig. 19.16).

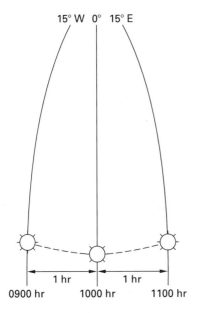

Figure 19.14 *Apparent Motion of the Sun*

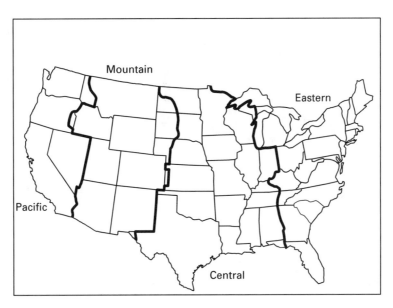

Figure 19.16 *United States Standard Time Zones*

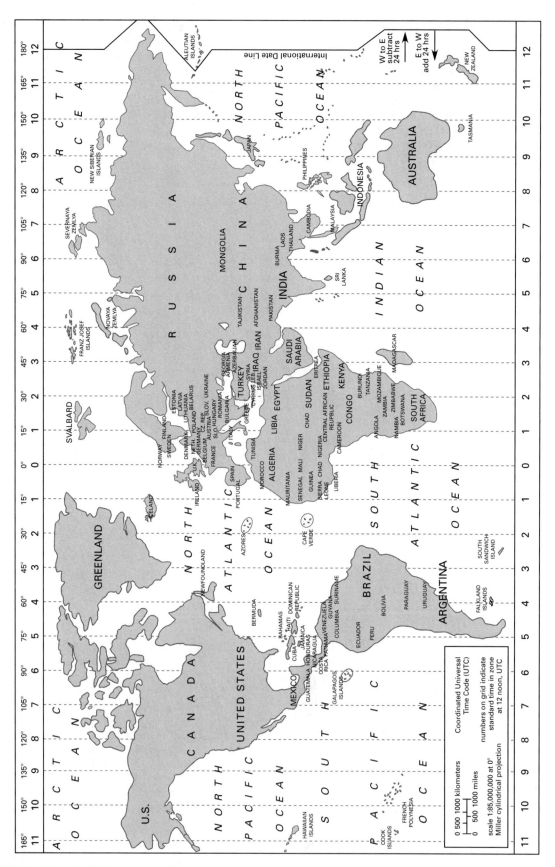

Figure 19.17 World Time Zone Map

Standard time zones in the United States are named Pacific, Mountain, Central, and Eastern. Standard time zone boundaries often run along state boundaries so that time is the same over a single state. In Fig. 19.17, time zones of the world are designated by letters of the alphabet.

Standard time in any zone is referred to a *local mean time* (LMT). It is clock time where the observer's position is located. It does not take into account daylight savings time.

Rather than establishing a reference meridian for measuring time in each time zone, it was decided to establish a single reference line for all parts on the earth. Standard time zone Z (see Fig. 19.17), which uses the Greenwich meridian as its basic time meridian, was chosen for computing data pertaining to mean solar time. Greenwich standard time is also *Greenwich mean time* (GMT). It is referred to as *Universal Time* in an ephemeris. *Central Standard Time* (CST) is standard time zone S.

Time can be expressed as the reading of the standard 24-hour clock at the Greenwich Observatory at the moment an observation is made on a celestial body; therefore, it is the same time throughout the world. Because the observer's watch is usually set to the local standard time (local mean time, or LMT), a conversion must be made from LMT to GMT. Data in an ephemeris are based on the Greenwich meridian and 0 h Greenwich mean time.

To convert LMT to GMT when the observer is located in west longitude, divide the value of the central meridian of the time zone in degrees of longitude by 15°. This equals the time zone correction in hours. The difference in time between the standard time zone of the observer's position and GMT must be added to the LMT to arrive at the Greenwich mean time of observation (see Fig. 19.18). If the result is greater than 24 hours, the amount over 24 hours is dropped and 1 day is added to obtain the Greenwich time and date. If the observer is located in east longitude, the difference is subtracted.

When observers sight the sun, it is obvious that they observe the apparent sun and not the mean sun on which their time is based. Therefore, they must convert mean time to apparent time, which is done by converting GMT at the point of observation to *Greenwich apparent time* (GAT). The observers first convert LMT to GMT. This is done by adding the time zone correction (see Fig. 19.18) to LMT. In the Central time zone, 6 hours would be added. To obtain GAT, the equation of time is added to GMT for observations in west longitude and subtracted from GMT for east longitude. The equation is found in an ephemeris, using the date and time of observation. In summary, for west longitude,

local mean time

    + time zone correction

        = GMT

           + equation of time 0 h (from ephemeris)

           + daily change (from ephemeris)

        = GAT

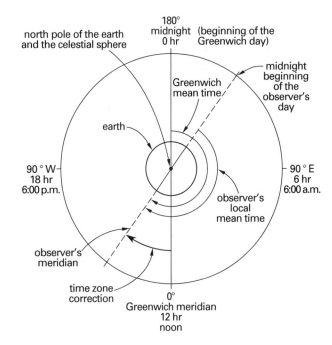

Figure 19.18 *Time Zone Correction*

An *hour angle* is any great circle on the celestial sphere that passes through the celestial poles. It corresponds to a meridian on the earth.

The *observer's meridian* is the great circle on the celestial sphere that passes through the celestial poles and the observer's zenith.

The *hour angle* of a celestial body is the angle at the celestial poles between the plane of the meridian of the observer and the plane of the hour angle of the celestial body. In Fig. 19.11, it is shown as the angle P; it is also known as the angle t.

The *Greenwich hour angle* (GHA) of a celestial body is the time that has elapsed since the body crossed the Greenwich meridian (projected on the celestial sphere).

The *local hour angle* (LHA) of a celestial body is the angle measured along the plane of the celestial equator from the meridian of the observer (zenith) to the meridian of the celestial body (projected on the celestial sphere).

For west longitude, where λ is the longitude of the observer's position,

$$LHA = GHA - W\lambda \qquad \text{19.1}$$

Example 19.1

The observer's longitude is 98°30'00" W, and the local mean time of observation is 09 h 00 m 00 s CST, 24 April 68. Determine the angle t for the solution of the *PZS* triangle.

Solution

	09 h	00 m	00 s	
+	06	00	00	time zone correction
				(90° W ÷ 15° = +6)
	15	00	00	GMT of observation
+		01	44	equation of time for 0 h GMT*
+		00	07	equation of time for partial day*
	15	01	15	GAT of observation
−	12	00	00	GHA measured from noon
	03	01	51	(GHA) = 45°27'45"
	45°	27'	45"	
	360°			(if necessary)
	405°	27'	45"	
−	98°	30'	00"	longitude of observer
	306°	57'	45"	LHA of the sun
	53°	02'	15"	angle t (360° − 306°57'45")

*from ephemeris

The following factors can be used in converting hours to degrees and degrees to hours.

$$24 \text{ h} = 360° \qquad 360° = 24 \text{ h}$$
$$1 \text{ h} = 15° \qquad 1° = 4 \text{ m}$$
$$1 \text{ m} = 15' \qquad 1' = 4 \text{ s}$$
$$1 \text{ s} = 15'' \qquad 1'' = 0.067 \text{ s}$$

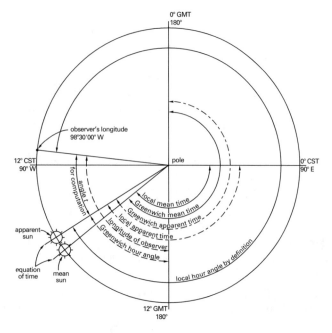

Figure 19.19 *Time, Hour Angle Relationship, and West Longitude*

As mentioned previously, Greenwich mean time, which is also Greenwich standard time, is referred to in an ephemeris as *Universal Time*, or UTC (Coordinated Universal Time). This time is broadcast by the radio station WWV of the National Bureau of Standards and can be received on receivers that are pretuned to WWV. Also, the time signals can be obtained by calling (303) 449-7111 at the caller's expense.

The time signals can be used to determine a more precise time called UT1. UT1 is obtained by adding a small correction called DUT (the difference between UTC and UT1; UT1 = UTC + DUT).

The DUT correction can be determined by listening carefully to the WWV time signal. Following a minute tone, there will be a number of double ticks. Each double tick represents a correction of 0.1 second and is positive for the first 7 seconds. Beginning with the ninth second, each double tick is a negative correction. For example, a voice on the radio will announce "Fifteen hours thirty-six minutes." Just after this will be a minute tone followed by the double ticks. This occurs for each minute.

The DUT correction changes 0.1 second periodically, but not uniformly; it does not change rapidly. It may remain constant for a week or more in some instances.

Ephemerides are based on UT1 time for observations on the sun. Data needed for azimuth calculations are tabulated in an ephemeris for each day of the year for 0 hour Universal Time, so that it is possible for the day of the month at Greenwich to be one day later than the date of the observation. At 6:00 P.M. CST it is midnight at Greenwich, so that for observations on Polaris after 6:00 P.M. CST, one day would be added to the local date to find the Greenwich date to enter the tables in an ephemeris. For most sun shots, Greenwich date and local date will be the same.

The *sidereal day* is defined by the time interval between successive passages of the vernal equinox over the upper meridian of a given location. The *sidereal year* is the interval of time required for the earth to orbit the sun and return to the same position in relation to the stars. Because the sidereal day is 3 minutes 56 seconds shorter than the solar day, this differential in time results in the sidereal year being 1 day longer than the solar year, or a total of 366.2422 sidereal days. And, because the vernal equinox is used as a reference point to mark the sidereal day, the sidereal time for any point at any instant is the number of hours, minutes, and seconds that have elapsed since the vernal equinox passed the meridian of the point.

The general steps for converting local mean time of observation to local hour angle (then to the interior angle t at the pole) from sidereal time are as follows.

(1) Greenwich mean time of observation is determined the same way as solar time is determined.

(2) Sidereal time for 0 h GMT plus the correction for GMT (from an ephemeris) determines the Greenwich sidereal time of observation.

(3) Greenwich sidereal time of observation minus the right ascension of the star (from an ephemeris) equals the Greenwich hour angle.

(4) Greenwich hour angle plus the observer's longitude if in east longitude (or minus the observer's longitude if in west longitude) is the local hour angle of the star.

(5) Angle t, the interior angle of the PZS triangle at the pole, equals the local hour angle when the star is the east. The specific steps performed in determining the local hour angle = angle t are as follows.

corrected watch time

+ time zone correction

= GMT

+ sidereal time for 0 h (from ephemeris)

+ correction for partial day (from ephemeris)

= Greenwich sidereal time

− right ascension from star (from ephemeris)

= GHA

− W longitude (+ for E longitude)

= LHA = angle t

In general, it can be stated that observations on the sun involve apparent solar time, while observations on the stars are based on sidereal time. The computations using either apparent solar time or sidereal time are similar in that they do nothing more than fix the location of both the celestial body and the observer in relation to the Greenwich meridian. Once a precise relationship has been established, it is a simple matter to complete the determination of azimuth to the celestial body.

7. METHODS AND TECHNIQUES OF DETERMINING AZIMUTH

There are two methods of determining azimuth by astronomical observations: the *altitude method* and the hour *angle method*. Both methods require a horizontal angle from an azimuth mark on the ground to the observed body (sun or star) in order to establish azimuth on the ground. The basic difference between the two methods is that the altitude method requires an accurate vertical angle measurement but does not require precise time; the hour angle method requires precise time but does not require a vertical angle measurement.

In both methods, it is extremely important that the instrument be leveled carefully. The vertical axis must be truly vertical; if it is not, the error caused by the inclination of the horizontal plane will not be eliminated

by a reversal of the telescope between sights. Centering the plate level bubbles on the instrument makes the vertical axis of the instrument truly vertical, provided that the plate levels are in perfect adjustment.

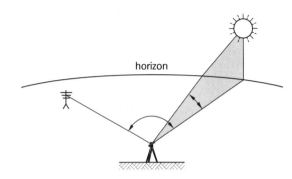

Figure 19.20 *Measuring Horizontal and Vertical Angles*

In the altitude method of determining azimuth, the PZS triangle is solved by using the three sides of the triangle. In addition to the horizontal angle from a ground point to the celestial body, three elements are necessary and must be determined: (1) the latitude for determining the colatitude side (colatitude equals 90° minus latitude), (2) the declination of the celestial body (angular distance from the celestial equator to the celestial body) for determining the polar distance side of the PZS triangle (polar distance equals 90° minus declination), and (3) the observed altitude (vertical angle) to the celestial body for determining the coaltitude side of the triangle (coaltitude equals 90° minus corrected altitude).

In the hour angle method of determining azimuth, the azimuth angle is determined from two sides and the included angle of the PZS triangle. The sides are the polar distance and the colatitude, as explained in the altitude method. The angle at the north celestial pole, the angle t, is determined as explained in Sec. 6.

The principle advantage of the altitude method is that precise time is not required. Until recent years, timepieces that make UT1 time possible in the field were not available. Because of this, the altitude method has been widely used.

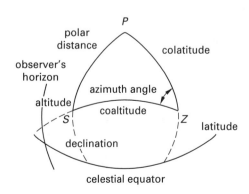

Figure 19.21 *Altitude Method*

A disadvantage of the altitude method is that a vertical angle is required for observation on both the sun and the stars, which makes it necessary to set both the horizontal and vertical crosshairs tangent to the sun simultaneously. Also, measuring a vertical angle introduces the necessity of making corrections for parallax and refraction.

Advantages of the hour angle method counter the disadvantages of the altitude method. Bringing the vertical crosshair tangent to the sun without concern for the horizontal crosshair is much less difficult than simultaneous tangency. Also, eliminating corrections for parallax and refraction (for sun sights) contributes to more accurate results.

Disadvantages of the hour angle method are the cost of timepieces and the additional training required to use them.

Both methods can be used for observation on either the sun or the stars. Both methods require the determination of the latitude and longitude of the point of observation. All in all, however, the hour angle method seems to be the preferred method.

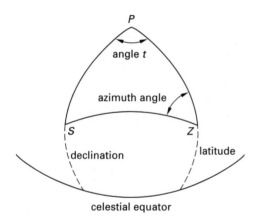

Figure 19.22 *Hour Angle Method*

In determining the azimuth of a line, the general equation is

$$\text{az line} = \text{az sun or star} + 360° - \text{angle right} \qquad 19.2$$

Example 19.2

Find the azimuth of a line when the azimuth to the sun is $78°31'24.''6$ and the angle right is $346°20'18.''1$.

Solution

$$\text{az line} = 78°31'24.''6 + 360° - 346°20'18.''1$$
$$= 92°11'06.''5$$

8. MAPS AND MAP READING

Topographic maps of various scales are available from the United States Geological Survey (USGS).

USGS quadrangle series maps cover areas bounded by parallels of latitude and meridians of longitude. Standard edition maps are produced at 1:24,000 scale in either 7.5 by 7.5 or 7.5 by 15 minute format. The 7.5 minute quadrangle map is satisfactory for determining latitude and longitude by scaling in determining azimuth.

The scale of the map (see Fig. 19.23), 1:24,000, is very convient because 24,000 in = 2000 ft exactly.

The east side of the Elm Mott map is bounded by a line representing the meridian of $97°00'$ west longitude; the west side is bounded by a line representing the meridian of $97°07'30''$ west longitude. The south side is bounded by a line representing the parallel of $31°37'30''$ north latitude; the north side is bounded by a line representing the parallel of $31°45'$ north latitude. Thus, the quadrangle formed is 7.5 minutes on each side. The east and west lines of the map are marked by ticks at 2.5 minute intervals ($31°40'$ and $31°42'30''$) north latitude, and the north and south lines are marked by ticks at 2.5 minute intervals ($97°02'30''$ and $97°05'$) west longitude. Connecting corresponding ticks on the east and west lines and connecting corresponding ticks on the north and south lines with lines divides the quadrant into nine 2.5' by 2.5' subquadrants.

If the latitude and longitude of Monument BM No. 498 are needed, first select the subquadrant that contains the monument and determine the lines of latitude and longitude that bound the subquadrant.

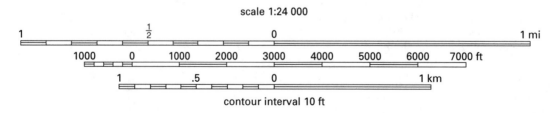

Figure 19.23 *Scale of the Elm Mott Map*

The borders of the Elm Mott map are not shown to scale in Fig. 19.24. The subquadrant that contains the monument is shown in Fig. 19.25.

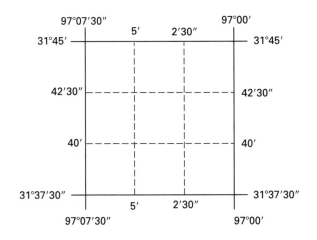

Figure 19.24 *Outline of the Elm Mott Map*

After the parallels and meridians for the subquadrant have been drawn (see Fig. 19.25), the geographic interval (angular distance between two adjacent lines) must be determined. Examination of the tick marks gives the interval. On the 7.5′ quad map, the interval is $2'30'' = 150''$. Any scale with 150 divisions may be used to find latitude and longitude. The twenty scale of an engineer's scale fits this requirement.

To find the longitude of the Monument BM No. 498, place the 0 mark of the twenty scale on meridian 97°00′ (east side because longitude increases from east to west) and the 150 mark of the scale on the meridian 97°02′30″ (west side) with the scale just above the monument. Then slide the scale downward, keeping the 0 and 150 marks on the lines, until the edge of the scale just touches the point of the monument. The reading of the scale will give the number of seconds west longitude from the meridian 97°00′. The scale reading is 66, so the longitude of the monument is 97°01′06″ west. Following the same procedure for latitude, the scale reading is 76, so the latitude of the monument is 31°43′46″ north.

Astronomy

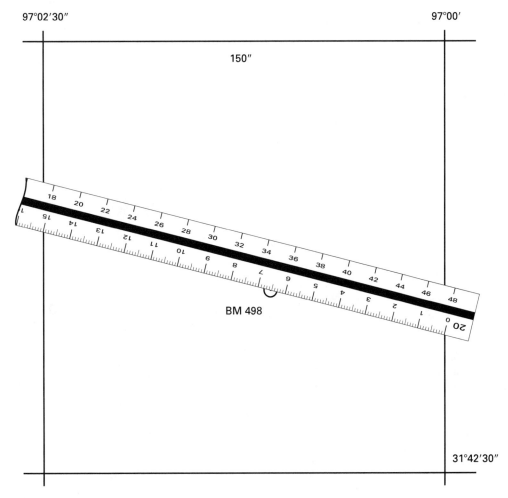

Figure 19.25 *Scaling Longitude*

9. LOCATING POLARIS

Ancient astronomers identified some groups of stars with mythological characters, animals, and everyday objects, and named the groups accordingly. These groups of stars are called *constellations*. Two of the most well-known constellations are Ursa Major and Ursa Minor. Ursa Major translates to Great Bear, and contains the prominent configuration of stars known as the *Big Dipper*. Ursa Minor is the Little Bear and contains the Little Dipper.

As mentioned in Sec. 6, because of the difference in solar time and sidereal time, a star apparently rises about 4 minutes earlier from night to night, or 2 hours earlier from month to month. Thus, at the same hour, night by night, a constellation seems to move slowly westward across the sky.

Because of the motion of the earth, the celestial sphere appears to rotate once a day, causing a constant movement of the constellations across the sky.

Stars that never set are said to be *circumpolar*. A star with declination greater than 90° minus an observer's latitude is circumpolar at that latitude. The stars in Ursa Major can be seen as far south as 30° south latitude, while the constellation Orion straddles the celestial equator and can be seen from anywhere in the world.

Polaris appears to move in a small, counterclockwise, circular orbit around the celestial north pole. Because Polaris stays so close to the north celestial pole, it is visible throughout most of the northern hemisphere. When the Polaris hour angle is 0 or 12 hours, the star is said to be in its *upper* or *lower culmination*. When the Polaris local hour angle is 6 or 18 hours, the star is said to be in its *western* or *eastern elongation*. When Polaris is near western or eastern elongation, it appears to move slowly and vertically.

Polaris is the brightest star in Ursa Minor, which is near Ursa Major and the constellation Cassiopeia. It is the end star of the three stars making up the handle of the Little Dipper. Polaris can also be identified with respect to the Big Dipper and Cassiopeia. The two stars forming the side of the bowl farthest from the handle of the Big Dipper are called the *pointer stars*. A line through the pointer stars toward Cassiopeia nearly passes through the celestial north pole. The distance from the nearest pointer star to Polaris is about five times the distance between the two pointer stars. Polaris and Cassiopeia are on the same side of the north celestial pole.

The vertical angle to Polaris is nearly the same as the latitude of the observer, depending on the position of Polaris in its orbit. In searching for Polaris, this angle can be set on the observing instrument. The telescope should first be focused on any bright star.

When the telescope is directed at Polaris, the observer will see two other stars nearby that are not visible to the naked eye. However, Polaris will be the only star visible when the crosshairs are lighted.

The orbit of Polaris is shown in Fig. 19.26, and the constellations that serve to identify Polaris are shown in Fig. 19.27.

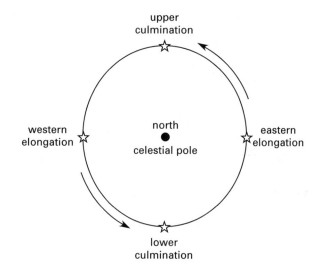

Figure 19.26 *Orbit of Polaris*

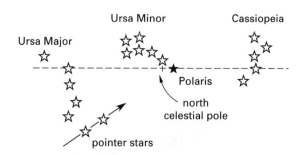

Figure 19.27 *Identification of Polaris*

10. A SIMPLER METHOD OF DETERMINING AZIMUTH

In determining azimuth from astronomical observations, the work involved has been made easier by present-day ephemerides. Determining the Greenwich hour angle as in Ex. 19.1 has been simplified.

Ephemerides tabulate the GHA and the declination of the sun and Polaris at 0 hour Universal Time in degrees, minutes, and seconds for each day of the year, thus eliminating some of the computations in Ex. 19.1. Interpolation for an exact time of day is, of course, still necessary. The LHA is determined as before (LHA = GHA - Wπ).

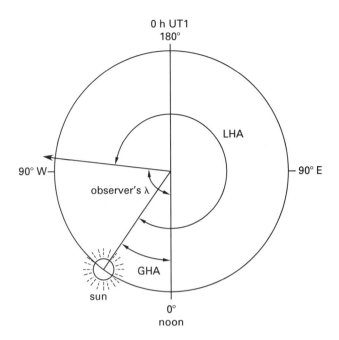

Figure 19.28 *Observations on Sun*

While single observations for azimuth can be made, it is better to increase accuracy by making at least six pointings on the sun or star and averaging the best of the computed azimuths.

There are two basic procedures in making a set of observations. The multiple foresight (MFS) procedure consists of a backsight on the mark, three foresights on the sun/star, three telescope direct, three foresights on the sun/star with the telescope reversed, then a final backsight on the mark with the telescope reversed. This method assumes that a sight on a stationary mark is more accurate than on a rapidly moving celestial body. The single foresight (SFS) procedure is a special case of the MFS procedure in that only one foresight is taken per backsight. An observation normally consists of three sets (i.e., six pointings on both the mark and the sun/star). This method is frequently used for Polaris and other large declination stars where the star motion is slow.

Once an observation has been completed, an azimuth is computed for each pointing on the celestial body. These azimuths are compared with each other with acceptable values being averaged.

Example 19.3

A solar observation is taken on June 6, 1997, by observing the left edge of the sun. All angles are turned to the right (clockwise).

Given data:

latitude	$= 36°04'00''\,$N
longitude	$= 94°10'08''\,$W
UTC when stopwatch was started	$= 13$ h 34 m 02 s,
	DUT $= -0.5$ s
stopwatch time of pointing	$= 0$ h 15 m 42.0 s

backsight on mark $= 180°00'05''$
foresight on sun $= 171°56'10''$
GHA, declination, and semidiameter are taken from the SOKKIA ephemeris table shown in Table 19.1.

The following symbols are defined as:

az	$=$	azimuth of sun
az$'$	$=$	azimuth of line
decl and δ	$=$	declination
ϕ	$=$	latitude
h	$=$	computed slope angle to sun
dH	$=$	semidiameter correction
DUT	$=$	UTC $-$ UT1 (correction)

Solution

Correction to stopwatch equals UT1 when stopwatch was started.

UTC $= 13$ h 34 m 02.0 s
DUT $= -\qquad 0.5$ s
UT1 $= \overline{13$ h 34 in $01.5}$ s (at stopwatch $= 0:00:00$)

From Table 19.1,

GHA 0 h	$= 180°21'40.1''$
GHA 24 h	$= 180°18'54.4''$
decl 0 h	$= 22°37'49.4''$
decl 24 h	$= 22°43'53.4''$
semidiameter	$= 0°15'47.1''$

$$\text{UT1} = 15 \text{ m } 42.0 \text{ s} + 13 \text{ h } 34 \text{ m } 01.5 \text{ s}$$
$$= 13 \text{ h } 49 \text{ m } 43.5 \text{ s}$$

$$\text{GHA} = \text{GHA 0 h} + (\text{GHA 24 h}$$
$$-\text{GHA 0 h} + 360)\left(\frac{\text{UT1}}{24}\right)$$
$$= 387°45'57.1''$$
$$= 27°45.57.1''$$

$$\text{LHA} = \text{GHA} - \text{W}\lambda$$
$$= -66°24'10.9''$$
$$= 293°35'49.1''$$

$$\text{decl} = \text{decl 0 h} + (\text{decl 24 h} - \text{decl 0 h})$$
$$\times \left(\frac{\text{UT1}}{24}\right) + (0.0000395)(\text{decl 0 h})$$
$$\times \sin(7.5 \text{ UT1})$$
$$= 22°41'19.1'' + 0°00'03.1''$$
$$= 22°41'22.3''$$

$$\text{az} = \tan^{-1}\frac{-\sin\text{LHA}}{\cos\phi\tan\delta - \sin\phi\cos\text{LHA}}$$
$$= 83°37'50.6''$$

when LHA is	correction	
	if az is positive	if az is negative
0° to 180°	180°	360°
180° to 360°	0°	180°

Table 19.1 *Sample Page from Ephemeris*

JUNE 1997
Greenwich Hour Angle for the Sun and Polaris for 0 Hour Universal Time

day	GHA (sun)			declination			eq. of time appt-mean		semi-diam.		GHA (Polaris)			declination			Greenwich transit		
	deg	ft	in	deg	ft	in	M	S	ft	in	deg	ft	in	deg	ft	in	H	M	S
1 Su	180	34	12.5	22	01	35.7	02	16.83	15	47.8	212	28	38.3	89	14	51.80	9	48	29.
2 M	180	31	53.4	22	09	37.1	02	07.56	15	47.7	213	27	29.5	89	14	51.53	9	44	34.
3 Tu	180	29	28.2	22	17	15.3	01	57.88	15	47.5	214	26	19.0	89	14	51.28	9	40	39.
4 W	180	26	57.4	22	24	30.2	01	47.83	15	47.4	215	25	07.5	89	14	51.05	9	36	45.
5 Th	180	24	21.2	22	31	21.6	01	37.42	15	47.3	216	23	55.4	89	14	50.84	9	32	50.
6 F	180	21	40.1	22	37	49.4	01	26.68	15	47.1	217	22	43.4	89	14	50.64	9	28	56.
7 Sa	180	18	54.4	22	43	53.4	01	15.63	15	47.0	218	21	32.0	89	14	50.46	9	25	01.
8 Su	180	16	04.4	22	49	33.5	01	04.29	15	46.9	219	20	21.4	89	14	50.29	9	21	06.
9 M	180	13	10.5	22	54	49.5	00	52.70	15	46.8	220	19	11.6	89	14	50.13	9	17	12.
10 Tu	180	10	13.1	22	59	41.4	00	40.87	15	46.7	221	18	02.4	89	14	49.96	9	13	17.
11 W	180	07	12.4	23	04	09.0	00	28.83	15	46.6	222	16	53.5	89	14	49.78	9	09	22.
12 Th	180	04	08.9	23	08	12.1	00	16.59	15	46.5	223	15	44.7	89	14	49.60	9	05	27.
13 F	180	01	02.9	23	11	50.8	00	04.19	15	46.4	224	14	35.4	89	14	49.41	9	01	33.
14 Sa	179	57	54.7	23	15	05.0	−00	08.35	15	46.3	225	13	25.3	89	14	49.22	8	57	38.
15 Su	179	54	44.7	23	17	54.5	−00	21.02	15	46.2	226	12	14.2	89	14	49.02	8	53	43.
16 M	179	51	33.2	23	20	19.3	−00	33.79	15	46.2	227	11	01.9	89	14	48.83	8	49	49.
17 Tu	179	48	20.5	23	22	19.4	−00	46.63	15	46.1	228	09	48.2	89	14	48.64	8	45	54.
18 W	179	45	07.0	23	23	54.8	−00	59.54	15	46.0	229	08	33.3	89	14	48.46	8	42	00.
19 Th	179	41	52.9	23	25	05.4	−01	12.48	15	45.9	230	07	17.6	89	14	48.31	8	38	06.
20 F	179	38	38.5	23	25	51.1	−01	25.43	15	45.9	231	06	01.6	89	14	48.17	8	34	11.
21 Sa	179	35	24.3	23	26	12.1	−01	38.38	15	45.8	232	04	46.0	89	14	48.06	8	30	17.
22 Su	179	32	10.3	23	26	08.3	−01	51.31	15	45.7	233	03	31.2	89	14	47.97	8	26	23.
23 M	179	28	57.0	23	25	39.7	−02	04.20	15	45.7	234	02	17.7	89	14	47.88	8	22	28.
24 Tu	179	25	44.5	23	24	46.3	−02	17.03	15	45.6	235	01	05.4	89	14	47.79	8	18	34.
25 W	179	22	33.2	23	23	28.2	−02	29.78	15	45.6	235	59	53.9	89	14	47.69	8	14	39.
26 Th	179	19	23.4	23	21	45.4	−02	42.44	15	45.5	236	58	42.4	89	14	47.58	8	10	45.
27 F	179	16	15.4	23	19	37.9	−02	54.97	15	45.5	237	57	30.1	89	14	47.45	8	06	50.
28 Sa	179	13	09.5	23	17	05.8	−03	07.36	15	45.5	238	56	16 4	89	14	47.31	8	02	56.
29 Su	179	10	06.1	23	14	09.1	−03	19.59	15	45.4	239	55	01.2	89	14	47.18	7	59	01.
30 M	179	07	05.5	23	10	48.0	−03	31.63	15	45.4	240	53	44.4	89	14	47.05	7	55	07.

Used with permission from 1997 *Celestial Observation Handbook and Ephemeris*, Dr. Richard L. Elgin, Dr. David R. Knowles, and Dr. Joseph Senne, published by SOKKIA Company.

Since LHA is between 180° and 360°, and az is positive, the normalized correction equals 0°.

az $= 83°37'50.6''$
R and rt $= 171°56'10'' − 180°00'05''$
$= −8°03'55''$
$= 351°56'05''$

$h = (\sin^{-1})(\sin\phi\sin\delta + \cos\phi\cos\delta\cos\text{LHA})$
$= 31°42'38''$

$dH = \dfrac{\text{semidiameter}}{\cos h}$
$= 0°18'33.3''$

left edge pointed D&R (down and right); therefore, correction dH is positive.

ang rt $= 351°56'05'' + 0°18'33.3''$
$= 352°14'38.3''$

azl $= \text{az} + 360 − \text{ang rt}$
$= 91°23'12.3''$

This same calculation procedure is used for each pointing on the sun.

Example 19.3 works equally well for Polaris and other stars. In the case of stars, the correction for semidiameter would be zero.

Computer software is now available that computes azimuths from observational data either using ephemeris tables or by generating the ephemeris internally. The latter permits azimuth calculations from the early 1900s to well into the twenty-first century.

Example 19.4

Given the following information, determine the azimuth of Polaris.

local time: June 7, 1997
UTC time: 01:56:31
latitude: N 31°38'20''
longitude: W 97°04'46''

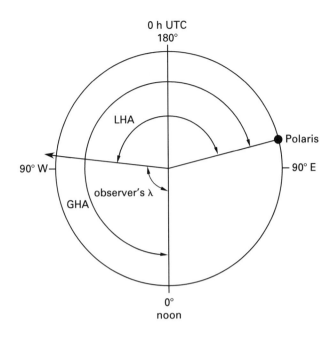

Observation on Polaris

11. GEODETIC AZIMUTH

The azimuth obtained from a celestial observation is known as *astronomical azimuth*. This can be converted to geodetic azimuth using Eq. 19.3.

$$\text{geodetic azimuth} = \begin{array}{c} \text{astronomic azimuth} \\ + \text{ Laplace correction} \end{array} \qquad 19.3$$

Except for mountainous areas, the Laplace correction is relatively small. Consequently, geodetic and astronomic azimuth are frequently considered to be the same. If Laplace corrections are necessary, they can be obtained from software furnished by the National Geodetic Survey (NGS).

Solution

Greenwich date is one day later than local date.

$$
\begin{aligned}
\text{GHA} \quad &= 251°18'13.8'' \text{(from ephemeris)} \\
\text{LHA} \quad &= \text{GHA} - \text{W}\lambda \\
&= 248°32'52'' - 97°04'46'' \\
&= 151°28'06.0'' \\
\text{decl} \quad &= 89°14'53.5'' \text{ (from ephemeris)} \\
\text{az to Polaris} &= 0°25'07.8'' \text{ (west of north)}
\end{aligned}
$$

Astronomy

20 Map Projections and State Coordinate Systems

Projections

1. GEODESY

Map makers have always had to face the problem of projecting the curved surface of the earth onto a plane surface. Representing the true shape of the lands and waters and the relative positions of points on the earth on a plane surface without distortion is impossible. *Geodesy* is the science of measuring the size and shape of the earth.

The earth is not a true sphere but a *spheroid*. Spinning on its axis, tilted $23^{1}/_{2}°$ to the plane of its orbit around the sun, it bulges at the equator. The equatorial diameter of 7927 mi is some 27 mi greater than the polar diameter of 7900 mi.

2. THE CLARKE SPHEROID OF 1866

The *Clarke Spheroid of 1866* is a theoretical spheroid representing the earth. It is the basis for the National Geodetic Survey (U.S. Department of Commerce) measurements and tables referenced in many state plane coordinate systems.[1]

3. GERARDUS MERCATOR

As discoveries of new lands were made, commerce increased accordingly. Ships needed charts that would guide them to their destination, but captains became aware that a straight line on a chart is not the same as a straight line on the globe of the earth. They needed charts on which a straight line drawn from port to port could be used to take them to their destination.

Gerardus Mercator, born Gerhard Kremer in Flanders in 1512, gave them what they wanted. Mercator wrote of his charts, "If you wish to sail from one port to another, here is a chart, and a straight line on it, and if you follow this line carefully you will certainly arrive at your destination. But the length of the line may not be correct. You may get there sooner or may not get there as soon as you expected, but you will certainly get there."

4. JOHANN HEINRICH LAMBERT

Probably the greatest contributor to modern cartography is *Johann Heinrich Lambert*, born in Alsace, France, in 1728. A philosopher and mathematician, Lambert demonstrated that maps could be made with truer shapes by using mathematics. His conic conformal projection is the basis of many state plane coordinate systems.

5. LATITUDE AND LONGITUDE

Dividing the earth into a huge grid are lines of latitude and longitude. Unlike the rectangular coordinate system in which distances from x- and y-axes are measured in linear units, the unit of measure for this grid system is the degree.

Latitude

Lines of *latitude* measure the distance from the equator to the north and south poles in degrees. As is shown in Fig. 20.1, the 30th parallel of north latitude, known as the *30th parallel*, is measured by an angle formed by a line from the center of the earth to a point on the equator and a line from the center of the earth to a point on the 30th parallel. All lines of latitude are parallel to the equator. Parallels of north latitude measure the 90° from the equator to the north pole. Parallels of south latitude measure from the equator to the south pole. Each degree of latitude represents about 69 mi on the surface of the earth.

---

[1]Formerly a component of the U.S. Coast and Geodetic Survey.

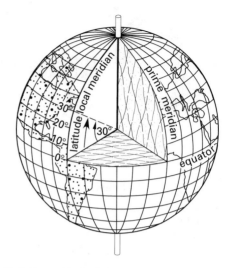

Figure 20.1 *Lines of Latitude*

Longitude

Measuring distance east and west around the earth at right angles to the equator are lines of *longitude*. These lines are also known as *meridians*, which are defined as great circles of the earth passing through both poles. They differ from lines of latitude in that they are not parallel but converge at the poles. Distance is measured east and west from a line of reference known as the *prime meridian*, which passes through the observatory in Greenwich, England. This reference line (longitude zero) is used by agreement by all nations. The measurement of longitude is either east or west from the prime meridian at Greenwich.

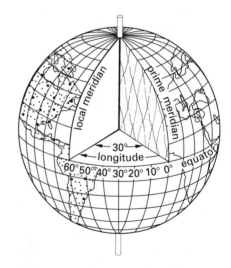

Figure 20.2 *Lines of Longitude*

Along the equator, each degree of longitude represents about 69 miles. Due to the convergence of the meridians, this distance becomes smaller toward the poles.

The National Geodetic Survey furnishes tables that give distances on the surface of the earth for $1°$ along any parallel of latitude.

6. PROPERTIES OF MAPS

Mercator was able to make a map that allowed sailors to plot a line on a chart from port to port and to follow that line to their destination. However, he admitted that his map would not scale to correct distances. To achieve one property for his map, he had to sacrifice another. The main properties of maps are shape, area, distance, and direction. Map makers may attain one of the properties and combinations of some of the others. Accurate representation of all properties is not possible if the map represents large areas such as continents or the whole world.

7. CONFORMAL MAPS

To show the true shape of the earth on a flat surface is highly desirable but impossible when considering large areas. If small areas such as cities, counties, or even states are shown, it may be possible to maintain their true shapes. The shape of a small area on the map will conform to the same area on the earth—thus, the word *conformal*. *Conformal projections* have the property that the scale at any point is the same in all directions.

8. MAP PROJECTIONS

A *map projection* is a representation of the surface of the earth on a flat sheet. Just as movie projectors "project" an image on a screen, map makers project points from a spherical surface onto a plane surface. Various methods are used in map projection depending on whether the map maker wants the map to be conformal (showing true shape), to be equal-area (having the area shown on the map in proportion to the area on the earth), or to show true distance or true direction. Map projections can be developed by using a cylinder, cone, or plane.

9. CYLINDERS AS DEVELOPABLE SURFACES

A flat sheet of paper can be rolled into a cylinder. If the sheet of paper is rolled around a globe, it will be in contact with the globe along the equator. If the equator is inked when the sheet is wrapped around the globe, a straight line will be printed on the sheet when it is unwrapped. The scale on this line will be exactly correct, but any point on the sphere that does not fall on the equator will not touch the sheet and will have to be "projected" onto the cylinder.

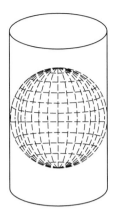

Figure 20.3 *Cylindrical Projections*

10. CONES AS DEVELOPABLE SURFACES

If a cone is formed with a sheet of paper and then placed over a sphere, the cone comes in contact with the sphere only along the parallel at $45°$ N latitude.[2] In other words, it is tangent to the sphere along the parallel. If the parallel is inked before the cone is placed, a line will be printed on the cone, but when the cone is unwrapped, the line will not be a straight line but rather an arc of a circle. The scale on this line will be exact, but points not on the $45°$ N parallel will have to be projected onto the cone. The $45°$ N parallel in this case will be known as the *standard parallel*.

11. PLANES AS DEVELOPABLE SURFACES

If a plane is passed parallel to the equator and tangent at the north pole, it will project points on the globe onto the plane. On this map, the parallels of latitude appear as concentric circles, and the meridians are straight lines radiating from the pole. Points can be projected from the center of the earth or from the opposite pole as shown in Fig. 20.5.

12. THE MERCATOR PROJECTION

The *Mercator projection* is known to most people. Mercator wanted navigators to be able to plot a line on a map and sail in that direction to their destination. He was successful, but in order to be so, he had to stretch the meridians and parallels on his projections, as can be seen in Fig. 20.6. On his map, meridians are straight parallel lines, as are parallels. Parallels and meridians are at right angles. Meridians are equally spaced but parallels are proportionally farther apart as the latitude increases. By comparing Figs. 20.5 and 20.6, it can be seen that the area around the north pole is greatly distorted in the Mercator projection, but the area along the equator is not.

---

[2]Cones can be made to contact the sphere at different parallels.

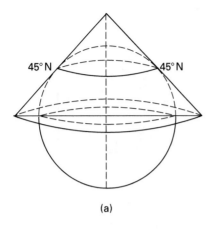

(a)

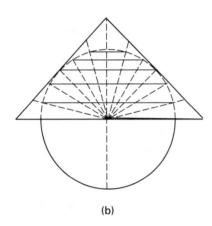

(b)

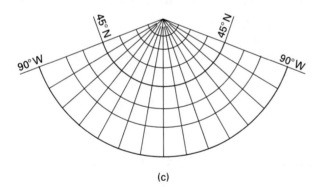

(c)

Figure 20.4 *Conical Projections*

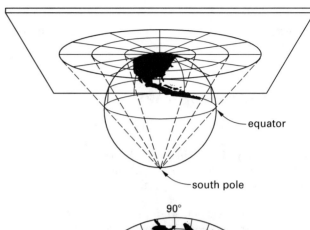

equator

south pole

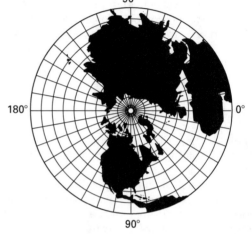

Figure 20.5 *Plane Projections*

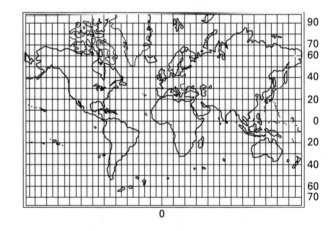

Figure 20.6 *The Mercator Projection*

13. TRANSVERSE MERCATOR PROJECTION

The *transverse Mercator projection* is a conformal cylinder projection. Instead of a cylinder wrapped around
the earth and tangent to it, a cylinder slightly less in
diameter than the globe is passed through the globe, at
right angles to the poles, so as to cut a band or strip
that produces a strip map. It is the basis for plane coordinate systems for many states that have a long axis
running north-south.

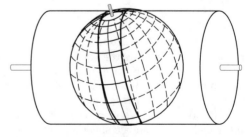

Figure 20.7 *Tranverse Mercator Projection*

14. THE LAMBERT CONIC PROJECTION

In many states that have their long axis orientated east-to-west, the basis for the state plane coordinate system is the *Lambert conformal projection*. Lambert's conformal projection is produced by mathematics. In theory, a cone, too small to cover one-half of the earth, is passed through the earth so that it intersects the earth at two parallels of latitude as shown in Fig. 20.8. The two parallels are known as the *standard parallels*.

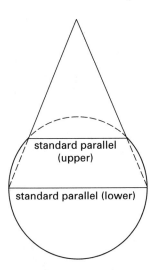

Figure 20.8 *Conic Two-Parallel Projection*

A Lambert conformal conic projection passing through the earth at the parallels of latitude N 33° and N 45° would produce the map projection of the United States shown in Fig. 20.9. Notice that the parallels of latitude appear as curved lines and the meridians as straight, converging lines.

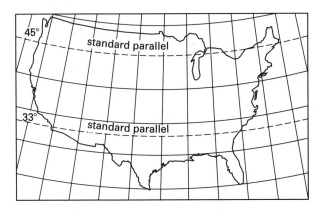

Figure 20.9 *Lambert Projection of the United States*

15. REDUCTION TO SEA LEVEL

Most map projection systems used for state plane coordinates project points from the spheroid at sea level. The distance between two points of known coordinates on the surface of the earth is not the same as the distance between two points when reduced to sea level. The difference in the two distances is a function of the elevation of the points.

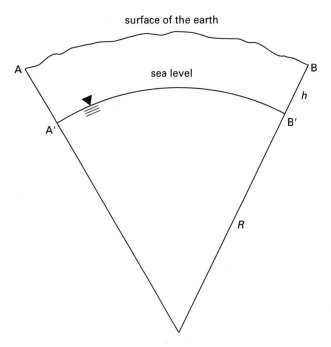

Figure 20.10 *Reduction to Sea Level*

In Fig. 20.10, points A and B are two points on the surface. A is projected along a radius of the earth to point A′ at sea level, and B is projected to point B′. If h is the mean elevation of AB and R is the mean radius of the earth, the *sea level factor* is given by Eq. 20.1 and tabulated in Table 20.1.

$$\frac{A'B'}{AB} = \frac{R}{R+h} \qquad 20.1$$

$$A'B' = \left(\frac{R}{R+h}\right) AB \qquad 20.2$$

For most computations, a value of R = 20,906,000 ft is sufficiently accurate.

Consider two points at a mean elevation of 500 ft that measure 5000.00 ft apart on the surface. Then, the sea level separation is

$$\begin{aligned} A'B' &= \left(\frac{R}{R+h}\right) AB \\ &= \left(\frac{20{,}906{,}000 \text{ ft}}{20{,}906{,}000 \text{ ft} + 500 \text{ ft}}\right) (5000.00 \text{ ft}) \\ &= 4999.88 \text{ ft} \end{aligned}$$

The correction at low elevations is negligible for short distances.

Table 20.1 *Multiplicative Factors for Reduction to Sea Level*

elevation (h)	sea level factor
500 ft	0.999976
1000 ft	0.999952
1500 ft	0.999928
2000 ft	0.999904
2500 ft	0.999880
3000 ft	0.999857
3500 ft	0.999833
4000 ft	0.999809
4500 ft	0.999785
5000 ft	0.999761
5500 ft	0.999737
6000 ft	0.999713
6500 ft	0.999689
7000 ft	0.999665
7500 ft	0.999641
8000 ft	0.999617
8500 ft	0.999593
9000 ft	0.999570
9500 ft	0.999546
10,000 ft	0.999522

16. REDUCTION FOR SCALE

To convert surface distance to grid distance, the surface distance must be multiplied by a *scale factor* that is a function of the latitude. Scale factors can be found in Table 1 of the NGS Projection tables. Latitude can be found in USGS topographic maps.

17. REDUCTION FOR CURVATURE OF EARTH (SECOND TERM)

A straight line on a sphere when projected on the Lambert cone will be a curved line. The angle formed by the straight line and the tangent to the curved line is very small and need not be considered for most traverses. For surveys over 5 mi in length, it may become a factor but it will not be discussed here. A formula for this correction can be found in the NGS Projection tables. The correction is known as the *second term*.

18. CONVERTING BETWEEN GEOGRAPHIC AND GRID POSITIONS

Horizontal control data furnished by the National Geodetic Survey show latitude and longitude as well as coordinates for triangulation stations. This would seem to indicate that there is no necessity to convert from geographic position to plane coordinates and vice versa. However, for surveys that cross from one zone to another (or from one state to another), it is convenient

to convert from grid position in one zone to geographic position, and then convert from geographic position to grid position in the other zone. This conversion is usually done with software.

19. NAD 83

NAD 83 (North American Datum of 1983) is a readjustment of the positions of some 250,000 control (survey) points. It is the first adjustment since 1927. NAD 27 involved the adjustment of 25,000 survey points. (Also, see Sec. 51.)

20. REASONS FOR THE ADJUSTMENT

The advent of EDM (Electronic Distance Measuring) instruments and GPS made it apparent to the NGS that there were distortions in NAD 27. These distortions were the main reasons for the readjustment. Other reasons were that length control was significantly deficient, that a number of azimuths were found to be inaccurate, that control in Alaska was not adequate, and that NAD 27 did not include Atlantic Seaboard control.

21. TERMS USED IN GEODESY

Understanding the difference between NAD 27 and NAD 83 requires some knowledge of mathematics and a vocabulary that includes terms such as *ellipse, ellipsoid, spheroid,* and *geoid*. A brief discussion of analytic geometry may shed some light on the meanings of these terms.

22. ANALYTIC GEOMETRY

Analytic geometry is the study of the line, the circle, the ellipse, the parabola, and the hyperbola. These geometric figures can be defined by mathematical equations.

In theory, the figures can be produced by passing a plane through, or tangent to, a cone. A plane tangent to a cone would form a line, the equation of which would be $Ax + By + C = 0$, as discussed in Chap. 12. If a plane were passed at right angles through a right circular cone, the intersection of the plane and the cone would form a circle, the equation of which is $x^2 + y^2 = r^2$, also discussed in Chap. 12.

23. THE ELLIPSE

If a plane were passed through a cone at an angle other than a right angle, the intersection of the plane and the cone would form an *ellipse* (see Fig. 20.11), which has the equation

$$\frac{x^2}{a^2} + \frac{y^2}{b^2} = 1 \qquad \text{20.3}$$

a is one-half the major axis and b is one-half the minor axis.

The flattening of an ellipse, which resembles the flattening of the earth at the poles, is given the symbol f.

$$f = \frac{a - b}{a} \qquad 20.4$$

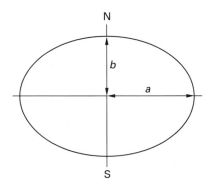

Figure 20.11 *Ellipse*

24. ELLIPSOID OF REVOLUTION

If the minor axis of an ellipse remained stationary and the ellipse revolved around this axis, it would form an *ellipsoid* of revolution (sometimes called a *spheroid*).

Because the earth is flattened slightly at the poles and bulges slightly at the equator, geodesists use the ellipsoid as the mathematical figure representing the earth. By varying the semimajor axis and f, they can produce an ellipsoid that very nearly matches the earth.

25. GRS 80 ELLIPSOID

All data have a reference ellipsoid. The original reference ellipsoid for NAD 83 is called *GRS 80* (Geodetic Reference System 80), for which the value of $a = 6\,378\,137$ m and $f = 1/298.2572221\ldots$. It is geocentric, meaning that its origin is at the mass center of the earth. The Clarke Ellipsoid (Spheroid) of 1866, which is the reference for NAD 27, has a value of $a = 6\,378\,206.4$ m and $b = 6\,356\,583.8$ m. The origin of this spheroid is not at the center of the earth. (Also, see Table 20.8.)

26. THE GEOID

The geoid surface coincides with mean sea level extended continuously through the continents. It is an uneven shape that does not describe a mathematical surface.

The direction of gravity is perpendicular to the geoid at every point. If a theodolite is oriented on the surface by a plumb line (direction of gravity) and a level bubble, the geoid is the surface of reference for astronomical observations.

27. DEFLECTION OF THE VERTICAL

In making astronomical observations, the plumb bob line is perpendicular to the geoid, but it is not perpendicular to the ellipsoid. The very small angle between the two perpendiculars is known as the *deflection of the vertical* (Fig. 20.12). Geodetic azimuth is astronomic azimuth plus the Laplace correction. The Laplace correction is the deflection of the vertical in the prime vertical component times the tangent of latitude at the station.

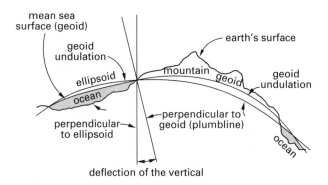

Figure 20.12 *Geoid-Ellipsoid Relationship*

Let

α = geodetic azimuth

α_A = astronomic azimuth

α_g = grid azimuth

ϕ = geodetic latitude of the station

η = deflection of the vertical in the prime vertical component

γ = the mapping angle, or convergence, for the Lambert projection and the Mercator projection

$\eta(\tan\phi)$ = Laplace correction

Then,

$$\alpha = \alpha_A + \eta \tan\phi \qquad 20.5$$

$$\alpha_g = \alpha - \gamma \qquad 20.6$$

28. FEATURES OF STATE PLANE COORDINATE SYSTEMS THAT CHANGED FROM 27 TO 83

The *State Plane Coordinate System of 1927* (SPCS 27), developed in the 1930s, is based on the North American Datum of 1927 (NAD 27), which uses the Clarke 1866 spheroid as a reference. Coordinates are given as X (east) and Y (north) in U.S. Survey foot units. Projections used to project from the earth's curved surface to a developable plane are the Lambert conformal conic projection and the Transverse Mercator projection. States that tend to be long in the east-west direction (e.g., Montana) use one or more Lambert projections (zones), while states prominent in the north-south direction (e.g., Idaho) use one or more Mercator projections. A few states (e.g., Florida) use both projections for different regions (see Table 20.2).

From 1975 to 1986, NAD 27 was readjusted using a simultaneous adjustment of all original survey data and a significant amount of very precise new data. A new reference surface, the *Geodetic Reference System of 1980* (GRS 80), was used that approximates the true shape of the earth worldwide better than the Clarke spheroid of 1866. This resulted in the *North American Datum of 1983* (NAD 83) and a subsequent change in the computed geodetic position (latitude and longitude) of control points on the earth's surface. It should be noted there is no precise direct mathematical conversion of geodetic position from NAD 27 to NAD 83. This is due primarily to the new adjustment rather than to the new reference surface (spheroid).

Since state plane coordinates are based on both geographic coordinates and the reference spheroid, implementation of NAD 83 dictates a new state plane coordinate system: SPCS 83. In converting to the new system, most states elected (or defaulted) to have the new system remain essentially the same as the old. That is to say, if the state was on the Lambert system with two zones, the new system remained on Lambert with the two zones covering the same geographical areas. A few states requested significant changes. As an example, Montana had three Lambert zones in SPCS 27 and only one zone for the entire state in SPCS 83.

There are two minor changes for all states in SPCS 83. The coordinate axes are designated N (north) and E (east), and units of measure are meters rather than feet. While meters are meters, feet are not necessarily feet. This is because there are two definitions of the foot: the U.S. Survey foot and the international foot. Conversion from meters to the two foot measurements can be made using the following *exact* equations.

$$\text{distance in U.S. survey feet} = (\text{distance in meters})$$
$$\times \left(\frac{3937}{1200}\right) \qquad 20.7$$

$$\text{distance in international feet} = \frac{\text{distance in meters}}{0.3048}$$
$$20.8$$

This results in only a very small difference between the two definitions. (In fact, the difference is two parts per million, or 0.02 ft in 10,000 ft, which is less than can be detected by almost any electronic distance meter.) When working with SPCS 83, however, extreme caution must be observed when converting coordinates in meters to feet. As an example, for a point near Maryville, Missouri, with coordinates of North 464 393.452 m and East 818 851.164 m, converting to survey feet results in 1,523,597.52 ft and 2,686,514.19 ft, whereas converting to international feet results in 1,523,600.56 ft and 2,686,519.57 ft—a difference of 3.04 ft and 5.38 ft.

All surveyors should be aware of the difference in definitions and be encouraged to always show the conversion factor when listing 1983 state plane coordinates in feet.

The official unit of measurement for NAD 83 is the meter. However, a number of states have passed legislation making either the international or the survey foot official for that state.

This may cause some confusion in the future since NAD 83 in meters made it easier to distinguish from NAD 27, which was in U.S. survey feet.

29. EXAMPLE CALCULATIONS USING THE STATE PLANE COORDINATE SYSTEM NAD 83

Two examples of working with state plane coordinates are presented in this chapter. The first is the Lambert conformal system for the State of Arkansas and the second is the transverse Mercator system for the State of Missouri.

30. LAMBERT SYSTEM

The *1927 system* was originally developed such that lines measured on the surface would change very little in length when projected onto the state plane grid (scale ratio or factor). Multiple zones were created so that any difference in these lengths would be less than one part in 10,000. For surveys of a low order of accuracy (the general rule for surveys prior to the 1970s), this enabled the surveyor to work on the system without applying any correction to the distances. Because of Arkansas' size, it has two separate but overlapping Lambert conformal projections or zones: north and south. Figure 20.13 illustrates these two zones.

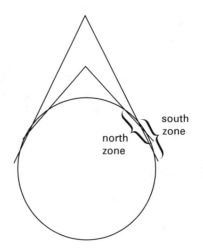

Figure 20.13 *Overlapping Projections*

The basics in developing any zone in either the 27 or 83 systems are the same; therefore, only one zone is discussed in the remainder of this section.

Table 20.2 State Zones for Plane Coordinates (NAD 83)

state	zone	projection	central meridian deg	ft
Alabama	E	TM	85	50
	W	TM	87	30
Alaska	1	TM	Oblique	
	2	TM	142	00
	3	TM	146	00
	4	TM	150	00
	5	TM	154	00
	6	TM	158	00
	7	TM	162	00
	8	TM	166	00
	9	TM	170	00
	10	L	176	00
Arizona	E	TM	110	10
	C	TM	111	55
	W	TM	113	45
Arkansas	N	L	92	00
	S	L	92	00
California	1	L	122	00
	2	L	122	00
	3	L	120	30
	4	L	119	00
	5	L	118	00
	6	L	116	15
Colorado	N	L	105	30
	C	L	105	30
	S	L	105	30
Connecticut		L	72	45
Delaware		TM	75	25
Florida	E	TM	81	00
	W	TM	82	00
	N	L	84	30
Georgia	E	TM	82	10
	W	TM	84	10
Hawaii	1	TM	155	30
	2	TM	156	40
	3	TM	158	00
	4	TM	159	30
	5	TM	160	10
Idaho	E	TM	112	10
	C	TM	114	00
	W	TM	115	45
Illinois	E	TM	88	20
	W	TM	90	10

state	zone	projection	central meridian deg	ft
Indiana	E	TM	85	40
	W	TM	87	05
Iowa	N	L	93	30
	S	L	93	30
Kansas	N	L	98	00
	S	L	98	30
Kentucky	N	L	84	15
	S	L	85	45
Louisiana	N	L	92	30
	S	L	91	20
	OF	L	91	20
Maine	E	TM	68	30
	W	TM	70	10
Maryland		L	77	00
Massachusetts	M	L	71	30
	IS	L	70	30
Michigan	N	L	87	00
	C	L	84	22
	S	L	84	22
Minnesota	N	L	93	06
	C	L	94	15
	S	L	94	00
Mississippi	E	TM	88	50
	W	TM	90	20
Missouri	E	TM	90	30
	C	TM	92	30
	W	TM	94	30
Montana		L	109	30
Nebraska		L	100	00
Nevada	E	TM	115	35
	C	TM	116	40
	W	TM	118	35
New Hampshire		TM	71	40
New Jersey		TM	74	30
New Mexico	E	TM	104	20
	C	TM	106	15
	W	TM	107	50
New York	E	TM	74	30
	C	TM	76	35
	W	TM	78	35
	LI	L	74	00
North Carolina		L	79	00
North Dakota	N	L	100	30
	S	L	100	30
Ohio	N	L	82	30
	S	L	82	30

state	zone	projection	central meridian deg	ft
Oklahoma	N	L	98	00
	S	L	98	00
Oregon	N	L	120	30
	S	L	120	30
Pennsylvania	N	L	77	45
	S	L	77	45
Rhode Island		TM	71	30
South Carolina		L	81	00
South Dakota	N	L	100	00
	S	L	100	20
Tennessee		L	86	00
Texas	N	L	101	30
	NC	L	98	30
	C	L	100	20
	SC	L	99	00
	S	L	98	30
Utah	N	L	111	30
	C	L	111	30
	S	L	111	30
Vermont		TM	72	30
Virginia	N	L	78	30
	S	L	78	30
Washington	N	L	120	50
	S	L	120	30
West Virginia	N	L	79	30
	S	L	81	00
Wisconsin	N	L	90	00
	C	L	90	00
	S	L	90	00
Wyoming	E	TM	105	10
	EC	TM	107	20
	WC	TM	108	45
	W	TM	110	05
Puerto Rico		L	66	26

Abbreviations

TM = transverse Mercator			S	= south
L = Lambert			SC	= south central
N = north			W	= west
NC = north central			WC	= west central
E = east			LI	= Long Island
EC = east central			IS	= island
C = central				
M = mainland				
OF = offshore				

Projections

Figure 20.14 illustrates a Lambert projection cone with the relationship between a geodetic distance L and a projection (grid) distance L'. Note that the ratio of L'/L (scale ratio or scale factor) is less than one between the standard parallels, is greater than one outside the parallels, and varies in a north-south direction. Also note that it does not vary in an east-west direction.

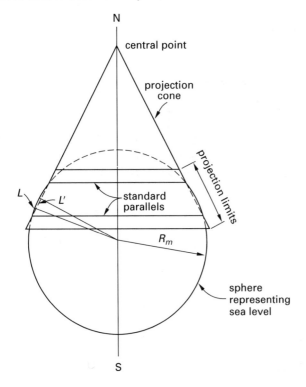

Figure 20.14 *Lambert Geodetic and Projection Distances*

Figure 20.15 illustrates a projection cone developed into a plane (cut and laid out flat). The central meridian is a line of longitude arbitrarily chosen such that it will pass through the center of the zone. It will not necessarily be the same degree of longitude for all zones within a state; however, for both Arkansas zones (north and south) it is at a longitude of $92°00'00''$. Note that the projected lines of longitude are straight, and lines of latitude are curved with the radius point (central point) being the apex of the cone. Also note the projected lines of latitude and longitude through a point and the corresponding R and γ dimensions. The grid declination angle γ was expressed as θ in the 27 Lambert system and $\Delta\alpha$ in the 27 Mercator system.

A rectangular coordinate system is superimposed onto the developed cone (see Fig. 20.16). The N-axis is placed parallel with the central meridian and the E-axis is placed such that it just touches the lower limit or at some offset distance (N_b) to the south. For Arkansas, N_b is 0.000 m for the north zone and 400,000.000 m for the south. The N-axis is placed far enough west of the central meridian (E_o) such that the entire area of interest will be in the upper right quadrant, ensuring all coordinate values will be positive. This distance is 400,000.000 m for both the north and south zones.

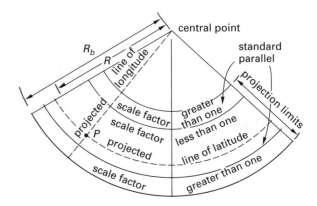

Figure 20.15 *Developed Projection Cone*

On Fig. 20.16, note that the lines of longitude (geodetic, or "true" north) are not parallel with the N-axis (grid north) except at the central meridian. This difference between geodetic and grid north is γ, sometimes referred to as *mapping angle*, grid declination, or variation. Also note that this angle varies in the geodetic east-west direction (a function of longitude) but does not vary with latitude. Also, γ is the angle measured at the central point from the central meridian to a projected line of longitude (see Fig. 20.15).

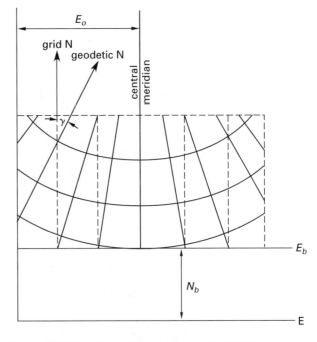

Figure 20.16 *Developed Projection Cone with Grid Overlay*

31. GEODETIC DISTANCE TO GRID DISTANCE (LAMBERT)

To convert geodetic distance to grid, the projection scale factor (scale ratio) is used.

$$\text{scale factor} = \frac{\text{grid distance}}{\text{geodetic distance}} \quad\quad 20.9$$

$$\text{grid distance} = (\text{geodetic distance})(\text{scale factor}) \quad 20.10$$

Scale factor is a function of latitude and can be obtained from the appropriate projection table. Depending on accuracy requirements and the extent of the project area in a north-south direction, the average latitude as obtained from a quad sheet is sufficient. For very precise surveys, a weighted average of scale factor for each line can be used. Note that for all practical purposes, scale factors did not change from SPCS 27 to SPCS 83 in most states.

32. SURFACE DISTANCE TO GRID DISTANCE (LAMBERT)

Converting a surface distance to grid involves multiplying the distance by both the scale factor and the sea level factor (see Eq. 20.13).

$$\text{grid distance} = (\text{surface distance})(\text{scale factor})$$
$$\times (\text{sea level factor}) \qquad \textit{20.11}$$

If an average elevation and latitude for the project area are being used, these two factors can be multiplied to obtain a single factor (combined scale factor, CSF), which is then multiplied by all of the measured distances—either slope or horizontal.

Example 20.1

A geodetic distance has been measured in Arkansas (north). What is the grid distance?

horizontal distance = 2640.00 ft
average elevation = 1400 ft
average latitude = 35°30′
83 north zone

Solution

scale factor = 0.99993594 $\quad\begin{bmatrix}\text{from Table 20.3, enter}\\\text{with latitude} = 35°30′\end{bmatrix}$

$$\text{sea level factor} = \frac{20{,}906{,}000}{20{,}906{,}000 + 1400}$$
$$= 0.99993304$$

combined
$$\text{scale factor} = (0.99993594)(0.99993304)$$
$$= 0.99986898$$

$$\text{grid distance} = (2640.00)(0.99986898)$$
$$= 2639.65 \text{ ft}$$

33. GEODETIC AZIMUTH TO GRID AZIMUTH (LAMBERT)

To convert geodetic azimuth to grid, use Eq. 20.12.

$$\text{grid azimuth} = \text{geodetic azimuth} - \gamma + \text{second term} \qquad \textit{20.12}$$

The second term can be neglected for most surveys; therefore, Eq. 20.12 reduces to

$$\text{grid azimuth} = \text{geodetic azimuth} - \gamma \qquad \textit{20.13}$$

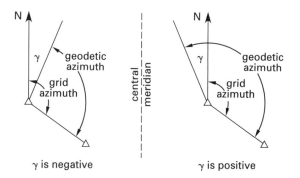

Figure 20.17 Grid Declination

γ (mapping angle or grid declination) is a function of longitude and can be computed from Eq. 20.14.

$$\gamma = (\text{longitude of central meridian}$$
$$- \text{longitude of station})\ell \qquad \textit{20.14}$$

The longitude of the central meridian (LCM) and ℓ are projection zone constants. They are listed in Table 20.4 as well as in projection tables.

Table 20.4 Zone Constants (Arkansas, north)

zone	LCM	ℓ
83 North	92°00′00″	0.581899128040
83 South	92°00′00″	0.559690686831

For most work, the longitude of the beginning point of the survey must be accurately known to convert geodetic azimuth to grid. Since this is usually a triangulation station or other point of known coordinates, the longitude is given or can be computed from the coordinates. Scaling longitude from a USGS quad sheet may be sufficient; however, for Arkansas a longitudinal scaling error of 5 in will result in approximately 3 in of error in γ.

Example 20.2

Determine the grid azimuth from the geodetic azimuth.

geodetic azimuth from tri
   station to point A = 241°12′37″
longitude of tri station = 94°10′15″2758
83 north zone

Solution

longitude of
   central meridian = 92°00′00″   [from Table 20.4]
ℓ = 0.581899128040
$\gamma = (92°00'00'' - 94°10'15'')$
   $\times (0.581899128040)$
   $= -1.26321°$
   $= -1°15'48''$
grid azimuth = $241°12'37'' - (-1°15'48'')$
   $= 242°28'25''$

Table 20.3(a) *Lambert Projection Tables*

Lambert conformal conic projection with two standard parallels
Plane coordinate projection tables

DATUM: NAD 83
The projection is ARKANSAS NORTH

The ellipsoidal constants are

$a = 6378137$ m

$f = 1/298.25722210$

The defining constants are

$\phi_b = 34°20'$ [latitude of grid origin]

$\gamma_{CM} = 92°0'$ [longitude of origin and central meridian, CM]

$\phi_s = 34°56'$ [southern standard parallel]

$\phi_n = 36°14'$ [northern standard parallel]

$E_o = 400000.0000$ m [easting coordinate of origin]
$N_b = 0.0000$ m [northing coordinate of origin]

The derived constants are

$\ell = 0.581899128040 = \sin\phi_o$

$K = 13112784.4998$ m [mapping radius at equator]

$R_b = 9062395.1981$ m [mapping radius at grid origin]

The Lambert coordinates (N, E) from geodetic positions (ϕ, γ) are

$\gamma = (\gamma_{CM} - \gamma)\sin\phi_o$

$E = R\sin\gamma + E_o$ [R from table]

$N = R_b - R\cos\gamma + N_b$

station	latitude longitude	R γ	$\sin\gamma$ $\cos\gamma$	E N
sample	35 35 0.00000	8923718.366 m	−0.0050780060	354685.304 m
1	92 30 0.00000	−0 17 27.41843	0.9999871068	138791.887 m

The geodetic positions from the Lambert coordinates are

$$\tan\gamma = \frac{E - E_o}{R_b - (N - N_b)}$$

$$R = \frac{R_b - (N - N_b)}{\cos\gamma}$$

$$\gamma = \gamma_{CM} - \frac{\gamma}{\ell}$$

[ϕ from table using R]

station	E N	$E - E_o$ $R_b - (N - N_b)$	R γ	latitude longitude
sample	450000.000 m	50000.000 m	8922535.2939 m	35 35 38.3891
2	140000.000 m	8922395.198 m	0 19 15.87038	91 26 53.6241

(Note: Use sufficient significant digits for trigonomic functions.)

Table 20.3(b) *Lambert Projection Tables*

Lambert Conformal Conic Projection Tables
ARKANSAS NORTH

latitude deg	ft	R (m)	difference	k	latitude deg	ft	R (m)	difference	k
34	20	9062395.198	30.81870	1.00017199	35	0	8988433.794	30.81644	0.99998756
34	21	9060546.076	30.81859	1.00016576	35	1	8986584.808	30.81644	0.99998466
34	22	9058696.961	30.81848	1.00015961	35	2	8984735.821	30.81644	0.99998185
34	23	9056847.852	30.81838	1.00015354	35	3	8982886.835	30.81644	0.99997911
34	24	9054998.749	30.81828	1.00014755	35	4	8981037.849	30.81644	0.99997646
34	25	9053149.652	30.81819	1.00014165	35	5	8979188.862	30.81645	0.99997390
34	26	9051300.561	30.81809	1.00013583	35	6	8977339.875	30.81646	0.99997142
34	27	9049451.475	30.81800	1.00013010	35	7	8975490.888	30.81647	0.99996902
34	28	9047602.395	30.81791	1.00012445	35	8	8973641.900	30.81649	0.99996670
34	29	9045753.320	30.81783	1.00011888	35	9	8971792.910	30.81650	0.99996447
34	30	9043904.250	30.81774	1.00011339	35	10	8969943.920	30.81652	0.99996233
34	31	9042055.186	30.81766	1.00010799	35	11	8968094.929	30.81655	0.99996026
34	32	9040206.126	30.81758	1.00010267	35	12	8966245.936	30.81657	0.99995828
34	33	9038357.071	30.81751	1.00009743	35	13	8964396.941	30.81660	0.99995639
34	34	9036508.020	30.81744	1.00009228	35	14	8962547.945	30.81663	0.99995458
34	35	9034658.974	30.81737	1.00008721	35	15	8960698.948	30.81666	0.99995285
34	36	9032809.932	30.81730	1.00008222	35	16	8958849.948	30.81670	0.99995121
34	37	9030960.894	30.81723	1.00007732	35	17	8957000.946	30.81674	0.99994965
34	38	9029111.861	30.81717	1.00007250	35	18	8955151.942	30.81678	0.99994817
34	39	9027262.830	30.81711	1.00006776	35	19	8953302.935	30.81682	0.99994678
34	40	9025413.804	30.81705	1.00006311	35	20	8951453.925	30.81687	0.99994547
34	41	9023564.781	30.81700	1.00005854	35	21	8949604.913	30.81692	0.99994425
34	42	9021715.761	30.81694	1.00005405	35	22	8947755.898	30.81697	0.99994311
34	43	9019866.744	30.81689	1.00004965	35	23	8945906.880	30.81702	0.99994205
34	44	9018017.731	30.81685	1.00004533	35	24	8944057.859	30.81708	0.99994108
34	45	9016168.720	30.81680	1.00004109	35	25	8942208.834	30.81714	0.99994019
34	46	9014319.712	30.81676	1.00003694	35	26	8940359.806	30.81720	0.99993939
34	47	9012470.707	30.81672	1.00003287	35	27	8938510.774	30.81726	0.99993867
34	48	9010621.703	30.81668	1.00002889	35	28	8936661.738	30.81733	0.99993803
34	49	9008772.702	30.81665	1.00002498	35	29	8934812.698	30.81740	0.99993748
34	50	9006923.704	30.81662	1.00002116	35	30	8932963.654	30.81747	0.99993701
34	51	9005074.707	30.81659	1.00001743	35	31	8931114.606	30.81755	0.99993663
34	52	9003225.711	30.81656	1.00001377	35	32	8929265.553	30.81763	0.99993633
34	53	9001376.718	30.81654	1.00001021	35	33	8927416.495	30.81771	0.99993611
34	54	8999527.726	30.81651	1.00000672	35	34	8925567.433	30.81779	0.99993598
34	55	8997678.735	30.81650	1.00000332	35	35	8923718.366	30.81787	0.99993594
34	56	8995829.745	30.81648	1.00000000	35	36	8921869.294	30.81796	0.99993597
34	57	8993980.756	30.81647	0.99999677	35	37	8920020.216	30.81805	0.99993609
34	58	8992131.768	30.81645	0.99999361	35	38	8918171.133	30.81814	0.99993630
34	59	8990282.781	30.81645	0.99999055	35	39	8916322.044	30.81824	0.99993659

Projections

Example 20.3

Determine the grid azimuth from the geodetic azimuth.

geodetic azimuth to backsight = $127°48'36''$
longitude of instrument station = $90°39'28''$
83 south zone

Solution

longitude of
central meridian = $92°00'00''$ [from Table 20.4]
$\ell = 0.559690686831$
$\gamma = (92°00'00'' - 90°39'28'')$
$\quad\quad \times (0.559690686831)$
$\quad = +0.75123°$
$\quad = +0°45'04''$
grid azimuth = $127°48'36'' - (+0°45'04'')$
$\quad\quad\quad = 127°03'32''$

34. ASTRONOMIC AZIMUTH TO GEODETIC AZIMUTH

Astronomic azimuth is based on the true shape and rotation of the earth; *geodetic azimuth* is based on a mathematical approximation of the earth's shape (e.g., GRS 80). Astronomic azimuth is obtained when making a celestial observation (sun or Polaris) and can be converted to geodetic azimuth from Eq. 20.15.

$$\text{geodetic azimuth} = \text{astronomic azimuth}$$
$$+ \text{ Laplace correction} \quad 20.15$$

The Laplace correction is less than 2 in for most of Arkansas and does not exceed 5 in. It may, however, exceed 20 in in the mountainous regions of some western states. If accuracy requires use of the correction, it is recommended that the NGS software be used (e.g., DEFLEC).

35. GEODETIC ANGLE TO GRID ANGLE

If the second term is neglected, the grid angle is the same as the geodetic (field) angle. Therefore, field angles can be used when computing grid azimuths.

36. PLANE COORDINATES TO GEOGRAPHIC COORDINATES (LATITUDE AND LONGITUDE)

To compute latitude and longitude from north and east coordinates, the following equations are used.

$$N' = N - N_b \quad\quad 20.16$$
$$E' = E - E_o \quad\quad 20.17$$

$$\tan \gamma = \frac{E'}{R_b - N'} \quad\quad 20.18$$

$$\text{longitude} = \text{LCM} - \gamma \quad\quad 20.19$$

$$R = \frac{R_b - N'}{\cos \gamma} \quad\quad 20.20$$

To obtain latitude, enter these equations into projection tables with R and interpolate.

N_b, E_o, R_b, and ℓ are constants for a zone. When working on the 83 system, coordinates in Eq. 20.20 must be calculated in meters. In solving Eqs. 20.12 through 20.20, computations must be significant to at least ten digits.

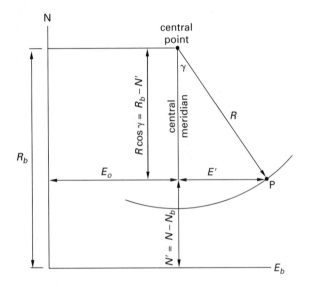

Figure 20.18 *Converting Plane Coordinates to Longitude and Latitude*

Example 20.4

Convert the plane coordinates to longitude and latitude.

N = 38 631.745 m
E = 201 012.881 m
83 north zone

Solution

$N_b = 0.000$ m   [from Table 20.3(a)]
$E_o = 400\,000.000$ m
longitude central meridian = $92°00'00''$
$R_b = 9\,062\,395.1981$ m
$\ell = 0.581899128040$

$N' = 38\,631.745$ m $- 0.000$
$\quad = 38\,631.745$ m

$E' = 201\,012.881$ m $- 400\,000.000$ m
$\quad = -198\,987.119$ m

$R_b - N' = 9\,062\,395.1981$ m $- 38\,631.745$ m
$\quad\quad\quad = 9\,023\,763.453$ m

Using Eqs. 20.18 and 20.20,

$$\tan \gamma = \frac{-198\,987.119 \text{ m}}{9\,023\,763.453 \text{ m}}$$

$$\gamma = -1.2632506048°$$

$$R = \frac{9\,023\,763.453 \text{ m}}{\cos(-1.26325060480°)}$$

$$= 9\,025\,957.164 \text{ m}$$

$$\text{longitude} = \text{LCM} - \frac{\gamma}{\ell}$$

$$= 92°00'00'' - \left(-\frac{1.26325060480°}{0.581899128040} \right)$$

$$= 92°00'00'' - (-2°10'15.2758'')$$

$$= 94°10'15.2758''$$

Enter the projection tables with R and interpolate for latitude.

The computed R is between $34°39'$ and $34°40'$ (see Table 20.3).

$$R = 9\,027\,262.830 \text{ m}$$

$$\text{difference} = 30.81711 \text{ m}$$

$$\text{[at latitude} = 34°39', \text{ Table 20.3]}$$

$$\Delta R = 9\,027\,262.830 \text{ m} - 9\,025\,957.164 \text{ m}$$

$$= 1\,305.666 \text{ m}$$

$$\Delta \text{ latitude} = \frac{1\,305.666 \text{ m}}{30.81711 \text{ m}}$$

$$= 42.3682''$$

$$\text{latitude} = 34°39'42.3682''$$

37. GEOGRAPHIC COORDINATES (LATITUDE AND LONGITUDE) TO PLANE COORDINATES

To compute north and east coordinates from latitude and longitude, the following equations are used.

$$\gamma = (\text{LCM} - \text{longitude})\ell \qquad 20.21$$

To obtain R from projection tables, enter with latitude and interpolate.

$$\text{N} = R_b - R \cos \gamma + N_b \qquad 20.22$$
$$\text{E} = R \sin \gamma + E_o \qquad 20.23$$

N_b, E_o, R_b, and ℓ are constants for a zone. When working on the 83 system, coordinates will be expressed in meters. In solving Eqs. 20.21 through 20.23, the calculator/computer must be capable of computing to at least 10 digits.

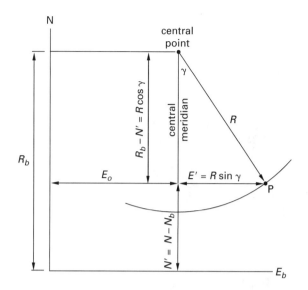

Figure 20.19 *Converting Longitude and Latitude to Plane Coordinates*

Example 20.5

Use Table 20.3 to convert the given latitude and longitude to plane coordinates.

latitude $= 34°44'47.7648''$
longitude $= 92°17'20.7595''$
83 north zone

Solution

$N_b = 0.000$ m   [from Table 20.3(a)]
$E_o = 400\,000.000$ m   [from Table 20.3(a)]
LCM $= 92°00'00''$ [from Table 20.4]
$R_b = 9\,062\,395.1981$ m   [from Table 20.3(a)]
$\ell = 0.581899128040$ [from Table 20.3(a)]

Enter latitude into the projection tables.

$$R = 9\,018\,017.731 \text{ m}$$

$$\text{difference} = 30.81685 \text{ m} \quad \text{[see Table 20.3]}$$

$$\Delta R @ (30.81685 \text{ m}, 47''7648)$$

$$= 1471.961$$

$$R = 9\,018\,017.731 \text{ m} - 1471.961 \text{ m}$$

$$= 9\,016\,545.770 \text{ m}$$

$$\gamma = (92°00'00'' - 92°17'20''7595)$$

$$\times (0.581899128040)$$

$$= -0.168226957°$$

Using Eqs. 20.22 and 20.23,

$$\text{N} = 9\,062\,395.1981 \text{ m} - 9\,016\,545.770 \text{ m}$$

$$\times \cos(-0.168226957) + 0.000$$

$$= 45\,888.293 \text{ m}$$

$E = (9\,016\,545.770 \text{ m})(\sin -0.168226957°)$

$\quad + 400\,000.000 \text{ m}$

$\quad = 373\,526.429 \text{ m}$

38. CONVERSION FROM SPCS 27 TO SPCS 83 (LAMBERT)

For all practical purposes, there is no accurate and convient means to convert from NAD 27 to NAD 83. The U.S. Army Corps of Engineers and NGS have software available (CORPSCON) to convert from one system to the other.

39. TRANSVERSE MERCATOR SYSTEM (MISSOURI)

The 1927 system was originally developed such that the lines measured on the surface would change very little in length when projected to the state plane grid (scale ratio or factor). In fact, multiple zones were created so that any difference in these lengths would be less than 1 part in 10,000. For surveys of a low order of accuracy (the general rule for surveys prior to the 1970s), this enabled the surveyor to work on the system without applying any correction to distances. Because of Missouri's size, Missouri has three separate but overlapping transverse Mercator conformal projections or zones: West, Central, and East. Figure 20.20 shows the three Missouri zones. Figure 20.21 illustrates a typical transverse Mercator projection.

The basics in developing each of these three zones in either the 27 or the 83 system are the same; therefore, only one zone will be discussed.

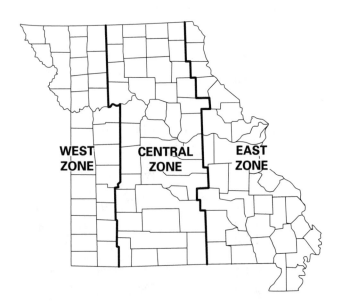

Figure 20.20 *Missouri State Plane Coordinate Zones*

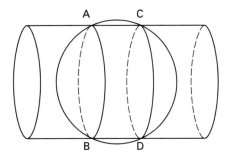

Figure 20.21 *Developable Cylinder*

Figure 20.22 illustrates a side view perpendicular to the central meridian of a transverse Mercator projection cylinder with the relationship between a geodetic distance L and a projection (grid) distance L'. Figure 20.23 is similar to Fig. 20.22 except the view is perpendicular to the axis of the cylinder.

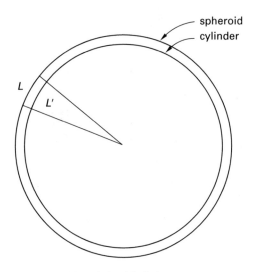

Figure 20.22 *View Along Axis of Cylinder*

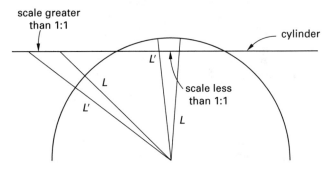

Figure 20.23 *View Perpendicular to Axis of Cylinder*

Figure 20.24 illustrates the cylinder after it has been developed (cut and laid flat). Note that the ratio of L'/L (scale ratio or scale factor) is less than one between the lines where the cylinder intersected the ellipsoid and greater than one outside these lines. This ratio (designated by the symbol k) varies in the east-west coordinate direction and does not vary in the north-south coordinate direction. The scale ratio at the central meridian can be made closer to one by increasing the size of the cylinder; however, the east-west extent of acceptable scale factors will be decreased.

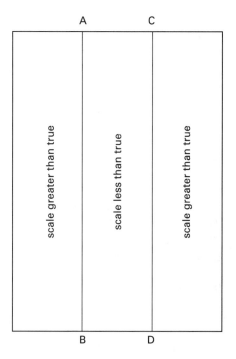

Figure 20.24 *Developed Cylinder*

Figure 20.25 illustrates a projection cylinder at a central meridian of 180° developed into a plane. Different zones are created by rotating the cylinder to a different central meridian. Also, the "diameter" of the cylinder can be increased or decreased to change the coverage in an east-west direction. For Missouri, the placement of the central meridian and the size of the cylinder (scale factor) for the west, central, and east zones are given in Table 20.5.

Note that the lines of latitude and longitude project as complex curved lines. On a Lambert conic projection, lines of longitude are straight and lines of latitude are circular arcs.

Table 20.5 *Missouri Transverse Mercator Zone Constants*

zone	longitude central meridian	latitude of grid origin	northing of grid origin	easting of central meridian (meters)	scale factor at central meridian
east	90°30′	35°50′	0.0	250,000	0.99993333 (1/15000)
central	92°30′	35°50′	0.0	500,000	0.99993333 (1/15000)
west	94°30′	36°10′	0.0	850,000	0.99994118 (1/17000)

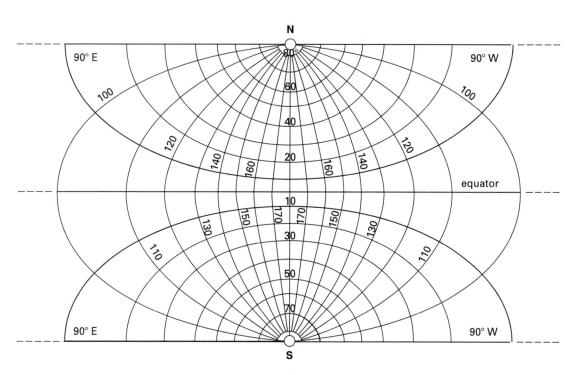

Figure 20.25 *Developed Mercator Projection Cylinder (Meridian = 180°)*

A rectangular coordinate system is superimposed onto the developed cylinder as shown in Fig. 20.26. The N-axis is placed parallel with the central meridian at some offset distance to the west (E_o), and the E-axis is placed at some latitude south of the area to be covered by the projection. The E_o for each of the three Missouri zones (see Table 20.5) have been established so that all the eastings in a given zone are greater than any in the zones to the east. This provides a convenient way of determining the zone for which the coordinates are given—that is, the coordinates in the central zone will be greater than 390 000 m, and those in the west zone will be greater than 740 000 m. Note that lines of longitude (geodetic or "true" north) are not parallel with the N-axis (grid north) except at the central meridian. This difference between geodetic and grid north is γ ($\Delta\alpha$ in the 27 system), or sometimes referred to as *mapping angle*, *grid declination*, or *variation*.

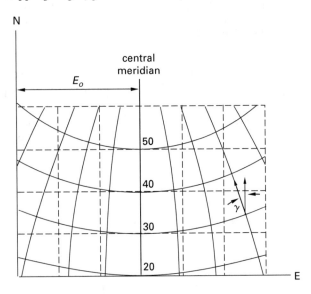

Figure 20.26 *Projection with Grid*

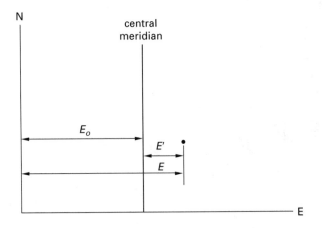

Figure 20.27 *Nomenclature for East-West Projections*

When working on a transverse Mercator projection, the east-west distance from the central meridian (E') is important. It can be obtained by simply subtracting E_o

from the easting coordinate value. While E' has a sign (+ if east of the central meridian and − if west), normally only the magnitude or absolute value of the term is necessary.

40. GEODETIC DISTANCE TO GRID DISTANCE (MERCATOR)

To convert geodetic distance to grid, the projection scale factor (scale ratio), k, is used in the following equations.

$$k = \frac{\text{grid distance}}{\text{geodetic distance}} \quad\quad 20.24$$

$$\text{grid distance} = (\text{geodetic distance})k \quad\quad 20.25$$

Scale factor is a function of how far the line is away from the central meridian in the east-west coordinate direction (E'). It can be computed for each zone using the following equations.

For the east and central zones,

$$k = 0.9999333 + (E')^2(1.231178 \times 10^{-14}) \quad 20.26$$

For the west zone,

$$k = 0.9999412 + (E')^2(1.231093 \times 10^{-14})$$
$$[E' = E - E_o \text{ and is in meters}] \quad\quad 20.27$$

Although these equations change slightly with latitude, the effect on k is small within a zone. In lieu of Eqs. 20.24 through 20.27, Tables 20.6 and 20.7 can be used.

Scale factor does not change rapidly with E'. Consequently, an approximate E' for the average location of the survey is sufficient in most cases. If an average E' is not readily available, it can be computed by scaling latitude and longitude from a quad sheet and using the following approximate equation.

$$E' = (111\,500 \text{ m})(\gamma_{CM} - \gamma)(\cos\phi) \quad\quad 20.28$$

(E' is approximate to within 100 m, ϕ is the latitude of the point, λ is the longitude of the point, and γ_{CM} is the longitude of the central meridian.)

Note that for all practical purposes, scale factors did not change from SPCS 27 to SPCS 83 in most states.

41. SURFACE DISTANCE TO GRID DISTANCE (MERCATOR)

Converting a surface distance to grid involves multiplying the distance by both the sea level factor and the scale factor.

$$\text{grid distance} = (\text{surface distance})(\text{sea level factor})$$
$$\times (\text{scale factor}) \quad\quad 20.29$$

Table 20.6 *Scale Factor—East and Central Zones—Missouri*

E' (m)	k	E' (m)	k	E' (m)	k
1000	0.9999333	41000	0.9999540	81000	1.0000141
2000	0.9999334	42000	0.9999550	82000	1.0000161
3000	0.9999334	43000	0.9999561	83000	1.0000181
4000	0.9999335	44000	0.9999572	84000	1.0000202
5000	0.9999336	45000	0.9999583	85000	1.0000223
6000	0.9999338	46000	0.9999594	86000	1.0000244
7000	0.9999339	47000	0.9999605	87000	1.0000265
8000	0.9999341	48000	0.9999617	88000	1.0000287
9000	0.9999343	49000	0.9999629	89000	1.0000309
10000	0.9999346	50000	0.9999641	90000	1.0000331
11000	0.9999348	51000	0.9999654	91000	1.0000353
12000	0.9999351	52000	0.9999666	92000	1.0000375
13000	0.9999354	53000	0.9999679	93000	1.0000398
14000	0.9999357	54000	0.9999692	94000	1.0000421
15000	0.9999361	55000	0.9999706	95000	1.0000444
16000	0.9999365	56000	0.9999719	96000	1.0000468
17000	0.9999369	57000	0.9999733	97000	1.0000492
18000	0.9999373	58000	0.9999747	98000	1.0000516
19000	0.9999378	59000	0.9999762	99000	1.0000540
20000	0.9999383	60000	0.9999777	100000	1.0000564
21000	0.9999388	61000	0.9999791	101000	1.0000589
22000	0.9999393	62000	0.9999807	102000	1.0000614
23000	0.9999398	63000	0.9999822	103000	1.0000639
24000	0.9999404	64000	0.9999838	104000	1.0000665
25000	0.9999410	65000	0.9999853	105000	1.0000691
26000	0.9999417	66000	0.9999870	106000	1.0000717
27000	0.9999423	67000	0.9999886	107000	1.0000743
28000	0.9999430	68000	0.9999903	108000	1.0000769
29000	0.9999437	69000	0.9999919	109000	1.0000796
30000	0.9999444	70000	0.9999937	110000	1.0000823
31000	0.9999452	71000	0.9999954	111000	1.0000850
32000	0.9999459	72000	0.9999972	112000	1.0000878
33000	0.9999467	73000	0.9999989	113000	1.0000905
34000	0.9999476	74000	1.0000007	114000	1.0000933
35000	0.9999484	75000	1.0000026	115000	1.0000962
36000	0.9999493	76000	1.0000044	116000	1.0000990
37000	0.9999502	77000	1.0000063	117000	1.0001019
38000	0.9999511	78000	1.0000082	118000	1.0001048
39000	0.9999521	79000	1.0000102	119000	1.0001077
40000	0.9999530	80000	1.0000121	120000	1.0001106

Table 20.7 *Scale Factor—West Zone—Missouri*

E' (m)	k	E' (m)	k	E' (m)	k
1000	0.9999412	41000	0.9999619	81000	1.0000219
2000	0.9999412	42000	0.9999629	82000	1.0000240
3000	0.9999413	43000	0.9999639	83000	1.0000260
4000	0.9999414	44000	0.9999650	84000	1.0000280
5000	0.9999415	45000	0.9999661	85000	1.0000301
6000	0.9999416	46000	0.9999672	86000	1.0000322
7000	0.9999418	47000	0.9999684	87000	1.0000344
8000	0.9999420	48000	0.9999695	88000	1.0000365
9000	0.9999422	49000	0.9999707	89000	1.0000387
10000	0.9999424	50000	0.9999720	90000	1.0000409
11000	0.9999427	51000	0.9999732	91000	1.0000431
12000	0.9999429	52000	0.9999745	92000	1.0000454
13000	0.9999433	53000	0.9999758	93000	1.0000476
14000	0.9999436	54000	0.9999771	94000	1.0000500
15000	0.9999439	55000	0.9999784	95000	1.0000523
16000	0.9999443	56000	0.9999798	96000	1.0000546
17000	0.9999447	57000	0.9999812	97000	1.0000570
18000	0.9999452	58000	0.9999826	98000	1.0000594
19000	0.9999456	59000	0.9999840	99000	1.0000618
20000	0.9999461	60000	0.9999855	100000	1.0000643
21000	0.9999466	61000	0.9999870	101000	1.0000668
22000	0.9999471	62000	0.9999885	102000	1.0000693
23000	0.9999477	63000	0.9999900	103000	1.0000718
24000	0.9999483	64000	0.9999916	104000	1.0000743
25000	0.9999489	65000	0.9999932	105000	1.0000769
26000	0.9999495	66000	0.9999948	106000	1.0000795
27000	0.9999502	67000	0.9999964	107000	1.0000821
28000	0.9999508	68000	0.9999981	108000	1.0000848
29000	0.9999515	69000	0.9999998	109000	1.0000874
30000	0.9999523	70000	1.0000015	110000	1.0000901
31000	0.9999530	71000	1.0000032	111000	1.0000929
32000	0.9999538	72000	1.0000050	112000	1.0000956
33000	0.9999546	73000	1.0000068	113000	1.0000984
34000	0.9999554	74000	1.0000086	114000	1.0001012
35000	0.9999563	75000	1.0000102	115000	1.0001040
36000	0.9999571	76000	1.0000123	116000	1.0001068
37000	0.9999580	77000	1.0000142	117000	1.0001097
38000	0.9999590	78000	1.0000161	118000	1.0001126
39000	0.9999599	79000	1.0000180	119000	1.0001155
40000	0.9999609	80000	1.0000200	120000	1.0001184

Projections

If an average elevation and E' for the project area are being used, these two factors can be multiplied to obtain a single factor, grid factor (GF), which is in turn multiplied by each of the field measured distances—either slope or horizontal.

Example 20.6

Convert the measured distance to grid distance.

horizontal distance = 2,640.00 ft (804.672 m)
average elevation = 1100 ft (335.28 m)
average $E = 276\,500$ m
83 east zone

Solution

$E_o = 250\,000$ m

$$E' = E - E_o$$
$$= 276\,500 \text{ m} - 250\,000 \text{ m}$$
$$= 26\,500 \text{ m}$$

From Table 20.6—east zone (enter with 26 500),
scale factor = 0.9999420

From Eq. 20.26,
scale factor = 0.99993333
$$+ \left((26\,500)^2 (1.231178 \times 10^{-14}) \right)$$
$$= 0.9999420$$

$$\text{sea level factor} = \frac{6\,372\,000 \text{ m}}{6\,372\,000 \text{ m} + 335 \text{ m}}$$
$$= 0.9999474$$

$$\text{grid factor} = (0.9999420)(0.9999474)$$
$$= 0.9998894$$

$$\text{grid distance} = (2640.00 \text{ ft})(0.9998894)$$
$$= 2639.71 \text{ ft}$$

Example 20.7

Convert the horizontal distance to grid distance.

horizontal distance = 2,640.00 ft (804.672 m)
average elevation = 1050 ft (320.04 m)
average latitude = 37°57′
average longitude = 91°46′
83 central zone

Solution

Use Eq. 20.28.

$$E' = (111\,500 \text{ m})(92°30' - 91°46')$$
$$\times \cos 37°57'$$
$$= 64\,477 \text{ m}$$

From the scale factor table—central zone (enter with 64 477 m),

scale factor = 0.9999845

$$\text{sea level factor} = \frac{6\,372\,000 \text{ m}}{6\,372\,000 \text{ m} + 320 \text{ m}}$$
$$= 0.9999498$$

$$\text{grid factor} = (0.9999845)(0.9999498)$$
$$= 9999343$$

$$\text{grid distance} = (2640.00 \text{ ft})(0.9999343)$$
$$= 2639.83 \text{ ft}$$

Example 20.8

Convert the horizontal distance to grid distance.

horizontal distance = 3120.00 ft (950.976 m)
average elevation = 720 ft (219.46 m)
average E = 750 400 m
83 west zone

Solution

$$E_o = 850\,000 \text{ m}$$

$$E' = E - E_o$$
$$= 750\,400 \text{ m} - 850\,000 \text{ m}$$
$$= -99\,600 \text{ m}$$

From the scale factor table—west zone (enter with 99 600),

scale factor = 1.0000633

$$\text{sea level factor} = \frac{6\,372\,000 \text{ m}}{6\,372\,000 \text{ m} + 219 \text{ m}}$$
$$= 0.9999656$$

$$\text{grid factor} = (1.0000633)(0.9999656)$$
$$= 1.0000289$$

$$\text{grid distance} = (3120.00 \text{ ft})(1.0000289)$$
$$= 3120.09 \text{ ft}$$

42. GEODETIC AZIMUTH TO GRID AZIMUTH (MERCATOR)

To convert geodetic azimuth to grid azimuth, Eq. 20.30 is used.

$$\text{grid azimuth} = \text{geodetic azimuth} - \gamma + \text{second term}$$
$$20.30$$

The second term can be neglected for most surveys; therefore, Eq. 20.30 reduces to

$$\text{grid azimuth} = \text{geodetic azimuth} - \gamma \qquad 20.31$$

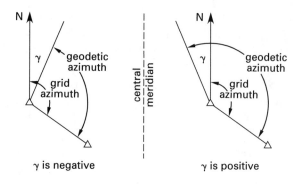

Figure 20.28 *Grid Declination*

Mapping angle (or *grid declination*) is a function of longitude and can be computed from Eq. 20.32.

$$\gamma = (\text{longitude of central meridian}$$
$$- \text{longitude of station}) \sin(\text{latitude of station})$$
$$20.32$$

(Longitude of the central meridian for each zone is given in Table 20.5)

For most work, the longitude of the beginning point of the survey must be accurately known to convert geodetic azimuth to grid. Since this is usually a triangulation station or other point of known coordinates, the longitude is given or can be computed from the coordinates. Scaling longitude from a USGS quad sheet may be sufficient; however, for Missouri a scaling error of 5 in of longitude will result in approximately 3 in of error in γ. Scaling latitude is not as critical but should be within about 10 in.

Example 20.9

Convert the geodetic azimuth to grid azimuth.

geodetic azimuth from
    UMR Stonehenge to point A = 241°12′37″
latitude of UMR Stonehenge = 37°57′22.84227″
longitude of UMR Stonehenge = 91°46′35.54261″
83 central zone

Solution

longitude of
    central meridian= 92°30′

$\gamma = (92°30′ - 91°46′35.54261″)$
        $\times \sin(37°57′22.84227″)$

$\gamma = 0.444972°$

    $= 0°26′42″$

grid azimuth $= 241°12′37″ - 0°26′42″$

            $= 240°45′55″$

43. CORRECTION TO THE γ TERM

The correction to the γ term was known as the *g term correction* in the NAD 27 plane coordinate tables. In Missouri, within a given zone, it will not exceed 0.5 arc-seconds. If significant, however, it can be included using the following revised formula for γ.

$\gamma = \arctan \big(\tan(\text{longitude of central meridian}$

    $- \text{longitude of station}) \sin(\text{latitude of station})\big)$

                                    20.33

44. SECOND TERM

The second term correction is also known as the *arc to chord correction* or *t-τ correction*. It represents the difference between the projected geodetic and grid azimuths. A good approximation of this value in arc-seconds can be obtained using the following formula.

$$t\text{-}\tau = (0.000283)D\cos(\text{az})(\gamma - \gamma_{\text{CM}})(\cos\phi) \quad \textit{20.34}$$

D equals the length of line in meters, ϕ equals the latitude of line midpoint, γ equals the longitude of line midpoint, γ_{CM} equals the longitude of central meridian, and az equals the initial grid azimuth.

In Missouri, this correction is less than 1 arc-second for lines up to 5 km in length and is added algebraically to grid azimuth. For more information on this second term, see the NOAA Manual NOS NGS-5. Because the term is small, field angles can generally be used when computing grid azimuths.

45. PLANE COORDINATES TO GEOGRAPHIC COORDINATES (LATITUDE AND LONGITUDE) AND VICE VERSA (MERCATOR)

These calculations are normally only performed when transferring coordinates from one zone to another. There is no direct conversion; consequently, the coordinates are converted to latitude and longitude using the constants for that zone, then the latitude and longitude are converted to plane coordinates using the constants for the other zone. It is recommended that software be used for these conversions. NOAA Manual NOS NGS-5 can be referred to, or Transverse Mercator tables (available from NGS) can be used for manual calculations.

46. CONVERSION FROM SPCS 27 TO SPCS 83 (MERCATOR)

For all practical purposes there is no accurate and convenient means to convert from NAD 27 to NAD 83. The U.S. Army Corps of Engineers and NGS have software available (CORPSCON) to convert from one system to the other.

47. THE GRID TRAVERSE

A *grid traverse* is a surface traverse that has been transformed to a plane coordinate system. Figure 20.28 shows an example of an open grid traverse. The steps for computation are as follows.

step 1: From an astronomic observation or other method, determine the grid azimuth of the first line.

step 2: Using the measured angles, determine the grid azimuths of all lines of the traverse.

step 3: Compute the sea level and scale factors discussed, using the same method explained earlier.

step 4: Apply these factors to the measured lines to get grid distance.

step 5: Compute latitude and departures of each line using grid azimuths and grid distances.

step 6: Starting with the initial coordinates at point A, compute the coordinates (northings and eastings) of each point.

48. CONVERTING LATITUDE AND LONGITUDE TO STATE PLANE COORDINATES

For every geographical position there is a corresponding set of plane coordinates and vice versa. For any given spheroid the geographical coordinates are fixed but the

Projections

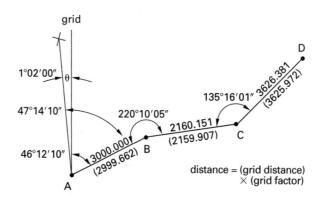

point	grid azimuth	grid distance	latitude	departure	N (m)	E (m)
A					402,012.851	611,415.272
	46°12′10″	2999.662	2076.090	2165.137		
B					404,088.941	613,580.409
	86°22′15″	2159.907	136.719	2155.576		
C					404,225.660	615,735.985
	41°38′16″	3625.972	2709.907	2409.165		
D					406,935.567	618,145.150

Figure 20.29 *Open Grid Traverse*

plane coordinates depend on the particular state or local system used. The mathematical procedure of this transformation is beyond the scope of this text, but software is available to the practicing surveyor for performing these computations. This software is available from the NGS, U.S. Army Corps of Engineers, or from private vendors. An example of output from software conversion is shown in Fig. 20.29.

49. CONVERTING POSITIONS FROM NAD 27 TO NAD 83 AND VICE VERSA

As stated previously, there is no direct mathematical relation between NAD 27 and NAD 83. Any conversion must be done using a data base of known plots. Accuracies can vary as much as 20 cm. The software for making this transformation is known as NADCON and is available from the NGS. CORPSCON, which is software from the U.S. Army Corps of Engineers, also uses NADCON in making this conversion. An example of output from software process is shown in Fig. 20.30.

50. THE HIGH-ACCURACY REFERENCE NETWORK (HARN)

In recent years, the NGS has begun establishing a network of high-accuracy points within various states. These points, primarily located using GPS, have an accuracy of about one in ten million with secondary points of about one in one million. For example, Missouri

has about 26 high-accuracy points with about 79 secondary stations set by the state. Since the densification is low, most states will continue to use NAD 83 as the official standard state coordinate system. An example of converting NAD 27 to NAD 83 (HARN) is shown in Fig. 20.31. Compare these results with those of Fig. 20.30.

51. THE NGS DATA BASE

NGS has placed most of its triangulation stations, azimuth marks, bench marks, and so on, into a single database that is available on CD-ROM. The data is organized by states and counties. A typical printout for an individual station is shown in Fig. 20.32.

52. THE UNIVERSAL TRANSVERSE MERCATOR (UTM) SYSTEM

The *Universal Transverse Mercator System* (UTM) has been designed as a worldwide plane coordinate system. The origin of the X-coordinate is the equator, going both north and south. There are 60 zones of 6° each of longitude beginning with the central meridian at 177 west longitude. The scale factor is the same for all zones and is 0.9996 at the central meridian. UTM differs from NAD 83 in that it uses different ellipsoids depending on what continent it is being used in.

*L,L to SPC NAD 83, meters, State = TXC   WACO   MAG   TEXAS   CENTRAL ZONE
Lat. = 31°34′35.09907″   Long. = 97°08′58.06753″   N = 3216034.451   E = 1002166.953
Convergence = 1°38′23.57″   Scale factor = 0.99993196   Long. = +W

Figure 20.30 *Example of CORPSCON Output*

COORDINATE CHANGES

GEODETIC VERSUS STATE PLANE

Point: WACO MAG 1943

Geodetic Coordinates

Latitude	Longitude	
31°34′34.50400″	97°08′57.06900″	— NAD 27
31 34 35.09907	97 08 58.06753	— NAD 83
+0.59507″	+0.99853″	
+60.13 ft	+86.38 ft	

This corresponds to a position shift of 105.25 ft.

State Plane Coordinates

Northing (Y)	Easting (X)	Map Angle	Scale Factor
708,691.42 ft	2,991,468.36 ft	+1°38′24.1″	0.9999319
3 216 034.451 m	1 002 166.953 m	+1°38′23.6″	0.9999320
(10,551,273.03 ft)*	(3,287,942.74 ft)*		

(*Note: Converted from meters to feet using the
U.S. Survey foot conversion value,
1 m = 3.280833333333... ft.)

Figure 20.31 *Example of NADCON Output*

COORDINATE CHANGES

GEODETIC AND STATE PLANE

Point: WACO MAG

Latitude	Longitude	
31°34′34.50400″	97°08′57.06900″	— NAD 27
31 34 35.08899	97 08 58.08559	— NAD 83
+0.58499″	+0.98989″	
18.018 m	26.806 m	
59.11 ft	87.95 ft	

This corresponds to a position shift of 105.97 ft.

State Plane Coordinates

Texas Central Zone NAD 27/NAD 83

Northing (Y)	Easting (X)	Map Angle	Scale Factor
708,691.42 ft	2,991,468.36 ft	+1°38′24.1″	0.9999319
3 216 034.127 m	1 002 166.486 m	+1°38′23.6″	0.9999320
(10,551,271.96 ft)*	(3,287,941.21 ft)*		
(10,551,293.07 ft)**	(3,287,947.79 ft)**		
(21.10)	(6.58)		

(*Note: Converted from meters to feet using the
U.S. Survey foot 1 m = 3.280833333333... ft.
**Converted from meters to feet using the
international foot 1 m = 3.2808398501... ft.)

Figure 20.32 *Example of NADCON Output (using HARN)*

```
DATABASE = Sybase, PROGRAM = datasheet, VERSION = 5.25
      Retrieval Date = JULY 8, 1997          Version = 5.25

Starting Datasheet Retrieval. . .
1            National Geodetic Survey,      Retrieval Date = JULY 8, 1997
      *******************************************************************************
BZ1087   DESIGNATION   —  WACO MAG
BZ1087   PID           —  BZ1087
BZ1087   STATE/COUNTY  —  TX/MCLENNAN
BZ1087   USGS QUAD     —  WACO WEST (1984)
BZ1087
BZ1087                        *CURRENT SURVEY CONTROL
BZ1087   _____
BZ1087
BZ1087*  NAD 83 (1993)  —  31 34 35.08899 (N)    097 08 58.08559 (W)    ADJUSTED
BZ1087*  NAVD 88        —        142.125 (meters)       466.29 (feet)   ADJUSTED
BZ1087   _____
BZ1087   LAPLACE CORR  —         4.44  (seconds)                        DEFLEC 96
BZ1087   GEOID HEIGHT  —       −26.91  (meters)                         GEOID 96
BZ1087   DYNAMIC HT    —       141.946 (meters)       465.70 (feet)     COMP
BZ1087   MODELED GRAV  —  979,383.4   (mgal)                            NAVD 88
BZ1087
BZ1087   HORZ ORDER    —  SECOND
BZ1087   VERT ORDER    —  FIRST     CLASS II
BZ1087
BZ1087
BZ1087
BZ1087.The horizontal coordinates were established by classical geodetic methods
BZ1087.and adjusted by the National Geodetic Survey in February 1996.
BZ1087
BZ1087.The orthometric height was determined by differential leveling
BZ1087.and adjusted by the National Geodetic Survey in June 1991.
BZ1087
BZ1087.The Laplace correction was computed from DEFLEC96 derived deflections.
BZ1087
BZ1087.The geoid height was determined by GEOID96.
BZ1087
BZ1087.The dynamic height is computed by dividing the NAVD 88
BZ1087.geopotential number by the normal gravity value computed on the
BZ1087.Geodetic Reference System of 1980 (GRS 80) ellipsoid at 45
BZ1087.degrees latitude (G = 980.6199 gals.).
BZ1087
BZ1087.The modeled gravity was interpolated from observed gravity values.
BZ1087
BZ1087;                    North        East       Units    Scale      Converg.
BZ1087;SPC TX C   —  3,216,034.127  1,002,166.486   MT   0.99993196  +1 38 23.6
BZ1087;UTM  14    —  3,494,971.168    675,605.308   MT   0.99998035  +0 58 09.3
BZ1087
BZ1087:              Primary Azimuth Mark                  Grid Az
BZ1087:SPC TX C   —  WACO TEXAS WATER CO NORTH TANK        237 17 45.9
BZ1087:UTM   14   —  WACO TEXAS WATER CO NORTH TANK        237 58 00.2
BZ1087
BZ1087| - - - - - - - - - - - - - - - - - - - - - - - - - - - - - - - - - - |
BZ1087|  PID      Reference Object               Distance      Geod. Az     |
BZ1087|                                                         dddmmss.s    |
BZ1087|  BZ1266  WACO M K T RAILROAD CO STACK    APPROX. 5.0 KM  0805044.8   |
BZ1087|  BZ1088  WACO MAG RM 1                   23.826 METERS   08654       |
```

Figure 20.33 *Sample NGS Data Sheet*

The UTM projection is not used for latitudes above 84° north or below latitudes of 80° south. Rather, the *Universal Polar Stereographic* (UPS) projection is used in the polar regions. The plane coordinates at the poles are 2 000 000.000 N and 2 000 000.000 E. The central meridian is taken as the Greenwich meridian and the international ellipsoid is used. The positive direction of the northings is directed north in the direction of the central meridian for both the north and south poles. Each polar region comprises a single zone.

Grid azimuths and distances are handled in the same way as any transverse Mercator system. Software is available to compute scale factor, geographical positions to plane coordinates, and so on.

Lists of the various reference ellipsoids and UTM zones are shown in Tables 20.8 and 20.9.

Table 20.8 *Common Reference Ellipsoids (Used with Universal Transverse Mercator Projections and Others)*[a]

number	name	year	country	eq. radius (m)	$1/F$
1	Clarke (NAD 27)	1866	USA (old)	6 378 206.4	294.978698214
2	WGS 72	1972	NASA, Dept. Def.	6 378 135.0	298.25972086
3	International (Hayford)	1924	Remainder of the world	6 378 388.0	297.0
4	Everest	1830	India, Burma, Pakistan, etc.	6 377 276.345	300.8017
5	Modified Everest		India, Burma, Pakistan, etc.	6 377 304.1	300.8017
6	Airy	1830	Great Britain	6 377 563.4	299.324965
7	Modified Airy		Great Britain	6 377 340.2	299.324965
8	Bessel	1841	Central Europe, Chile, Indonesia	6 377 397.2	299.152813
9	Krasovsky	1940	Russia	6 378 245.0	298.3
10	Australian National	1965	Australia	6 378 160.0	298.25
11	Clarke	1880	Most of Africa, France	6 378 249.145	293.465
12	GRS 80 (NAD 83)	1980	USA (new)	6 378 137.0	298.25722210
13	WGS 84	1984	World	6 378 137.0	298.25722356

Legend: NAD 27 - North American Datum (1927)
GRS 80 - Geodetic Reference System (1980)
WGS 84 - World Geodetic System (1984)

[a]Incomplete list

Table 20.9 *Universal Transverse Mercator*

UTM Zone numbers and corresponding longitude of central meridians

	western hemisphere				eastern hemisphere			
zone	central meridian	zone	central meridian		zone	central meridian	zone	central meridian
1	177	16	87		31	−03	46	− 93
2	171	17	81		32	−09	47	− 99
3	165	18	75		33	−15	48	−105
4	159	19	69		34	−21	49	−111
5	153	20	63		35	−27	50	−117
6	147	21	57		36	−33	51	−123
7	141	22	51		37	−39	52	−129
8	135	23	45		38	−45	53	−135
9	129	24	39		39	−51	54	−141
10	123	25	33		40	−57	55	−147
11	117	26	27		41	−63	56	−153
12	111	27	21		42	−69	57	−159
13	105	28	15		43	−75	58	−165
14	99	29	09		44	−81	59	−171
15	93	30	03		45	−87	60	−177

Note: Each central meridian is in the center of a 6° wide zone.
Scale factor is 0.9996 at each central meridian.

PRACTICE PROBLEMS

1. The axis of the earth is tilted to the plane of its orbit by which of the following?

(A) $18\frac{1}{2}°$

(B) $23\frac{1}{2}°$

(C) $27\frac{1}{2}°$

(D) $28\frac{1}{2}°$

2. The earth is a spheroid; its equatorial diameter is greater than its polar diameter by

(A) 27 mi

(B) 37 mi

(C) 105 mi

(D) 127 mi

3. Mercator's map projections, in theory, use a

(A) plane

(B) cone

(C) cylinder

(D) ellipse

4. Lambert's map projections, in theory, use a

(A) cone

(B) cylinder

(C) plane

(D) ellipse

5. Lines of latitude

(A) measure the arc distance from the equator to the north and south poles in degrees

(B) measure the distance east and west around the earth at right angles to the equator

6. Lines of longitude

(A) measure the arc distance from the equator to the north and south poles in degrees

(B) measure the distance east and west around the earth at right angles to the equator

7. Properties of maps include

(A) shape

(B) area

(C) distance

(D) direction

(E) all of the above

8. A conformal map is one in which

(A) the shape of a small area on the map conforms to the same shape on the earth

(B) the meridians converge toward the poles

(C) the meridians are parallel

(D) parallels of latitude are straight lines

9. On Mercator's maps projected on a cylinder,

(A) meridians converge

(B) parallels are arcs of circles

(C) meridians and parallels are at right angles

(D) parallels are uniformly spaced

10. The Lambert conic two-parallel projection

(A) uses two cones

(B) uses one cone to pierce the earth at two parallels

(C) has parallel lines of longitude

(D) includes all of the above

11. Maps that use a plane parallel to the equator tangent at the north pole for projection

(A) show parallels as concentric circles

(B) show parallels equally spaced

(C) show meridians converging at the poles

(D) show all of the above

12. In order for distortion to be no more than one part in ten thousand on the Lambert projection, limiting parallels may not be more than

(A) 108 mi apart

(B) 158 mi apart

(C) 168 mi apart

(D) 268 mi apart

13. The mapping angle θ, Lambert projection, varies with

(A) longitude

(B) latitude

(C) latitude and longitude

(D) all of the above

14. The scale factor, Lambert projection, varies with

(A) longitude

(B) latitude

(C) latitude and longitude

(D) all of the above

15. Greatest accuracy is obtained when the distance between the standard parallels is

(A) two-thirds the distance between the limiting parallels

(B) one-third the distance between the north and south limits of the map

(C) two-thirds the distance between the standard parallels and the limiting parallel

16. Determine the mapping angle θ at longitude $95°52'18.352''$, central zone, Texas.

17. Determine the grid azimuth from triangulation station No. 2, north central zone, Texas, to station A from the following information.

$$\text{station no. 2: latitude} = 33°42'22.770''$$
$$\text{longitude} = 101°31'14.604''$$
$$\text{azimuth to azimuth mark} = 304°30'12''$$
$$\text{angle right to A} = 112°40'22''$$

18. Using the combined scale factor, determine the grid distance between two points on the surface that are 2156.35 ft apart at average latitude $31°42'$, central zone, Texas, and average elevation of 650 ft.

19. Determine the x- and y-coordinates of a station at latitude $31°36'31.177''$, longitude $97°13'22.471''$, central zone, Texas.

20. Determine the latitude and longitude of a station at central zone, Texas, $x = 1,151,277.27$, $y = 697,447.02$.

21. Using Horizontal Control Data for triangulation station PRICE and field measurements shown as follows for the traverse ABCDEA, compute the coordinates of A, B, C, D, and E and the area of ABCDEA. Average elevation of the traverse is 460 ft.

point	angle right	surface distance (ft)
az mark PRICE	232°02'18''	
		485.22
A	272°13'44''	
		560.10
B	95°18'22''	
		484.18
C	65°13'08''	
		375.51
D	216°19'30''	
		311.54
E	67°05'20''	
		449.83
A	96°03'40''	

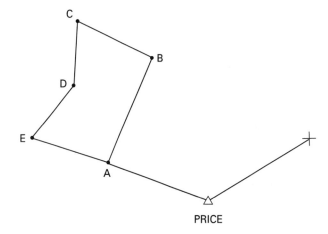

SOLUTIONS

1. *The answer is B.*

2. *The answer is A.*

3. *The answer is C.*

4. *The answer is A.*

5. *The answer is A.*

6. *The answer is B.*

7. *The answer is E.*

8. *The answer is A.*

9. *The answer is C.*

10. *The answer is B.*

11. *The answer is D.*

12. *The answer is B.*

13. *The answer is A.*

14. *The answer is B.*

15. *The answer is A.*

16.

$$\theta(96°52') = +1°47'07.9349''$$
$$(18.352'')(0.51505889) = \underline{9.4524''}$$
$$\theta(96°52'18.352'' = \boxed{+1°46'58.4825''}$$

17.

$$\theta(101°31') = -2°11'26.4032''$$
$$(14.604'')(0.54539441) = \underline{7.9649''}$$
$$\theta(\text{sta no. 2}) = -2°11'34''$$

geodetic az (N) to mk $= 124°30'12''$
angle right to A $= \underline{112°40'22''}$
geodetic az to A $= 237°10'34''$
$\theta = \underline{2°11'34''}$
grid az to A $= 239°22'08''$

18. grid distance

$$= \frac{(20{,}906{,}000 \text{ m})(0.9999559)(2156.35 \text{ ft})}{20{,}906{,}650 \text{ m}}$$

$$= \boxed{2156.19 \text{ ft}}$$

19. From Sec. A of App. I,

$$C = 2{,}000{,}000.00 \text{ m}$$
$$R_b = 35{,}337{,}121.23 \text{ m}$$
$$\ell = 0.5150588857$$
$$R(31°36') = 34{,}633{,}982.12 \text{ m}$$
$$(31.177'')(101.04000) = \underline{3{,}150.12 \text{ m}}$$
$$R(\text{sta D}) = 34{,}630{,}832.00 \text{ m}$$

$$\theta(97°13') = +1°36'18.9607''$$
$$(22.471'')(0.51505889) = \underline{11.5739''}$$
$$\theta(\text{sta D}) = +1°36'07.3868''$$

$$X = R\sin\theta + C$$
$$= (34{,}630{,}832.00 \text{ m})\sin(1°36'07.3868'')$$
$$\quad + 2{,}000{,}000 \text{ m}$$
$$= 2{,}968{,}189.30 \text{ m}$$
$$Y = R_b - R\cos\theta$$
$$= 35{,}337{,}121.23 \text{ m}$$
$$\quad - (34{,}630{,}832.00 \text{ m})\cos(1°36'07.3868'')$$
$$= 719{,}825.92 \text{ m}$$

The coordinates are

$$\boxed{x = 2{,}968{,}189.30 \text{ m}}$$

$$\boxed{y = 719{,}825.92 \text{ m}}$$

20. From App. I,

$$C = 2{,}000{,}000.00 \text{ m}$$

$$\text{central meridian} = 100°20'00.00''$$

$$R_\mathrm{D} = 35{,}337{,}121.23 \text{ m}$$

$$\ell = 0.5150588857$$

$$\tan\theta(\text{sta B}) = \frac{x - c}{R_b - y}$$

$$= \frac{1{,}151{,}277.27 \text{ m} - 2{,}000{,}000.00 \text{ m}}{35{,}337{,}121.23 \text{ m} - 697{,}447.02 \text{ m}}$$

$$= -0.024501464 \text{ m}$$

$$\theta(\text{sta B}) = -1.403549662°$$

$$= -1°24'12.7789''$$

$$R(\text{sta B}) = \frac{R_b - y}{\cos\theta}$$

$$= \frac{35{,}337{,}121.23 \text{ m} - 697{,}447.02 \text{ m}}{\cos\theta}$$

$$= 34{,}650{,}070.13 \text{ m}$$

$$R(31°33') = 34{,}652{,}169.12 \text{ m}$$

$$R(31°33') - R(\text{sta B}) = 34{,}652{,}169.12 \text{ m}$$
$$- 34{,}650{,}070.13 \text{ m}$$
$$= 2098.99 \text{ m}$$

$$\frac{2098.99 \text{ m}}{101.03833} = 20.774''$$

$$\phi = 31°33'00.000'' + 20.774''$$
$$= 31°33'20.774''$$

$$\Delta\lambda = \frac{\theta}{\ell} = \frac{-1.403549662}{0.5150588857}$$
$$= -2°43'30.099''$$

$$\lambda = 100°20'00.000''$$
$$- (-2°43'30.099'')$$
$$= 103°03'30.099''$$

$$\boxed{\text{latitude} = 31°33'20.774''}$$

$$\boxed{\text{longitude} = 103°03'30.099''}$$

21. combined scale factor $= \left(\dfrac{20{,}906{,}000 \text{ m}}{20{,}906{,}000 \text{ m} + 460 \text{ m}}\right)$
$$\times (0.9999329)$$
$$= 0.9999109$$

Field measurements and computations for coordinates are as shown.

point	right angle	grid az	surface distance (ft)	grid distance (ft)	latitude (ft)	departure (ft)
az mark		242°35'46''(S) 62°35'46''				
PRICE	232°02'18''					
		294°38'04''	485.22	485.18	+202.24	−441.02
A	272°13'44''					
		26°51'48''	560.10	560.05	+499.61	+253.07
B	95°18'22''					
		302°10'10''	484.18	484.14	+257.77	−409.81
C	65°13'08''					
		187°23'18''	375.51	375.48	−372.36	−48.28
D	216°19'30''					
		223°42'48''	311.54	311.51	−225.16	−215.27
E	67°05'20''					
		110°48'08''	449.83	449.79	−259.74	+420.47
A	96°03'40''				−12	0.18

$$\text{ratio of error} = \frac{0.22 \text{ ft}}{2186 \text{ ft}} = 1{:}9900$$

point	balanced latitude	balanced departure	state plane coordinates y	x
PRICE			711,233.43	3,014,994.39
	+202.24	−441.02		
A			711,435.67	3,014,553.37
	+499.58	+253.03		
B			711,935.25	3,014,806.40
	+257.75	−409.84		
C			712,192.99	3,014,396.56
	−372.38	−48.31		
D			711,820.61	3,014,348.25
	−225.18	−215.30		
E			711,595.43	3,014,132.95
	−159.76	+420.42		
A			711,435.67	3,014,553.37

Area computations:

In computing the area of the traverse, subtract 711,000 from each y-coordinate and 3,014,000 from each x-coordinate.

point	coordinates y	x		area
A	435.67	553.37	(553.37)(595.43 − 935.25)	−188,046
B	935.25	806.40	(806.40)(435.67 − 1192.99)	−610,703
C	1192.99	396.56	(396.56)(935.35 − 820.61)	45,462
D	820.61	348.25	(348.25)(1192.99 − 595.43)	208,100
E	595.43	132.95	(132.95)(820.61 − 435.67)	51,178
				−494,009

$$\text{sea level area} = \frac{494,009 \text{ ft}^2}{2}$$

$$= 247,005 \text{ ft}^2 \quad (5.6705 \text{ ac})$$

$$\text{surface area} = \frac{247,005 \text{ ft}^2}{(0.9999109)^2}$$

$$= 247,049 \text{ ft}^2 \quad (5.6715 \text{ ac})$$

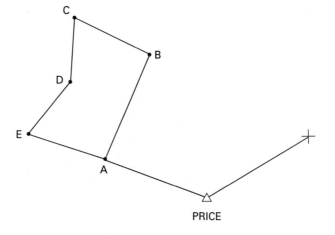

21 Property Law

Law

Part 1: Colonization History

1. ENGLISH COMMON LAW

English common law consists of those ideas of right and wrong determined by court decisions over many centuries. Such ideas have been accepted by generations trying to establish rules to meet social and economic needs. Sir William Blackstone, eighteenth century author of *Commentaries on the Law of England*, called it "unwritten law" in the sense that it was not enacted by a legislative body. It was legal custom expressed by the decisions of judges.

English common law was evolutionary. It changed slowly. Judges made decisions based on former decisions, but they also modified their decisions to reflect changing times.

Some of the important parts of our present law that came from English common law are the grand jury, trial by jury, freedom of press, habeas corpus, and oral testimony.

2. STATUTE LAW

Laws enacted by legislative bodies are known as *statute laws*. In contrast with common law, it is written law. Laws of France, Germany, Spain, and other countries in the continent of Europe are largely statute laws. After the French Revolution, France adopted the Code Napoleon to clarify its laws.

3. COLONIAL LAW

Colonial law was not evolutionary because there was nothing on which to base precedence. Therefore, settlers of the original thirteen colonies adopted English common law. But they did not adopt it in its entirety. It did not fit their new social economic environment entirely. Many of their laws were statutory (written).

Furthermore, the same parts of English common law were not adopted in each of the colonies. The parts of the law that seemed to fit the needs of the particular situation in each colony were the parts that were adopted.

4. SPAIN AND FRANCE IN THE NEW WORLD

Spain acquired title to land in the New World by grant from Pope Alexander VI in 1493. The grant conveyed all lands not held by a Christian prince on Christmas day of 1492 from a meridian known as the *Line of Demarcation*, 100 leagues west of the Azores and Cape Verde Islands. Later, a treaty with Portugal moved the Line of Demarcation 270 leagues west, with Portugal to have rights to the east and Spain to the west. Possession came from conquest, and by 1600, the territory extended from New Mexico and Florida on the North to Chile and Argentina on the south. The first seat of government was at Santa Domingo.

In 1511, Don Diego Velasquez led an expedition for the conquest of Cuba that was accomplished without serious opposition. Velasquez was appointed Governor of the island. During his rule, he promoted settlement of the land by Spaniards.

Hernandez de Cordova set out in 1517 from Cuba on an expedition to the Bahama Islands to obtain Indian slaves, but a storm drove him off his course. Three weeks later, he landed on the coast of Yucatan. Cordova returned to Cuba with tales of a more advanced civilization than previously found in the New World. And of great interest to Velasquez, he brought tales of gold and fine cotton garments. In 1518 Velasquez sent his nephew, Juan de Grijalva, on an expedition to explore the coast of Mexico. Grijalva also landed on Yucatan and was impressed by the advanced civilization. He returned to Cuba with many gold ornaments he had received in trade.

After approval from Spain, Velasquez decided on the conquest and colonization of the new land. He chose Hernando Cortes as the commander of his expedition.

After many skirmishes and battles, Cortes reached the capital of the Aztecs, Tenochtitlan, now Mexico City, on November 8, 1519. He quickly conquered the forces of Montezuma, the Aztec emperor, and placed him under house arrest where he was treated very cordially by Cortes and allowed to retain his many luxuries.

After the conquest of the new land, Spanish statute law was introduced. Legislation for this new land was codified in *Recopilacion de las Leyes de Indias* in 1680. The Crown of Spain had complete authority that was administered through the Minister of the Indies, and all the land belonged to the king of Spain.

Spain ruled Mexico for 300 years. However, in 1821, the people of Mexico revolted and declared their independence from Spain. Augustin de Iturbide, leader of the revolt, was crowned as Augustine I, Emperor of Mexico in 1822. In 1824, he was deposed by Lopez de Santa Anna who established a constitutional government.

France established Quebec in 1608. From there, settlement moved south along the Mississippi to its mouth. René Robert Cavalier, Sieur de la Salle, claimed all the Mississippi Valley for France and named it Louisiana in honor of Louis XIV.

Napoleon, in 1803, sold all of Louisiana to the United States, "with the same extent that it now has in the hands of Spain, and that it had when France possessed it." President Thomas Jefferson claimed Texas as part of the purchase but Spain protested vigorously. The boundary between Texas and the United States was established by treaty between the king of Spain and the United States in 1819.

An act barring emigration from the United States and other restrictive land laws caused unrest in Texas, which resulted in Texas winning its independence from Mexico. In 1836, the Republic of Texas was established.

Part 2: A Brief History of Property Law

5. EARLY HISTORY OF PROPERTY LAW

The earliest record of property ownership goes back to the Babylonians in 2500 B.C. The Bible also furnishes many references to property ownership. Numerous references are found in the book of Genesis, including a passage relating to the purchase of land by Abraham on which to bury his wife, Sarah.

In the book of Jeremiah is "Thou shall not remove thy neighbor's landmark which they of old times have set in thine inheritance which thou shall inherit in the land that the Lord thy God giveth thee to possess it." Also in the book of Jeremiah is "Cursed be to he that removeth his neighbor's landmark and all the people shall say Amen."

Historical records show that in about 1400 B.C. the king of Egypt divided the land into squares of equal size and gave each Egyptian one square. The king in turn levied taxes on each person. This land was in the fertile Nile valley where the river overflowed and destroyed parts of these plots. The owner of the destroyed land was required to report his loss to the king and request a reduction of taxes. The need to determine the actual losses of property resulted in the beginning of surveying in this part of the world. These early surveyors are referred to as *rope stretchers* in the Bible. Drawings on the walls of tombs show these rope stretchers accompanied by officials who recorded the measurements.

The Greeks and Romans also recognized individual property ownership and the Romans used taxes on land to support the cost of government.

At the beginning of the Christian era in Europe, the ownership of land was usually determined by conquest. After conquest, the ruler took over all the land and earlier land titles were extinguished. Often, sovereign rights were vested in the ruler by the Pope, but in non-Christian lands, the practice of the ruler having rights to all the land was much the same. The land belonged to the Sovereign, often referred to as the *Crown*.

6. FEUDAL SYSTEM

In the 11th century Great Britain was conquered by William the Conquerer (William I). He claimed all the land and ruled over all the people. Later, he introduced the feudal system, which was the social and political system in both Great Britain and Europe during the 11th, 12th, and 13th centuries. The feudal system was the basis of real property law in medieval times. Many of its principles found their way into American law.

In England, the system was an arrangement between the king, noblemen, and vassals. The arrangement included an intricate set of rules for the tenure and transfer of real property. The noblemen ruled over the tenants (*vassals*) of the land. These landlords protected their vassals but expected them to pay rent on the land they used and to pay allegiance to their lord. Often, they were required to fight for their lord. The land was in possession of the lord and could not be sold. It passed to the eldest son by inheritance. Vassals had no chance to own property.

7. COMMON LANDS

Almost all the peasant class at the time of William the Conqueror was engaged in farming. Tenant farmers acquired strips of land from the lord for row crops and for producing hay for the cattle. While the crops were growing, these farmers needed pasture for their cattle, and they acquired from their lord *right in common* on land used for permanent pasture. These lands eventually became known as *commons*, and the word was brought to this country by early settlers. In the United States, "commons" means a park.

8. DOMESDAY BOOK

In 1086, William the Conqueror made a survey for tax purposes that included every farm, every farm owner, and all common land. This was known as the *Domesday Book*. Besides being useful for collecting taxes, it allowed him to demand allegiance from everyone in his kingdom. The Domesday Book was a complete record for doomsday, the modern spelling.

A modern edition of the Domesday Book, which lists these common lands, has recently been published. Over the centuries, they have become private property subject to certain rights by claimants to rights in common. Parliament, with the new book, is trying to pin down just what may be done to this land (which comprises 4% of England).

9. TREND TO PRIVATE OWNERSHIP

As people turned from agriculture to crafts and trade in villages and cities, they curbed the power of kings and began to think of ownership of land without fealty to the lord, without obligations of service, and with the right to dispose of the land as they saw fit. This is known as *fee ownership*.

Early grants of the kings were not recorded. The grantee received a packet of papers as evidence of his ownership. These documents became so voluminous that loss of them became common. Parliament began to take steps to correct the confusion caused by lost or stolen documents. Laws were established that abolished the practice of passing title from father to eldest son, and steps were taken to better describe the property transferred.

10. MAGNA CARTA

In 1215, powerful English noblemen forced the king of England to sign the *Magna Carta*, a document that forced the king to share authority with the nobles. By 1700, Parliament had gained supremacy over the king.

11. STATUTE OF FRAUDS

In 1677, the English Parliament passed the *Statute of Frauds*, which, among other things, prohibited any transfer of land or any transfer of interest in land by oral agreement. All conveyances were required to be in writing, a requirement that is very much a part of our present-day law.

12. PROPERTY LAW IN THE UNITED STATES

Ideas of property ownership in existence in England at the time were brought to this country by English settlers. Lands in America were granted to settlers by kings and queens of England. The idea of ownership by the sovereign still prevailed, and some tribute to the crown was still required.

The settlers themselves still had the idea of ownership by the conqueror as they displaced the Native American Indians from their home and lands with little compensation to them.

Because land was cheap in the early days of our country, exact descriptions and exact locations were not necessary. Some grants to the original colonists extended from the Atlantic to the Pacific.

13. *STARE DECISIS* (PRECEDENT)

The doctrine of *stare decisis* established the principle that when a court hands down a decision regarding certain facts, it will adhere to that decision in deciding all future cases where the facts are substantially the same.

14. TYPES OF PROPERTY

Property is divided into two classes: real property and personal property. *Real property* is immovable and can

be recovered. It has been defined as "the interest that a man has in lands, tenements, or hereditaments." These terms include land, buildings, trees, and the right to use them. Anything that grows on the land or any structure that is fixed to the land is real property.

Real property law is, for the most part, state law rather than federal law. It, therefore, varies among the states.

Personal property is movable and often cannot be recovered. Action to recover such things as money and valuable goods is often taken against the person who removed them illegally.

Part 3: Title

15. DEFINITION

Title is the right to own real property and the evidence of that right. Right to ownership is not enough, however. There must also be possession of property. Title, then, is the outward evidence of the right to ownership.

16. CLEAR TITLE, GOOD TITLE, MERCHANTABLE TITLE

The terms *clear title, good title,* and *merchantable title* are essentially synonymous. Clear title means the property is free from encumbrances. Good title is a title free from litigation.

17. RECORD TITLE

A title entered on the public records is referred to as a *record title*.

18. COLOR OF TITLE

Any written instrument, such as a forgery, that appears to convey title but in fact does not, establishes *color of title*. A consecutive chain of transfers of title down to the person in possession in which one or more of the written instruments is not registered may also establish color of title.

19. CLOUD ON TITLE

A claim on land that would, if valid, impair the title to the land creates a *cloud on title*. The claim may be any encumbrance such as a lien, judgment, tax-levy, mortgage, or conveyance.

20. CHAIN OF TITLE

The change in ownership of a piece of property in sequence is known as the *chain of title*. Any defective conveyance of title in the chain adversely affects the title from that point on.

21. ABSTRACT OF TITLE

Before buying real property, a buyer should institute a search of title—a review of all documents affecting the ownership of the property to determine if the person selling the property has a good and clear title. A compilation of abstracts of deeds, deeds of trust, or any other estate or interest, together with all liens or liabilities that affect the title to the property, may be obtained from an abstract or title company. This condensed history of the title to the land in chronological order is known as an *abstract of title* or simply an *abstract*.

22. ATTORNEYS' OPINION

After attorneys secure an abstract of title, they examine the various transfers of title and write an opinion for their client as to whether they think the grantor has a good and clear title to the property. The attorneys do not guarantee the title; they merely state their opinion from the facts shown in the abstract. They cannot guarantee that there has not been fraud or forgery.

Attorneys may point out errors that were made in the execution of conveyances but that, in their opinion, will not affect the title. For instance, a deed was dated October 5, 1938, and the acknowledgment was dated October 4, 1938. The attorneys' opinion was that the instrument had been recorded for more than ten years and the acknowledgment was cured by limitation. They further stated that, if necessary, an affidavit could be secured from the notary public involved to the effect that a stenographic error had been made in dating the acknowledgment.

23. AFFIDAVIT

An *affidavit* is a statement made under oath in the presence of a notary public or other authorized person. In the case of the misdated acknowledgment mentioned in Sec. 22, the notary public made a sworn statement, in the presence of another notary public, to the effect that the acknowledgment was actually made after the grantor had signed the instrument.

24. TITLE INSURANCE POLICY

In recent years, the practice of preparing an abstract of title and the practice of submitting an opinion on the title has been replaced by the issuance of a *title insurance policy*, often referred to as a *title policy*. Title insurance policies assure purchasers of real property that they have good title to the land they have purchased. These policies are issued by title abstract companies operating under the insurance laws of the state.

The amount the assureds are guaranteed is usually limited to the amount they are paying for the property. If the property is enhanced in later years, the assureds would receive no extra compensation for the increased value of the property if their title were defeated.

The policy is usually issued subject to certain exceptions: taxes, easements, encumbrances, oil royalties, and so on.

25. HOMESTEAD RIGHTS

Many of the colonists in Stephen F. Austin's colony had left the United States because of a financial crisis of the time. Austin knew they needed time to establish themselves in Texas, and he appealed to the legislature of Coahuila-Texas for legislation to protect them from property seizure for old debts in the United States. A bill was passed in 1829 that exempted, without limitation, lands acquired by virtue of a colonization law from seizure for debts incurred before the acquisition of the land. The law had some precedence in the laws of Spain in the 15th century.

In 1839, the Third Congress of the Republic of Texas passed a law that read as follows:

> Be it enacted... that from and after the passage of this act, there shall be reserved to every citizen or head of a family in this Republic, free and independent of the power of a writ of fire facies, or other execution issuing from any court of competent jurisdiction whatever, fifty acres of land or one town lot, including his or her homestead, and improvements not exceeding five hundred dollars in value, all household and kitchen furniture (provided it does not exceed in value two hundred dollars), all implements of the husbandry (provided they shall not exceed fifty dollars in value), all tools, apparatus and books belonging to the trade or profession of any citizen, five milch cows, one yoke of work oxen or one horse, twenty hogs, and one year's provisions...

This was the first law of its kind. It has since been adopted by most of the states, with realistic revisions in the protected quantities.

Part 4: Transfer of Ownership of Real Property

26. CONVEYANCE

A *conveyance* is a written instrument that transfers ownership of property. It includes any instrument that affects the ownership of property. The term not only refers to a written document, but also means a method of transfer of property.

Changes in title law in early U.S. history were accompanied by changes in methods of conveying land. Deeds had to be in writing, but they were shortened and simplified. No longer was it necessary that deeds be written by lawyers learned in English law.

27. ESTATE

An *estate* in real property is an interest in real property. It can be complete and inclusive without limit or duration; it can be partial and of limited duration; it can be for the life of one person or for the life of several; it can include surface and all minerals below, or surface and no minerals below, or minerals but not the surface. It can be acquired in many ways: by purchase, by inheritance, by power of the state, or by gift.

Several people may hold an interest in the same property. Consider a person who buys a house with a mortgage and then leases the house to someone else. The lessee, the owner, the mortgagee, and various taxing agencies have an interest in the house.

28. FEE

The word *fee* comes from the feudal era and refers to an estate in land. The true meaning of the word is the same as that of "feud" or "fief." Under the feudal system, a freehold estate in lands came from a superior lord as a reward for services and on the condition that services would be rendered in the future. A fee and a freehold estate are the same.

29. FEE TAIL

Under the feudal system, a fee or freehold estate was passed on to the eldest son on the death of the fee holder. An estate in which there is a fixed line of heirs to inherit the estate is known as a *fee tail*.

30. ESTATE IN FEE SIMPLE ABSOLUTE

In U.S. law, an *estate in fee simple absolute* (also called an *estate in fee simple*) is the highest type of interest. It is an estate limited absolutely to a person and the heirs, and assigns forever without limitation. In other words, a person who owns a parcel of land "in fee" can hold it, sell it, or divide it without limitations.

31. DEED

The most important document in the transfer of ownership of real property is the *deed*, which is evidence in writing of the transfer of an estate. A deed is a formal document. It needs not only to be in writing but also to be written by a person versed in the law. Deeds are of two principle types: warranty deeds and quitclaim deeds.

Warranty Deeds

In a *warranty deed*, the grantors proclaim that they are the lawful owners of the real estate and bind themselves, bind their heirs, and assigns to warrant and forever defend the property unto the grantees and their heirs, and assigns against every person who lawfully claims it or any part of it. The warranty deed is the instrument used to convey an estate in fee simple absolute (in fee).

Quitclaim Deeds

The *quitclaim deed* passes on to the grantees whatever interest the grantors have. If the grantors have a complete title, they pass on a complete title. If their title is incomplete, they pass on whatever interest they have.

32. ESSENTIALS OF A DEED

Because a deed is evidence of the transfer of an estate, the evidence must be clear and concise. The wording of the deed must clearly state the intent of the parties involved in the transfer. It is not sufficient that the grantors and grantees understand the terms of the transfer. In order to protect the rights of the real property owners and to establish an orderly method of transfer of real property, state legislatures and courts have adopted requirements for conveyance of such interest.

- A deed must be in writing. As previously mentioned, this requirement originated in the *statute of frauds* and now is found in the statutes of all states.

- A deed must be in legal terminology.

- Parties to a deed must be competent. A person of unsound mind or a minor cannot execute a deed.

- There must be a grantor and a grantee, and they must be clearly identified.

- There must be a valid consideration, although the total amount of the consideration need not be shown. Deeds containing the phrase "ten dollars and other consideration" provide evidence that the grantor received remuneration for the property.

- A deed must contain a description of the property being conveyed and clearly show the interest conveyed.

- A deed must be signed. In the case of joint ownership by husband and wife, both must sign.

- A deed must be acknowledged. The signer or signers of the deed must sign in the presence of a registered notary public who must know the identity of the signer or signers. The notary must sign the acknowledgment and affix a seal to it.

- A deed must be delivered. Centuries ago, land was conveyed by a ceremony known as *livery of seisin*. Parties to the transfer of ownership met on the property to be conveyed and performed such acts as handing over twigs and soil, driving stakes in the ground and shouting. The ceremony was practiced in England as late as 1845. Today, delivery of the deed is considered to be the delivery of the property.

33. RECORDING DEEDS

It is important that deeds be recorded in order to constitute notice to the public. Unrecorded deeds may be valid, but to avoid future controversy, deeds should be recorded as soon after execution as possible. It is not necessary for the grantor to actually carry the deed to the grantee.

34. PATENT

A *patent* is a conveyance or deed from the sovereign for the sovereign's interest in a tract of land. Most, but not all, land in the United States was patented by the United States. The original thirteen colonies received grants from the king of England. Owners of land in Texas have received patents (grants) from the king of Spain, the Republic of Mexico, the Republic of Texas, and the State of Texas. The lands of Texas have never come under the ownership of the United States, and no patents have been conveyed from that source.

35. WILL

A *will* is a declaration of a person's wishes for the distribution of his or her property after death. These wishes are carried out by a probate court. A *devise* transfers real property, whereas a *bequest* transfers personal property. The devisee is the person receiving the real property. The *probate court* will distribute the property according to the wishes of the *testator* (the deceased) if a will exists. If no will exists, the court will distribute the property in accordance with the law of descent and distribution. Widows, widowers, and children come first in this succession.

A will may devise certain property to certain individuals, or it may devise an entire estate to several heirs. In the latter case, the heirs will own the undivided property jointly.

Before property can be transferred under the terms of the will, the heirs must submit the will to a probate court or a county court that has probate jurisdiction. If the will designates an *executor*, the court will recognize him or her. If no will exists, the court will appoint an administrator. Heirs of an estate must then file an inventory of the property of the estate. Public notice must be given to creditors of the estate and these claims must be paid, if valid. State and federal taxes must also be paid before final settlement of the estate.

In Louisiana, the *forced heirship law*, based on the Napoleonic Civil Code, decrees that children are entitled to a testator's property regardless of the terms of the will. Under this law, children are entitled to one-fourth of the estate. If there are no children, parents are entitled to this one-fourth.

36. HOLOGRAPHIC WILL

A *holographic will* is a will in the handwriting of the deceased.

37. EASEMENT

An *easement* is the right that the public or an individual has in the lands of another. An easement does not give the grantee a right to the land—only a right to use the land for a specified purpose. The owner of the land may also use it for any purpose that does not interfere with the specified use by the grantee.

Utilities wishing to install power lines, underground pipe, canals, drainage ditches, and so on, sometimes do not require fee title to land but need only the use of the land to install and maintain the facility. The owner of the land retains title to it, subject to the terms of the easement.

38. LEASE

A *lease* must be for a certain term, and there must be a consideration. It is a contract for exclusive possession of lands or tenements though use may be restricted by reservations. The person who conveys is known as the *lessor* and the person to whom the property is conveyed is known as the *lessee*. Both parties must be named in the lease.

In many cases a tenant holds real estate without a lease, paying rent each week, month, or year. This is known as *tenancy without lease*.

In general, whatever buildings or improvements stand upon the land and whatever grows upon the land belongs to the landlord. Under a lease, the tenant is entitled to the crops of annual planting.

39. SHARECROPPER'S LEASE

A lease of farmland wherein the landlord and tenant each receive a predetermined share of the total income from crops on the land is known as a *sharecropper's lease*. The share to each is usually determined by custom in certain areas, but it can be set at any figure by agreement between the two.

40. OIL LEASE

An oil company or private individual may enter into an agreement with a land owner to remove oil, gas, or other minerals from the land and to share the profits from the sale of the minerals with the landowner. This is known as a *mineral lease* or *oil lease*. The shares are usually set by the company that removes the minerals, and this share has been accepted by custom.

An oil lease is for a definite number of years (often five), stipulating rental on a per-acre, per-year basis. In addition to the yearly rental, the agreement usually includes a bonus paid by the oil company at the beginning of the lease period. This also is usually on a per-acre basis. If drilling has not commenced by the end of a specified period, the lease expires.

41. MORTGAGE

A *mortgage* is a conditional conveyance of an estate as a pledge for the security of a debt. People borrowing money to purchase property guarantee that they will repay the lender by making a conditional conveyance to the lender. If they repay the loan as specified, the mortgage becomes null and void. If they do not pay the loan as specified (a *default*), they must deliver the property to the lender.

42. DEED OF TRUST

A *deed of trust* is a mortgage that gives the creditor the right to sell property, in case of default, through a third person known as the *trustee*. Early American law regarding mortgages included a complicated system of equitable foreclosure to give the debtor protection. It included the debtor's "equity," which gave him the right to redeem his land after it had been foreclosed on. This "equity" created difficulties for the lender, and in time, laws in many states were modified so that if the debtor agrees in advance, the creditor can sell the property through a third person, known as the *trustee*, without going through court, in case of default.

43. CONTRACT OF SALE

Often the sale of a large estate involves many complexities that are time consuming for the parties involved and their attorneys. A thorough examination of the complexities of the transaction in advance can save time that otherwise might be spent in court settling a dispute.

In order that buyers may express their intent to buy and sellers may express their intent to sell, the two parties may enter into a *contract of sale*, which describes the property involved and the terms of the sale and specifies a date, not later than which the transfer of property must be completed. This contract usually stipulates that the sellers will furnish a good and merchantable title to the buyers by a warranty deed, and that if the sellers cannot furnish such a deed, the contract is null and void. The contract also often provides for an *escrow fund*, which the buyers will forfeit if they do not carry out the terms of the contract.

In some instances, real property is sold by contract of sale and all payments are made before the deed is executed. Throughout the period of the payments, the title remains in the name of the sellers, and their names appear on the tax roll as owners of the property. Contracts of sale are frequently not recorded.

44. UNWRITTEN TRANSFERS OF LAND OWNERSHIP

The statute of frauds requires that all conveyances of real property be in writing. But, if that statute would deprive the rightful owners of their property, then the law may be set aside and an unwritten transfer of real property may take place.

This transfer may take place by expressed or implied agreement such as by the principle of recognition and acquiescence over a long period of time, by dedication, by adverse (hostile) relationships, or by acts of nature.

A legal unwritten transfer of title supersedes written title and will extinguish written title. Evidence to prove the location of a written title will not overturn a legal unwritten title.

45. RECOGNITION AND ACQUIESCENCE

Acquiescence in a boundary line is evidence from which it may be inferred that the parties by agreement established a line as the true line. From such acquiescence, a jury or court may find that the line used is the true line. Acquiescence in a line other than the true line will not support a finding of an agreement establishing the line as the boundary when there is no evidence of agreement other than acquiescence and where it is shown that the use of the line resulted, not from agreement, but only from a mistaken belief of the parties that it was the true line.

46. DEDICATION

Dedication is the giving of land or rights in land to the public. It must be given voluntarily, either expressed or implied. It may be written or unwritten, but there must be acceptance of the dedication. A consideration is not necessary.

Common law dedication may be expressed, as when the intention to dedicate is expressed by a written document or by an act that makes the intent obvious. It may be implied, as when some act or acts of the donor make it reasonable to infer that he or she intended to dedicate.

Dedication made in accordance with the provisions of a statute is called *statutory dedication* and usually requires that the donor sign and acknowledge the dedication.

Developers of a subdivision may subdivide a tract of land, lay out streets, lay sewer lines and water lines, and pave streets. They may then turn over the use of these facilities to the public. The facilities must be accepted for use by the public by the state, city, town, or other governing body. Other examples of land dedicated to the public include parks, cemeteries, and schools.

47. ADVERSE POSSESSION AND TITLE BY LIMITATION

Transfer of property may occur without the agreement of the owner by the method known as *adverse possession*. Adverse possession is the acquisition of title to property belonging to another by performing certain acts. The rights to acquire property in this manner are often referred to as *squatter's rights*.

Requirements for transfer of title by adverse possession vary among the states but are essentially the same.

- Possession of the land by the person claiming it from another must be such that the owner will be aware of the possession if he or she visits the property.

- Possession must be open and notorious. Possession so open, visible, and notorious that it will raise the presumption of an adverse claim is the equivalent of actual knowledge. The land must be occupied in a straightforward, not clandestine, manner.

- Possession must be continuous. Statutes vary among the states as to the period necessary to establish title from adverse possession, but the land in question must be held continuously for the period required by statute.

- Possession is required to be exclusive. This means that the person making the claim cannot share the possession with the owner or others. He or she must have complete control of the property.

- The possession must be hostile. The claimant must possess the land as if he or she were the owner in defiance of the owner.

48. ADVERSE POSSESSION USED TO CLEAR TITLE

In modern times, people seldom squat on land with the intention to acquire title by adverse possession (although there have been many instances in the past when this has occurred). The importance of adverse possession today is in its use to clear up defects in title or to settle boundary disputes between adjacent land owners.

Honest differences may occur between adjacent owners as to where the boundary between them actually is. Monuments and landmarks may be obliterated; changes in the location of fences, ditches, and roadways may have occurred. Adverse possession provides a means of clarifying an obliterated boundary line.

49. RIGHT OF THE STATE AGAINST ADVERSE POSSESSION

Title to state or public land generally cannot be acquired by adverse possession.

50. TRESPASS TO TRY TITLE ACTION

The action usually taken by the record owner of land against a person in adverse possession of land is known as *trespass to try title*. The record owner brings suit against the person in possession for recovery of the land and for damages for any trespass committed. If the court rules in favor of the plaintiff, the person in possession is evicted, but if the court rules in favor of the defendant, the defendant acquires a good title to the land.

51. PRESCRIPTION

The method of obtaining easement rights from long usage is known as *prescription*. A person may travel across a tract or parcel of land for a period of time required by the statute of limitations and acquire a right to continue the act of using the land. The act of using the land must have been open, continuous, and exclusive for the period of time required.

A highway right of way can be acquired by the state if it has been used by the public for a long period of time. As with individual acquisition, the use must be open and continuous for the required period of time.

52. RIGHT OF EMINENT DOMAIN

When the owner of land refuses to sell and the improvement is of public character, the law allows that land shall be taken under what is called the right of *eminent domain*. Eminent domain gives the state, or others delegated, the right and power to condemn private property for public use.

The constitution of the United States and state laws limit eminent domain. Owners are guaranteed adequate compensation for their property and they may not be deprived of their property without *due process of law*.

The power to exercise eminent domain must be authorized by the state legislature by statute, and the legislature may delegate this power to such agencies as it deems proper. Counties, incorporated cities and towns, water districts, and school districts have been delegated the power of eminent domain. The power is also given to private corporations that are engaged in public service.

Owners of condemned property must be fully compensated for the property. When only a part of their property is taken, they are entitled to compensation for *consequential damage*. (A highway that cuts off access to a watering tank for cattle might create consequential damage.)

The owner of the land is entitled to know the precise boundaries of the land to be condemned. It is the obligation of the agency executing the acquisition to furnish an adequate description of the boundaries.

Before the right of eminent domain can be exercised, it is essential that no purchase agreement be reached between the parties. It is necessary that the state or city or other governing body make the owners an offer that they refuse, and that the owners shall name their price, which the state or city refuses, or else that the owners refuse to name the price. There must be a definite failure to agree. After disagreement, the state or other body must initiate *condemnation proceedings*.

53. ENCROACHMENT

An *encroachment* is a gradual, stealthy, illegal, acquisition of property. By moving a fence a small amount over a period of years, an adjoining owner may acquire from the lawful owner a strip of land.

54. ACTION TO QUIET TITLE

Where the boundary between adjacent landowners is not clear or where there is a dispute over the location of the boundary line, one of the parties can sue the other to determine the location of the line. Either party may employ a surveyor as an expert witness. The judgment in the lawsuit becomes a public record and will be reflected in abstracts of title.

55. COVENANT

An agreement on the part of the grantee to perform certain acts or to abstain from performing certain acts regarding the use of property that has been conveyed to him or her is known as a *covenant*. Developers of residential property, in order to assure buyers that their neighborhood will be pleasing to the eye and pleasant to live in, require the buyer to accept certain restrictions as to the use of property he or she buys. These restrictions include such things as type of building construction, minimum distance between house and property line, minimum number of square feet in floor plan, use of the property, and kinds of animals allowed on premises. These covenants are sometimes called *deed restrictions*.

56. LIEN

A *lien* is a claim or charge on property for payment of a debt or obligation. It is not the right of possession and enjoyment of property, but it is the right to have the property sold to satisfy a debt. Mortgages and deeds of trust constitute liens.

57. TAX LIEN

States, counties, and other governing bodies impose taxes on real property, which gives them a first lien on the property. Failure to pay taxes gives the governing body the right to have the property sold to satisfy the tax debt. Failure to pay income taxes gives the federal government the same right. Before purchasing the real estate, the buyer may obtain a *tax certificate* in which the tax collector certifies that there are no unpaid taxes on the property up to a certain date.

58. PROMISSORY NOTE

A *promissory note* is the written promise of the borrower to pay the lender a sum of money with interest. The principal sum, the interest rate, and a schedule of dates of payment are included on the face of the note. Also included in some notes is a listing of the *security* for the note—the property that is to be mortgaged to guarantee payment of the note. The same information as to principal, interest, and schedule of payments shown on the note is, where applicable, shown in the deed of trust, and reference is made in the deed of trust to the promissory note between the two parties involved.

A deed of trust does not necessarily accompany all promissory notes, and a promissory note need not list any security. Lenders may, if they wish, lend simply on a person's *personal note*, which is a promise to pay. But this does not prevent lenders from taking legal action to collect the amount of the note or obtaining property of equal value in the event the borrower does not pay the note.

Part 5: Ownership of Beds of Rivers and Streams

59. COMMON LAW

Both English common law and Spanish civil law pertaining to ownership of the beds of rivers and streams are based on Roman civil law and follow the same rules in certain particulars. American common law has not followed English common law exactly because there are variations in the interpretation of English common law.

It appears that early English law was concerned only with the water in streams and the public right to use it, such as for fishing. Ownership of the beds of streams is not clear except for the beds of streams that are affected by the tide. Under English common law, title to the beds of streams is retained by the sovereign insofar as the waters of the stream are affected by the tide. Ownership beyond the point where the waters of the stream are affected is not clearly defined, and *navigability in fact* of streams seems to be of no consequence.

One theory as to the reason for the lack of classification of streams as navigable or nonnavigable is that in England there were no navigable streams except where the tide affected them.

It has been stated previously that, for the most part, property law is state law. This is also true in the matter of ownership of the beds of streams and the water in streams.

Differences between states pertain mostly to ownership of the beds of navigable streams. Under both English common law and Spanish law, grants bordering on nonnavigable streams extend to the center of the stream. In other words, beds of nonnavigable streams are privately owned. But the beds of navigable streams are owned by the state.

Some states consider streams to be navigable in law if navigable in fact. Other states have defined a navigable stream by a legislative act. Some states, such as Mississippi, do not claim streams above tidal effect even if stream is navigable in fact.

60. SPANISH AND MEXICAN LAW

In Spanish law, the sovereign owns both the water and beds of *perennial streams* whether they are navigable or not. The beds of nonperennial streams, called *torrential streams*, belong to the adjacent owners, each owning to the center of the stream. Perennial streams are considered to have continuous flow, except in periods of drought. Nonperennial streams flow after heavy rains or melting snow.

61. FEDERAL LAW

While the waters and beds of navigable streams are the property of the states, the federal government has absolute supremacy over navigation on navigable streams within a state.

The federal government can take action against a state or individual for any acts that might interfere with or diminish the navigability of streams. To ensure commerce between the states, state legislatures are barred from enacting legislation that would interfere.

62. ISLANDS

If an island is on one side of a nonnavigable stream—between one bank and the middle of the stream—the island belongs to the riparian owner on that side. If an island is formed in the middle of a nonnavigable stream, the island is owned by both riparian owners to the middle of the stream. This is true only if the island emerged after date of statehood. Otherwise, islands were *not* conveyed with submerged lands to state via statehood. Rather, they were subject to disposal to put interests by federal government. Therefore, most islands in navigable streams are privately owned.

63. THE GRADIENT BOUNDARY

Discovery of oil early in the 20th century in north central Texas caused a dispute over the boundary between Texas and Oklahoma along the Red River. Texas claimed that the boundary was in the center of the stream while Oklahoma claimed that the boundary was on the south bank of the river.

The U.S. Supreme Court held that the boundary is on and along the south bank at the mean level attained by the waters of the river when they reach and wash the bank without overflowing it. The boundary is a gradient of the flowing water in the river. It is located midway between the lower level of the flowing water that just reaches the cut bank and that higher level of it that just does not overtop the cut bank.

The *gradient boundary*, as defined, was established in Texas as the boundary between state-owned beds of navigable streams and adjoining private lands.[1]

64. MEANDER LINES

Surveyors run *meander lines* in order to plat a stream. Where a grant or patent calls for the shore of a stream or for the middle of the stream, it is this line that is the boundary and not the meander line. Meander lines, run for patents to U.S. lands, were used in computing the acreage to be paid for by the grantee. Often no charge was made for land between the meander line and the stream itself.

Part 6: Ownership of Tidelands and Lake Beds

65. TIDES

Waters of the oceans are attracted by the sun, moon, and planets. This attraction causes the rise and fall of the surface of the sea (the tide). When the sun, moon, and earth are in line and pulling in the same direction, the tides are highest and called *spring tides*. When the attraction of the sun and moon are at 90° to each other, the tides are lowest and are called *neap tides*.

Along the Pacific and Atlantic coasts, there are two high and two low tides during a tidal day. Along the western portion of the Gulf of Mexico there is usually one high and one low tide each tidal day.

---

[1]The method of locating the "gradient boundary" is defined in *The Gradient Boundary* by Arthur A. Stiles (*Texas Law Review*, vol. 30, no. 3, p. 305).

66. TIDAL WATERS

Waters in which the tide ebbs and flows are known as *tidal waters* or *coastal waters*. Saltwater flats that are alternately covered and uncovered as the tide ebbs and flows are also considered to be covered by tidal waters. A saltwater marsh that is not an integral part of a bay but that is affected by the ebb and flow of the tide only because a ditch or ditches that have been excavated for the purpose of drainage are not considered to be tidal waters.

Rivers and streams are considered to be tidal waters to the extent that the waters in these streams are affected by the ebb and flow of the tide.

67. TIDELANDS

The waters of bays, inlets, bayous, and arms of the Atlantic, Pacific, and the Gulf of Mexico and the soil laying beneath them are owned by the sovereign, the state.

Under common law, the line between the seashore and the upland estate (between state land and private land) is the line of mean high tide. Mean high tide is the average height of the high water over a 19-year period. Under Spanish law the shore of the sea is that ground that is covered by the "highest wave of winter."

68. OWNERSHIP OF BEDS OF LAKES

The common law rule is that ownership of the bed of a nonnavigable lake is with the littoral owner and extends to the center of the lake unless the conveyance specifies otherwise. Beds of navigable lakes belong to the state, but to be navigable the lake must be navigable in fact to the extent that it is used as a highway for commerce and not for pleasure boating only.

In a decision to determine who had the right to fish in Stanmire Lake in Leon County, Texas, the court in its deliberations defined the meaning of "navigable in fact" and in determining who had the right to fish in the lake, first determined who owned the bed of the lake. In the suit *Taylor Fishing Club* v. *Hammett*, Taylor Fishing Club brought suit to enjoin John Hammett from fishing in Stanmire Lake.

Taylor Fishing Club owned all of Stanmire Lake except a small portion in the north end of the lake that was owned by Hammett. The part of the lake owned by Taylor Fishing Club was fenced but Hammett insisted that he and his friends had the right to enter the lake by boat on Hammett's premises, cross the partition fence, and fish in the part of the lake owned by Taylor Fishing Club.

Field notes in the patents issued by the state for the part of the lake owned by the fishing club cross the lake and show an unmistakable intention to convey the bed of the lake.

In considering the case, the court first set out to determine whether the lake was navigable or nonnavigable and commented that the rule is that

streams or lakes which are navigable in fact must be regarded as navigable in law; that they are navigable in fact when they are used, or are susceptible of being used, in their natural or ordinary condition, as highways for commerce, over which trade and travel are, or may be, conducted in the customary modes of trade and travel on water, and further that the navigability does not depend on the particular mode in which such use is or may be had ... whether by steamboats, sailing vessels, or flat boats ... nor an absence of occasional difficulties, as in navigation, but on the fact, if it is a fact, that the stream in its natural and ordinary condition affords a channel for useful commerce. (*United States* v. *Holt State Bank*, 270 U.S. 46, S Ct. 197, 199, 70 L. Ed. 465).

The court concluded that the lake was useful for fishing and as a pleasure resort but that it is generally held that a lake that is chiefly valuable for fishing or pleasure boats of a small size is not navigable. It was the opinion of the court that Stanmire Lake was not a navigable lake.

The court then considered the pleading by Hammett that under the common law, a riparian landowner whose land abuts on a nonnavigable lake and whose field notes call for the lake as boundary line, implicitly owns the land under the water to the center of the lake and that all riparian owners whose lands abut on such a lake have a right to the joint use of the entire lake for fishing and boating.

The court concluded that the rule applied as a general proposition, but did not apply in this case because Taylor Fishing Club, by specific grant from the state, owned the land under a definite and specific portion of the lake and has a right to control that part of the surface of the lake above its land, including the right to fish in or boat upon the waters. It granted the injunction to prevent Hammett and friends from entering the part of the lake controlled by Taylor Fishing Club.

The court further commented that the decision was not a variance with the decision of the Supreme Court in *Diversion Lake Club* v. *Heath*, 186, S.W.2d 441, 443, because in that case the court held that "the general rule is well established by the authorities that the right to fish in a stream, whether belonging to the public in common or exclusively to the owners of the land bordering the stream, is determined by the ownership of the bed."

In the case of *Diversion Lake Club* v. *Heath*, a lake was formed by impounding water behind a dam on a navigable stream, which established that the waters were public waters.

Law

69. OWNERSHIP OF OFFSHORE SUBMERGED LANDS

Until 1947, coastal states claimed ownership of offshore submerged lands. In a suit brought against California, Louisiana, and Texas by the the United States, the U.S. Supreme Court held that these lands belonged to the federal government. In 1953, by an act of Congress, these lands were returned to the states.

Part 7: Riparian and Littoral Rights

70. RIPARIAN AND LITTORAL OWNERS

Persons who own land abutting on a body of moving water are known as *riparian owners*, and they have certain rights to, or in, the water. Their rights rest solely in the fact that their land abuts on the water.

Persons whose land does not abut on a body of water, even though there is a very small tract of land between their land and the water, are not considered to be riparian owners.

Persons who own land on a body of water not in motion, such as a pond or lake, are also known as riparian owners. They also have rights in the body of water. Persons who own land on a body of water such as a gulf or ocean are known as *littoral owners*.

Riparian Rights

The riparian owner may use the water in the stream for any reasonable use to which the stream is adapted. These include domestic water, water for animals, fishing, boating, and swimming. Many of these are rights in common, and one owner may not interfere with another's use of the water. Riparian owners have the right to use water for irrigation of crops within the limits of the amount of water available. *Riparian rights* also include rights to stream beds and to the alluvium below the water.

Littoral Rights

Littoral rights are very much the same as riparian rights. Littoral owners on the ocean or gulf have the right to use the water and beach for bathing and boating, but they do not have exclusive rights.

71. NATURAL CHANGES IN STREAMS, LAKES, AND TIDELANDS

Rivers are constantly changing their boundaries with the rise and fall of the water. Lakes also change their shoreline with the rise and fall of the surface of the water, and the shoreline of tidelands changes with the tide and storm waters.

Both English common law and Spanish law follow Roman law in regard to changing boundaries in streams, lakes, and tidal waters. So courts have not found it necessary to distinguish between English common law and Spanish law in cases of this kind.

72. EROSION

Erosion is the gradual wearing away of the soil by operation of the action of water, wind, or other elements.

73. ALLUVIUM

Rivers carry silt, pebbles, and rocks as they travel to the sea. As the flow of the river rises and falls, the amount of material it carries rises and falls. The deposits made by water on a shore are known as *alluvium*.

74. ACCRETION

When land is formed slowly and imperceptibly by alluvium, the process is known as *accretion*. The buildup of alluvium may be on the banks of rivers and streams, or on the shores of lakes and tidal waters.

75. RELICTION

The gradual withdrawal of water that leaves land uncovered, such as the shore of a lake gradually receding, is known as *reliction*.

76. AVULSION

The sudden and perceptible change of a course of a river or stream forming a new channel across a horseshoe bend is known as *avulsion*.

77. BOUNDARY CHANGES CAUSED BY ACCRETION, EROSION, RELICTION, AND AVULSION

Where accretion, erosion, or reliction occurs along a river or stream, the boundary of the riparian owner does not remain fixed but changes with the change in the gradient boundary. This means that where accretion occurs, the riparian owner, along whose land the accretion is joined, gains this land.

Where erosion occurs, the riparian owner loses the land that may have formed accretion downstream.

Where reliction occurs, the boundary of the riparian or littoral owner changes with the boundary of the stream or with the shoreline of a lake or of tidal waters. Land left uncovered by reliction is owned by the riparian or littoral owner.

When avulsion occurs along a navigable stream, the owner of the land across which the new channel was formed loses title to the bed of the new channel to the state, but ownership of the land between the new and old channels does not change. Title to the bed of the old channel passes to the riparian owners from the state, unless the stream uses both channels, in which case the state has title to the bed of both channels.

Part 8: Metes and Bounds Surveys

78. HISTORY

Thirty states of the United States were subdivided into rectangular tracts by a system known as *The U.S. System of Rectangular Surveys* (see Part 10) before the sale of the land to settlers. Prior to this, surveys were made by the *metes and bounds* method. To survey a grant of land the surveyor went on the ground and went around the perimeter of the grant and measured and recorded the direction of each side of the perimeter and the length of each side. These surveys were made by citizens, or subjects, of England, Spain, France, Mexico, the Republic of Texas, as well as the United States.

In 1493, Spain acquired from Pope Alexander VI title to land in the New World, which included territory from Florida along the Gulf coast to New Mexico. The Kingdom of New Spain was set up in 1529 in what is now Mexico City.

France established Quebec in 1608 and later claimed all of the Mississippi River basin, which is called Louisiana. All of Louisiana was sold to the United States in 1803.

The people of New Spain declared their independence from Spain and established the Republic of Mexico in 1821 with the capital at Mexico City.

Texas declared its independence from Mexico on March 2, 1836, and established the Republic of Texas. Ten years later, Texas was annexed to the United States as a state. The Republic of Texas adopted the Common Law of England, with modifications. The State of Texas adopted the laws of the Republic.

The laws of Spain and Mexico were based on *Las Siete Partidas* (The Seven Parts) from the 13th century. Laws of Louisiana were based on the Code Napoleon. Roman law had influenced both, as did English common law.

79. BOUNDARY

A *boundary* is any marking or boundary line, natural or artificial, dividing two parcels of land.

80. METES AND BOUNDS DESCRIPTIONS

"Metes and bounds" refers to the measurements of the limits, or boundaries, of a tract of land. This type of description identifies a beginning point and then describes each course of the tract in sequence in either a clockwise or counterclockwise direction and returns to the beginning point. The description includes not only the direction and distance of each course but also includes calls for monuments and adjoiners. Calls for course and distance are referred to as *metes* and calls for objects and adjoining boundaries are referred to as *bounds*.

81. CORNER

A *corner* is a point of change of direction.

82. BEGINNING CORNER

The beginning corner of a tract of land is the corner where the metes and bounds description begins.

83. MONUMENT

Objects referred to in a metes and bounds description are known as *monuments*. Natural monuments, such as rivers, lakes, oceans, bays, and large boulders, are formed by nature. Artificial monuments, such as iron pipes, steel stakes, wooden stakes, or fences, are made by humans.

84. FIELD NOTES

The term *field notes* refers to notes made by surveyors in the field while making a survey, describing by course and distance and by natural or artificial marks found or made by them, the running of the lines and the making of the corners. The field notes constitute a description of the survey that is the substance and consists of the actual acts of the surveyors.

85. CALLS

A *call* is a phrase in the written description of the location of a parcel of land contained in the body of a conveyance. Calls start with the word "BEGINNING," for the first call, and "THENCE" for each succeeding call, written in capital letters with each call separated from another as a paragraph.

Locative Call

Calls that give the exact location of a point or line are known as *locative calls*. A call for a monument referenced to witness trees or other objects is a locative call.

Law

Directory Call

A call leading to the beginning corner is known as a *directory call*.

Passing Call

A call that refers to a creek, highway, fence, tree, or other object that is crossed or passed in a survey is called a *passing call*. It does not serve as a locative call but better identifies the location of the land.

Call for a Monument

A *call for a monument* describes a natural or artificial object at the corner of a survey or describes a course of the survey as being along a natural or artificial monument, such as a stream or highway.

Call for Adjoiner

Calls for the lines and corners of an adjoining tract are known as *calls for adjoiner*.

Call for Direction and Distance

A call giving the bearing and distance of a line in a survey is known as a *call for direction and distance*.

Call for Area

A *call for area* gives the area contained in the survey.

86. SENIOR RIGHTS AND JUNIOR RIGHTS

Many original metes and bounds surveys resulted in overlaps of surveys and gaps between surveys, which are known as *vacancies*. It became the rule of law in the early history of the United States that where an overlap occurred, the holder of the first grant or patent, the senior awardee, retained ownership of the overlap area, and the holder of the latter grant or patent, the junior awardee, lost the area. These rights are known as *senior rights* and *junior rights*. A common expression explains these rights: "The first deed is the best deed."

Where two parties have title to the same land, the party holding the senior conveyance has the right of possession. As an example, Smith owns 200 ac of land, or thinks he owns that amount. He sells one half of his land to Jones by a metes and bounds description that calls for 100 ac. Later, Smith sells the other half of his land to Brown, but finds that he had only 195 acres originally. He cannot sell more than 95 ac to Brown; Jones has the senior deed and is entitled to the full 100 ac that Smith sold to him. Smith cannot recover any of Jones's land. This principle in law assures the first buyer that his land cannot be taken from him as it was conveyed. It is the basis of senior rights and junior rights.

Regardless of how great or small a deficiency may be, the first deed holder cannot lose land conveyed to him because of some error that was discovered afterward.

There can be no private overlapping of acreage where junior survey expressly calls to begin at a senior survey (*Stanolind Oil & Gas Co.* v. *State*, 101 S.W.2d 801).

Where the junior survey adjoins and begins at a corner of the senior survey, the calls for distance and acreage of the junior survey in conflict with the senior survey must yield to the senior survey. A call for an adjoiner is like a call for artificial objects (*Kirby Lumber Co.* v. *Gibbs Bros. & Co.*, 14 S.W.2d 1013).

When all calls of a junior survey are calls for adjoining senior surveys, the lines of the junior survey follow the lines of the adjoining survey as they were actually located on the ground, not as they were supposed to have been located (*Peal* v. *Luling Oil & Gas Co.*, 157 S.W.2d 848).

When the north line of a junior survey was called to be identical with the south line of an adjoining senior survey, the proper location on the ground of the north line of the junior survey is the south line of the senior survey, regardless of where it may be (*Leone Plantation* v. *Roach*, 187 S.W.2d 674).

87. STEPS IN RESURVEYING LAND

Evidence to prove the location of a written title will not overturn a legal unwritten title. Title acquired by adverse possession does not depend on the calls of a survey. Therefore, the first step in arriving at the true location of land is to first determine who is in possession of the land. If the land is in possession of someone other than the record title holder, the record title holder may have to have the person in adverse possession removed from the land by court action. This action is brought about by a trespass to try title suit.

In a *trespass to try title suit*, the plaintiff (the record title holder) has the burden of proof to recover the land. He or she must show a regular chain of title that he or she is the lawful owner of the land, and he or she has the burden to locate on the ground the land sued for.

The *second step* in determining the true location of land is to determine whether there could be an overlap involving a senior conveyance.

The *third step* in locating land is retracing the original survey.

Part 9: Retracing the Original Survey

88. INTENT

The controlling factor in locating the boundaries of a tract of land is the *intent* of the parties to the conveyance. The parties' intent with respect to the boundaries is to be ascertained from the face of the conveyance in light of surrounding circumstances.

The primary question is "Where did the surveyor intend the boundaries to be located?" The intent of the surveyor is considered to be the intent of the parties, but the intent of the surveyor must be determined by their acts in locating the boundaries as expressed in their field notes. It is not confined to their minds.

89. FOLLOW THE FOOTSTEPS OF THE ORIGINAL SURVEYOR

It is a fundamental principle of law that boundaries are to be located in a resurvey where the original surveyors ran lines and called for them to be located in their field notes. The primary objective in locating a survey is to follow the footsteps of the surveyors, that is, to trace the lines on the ground as the surveyors actually ran them in making the survey.

There is one exception to the rule: The footsteps will not be followed in their full extent where the monuments are established along a meander line and the field notes call for the river. In such cases, the call for the river will control and the lines will be extended beyond the monuments so that they intersect the river.

90. SIGNIFICANCE OF CALLS

In resurveying, if the lines and corners of a tract of land can be definitely located from the calls in the field notes of the original surveyors, as found in the conveyance, and there is no uncertainty as to the location and no ambiguity in the calls, then they should there be established. But when calls of the conveyance are inconsistent and the location is uncertain, certain rules must be followed.

Inconsistency between two calls is usually the result of a mistake in one of them. If the mistake can be found, that call will be rejected and the other adopted.

91. HARMONY OF CALLS

All the calls in a description should be considered together, and when they cannot all be reconciled, as few of them should be disregarded as possible. Calls that are the most reliable and certain, from the evidence, are regarded as controlling. The primary purpose is to determine where on the ground the surveyors actually ran the line, or where the grantor intended it to be located when there was no actual survey. Where there was an actual survey, the inquiry is not where the line ought to have been, but where, in fact, it was located. The footsteps of the surveyors must be followed, if ascertainable, in locating the true line. When they are found and identified they control (*Stafford* v. *King*, 30 Tex. 257).

92. BEGINNING CORNER

A survey may be traced by beginning at any well-established corner in the survey, but it must be traced to its beginning corner. The beginning corner is of no greater weight or importance than any other corner that is well established and identified. All corners are of equal importance.

93. CONFLICTING CALLS

Boundary calls that will give effect to and carry out the intention of the parties are given controlling effect; calls inconsistent with this intention are rejected as false, regardless of their comparative dignity. But when the intention of the grantor is not expressed or is ambiguous, controlling effect will be given to the calls that are regarded as most reliable, most material, and most certain, and therefore, as having a higher dignity and importance, since they are presumed to be the most prominent in the mind of the grantor.

In the absence of a clear intention to the contrary, the order of priority for conflicting calls is as follows.

(1) calls for natural objects such as rivers, creeks, springs, mountains

(2) calls for artificial objects such as stakes, marks on trees, and marked lines

(3) calls for course and distance

(4) calls for quantity

This rule for the order of priority of calls is a *rule of evidence*. Every rule of evidence for boundary location is for the purpose of ascertaining true location. It is not a hard and fast rule, but a rule to determine which call or calls was made by mistake. There is no positive law giving to one call more weight or importance than another.

94. CALLS FOR NATURAL OBJECTS

Calls for natural objects are more certain and less subject to change or error than calls for course and distance and calls for quantity.

95. CALLS FOR ARTIFICIAL OBJECTS

Calls for rocks set in place, fences, marked lines and corners, and adjoining surveys are *calls for artificial objects* and have precedence over calls for course and distance and calls for quantity.

96. *STAFFORD* v. *KING*

A widely quoted court decision has clarified many questions in boundary location and is considered a classic in its clarity. The decision in *Stafford* v. *King* (30 Tex

257) was handed down in 1867 by the Supreme Court of Texas.

Mrs. Stafford, formerly Mobley Rhone, held a patent to 640 acres in Cherokee County, Texas. The land, as described in the field notes, was held by Adam King. Mrs. Stafford brought trespass to try title suit against King.

The field notes in Mrs. Stafford's patent started at the southeast corner of the John R. Taylor Survey, which was a well established and known point.

The first line called for was a directory call beginning at the Taylor corner and running east a distance of 3160 varas to the point of beginning of the Stafford patent.

From the point of beginning, the field notes called for various natural objects, including two creeks which were crossed twice, a lake, and a spring.

When a survey was run, none of the objects called for in the field notes were found. However, it was ascertained by the court that another patent, just west of the Stafford patent, had been awarded to Franklin, whose field notes had exactly the same calls as those found in the Stafford field notes, with one exception: The directory call from the southeast corner of Taylor to the beginning of Franklin was for 750 varas.

The surveyor testified that if the directory call in Stafford's field notes had been for 750 varas instead of 3160 varas, the resurvey would have followed as nearly perfect as any survey he had ever followed.

The survey would have followed in the footsteps of the original surveyor.

The decision of the court was as follows.

> It is the duty of the surveyor to run around the land located and intended to be embraced by the survey and patent, and see that such objects as designated on it will clearly point out and identify the locality and the boundaries of the tract, and to extend a correct description of these objects (natural and artificial with courses and distances) into the field notes of the survey, in order that they may be inserted in the patent, which will afford the owner, as well as other persons, the means of identifying the land that was in fact located and surveyed for the owner; and until the reverse is proved, it will be presumed that the land was thus surveyed and boundaries plainly marked and defined. And if any object of a perishable nature, called for in the patent, be not found, the presumption will be indulged that it is destroyed or defaced; that if it be established, by undoubted evidence, that the land was not in fact surveyed, yet, as the omission was the fault of the government officer and not the owner, it would seem extremely unjust to deprive him of the land, by holding the patent to be void, if the land can, by any reasonable evidence, be identified. And if course and distance alone, from a defined beginning-point will with reasonable certainty, locate and identify the land, that will be held sufficient. Then we

must conclude that the position of the appellant, that a patent without a survey having been made of the land, should be held void, cannot be sustained.

The main point in this case appears to be whether the Mobley Rhone patent for 640 acres of land in fact covers any of the land claimed by the defendant below.

As has been intimated, it is the purpose of the government (Texas) and the locator to select a particular tract of land and designate it from the mass of the public domain. And hence the directions given by law to run around the land—in fact, point out and define upon it such natural objects, or plain artificial marks, with course and distance, by which the land can at all times be easily found and identified. Natural objects are mountains, lakes, rivers, creeks, rocks and the like. Artificial objects are marked lines, trees, stakes, etc. A description of these objects and marks of identity should be faithfully transferred into the field notes, and thence into the patent to serve the purpose of aforesaid; and in all future controversies in respect to the locus of boundaries of the tract, recourse must be had to these calls and when they are all found and established in conformity with those set force in the patent, the conclusion is almost irresistible that the tract of land covered by the patent is identified, and there can be little or no room for controversy about the boundaries of land; but when all the calls of the patent cannot be found, or if found to be inconsistent with others, in whole or in part, and leading to a different result or confusion, then it becomes important to look to the rule of law that must govern the actions of the court and jury, in respect to the character and weight of evidence to be considered by them in fixing upon and establishing the true boundaries of the survey.

It has been often said by this court that the general rules are that location should be governed, first, by natural objects or boundaries, such as rivers, lakes, creeks, etc.; second, artificial marks, such as marked trees, lines stakes, etc., and third, course and distance.

The true and correct location of the land is ascertained by the application of all or any of these rules to the particular case, and when they lead to contrary results or confusion, that rule must be adopted which is most consistent with the intention apparent on the face of the patent, read in the light of surrounding facts and circumstances.

The rule stated by Chief Justice Marshall, in *Newson v. Pryor*, is "that the most material and most certain calls shall control those which are less material and less certain. A call for a natural object, as a river, a known stream, a spring, or even a marked tree, shall control both course and distance."

Of all these indicia of the locality of the true line, as run by the surveyor, course and distance are regarded as the most unreliable, and generally distance more than course, for the reason that chain carriers may miscount and report distances inaccurately, by mistake or design. At any rate, they are more liable to err than the compass. The surveyor may fall into an error in making out the field notes, both as to course and distance (the former no more than the latter), and the commissioner of the general land office may fall into a like error by omitting lines and calls, and mistaking and inserting south for north, east for west. And this is the work of the officers themselves, over whom the locator has no control. But when the surveyor points out to the owner rivers, lakes, creeks, and marked trees, and lines on the land, for the lines and corners of his land, he has the right to rely upon them as the best evidence of his true boundaries, for they are not liable to change and the fluctuations of time, or accident by mistake, like calls for course and distance, and hence the rule that when the course and distance, or either of them, conflict with natural or artificial objects called for, they must yield to such objects, as being more certain and reliable.

There is an intrinsic justice and propriety in this rule, for the reason that the applicant for land, however unlearned he may be, needs no scientific education to identify and settle upon his land, when the surveyor, who is the agent of the government, authoritatively announces to him that certain well-known rivers, lakes, creeks, springs, marked corners and lines constitute the boundaries of his land. But it would require some scientific knowledge and skill to know that the courses and distances called for are true and correct, and with the aid of the best scientific skill mistakes and errors are often committed in respect for the calls for course and distance in the patent. The unskilled are unable to detect them, and the learned surveyor is often much confused.

Although course and distance under certain circumstances may become more important than even natural objects—as when, from the face of a patent, the natural calls are inserted by mistake or may be referred to by conjecture and without regard to precision, as in the case of descriptive calls—still they are looked upon and generally regarded as mere pointers or guides that lead to the true lines and corners of the tract, as, in fact, surveyed at first.

This identification of the actual survey as made by the surveyor is the desideratum of all these rules. The footsteps of the surveyor must be followed, and the above rules are found to afford the best and most unerring guides to enable one to do so.

There is another rule to be observed in estimating these natural and artificial calls. They are divided into two classes: *descriptive* or *directory* and *special locative calls*. The former, though consisting of rivers, lakes, and creeks, must yield to special locative calls for the reason that the latter, consisting of the particular objects upon the lines or corners of the land, are intended to indicate the precise boundary of the land, about which the locator and surveyor should be, and are presumed to be, very particular; while the former are called for without any care for exactness

and merely intended to point out or lead a person to the region or neighborhood of the tract surveyed, and hence not considered as entitled to much credit in locating the particular boundaries of the land when they come in conflict with special locative calls, and must give way to them.

In this case the southeast corner of the John R. Taylor survey appears to have been notorious in the neighborhood and well established. The line calling to run east 3160 varas, to the beginning corner of the Mobley Rhone survey of 640 acres when extended that distance, is unsupported by any natural or artificial call mentioned in the patent or proved to have been made by the surveyor. When that line was run out by Armstrong, the distance and course called for, thence around the tract, the courses and distances called for in the patent, for the four corners and lines of the tract, not one single natural or artificial object called for in the patent could be found on the ground at all corresponding with it, as described in the patent. But it is contended by the appellee that no survey was in fact made by the surveyor, and the tract was surveyed by protraction, beginning at the Taylor southeast corner. To support the truth of this supposition, it must be admitted that the surveyor did not perform his duty; that the locative calls of the patent were never in fact made or found, and were falsely placed in the patent by the surveyor, or that they have, since the designation of them, been destroyed or defaced and cannot now be found; while it must be admitted that Mud Creek and Camp Creek, called for in the patent, are not liable to destruction, if it be admitted that all of the marked trees may have been destroyed or defaced by accident or design.

There is no evidence adduced, except the admission of the plaintiff himself, that the survey was in fact not made, or that it was made by protraction; while on the contrary, it has been most satisfactorily proved that if the line east from Taylor's corner be stopped at 750 varas from that point, there following the course and distance called for, we find every call of the patent on the ground, natural or artificial, Mud and Camp Creeks are crossed, each twice, and a spring reached at the very point called for in the patent, the course and bearing trees also corresponding to the patent, fortified by course and distance. It is true no witness testified that these were the lines actually run by the surveyor for the Mobley Rhone survey. Some evidence tends to show that it was surveyed by Reason Franklin. It may have been surveyed at the time of Franklin, but, like the Irby survey, which was transferred to Taylor, this may have been applied to the Mobley Rhone claim. The wonderful coincident of the locative calls of the patent with the objects found here on the ground would leave but little doubt that they were the true locative calls of the Mobley Rhone survey, and that there was a mistake in the length of the descriptive call of 3160 varas, and its true length is 750 varas. The distance of the line is the least important or material of all the calls of the patent. It is directory or descriptive in its character, and forms no part of the boundary of the survey, and evidently was intended only to point and direct a person to the neighborhood in which the special locative calls could be found, and about its accuracy the parties may not be been very particular. And in this view we are of the opinion that call should yield to the more specific calls found on the ground. It will be observed that if we hold that this line is correctly stated to be 3160 varas, we do not find one solitary other call of the patent on the ground; and we are asked to presume that the report of the calls in the patent was false and that no survey was in fact ever made on the land; while on the other hand, if we hold that this descriptive call of 3160 varas was a mistake in distance, and that the true length is only 750 varas, then we find every corner, bearing tree, line and the two creeks called for, all corresponding with the calls in the patent, in respect to identity and course and distance; and from the rules laid down it seems that we must hold the true length of that line to be 750 varas, and force it to yield to the more material and locative calls of the patent found upon the ground.

It is said that this same tract was surveyed for and patented to Franklin. This, we conceive, can be of but little importance in fixing upon the boundaries of the Rhone survey. If it amounts to anything, it only proves that the two patents have been issued for the same tract of land, which, by the by, has been very often done, but not often, perhaps, with lines corresponding so literally. But we do not think that at all affects the question of boundary involved in this case; because Mobley Rhone has secured a patent upon Franklin's land, we cannot see how that can furnish any reason why you should take that of any other person in its stead.

The line from Taylor's southeast corner we regard as a descriptive or direct call; and if found to be in conflict with any of the locative calls found and identified upon the ground, then it must yield to them, as being more material and important.

We believe the charge of the court gave too much importance to the descriptive call of 3160 varas over those of the locative character; and that it did not present to the jury properly the rules that should govern them in establishing the boundaries of the Rhone survey, and that it erred in refusing the charge asked by the defendant in that respect, and should have granted the defendant a new trial on her motion.

97. CALLS FOR COURSE AND DISTANCE

A *call for course and distance* is the least reliable of all calls except a call for quantity; and a call for distance is less reliable than a call for course. A call for course and distance is a guide to the true lines and corners of the survey.

A call for course and distance will ordinarily yield to a call for adjoiner, except where it is proven that a call for adjoiner was made by conjecture or under mistaken belief as to location of the adjoiner.

If a survey cannot be located on the ground from natural or artificial objects called for in the field notes, then it can be located by course and distance from the nearest recognized and established corner that is included in the field notes.

98. CALL FOR QUANTITY

Calls for quantity yield to calls for course and distance and are considered descriptive, except where conveyance is by the acre.

99. LOT AND BLOCK NUMBER

A description by *lot and block number* is more certain than a description by calls for course and distance.

100. SUBDIVISIONS

Lots in a *subdivision* created at the same time—when the plat is filed—have equal rights, regardless of when the lots were sold.

101. EXCESS AND DEFICIENCY

When a tract of land is subdivided and platted, any *excess* or *deficiency* in lot dimensions when the plat dimensions are compared to actual measurement between monuments is apportioned to the lots in proportion to their respective lengths, unless the lot lines are marked on the ground.

102. PAROL AGREEMENT

Where there is a disputed boundary line, adjoining land owners may enter into a *parol* (oral) *agreement* fixing the line. If the agreement is fairly and honestly made, it will be recognized as binding on them and persons claiming under them, even though it is later found that the line as fixed was not the true line. For the agreement to be valid, there must have been a dispute, doubt, or uncertainty as to the true location of the boundary. For the agreement to be binding, the parties must erect physical monuments or otherwise mark the line.

The agreement is not invalidated by the statute of frauds because it is not a conveyance of land but a settlement of a dispute as to the boundary. But if the parties agree on a location that they know is not the true one, the agreement is invalid under the statute.

103. EVIDENCE

Evidence is a collection of facts that may or may not be proof. Rules of evidence are applied for the purpose of ascertaining the true location.

104. BURDEN OF PROOF

Generally, the *burden of proof* lies with the plaintiff. The plaintiff must establish the boundaries that he or she alleges are the true boundaries. He or she must show and establish lines and corners.

105. PRESUMPTIONS

If a surveyor certifies that he or she surveyed a tract of land and that the lines and corners are as he or she has shown in the field notes, then it will be presumed that the surveyor did his or her duty and actually did survey the land and found the lines and corners as recorded in the field notes. The presumption will continue until the contrary is proved. It will also be presumed that the surveyor's work is accurate and truthful until the contrary is proved. It will not be presumed that a surveyor has done what it is unlawful for him or her to do.

106. ADMISSIBILITY

In general, the field notes of the original surveyor are *admissible*, but they are not conclusive if, after a long period of time, evidence left by the original surveyor has been destroyed. In this case, many facts that the rules of evidence would exclude in other areas are permitted to go to the jury if they tend to locate the true boundary.

107. HEARSAY EVIDENCE

Evidence about what someone else has said or written is generally considered as untrustworthy and not acceptable as evidence. However, if the person who made the statement is dead, the secondhand evidence may be admissible if the statement was made under certain conditions due to a long lapse of time where better evidence cannot be found.

108. PRIMA FACIE EVIDENCE

Prima facie evidence is evidence sufficient to establish a fact unless rebutted.

109. EXTRINSIC EVIDENCE

Extrinsic evidence is evidence not contained in the body of a document but must be derived from outside sources.

110. TESTIMONY OF SURVEYORS AND ASSISTANTS

The surveyor does not decide where the boundary is—that is a matter for the court—but the testimony of the surveyor is admissible in boundary dispute cases. The surveyor may testify as to the lines he or she ran on the ground and the observations he or she made while making the survey. The surveyor may explain an erroneous call and inconsistencies in the field notes, but may not give his or her opinion as to whether or not a survey was an office survey.

When qualified as an expert witness, the surveyor can testify as to the identity and dignity of calls, but the surveyor's opinion as to the location of lines and corners of a survey is not admissible; that is a matter for the court to decide.

Testimony of a witness who was with the surveyor at the time the surveyor found a marked line about which the surveyor testified, is admissible to show when, how, and under what circumstances the surveyor gained the knowledge.

111. BEST EVIDENCE RULE

Oral evidence of the contents of a written instrument is not admissible if the instrument itself is available. The surveyor cannot testify about the description in a written instrument unless the instrument has already been introduced.

112. MAPS AS EVIDENCE

Maps and diagrams are admissible as evidence to illustrate and explain the testimony of the witness, but a map cannot be admitted as evidence of the conclusion of the witness.

113. REPUTATION

Evidence of reputation may be admitted in boundary disputes only if the boundary in question is an ancient one and was of such interest in the neighborhood as to have provoked discussion and attracted general attention. The reputation must have been formed and in existence before the inception of the controversy under adjudication. Such evidence is admissible of necessity because of the lack of other satisfactory proof.

114. THE SURVEYOR'S REPORT

The surveyor's report is a detailed account of the surveyor's activities in making a survey for a client. It should include for whom, where, when, reasons for decisions, type of instrument used, and other comments.

An example of a surveyor's report follows.

Surveyor's Report

This report covers the survey of the Ralph Cramden 615.03 ac tract of land in Williams County, Texas, said tract being a part of the James Jone Survey no. 9 and the Frank Smith Survey no. 3. Said tract being all of the land conveyed by Ed Norton to said Ralph Cramden by deed dated April 1, 1982.

The sole purpose of this survey was to determine the location of boundary lines of land conveyed to Ralph Cramden in said deed, to mark any corners found, and to reestablish any missing corners so that a fence may be erected along the boundaries of said tract.

Title information and patent notes of the James Jones Survey and the Frank Smith Survey from the General Land Office in Austin, Texas, were obtained from the client.

Field work was commenced April 6, 1982, and was completed April 10, 1982. All work on the property was tied to a closed traverse around the perimeter of the property. The traverse was run with a Wild T-2 theodolite and a Hewlett-Packard Model 3800 EDM. All bearings are based on geodetic (true) north as determined from NGS Triangulation Station Maud and azimuth mark of said station.

With a working sketch of the original survey and a subdivision of said survey in hand, I and the field party went on the ground. We were able to locate the stones called for at the northeast and southeast corners of said James Jones Survey no. 9 and the stones called for at the northwest and southwest corners of Frank Smith Survey no. 3.

An exhaustive search was made for the mound and pits called for at the southwest and northwest corners of the said James Jones Survey, but no evidence or traces could be found.

We then proceeded to make necessary calculations for reestablishing said lost corners. It was determined from our survey that the east line of senior James Jones Survey is South 1920.00 varas (called South 1900 varas). Since the calls were for right-angle bearings off this line, we reestablished the northwest corner of this survey by setting a 1 in iron pin West 1900.00 varas from the stone at the northeast corner. We then set a 1 in iron pin West 1900.00 varas from the stone at the southeast corner. This made the west line of the James Jones Survey North 1920.00 varas (called North 1900 varas).

Since the junior Frank Smith Survey calls to begin at the northwest corner of the James Jones Survey and run south to the southwest corner of the James Jones Survey, we were able to determine that the stone found for the southwest corner of the Frank Smith Survey is actually West 1850.00 varas (called West 1900 varas).

The north line of the Frank Smith Survey from the stone found at the northwest corner to the iron pin set by us for the northwest corner of the senior James Jones Survey is actually North 89°41′28″ East 1860.01 varas (called East 1900 varas).

Since we had reestablished the original positions of these surveys, we then began locating Mr. Ralph Cramden's deed.

Since the deed to Ralph Cramden had no description other than "the 620 acres owned by me" (Ed Norton), out of the said surveys, we then had to refer to the deeds from James Jones to Ed Norton and from Frank Smith to Ed Norton, both being dated November 10, 1902.

The deed from James Jones to Ed Norton calls for the West 300 ac of the James Jones Survey. From this description, we calculated the east line of this tract to be 882.09 varas east of and parallel to the west line of said James Jones Survey. One inch iron pins were set for the northeast and southeast corners of this tract. In the deed from Frank Smith to Ed Norton, the conveyance was for the East One Half (*Edfrac*12) of the Frank Smith Survey. From this description, we bisected the north line and south line of said Frank Smith Survey, thereby reestablishing the west line of this tract, found to be 315.03 ac of land. One inch iron pins were then set for the northwest and southwest corners of this tract.

It should be noted that the actual acreage owned in these surveys by Ralph Cramden is five acres short of his deed call due to the Frank Smith Survey being short of the intended call by 10.22 ac. For more details, see the plat and field notes filed with this report.

115. SUMMARY OF RETRACING THE ORIGINAL SURVEY

- It is a rule of evidence that when there is a doubt as to whether a survey was made or not, that it be presumed that the law was complied with until it can be proved that it was not.

- If an overlap occurs the holder of the first grant or patent (senior) retains ownership of the area overlapped and the holder of the latter grant or patent (junior) loses the area overlapped.

- Where two parties have title to the same land, the party holding the senior conveyance has the right of possession.

- Calls for course distance are referred to as "metes" and calls for objects and adjoining boundaries are referred to as "bounds."

- Evidence to prove the location of a written title will not overturn a legal unwritten title.

- In locating boundaries, they should be located where the parties to the conveyance intended them to be.

- Intention of the surveyor is considered to be the intent of the parties, but intentions of the surveyor must be determined by his or her acts as shown in his or her field notes.

- Primary objective in locating a survey is to follow the footsteps of the surveyor.

- Natural objects and artificial objects called for in the field notes would be searched for, and if found, there can be little room for controversy.

- If natural objects and artificial objects called for in the field notes cannot be found, then the rule of law must control.

- General rules as to controlling calls are: (1) natural objects, (2) artificial objects, (3) course and distance, and (4) area.

- True location of land is ascertained by application of all, or any, rules of law.

- When contradiction occurs, that rule must be adopted that is most consistent with intention of the parties, apparent upon the face of the conveyance, read in the light of surrounding circumstances.

- Most material and certain calls control those that are less material and less certain.

- Course and distance are the most unreliable calls; distance is less reliable than course.

- Under certain conditions, course and distance may control, but generally they are but guides to calls for natural or artificial objects.

- The actual identification of the survey, the footsteps of the surveyor on the ground, should always be followed, by whatever means.

Part 10: Surveys of the Public Lands Under the U.S. System of Rectangular Surveys

116. GENERAL

In 1785, Congress enacted a law that provided for the subdivision of the public lands into townships 6 mi² with townships subdivided into 36 sections, most of which are 1 mi on a side. Sections were subdivided into half-sections, quarter-sections, and quarter-quarter sections (the quarter-quarter section being 40 ac in area). Thirty states of the United States were subdivided into tracts by this system known as the U.S. System of Rectangular Surveys.

The other states did not pass title to vacant lands to the United States. These states are Texas, West Virginia, Kentucky, Tennessee, the Colonial States, and the other New England and Atlantic Coast states except Florida.

117. QUADRANGLES

A *quadrangle* is approximately 24 mi² and consists of 16 townships. Quadrangles were laid out from an initial point through which was established a *principle meridian* and a baseline extending east and west that is a true parallel of latitude. All north-south township lines are true meridians, and all east-west township lines are circular curves that are parallels of latitude. Because of the convergence of meridians, quadrangle corners do not coincide except along the principal meridian.

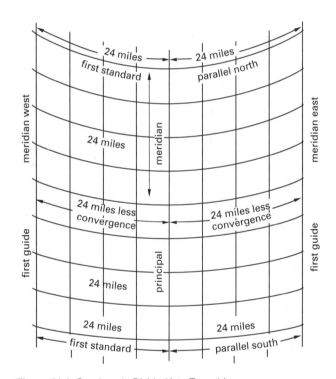

Figure 21.1 *Quadrangle Divided into Townships*

118. SUBDIVISION OF TOWNSHIPS

As explained in Sec. 116, in 1785 townships were divided into sections numbered from 1 to 36 beginning in the northeast corner and ending in the southeast corner as shown in Fig. 21.2. As many sections as possible with 1 mi on a side (640 ac) were laid out in the township. But, due to convergence of the east and west boundaries of a township, it was impossible for all 36 sections to be 1 mi on each side. East and west section boundaries were laid out parallel and not as true meridians. They were laid out parallel to the east boundary of the township. This made it impossible for all sections to be 1 mi on a side and at the same time coincide with the township lines.

To produce as many sections as possible 1 mi on a side, the sections along the north township line and the west township line were of varying dimensions to compensate for errors and the convergence of the west township line. The errors were actually thrown into the north one-half of the sections along the north township line and into the west one-half of the sections along the west township line. Thus, sections 1–6, 7, 18, 19, 30, and 31 were not regular sections. When a section was limited by a lake, river, or old survey, part of it was eliminated, but the existing section was numbered as if the whole section were laid out.

119. SUBDIVISION OF SECTIONS

Sections may be divided into half-sections, quarter-sections, half-quarter sections, or quarter-quarter sections.

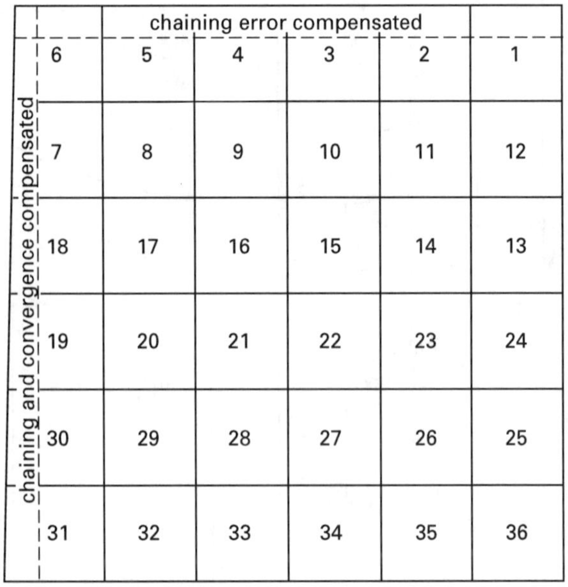

	chaining error compensated				
6	5	4	3	2	1
7	8	9	10	11	12
18	17	16	15	14	13
19	20	21	22	23	24
30	29	28	27	26	25
31	32	33	34	35	36

(left axis label: chaining and convergence compensated)

Figure 21.2 *Township Subdivided into Sections*

Part 11: Restoring Lost Corners in U.S. System of Rectangular Surveys

120. JURISDICTION

The U.S. Bureau of Land Management, under the supervision of the Secretary of the Interior, has complete jurisdiction over the survey and resurvey of the public lands of the United States.

After title to a piece of land is granted by the United States, jurisdiction over the property passes to the state. The federal government retains its authority only with respect to the public lands in federal ownership. Where the lands are in private ownership, it is a function of the county or local surveyor to restore lost corners and to subdivide the sections. Disputes concerning these questions must come before the local courts unless settled by joint survey or agreement. It should be understood, however, that no adjoining owner can make a valid encroachment upon the public lands.

121. RESURVEYS

Public and privately owned lands may both be resurveyed by the Bureau of Land Management in certain cases, under the authority of an act of Congress approved March 3, 1909, and amended June 25, 1910:

> That the Secretary of the Interior may, in his discretion, cause to be made, as he may deem wise under the rectangular system now provided by law, such resurveys or retracements of the surveys of public lands as, after full investigation, he may deem essential to properly mark the boundaries of the public lands remaining undisposed of.

The 1909 act is generally invoked where the lands are largely in federal ownership and where there may be extensive obliteration or other equally unsatisfactory conditions.

Another act of Congress approved September 21, 1918, provides authority for the resurvey by the government of townships previously ineligible for resurvey by reason of the disposals' being in excess of 50% of the total area.

The 1918 act may be invoked where the major portion of the township is in private ownership, where it is shown that the need for retracement and remonumentation is extensive, and especially if the work proposed is beyond the scope of ordinary local practice. The act requires that the proportionate costs be carried by the landowners.

122. PROTECTION OF BONA FIDE RIGHTS

Under the laws discussed in Sec. 121, and in principle as well, it is required that no resurvey or retracement shall be so executed as to impair the bona fide rights or claims of any claimant, entryman, or owner of land so affected.

Likewise in general practice, local surveyors should be careful not to exercise unwarranted jurisdiction, nor to apply an arbitrary rule. They should note the distinction between the rules for original surveys and those that relate to retracements. The disregard for these principles, or for acquired property rights, may lead to unfortunate results.

In unusual cases where the evidence of the survey cannot be identified with ample certainty to enable the application of the regular practices, the surveyors may submit their questions to the proper state office of the

Bureau of Land Management, or to the Director of the Bureau of Land Management.

123. ORIGINAL SURVEY RECORDS

The township plat furnishes the basic data relating to the survey and the description of all areas in the particular township. All title records within the area of the former public domain are based upon a government grant or patent, with description referred to an official plat. The lands are identified on the ground through the retracement, restoration, and maintenance of the official lines and corners.

The plats are developed from the field notes. Both are permanently filed for reference purposes and are accessible to the public for examination or making of copies.

Many supplemental plats have been prepared by protraction to show new or revised lottings within one or more sections. These supersede the lottings shown on the original township plat. There are also many plats of the survey of islands or other fragmentary areas of public land that were surveyed after the original survey of the township.

These plats should be referred to as governing the position and a description of the subdivision should be shown on them.

124. RESURVEY RECORDS

The plats and field notes of resurveys that become a part of the official record fall into two principal classes according to the type of resurvey.

A *dependent resurvey* is a restoration of the original survey according to the record of that survey, based upon the identified corners of the original survey and other acceptable points of control, and the restoration of lost corners in accordance with proportional measurement as described herein. Normally, the subdivisions shown on the plat of the original survey are retained on the plat of the dependent resurvey, although new designations and areas for subdivisions still in public ownership at the time of the resurvey may be shown to reflect true areas.

An *independent resurvey* is designed to supersede the original survey and creates new subdivisions and lottings of the vacant public lands. Provision is made for the segregation of individual tracts of privately owned lands, entries, or claims that may be based upon the original plat, when necessary for their protection, or for their conformation, if feasible, to the regular subdivisions of the survey.

125. RECORDS TRANSFERRED TO STATES

In those states where the public land surveys are considered as having been completed, the field notes, plats, maps, and other papers relating to those surveys have been transferred to an appropriate state office for safekeeping as public records. No provision has been made for the transfer of the survey records to the State of Oklahoma, but in the other states the records are filed in offices where they may be examined and copies made or requested.

126. GENERAL PRACTICES

The rules for the restoration of lost corners have remained substantially the same since 1883, when they were first published. These rules are in harmony with the leading judicial opinions and the most approved surveying practice. They are applicable to the public land rectangular surveys and to the retracement of those surveys (as distinguished from the running of property lines that may have legal authority only under state law, court decree, or agreement).

In the New England and Atlantic coast states except Florida, and in Pennsylvania, West Virginia, Kentucky, Tennessee, and Texas, jurisdiction over the vacant lands remained in the states. The public land surveys were not extended in these states, and it follows that the practices that are outlined herein are not applicable there, except as they reflect sound surveying methods.

The practices outlined herein are in accord with the related provisions of the BLM *Manual of Surveying* (1973). They have been segregated for convenience to separate them from the instructions pertaining only to the making of original surveys.

For clarity, the practices, as such, are set in boldface type. The remainder of the text is explanatory and advisory only, the purpose being to exemplify the best general practice.

In some states, the substance of the practices for restoration of lost or obliterated corners and subdivision of sections as outlined herein has been enacted into law. It is incumbent on the surveyor engaged in practice of land surveying to become familiar with the provisions of the laws of the state, both legislative and judicial, that affect his or her work.

127. GENERAL RULES

The general rules followed by the Bureau of Land Management, which affect all public lands, are summarized in the following paragraphs.

- **The boundaries of the public lands, when approved and accepted, are unchangeable.**

- **The original township, section, and quarter-section corners must stand as the true corners that they were intended to represent, whether in the place shown by the field notes or not.**

- Quarter-quarter section corners not established in the original survey shall be placed on the line connecting the section and quarter-section corners, and midway between them, except on the last half-mile of section lines closing on the north and west boundaries of the township, or on the lines between fractional or irregular sections.

- The center lines of a section are to be straight, running from the quarter-section corner on one boundary to the corresponding corner on the opposite boundary.

- In a fractional section where no opposite corresponding quarter-section corner has been or can be established, the center line must be run from the proper quarter-section corner as nearly in a cardinal direction to the meander line, reservation, or other boundary of such fractional section, as due parallelism with the section boundaries will permit.

Corners established in the public land surveys remain fixed in position and are unchangeable, and lost or obliterated corners of those surveys must be restored to their original locations from the best available evidence of the official survey in which such corners were established.

128. RESTORATION OF LOST OR OBLITERATED CORNERS

The restoration of lost corners should not be undertaken until after all control has been developed. Such control includes both original and acceptable collateral evidence. However, the methods of proportionate measurement will be of material aid in the recovery of evidence.

An existent corner is one whose position can be identified by verifying the evidence of the monument, or its accessories, by reference to the description that is contained in the field notes, or where the point can be located by an acceptable supplemental survey record, some physical evidence, or testimony.

Even though its physical evidence may have entirely disappeared, a corner will not be regarded as lost if its position can be recovered through the testimony of one or more witnesses who have a dependable knowledge of the original location.

An obliterated corner is one at whose point there are no remaining traces of the monument, or its accessories, but whose location has been perpetuated, or the point that may be recovered beyond reasonable doubt, by the acts and testimony of the interested landowners, competent surveyors, or other qualified local authorities, or witnesses, or by acceptable record evidence.

A position based upon collateral evidence should be duly supported, generally through proper relation to known corners, and agreement with the field notes regarding distances to natural objects, stream crossings, line trees, and off-line tree blazes, and so on, or unquestionable testimony.

A lost corner is a point of survey whose position cannot be determined, beyond reasonable doubt, either from traces of the original marks or from acceptable evidence or testimony that bears upon the original position, and whose location can be restored only by reference to one or more interdependent corners.

If there is some acceptable evidence of the original location of the corner, that position will be employed.

The decision that a corner is lost should not be made until every means has been exercised that might aid in identifying its true original position. The retracements, which are usually begun at known corners and run according to the record of the original survey, will indicate the probable position for the corner, and show what discrepancies may be expected. Any supplemental survey record or testimony should then be considered in light of the facts thus developed. A line will not be regarded as doubtful if the retracement affords recovery of acceptable evidence.

In cases where the probable position for a corner cannot be made to harmonize with some of the calls of the field notes, due to errors in description or to discrepancies in measurement developed in the retracement, it must be ascertained which of the calls for distances along the line are entitled to the greater weight. Aside from the technique of recovering traces of the original marks, the main problem is one that treats with the discrepancies in alignment and measurement.

Existing original corners cannot be disturbed; consequently, discrepancies between the new and the record measurements will not in any manner affect the measurements beyond the identified corners, but the difference will be distributed proportionally within the several intervals along the line between the corners.

129. PROPORTIONATE MEASUREMENT

The ordinary field problem consists of distributing the excess or deficiency in measurement between existent corners in such a manner that the amount given to each interval shall bear the same proportion to the whole difference as the record length of the interval bears to the whole record distance. After having applied the proportionate difference to the record length of each interval, the sum of the several parts will equal the new measurements of the whole distance.

A proportionate measurement is one that gives concordant relation between all parts of the line—that is, the new values given to the several parts, as determined by the remeasurement, shall bear the same relation to the record lengths as the new measurement of the whole line bears to that record. Lengths of proportioned lines are comparable only when reduced to their cardinal equivalents.

Discrepancies in measurement between those recorded in the original survey and those developed in the retracements should be carefully verified with the object by placing each such difference properly where it belongs. This is quite important at times, because, if disregarded, the result may be the fixing of a corner position where it is obviously improper. Accordingly, wherever possible, the manifest errors in the original measurements should be segregated from the general average difference and placed where the blunder was made. The accumulated surplus or deficiency that then remains is the quantity that is to be uniformly distributed by the methods of proportionate measurement.

130. SINGLE PROPORTION

The term "single proportionate measurement" is applied to a new measurement made on a line to determine one or more positions on that line.

In *single proportionate measurement*, the position of two identified corners controls the direction of the line between those corners, and intermediate positions on that line are determined by proportionate measurement between those controlling corners. The method is sometimes referred to as a *two-way proportion*. Examples are: a quarter-section corner on the line between two section corners; all corners on standard parallels; and all intermediate positions on any township boundary line.

131. DOUBLE PROPORTION

The term "double proportionate measurement" is applied to new measurement made between four known corners, two each on intersecting meridional and latitudinal lines, for the purpose of relating the intersection of both.

By *double proportionate measurement*, the lost corner is reestablished on the basis of measurement only, disregarding the record directions. An exception will be found in those cases where there is some acceptable survey record, some physical evidence, or testimony that may be brought into the control. The method may be referred to as a *four-way proportion*. Examples are a corner common to four townships, or one common to four sections within a township.

The double proportionate measurement is the best example of the principle that existent or known corners to the north and to the south should control any intermediate latitudinal position, and that corners east and west should control the position in longitude.

As between single or double proportionate measurement, the principle of precedence of one line over another of less original importance is recognized in order to harmonize the restoring process with the method followed in the original survey, thus limiting control.

132. STANDARD PARALLELS AND TOWNSHIP BOUNDARIES

Standard parallels will be given precedence over other township exteriors, and ordinarily the latter will be given precedence over subdivisional lines; section corners will be relocated before the position of lost quarter-section corners can be determined.

To restore a lost corner of four townships, a retracement will first be made between the nearest known corners on the meridional line, north and south of the missing corners, and upon that line a temporary stake will be placed at the proper proportionate distance; this will determine the latitude of the lost corner.

Next, the nearest corners on the latitudinal line will be connected, and a second point will be marked for the proportionate measurement east and west; this point will determine the position of the lost corner in departure (or longitude).

Then, through the first temporary stake run a line east or west, through the second temporary stake a line north or south, as relative situations may determine; the intersection of these two lines will fix the position for the restored corner.

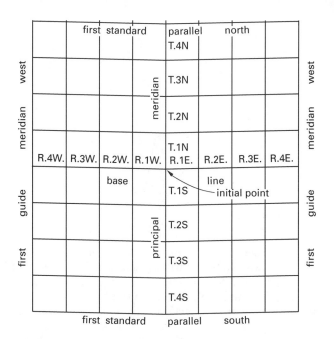

Figure 21.3 *Standard Parallels and Township Boundaries*

In Fig. 21.4, points A, B, C, and D represent four original corners. Point E represents the proportional measurement between A and B; and similarly, F represents the proportional measurement between C and D. Point X satisfies the first control for latitude and the second control for departure.

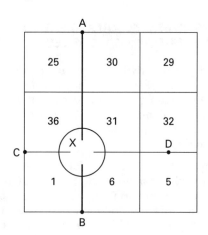

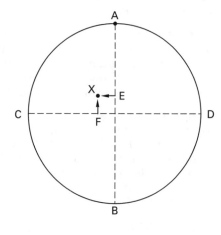

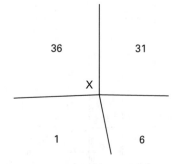

Figure 21.4 *Lost Township*

A lost corner of four townships should not be restored, nor the township boundaries reestablished, without first considering the full field note record of the four intersecting lines and the plats of the township involved. In most cases there is a fractional distance in the half-mile to the east of the township, corner, and frequently in the half-mile to the south. The lines to the north and to the west are usually regular—that is, quarter-section

and section corners at normal intervals of 40.00 and 80.00 chains—but there may be closing-section corners on any or all of the boundaries. So, it is important to verify all of the distances by reference to the field notes.

133. INTERIOR CORNERS

A lost interior corner of four sections will be restored by double proportionate measurement.

When a number of interior corners of four sections, and the intermediate quarter-section corners, are missing on all sides of the one sought to be reestablished, the entire distance between the nearest identified corners both north and south, east and west, must be measured. Lost section corners on the township exteriors, if required for control, should be relocated.

134. RECORD MEASURMENT

Where the line has not been established in one direction from the missing township or section corner, the record distance will be used to the nearest identified corner in the opposite direction.

Thus, in Fig. 21.4, if the latitudinal line in the direction of point D has not been established, the position of point F in departure would have been determined by reference to the record distance from point C. Point X would then be fixed by cardinal offsets from points E and F as already explained.

Where the intersecting lines have been established in only two of the directions, the record distances to the nearest identified corners on these two lines will control the position of the temporary points; then from the latter the cardinal offsets will be made to fix the desired point of intersection.

135. TWO SETS OF CORNERS

In many surveys the field notes and plats indicate two sets of corners along township boundaries and section lines where parts of the township were subdivided on different dates. In such cases, there are usually corners of two sections at regular intervals, and closing section corners established later upon the same line. The quarter-section corners on such lines usually are controlling for one side only.

In the more recent surveys, where the record calls for two sets of corners, those that are the corners of the two sections first established and the quarter-section corners relating to the same sections will be employed for the retracement, and they will govern both the alignment and the proportional measurements along that line. The closing section corners set at the intersections will be employed in the usual way—that is, to govern the direction of the closing lines.

136. RESTORATION BY SINGLE PROPORTION

The method of single proportionate measurement is generally applicable to the restoration of lost corners on standard parallels and other lines established with reference to definite alignment in one direction only. Intermediate corners on township exteriors and other controlling boundary lines are to be included in this class.

To restore a lost corner by single proportionate measurement, a retracement will be made connecting the nearest identified regular corners on the line in question. A temporary stake will be set on the trial line at the original distance. The total distance and the falling at the objective corner will be measured.

On meridional township lines, an adjustment will be made at each temporary stake for the proportional distance along the line. The temporary stake will be set over to the east or to the west for falling, counting its proportional part from the point of beginning.

On east-and-west township lines and on standard parallels, the proper adjustment should be made at each temporary stake for the proportional distance along the line for the falling. (The temporary stake will either be advanced or set back for the proportional part of the distance between the record distance and the new measurement. It will then be set over for the curvature of the line, and eventually corrected for the proportional part of the true falling.)

The adjusted position is thus placed on the true line that connects the nearest identified corners, and at the same proportional interval from either as existed in the original survey. Any number of intermediate lost corners may be located on the same plan by setting a temporary stake for each when making the retracement.

Lost standard corners will be restored to their original positions on a baseline, standard parallel or correction line, by single proportionate measurement on the true line connecting the nearest identified standard corners on opposite sides of the missing corner or corners, as the case may be.

The term *standard corners* will be understood to mean all corners that were established on the standard parallel during the original survey of that line, including, but not limited to, standard township, section, quarter-section, meander, and closing corners. Closing corners, or other corners purported to be established on a standard parallel after the original survey of that line, will not control the initial restoration of lost standard corners.

Corners on baselines are to be regarded as the same as those on standard parallel. In the older practice, the term *correction line* was used for what later has been called the *standard parallel*. The corners first set in the running of a correction line will be treated as original standard corners. Those that were set afterwards at the intersection of a meridional line will be regarded as closing corners.

All lost section and quarter-section corners on the township boundary lines will be restored by single proportionate measurement between the nearest identified corners on opposite sides of the missing corner, north and south on a meridional line, or east and west on a latitudinal line, after the township corners have been identified or relocated.

An exception to this rule will be found in the case of any exterior the record of which shows deflections in alignment between the township corners.

A second exception to this rule is found in those occasional cases were they may be persuasive proof of a deflection in the alignment of the township boundary, even though the record shows the line to be straight. For example, measurements east and west across a range line, or north and south across a latitudinal township line, counting from a straight-line exterior adjustment, may show distances to the nearest identified subdivisional corners to be materially long in one direction and correspondingly short in the opposite direction as compared to the record measurements. This condition, when supported by corroborative collateral evidence as might generally be expected, would warrant an exception to the straight-line or two-way adjustment under the rules for the acceptance of evidence—that is, the evidence outweighs the record. The rules for a four-way or double proportionate measurement would then apply, provided there is conclusive proof.

All lost quarter-section corners on the section boundaries within the township will be restored by single proportionate measurement between the adjoining section corners, after the section corners have been identified or relocated.

This practice is applicable in the majority of the cases. However, in those instances where other corners such as meander corners, sixteenth-section corners, and so on, were originally established between the quarter-section and the section corners, such minor corners, when identified, will exercise control in the restoration of lost quarter-section corners.

Lost meander corners, originally established on a line projected across the meanderable body of water and marked upon both sides, will be relocated by single proportionate measurement, after the section or quarter-section corners upon the opposite sides of the missing meander corner have been duly identified or relocated.

Under ordinary conditions, the actual shore line of a body of water is considered the boundary of lands included in an entry and patent, rather than the meander line returned in the field notes. It follows that the restoration of a lost meander corner would be required only infrequently. Under favorable conditions, a lost meander corner may be restored by treating the shore-line as an identified natural feature controlling the measurement to the point for the corner. This is particularly applicable where it is evident that there has been no change in the shore line.

A lost closing corner will be reestablished on the true line that was closed upon, and at the proper proportional interval between the nearest regular corners to the right and left.

To reestablish a lost closing corner on a standard parallel or other controlling boundary, the line that was closed upon will be retraced, beginning at the corner from which the connecting measurement was originally made. A temporary stake will be set at the record connecting distance, and the total distance and falling will be noted at the next regular corner on that line on the opposite side of the missing closing corner. The temporary stake will then be adjusted as in single proportionate measure.

A closing corner not actually located on the line that was closed upon will determine the direction of the closing line, but not its legal terminus. The correct position is at the true point of intersection of the two lines.

137. IRREGULAR EXTERIORS

Some township boundaries, not established as straight lines, are termed *irregular* exteriors. Parts were surveyed from opposite directions, and the intermediate portion was completed later by random and true line, leaving a fractional distance. Such irregularity follows some material departure from the basic rules for the establishment of original surveys. A modified form of single proportionate measurement is used in restoring lost corners on such boundaries. This is also applicable to a section line or township line that has been shown to be irregular by a previous retracement.

To restore one or more lost corners or angle points on such irregular exteriors, a retracement between the nearest known corners is made on the record courses and distances to ascertain the direction and length of the closing distance. A temporary stake is set for each missing corner or angle point. The closing distance is then reduced to its equivalent latitude and departure.

On a meridional line, the latitude of the closing distance is distributed among the courses in proportion to the latitude of each course. The departure of the closing distance is distributed among the courses in proportion to the length of each course. That is, after the excess or deficiency of latitude is distributed, each temporary stake is moved east or west an amount proportional to the total distance from the starting point.

On a latitudinal line, the temporary stakes should be placed to suit the usual adjustments for curvature. The departure of the closing distance is distributed among the courses in proportion to the departure of each course. Then, each temporary stake is moved

north or south an amount proportional to the total distance from the starting point.

Angle points and intermediate corners are treated alike.

138. ONE-POINT CONTROL

Where a line has been terminated with measurement in one direction only, a lost corner will be restored by record bearing and distance, counting from the nearest regular corner, the latter having been duly identified or restored.

Examples will be found where lines have been discontinued at the intersection with large meandrous bodies of water, or at the border of what was classified as impassable ground.

139. INDEX ERRORS FOR ALIGNMENT AND MEASUREMENT

Where the original surveys were faithfully executed, it is to be anticipated that retracement of many miles of the lines in a given township will develop a definite and consistent difference in measurement and in a bearing between original corners. Under such conditions, it is proper that allowance be made for the average differences in the restoration of a lost corner where control is lacking in one direction. The adjustment will be taken care of automatically where there is a suitable basis for proportional measurement.

140. SUBDIVISION OF SECTIONS

The ordinary unit of administration of the public lands under the rectangular system of surveys is the quarter-quarter-section of 40 ac. Usually the sections are not subdivided on the ground in the original survey. The boundaries of the legal subdivisions generally are shown by protraction on the plats.

On the plat of the original survey of a normal township, it is to be expected that the subdivision of sections will be indicated (by protraction) according to standard procedures.

The sections bordering the north and west boundaries of the township, except section 6, are subdivided into two regular quarter-sections, two regular half-quarter sections, and four fractional quarter-quarter units, which are usually designated as lots. In section 6, the subdivision will show one regular quarter-section, two regular half-quarter sections, one regular quarter-quarter section, and seven fractional quarter-quarter units. This is the result of the plan of subdivision, whereby the excess or deficiency in measurement is placed against the north and west boundaries of the township.

The plan of subdivisions and controlling measurements employed is illustrated in Fig. 21.6.

In a normal section that is subdivided by protraction into quarter-sections, it is not considered necessary to indicate the boundaries of the quarter-quarter sections on the plat. Such subdivisions are aliquot parts of the quarter-sections, based on midpoint protraction.

Sections that are invaded by meanderable bodies of water or by private claims that do not conform to the regular legal subdivisions are subdivided by protraction into regular and fractional parts as nearly as practicable in conformity with the uniform plan already outlined. The meander lines, and the boundary lines of the private claims, are platted according to the field note record. The subdivision-of-section lines are terminated at the meander line or claim boundary, as the case may be, but their positions are controlled precisely as though the section had been completed regularly. For the purpose of protracting the subdivisional lines in a section whose boundary lines are partly within the limits of a meanderable body of water or private claim, the fractional section boundaries are completed in theory.

The protracted position of the subdivision-of-section line is controlled by the theoretical points so determined (see Fig. 21.5).

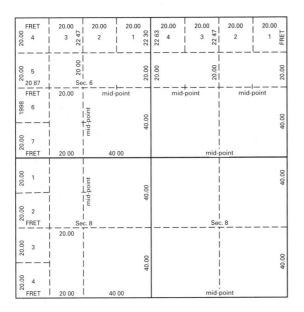

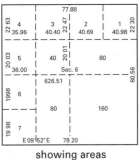
showing areas

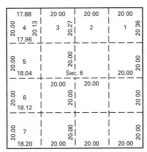
showing calculated distances

Figure 21.5 *Examples of Subdivision by Protraction*

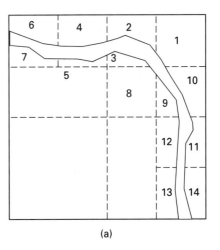
(a)

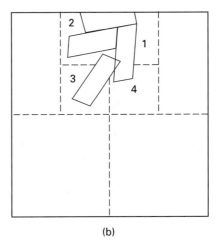
(b)

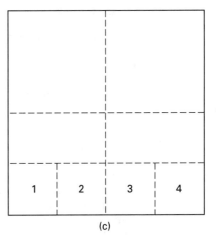
(c)

Figure 21.6 *Examples of Subdivision of Fractional Sections*

141. ORDER OF PROCEDURE IN SURVEY

The order of procedure is first to identify or reestablish the corners on the section boundaries, including determination of the points for the necessary one-sixteenth section corners. Next, fix the boundaries of the quarter-section. Finally, form the quarter sections or small tracts by equitable and proportionate division.

142. SUBDIVISION OF SECTIONS INTO QUARTER-SECTIONS

To subdivide a section into quarter-section, run straight lines from the established quarter-section corners to the opposite quarter-section corners. The point of intersection of the lines thus run will be the corner common to the several quarter-sections, or the legal center of the section.

Upon the lines closing on the north and west boundaries of a regular township, the quarter-section corners were established originally at 40 chains to the north or west of the last interior section corners. The excess or deficiency in measurement was thrown into the half-mile next to the township or range line, as the case may be. If such quarter-section corners are lost, they should be reestablished by proportionate measurement based upon the original record.

Where there are double sets of section corners on township and range lines, the quarter-section corners for the sections south of the township line and east of the range line usually were not established in the original surveys. In subdividing such sections, new quarter-section corners are required. They should be placed to suit the calculations of the areas that adjoin the township boundary, as indicated upon the official plat, adopting proportional measurements where the new measurements of the north or west boundaries of the section differ from the record distances.

143. SUBDIVISION OF FRACTIONAL SECTIONS

The law provides that where opposite corresponding quarter-section corners have not been or cannot be fixed, the subdivision-of-section lines shall not be ascertained by running from the established corners north, south, east, or west, as the case may be, to the water course, reservation line, or other boundary of such fractional section, as represented upon the official plat.

This law presumes that the section lines are either due north and south, or east and west lines, but usually this is not the case. Hence, to carry out the spirit of the law, it will be necessary in running the center lines through fractional sections to adopt mean courses, where the section lines are not on due cardinal, or to run parallel to the east, south, west, or north boundary of the section, as conditions may require, where there is no opposite section line.

144. SUBDIVISION OF QUARTER-SECTIONS

Preliminary to the subdivision of quarter-sections, the quarter-quarter, or sixteenth-section, corners will be established at points midway between the section and quarter-section

corners, and the center of the section, except on the last half mile of the lines closing on township boundaries, where they should be placed at 20 chains, proportionate measurement, counting from the regular quarter-section corner.

The quarter-quarter, or sixteenth-section, corners having been established as directed previously, the center lines of the quarter-section will be run straight between opposite corresponding quarter-quarter, or intersection of the lines thus run will determine the legal center of a quarter-section (see Fig. 21.7).

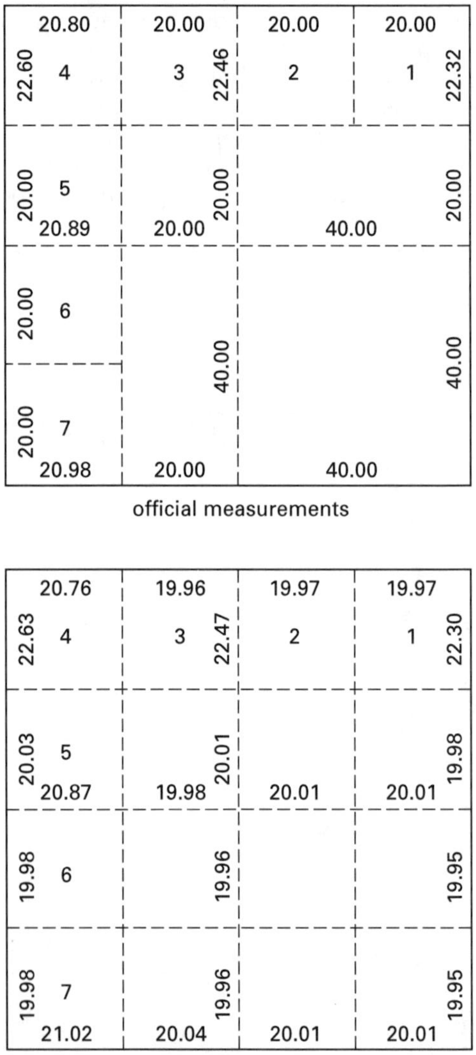

official measurements

remeasurements

Figure 21.7 *Example of Subdivision by Survey*

145. SUBDIVISION OF FRACTIONAL QUARTER-SECTIONS

The subdivisional lines of fractional quarter-sections will be run from properly established quarter-quarter, or sixteenth-section corners, with courses governed by the conditions represented upon the official plat, to the lake, watercourse, reservation, or other irregular boundary that renders such sections fractional.

What has been written on the subject of subdivision of sections relates to the procedure contemplated by law, and refers to the methods to be followed in the initial subdivision of the areas, prior to development and improvement. Care should be exercised to avoid disturbing satisfactory improvements such as roads, fences, or other features making subdivision-of-section lines and that may define the extent of property rights.

146. RETRACEMENTS

When it is necessary to retrace the lines of the rectangular public-land surveys, the first step is to assemble copies of field notes and plats and determine the names of the owners who will be concerned in the retracement and survey. A thorough search and inquiry with regard to the record of any additional surveys that have been made since the approval of the original survey should be made. The county surveyor, county clerk, register of deeds, practicing engineers and surveyors, landowners, and others who may furnish useful information should be consulted as to such features.

The matter of boundary disputes should be carefully reviewed, particularly as to whether claimants have based their locations upon evidence of the original survey and a proper application of surveying rules. If there has been a boundary suit, the record testimony and the court's opinion and decree should be carefully examined insofar as these may have a bearing upon the problem at hand.

The law requires that the position of original corners must not be changed. There is a penalty for defacing corner marks, or for changing or removing a corner. The corner monuments afford the principle means for identification of the survey, and accordingly, the courts attach the greatest weight to the evidence of their location. Discrepancies that may be developed in the directions and lengths of lines, as compared with the original record, do not warrant any alteration of a corner position.

Obviously, on account of roadways or other improvements, it is frequently necessary to reconstruct a monument in some manner in order to preserve its position. Alterations of this type are not regarded as changes in willful violation of the law, but rather as being in complete accordance with the legal intent to safeguard the evidence.

Therefore, whatever the purpose of the retracement may be—if it calls for the recovery of the true lines of the original survey, or for the running of the subdivisional lines of a section—the practices outlined require some or all of certain definite steps.

step 1: Secure a copy of the original plat and field notes.

step 2: Secure all available data regarding subsequent surveys.

step 3: Secure the names and contact the owners of the property adjacent to the lines that are involved in the retracement.

step 4: Find the corners that may be required:

- first, by the remaining physical evidence;

- second, by collateral evidence, supplemental survey records, or testimony, if the original monument is regarded as obliterated, but not lost, or;

- third, by application of the rules for proportionate measurement, if lost.

step 5: Reconstruct the monuments as required, including the placing of reference markers where improvements of any kind might interfere, or if the site is such as to suggest the need for supplemental monumentation.

step 6: Note the procedure for the subdivision of sections where these lines are to be run.

step 7: Prepare and file a suitable record of what was found, the supplemental data that were employed; a description of the methods used, the direction and length of lines, the new markers, and any other facts regarded as important.

A knowledge of the practices and instructions in effect at the time of the original survey will be helpful. These should indicate what was required and how it was intended that the original survey should be made.

The data used in connection with the retracements should not be limited to the section or sections under immediate consideration. They should also embrace the areas adjacent to those sections. The plats should be studied carefully. Fractional parts of sections should be located on the ground as indicated on the plats.

147. DOUBLE SETS OF CORNERS

The methods of determining true meridian and true latitudinal curve were developed many years after the inception of the rectangular survey system. Without these refinements, accumulated discrepancies were bound to develop.

As a result, in order to maintain rectangularity in some of the older surveys, two sets of corners were established on the township boundaries. The section and quarter-section corners established in the survey of the boundary itself are the corners to be adopted in retracement and for control of proportionate measurements. These corners control the subdivisions on only one side of the township boundary. The second set of corners on these boundaries are the closing section corners for the subdivisional surveys on the opposite side of the boundary. The descriptions of these closing corners, and the connecting distances to the regular township boundary corners, will be found in the field notes of the subdivisional survey in which they were established. These closing section corners should be considered and evaluated as evidence in the solution of the whole problem.

Where the section corners of the township boundaries are of minimum control, the quarter-section corners have the same status for the same side of the boundary. In the older surveys, quarter-section corners usually were not established for the opposite side of the boundary. Subsequent to 1919, it was the practice to establish the second set of quarter-section corners. These are at midpoint for distances between the closing section corners, except where the plan of subdivision dictates otherwise, in which event the quarter-section corner is placed at 40.00 chains from the controlling closing corner.

These conditions merit careful study of the plats to the end that the subdivisions shown on the plats be given proper protection. The plats will indicate whether these quarter-section corners should be at midpoint between the closing corners, or if they should be located with regard to a fractional distance. The surveyor should make sure that the position is determined for all corners necessary for control in his or her work.

There is nothing especially different or complicated in the matter of one or two sets of corners on the township boundary lines. It is merely a matter of assembling complete data and of making a proper interpretation of the status of each monument.

The same principles should be applied in the consideration of the data of the subdivisional surveys, where for any of the several causes, there may be two sets of interior corners.

148. THE NEEDLE COMPASS AND SOLAR COMPASS

Very simple needle-compass equipment in the hands of persons skilled as surveyors, coupled with natural woodcraft and faithfulness in doing their work, satisfied the requirements of the early public-land surveys.

A large proportion of surveys made prior to 1890 are of the needle-compass type. It should be noted that retracements may be made—that is, the evidence of the marks can be developed—by needle-compass methods if properly employed. Some surveyors maintain that you can "follow the steps" of the original surveyor more closely by use of the needle compass than by more precise methods.

In addition to the uncertainties of local attraction and temporary magnetic disturbances, the use of the needle compass is exceedingly unreliable in the vicinity of power lines, pipelines, steel rails, steel-framed structures of all kinds, wire fences, and so on. Its use is now much more restricted because of these improvements. The needle compass is rapidly becoming obsolete because it fails to satisfy the present need for more exact retracements.

149. EXCESSIVE DISTORTION

The needle-compass surveys, before being discontinued, had extended into the region of magnetic ore deposits of the Lake Superior watershed in northern Michigan, Wisconsin, and Minnesota. Here many townships were surveyed, and the lands patented, in which the section boundaries are now found to be grossly distorted. There is no way in which to correct these lines, nor to make an estimate (except by retracement), of the extent of the irregularities, which involve excessive discrepancies both in the directions and lengths of lines.

Considerable experience is required in retracing and successfully developing the evidence of the lines and corners in these areas of excessive distortion. However, the procedure for restoration of lost corners and for the subdivision of sections is the same as in areas of more regularity.

Another feature to be considered in connection with the retracements is that the record may show that one surveyor ran the south boundary, a second the east boundary, and others the remaining exteriors and subdivisional lines. All of these lines may be reported on cardinal, but may not be exactly comparable—that is, the east boundary may not be truly normal to the south boundary, and so on. It was customary to retrace one or more miles of the east boundary of the township to determine the "variation" of the needle. This value was then adopted in the subdivision of the township. It follows that the meridional section lines should be found to be reasonably parallel with the east boundary. Under that plan of operation, it should be anticipated that the latitudinal section lines should be found to be reasonably parallel with the south boundary. However, discrepancies in measurement on the meridional lines frequently affect such parallelism. For these reasons, the index corrections for bearings may not be the same for the east and west as for the north and south lines. The two classes should be considered separately in this respect.

Before 1900, most lines were measured with the Gunter's link chain. The present surveyor must realize the difficulties of keeping that chain at standard length and the inaccuracy of measuring steep slops by this method. It is to be expected that the retracements will show various degrees of accuracy in the recorded measurements, which were intended to reflect true horizontal distances.

150. INDEX ERRORS

Where the original surveys were faithfully made, there will generally be considerable uniformity in the directions and lengths of the lines. Frequently, this uniformity is so definite as to indicate *index errors* which, if applied to the record bearings and distances, will place the trial lines in close proximity to the true positions and aid materially in the search for evidence. With experience, the present surveyor will become familiar with the work of the original surveyor and know about what to expect in the way of such differences.

151. COLLATERAL EVIDENCE

The identified corners of the original survey constitute the main control for the surveys to follow. After those corners have been located, and before resorting to proportionate measurement for restoration of lost corners, the other calls of the field notes should be considered. The recorded distances to stream crossings or to other natural objects that can be identified often lead to the position for a missing corner. At this stage, the question of acceptance of later survey marks and records, the location of roads and property fences, and the reliability of testimony are to be considered.

A line tree, connection to some natural object, or an improvement recorded in the original field notes that can be identified may fix a point of the original survey. The calls of the field notes for the various items of topography may assist materially in the recovery of the lines. The mean position of a blazed line, when identified as the original line, will identify a meridional line for departure or a latitudinal line for latitude. These are matters that require the exercise of considerable judgment.

152. ORIGINAL MARKS

Original line-tree marks, off-line tree blazes, and scribe marks on bearing trees and tree corner-monuments whose age exceeds 100 years are found occasionally. Such marks of later surveys are recovered in much greater number. Different surveyors used distinctive marks. Some surveyors used hacks instead of blazes, and some used hacks over and under the blazes; some employed distinctive forms of letters and figures. All of these will be recognized while retracing the lines of the same survey and will serve to verify the identification of the work of a particular surveyor.

The field notes give the species and the diameter of the bearing trees and line trees. Some of the smooth-barked trees were marked on the surface, but most of the marks were made on a flat smoothed surface of the live wood tissue. The marks remain as long as the tree is sound. The blaze and marks will be covered by a gradual overgrowth, showing a scar for many years. The overgrowth will have a lamination similar to the annual rings of the tree, which may be counted in order to verify the date

of marking, and to distinguish the original marks from later marks and blazes. On the more recent surveys, it is to be expected that the complete quota of marks should be found, clear cut and plainly legible. This cannot be expected in the older surveys, however.

It is advisable not to cut into a marked tree except as necessary to secure proof. The evidence is frequently so abundant, especially in the later surveys, that the proof is conclusive without inflicting additional injury that would hasten the destruction of the tree.

Finding original scribe marks, line-tree hacks, and off-line tree blazes furnishes the most convincing identification that can be desired.

It is not intended to disturb satisfactory local conditions with respect to roads and fences. Surveyors have no authority to change a property right that has been acquired legally. On the other hand, they should not accept the location of roads and fences as evidence of the original survey without something to support these locations. This supporting evidence may be found in some intervening survey record, or the testimony of individuals who may be acquainted with the facts.

153. RULES ESTABLISHED BY STATE LAW OR DECISIONS

Other factors that require careful consideration are the rules of the state law and the state court decisions, as distinguished from the methods followed by the Bureau of Land Management. Under state law, property boundaries may be fixed by agreement between owners, acquiescence, or adverse possession. Such boundaries may be defined by roads, fences, or survey marks, disregarding exact conformation with the original section lies. The rights of adjoining owners may be limited to such boundaries.

In many cases, due care has been exercised to place the property fences on the lines of legal subdivision. It has been the general practice in the prairie states to locate the public roads on the section lines. These are matters of particular interest to the adjoining owners. It is reasonable to presume that care and good faith were exercised in placing such improvements with regard to the evidence of the original survey in existence at the time. Obviously, the burden of proof to the contrary must be borne by the party claiming differently. In many cases, at the time of construction of a road, the positions for the corners were preserved by subsurface deposits of marked stones or other durable material. These are to be considered as exceptionally important evidence of the position of the corner, when duly recovered and verified.

The replacement of those corners that are regarded as obliterated, but not lost, should be based on such collateral evidence as has been found acceptable. All lost corners can be restored only by reference to one or more interdependent corner.

154. ADEQUATE MONUMENTATION ESSENTIAL

Surveyors will appreciate the great extent to which their successful retracements have depended upon an available record of the previous surveys, and upon the markers that were established by those who preceded them. The same appreciation will apply in subsequent retracements. It is essential to the protection of the integrity and accuracy of the work, the reputation of the surveyors, and the security of the interested property owners, that durable new corner markers be constructed in all places where required, and that a record of the survey as executed be filed.

The preferred markers are of stone; concrete block; glazed sewer-tile filled with concrete, cast-iron or galvanized-iron pipe; and similar durable materials. Many engineers and surveyors, counties, and landowners employ specially designed markers with distinctive lettering, including various cast-iron plates or bronze tablets.

The Bureau of Land Management has adopted a standard monument for use on the public-land surveys. This is made of wrought-iron pipe, zinc-coated, $2^{1}/_{2}$ in diameter and 28 in long, with one end split and spread to form flanges or foot plates. A brass cap is securely attached to the top on which appropriate markings for the particular corner are inscribed by use of steel dies.

Frequently, on account of roadway or other improvements, it is advisable to set a subsurface marker and, in addition, to place a reference monument where it may be found readily, selecting a site that is not likely to be disturbed.

155. MEANDER LINES AND RIPARIAN RIGHTS

The traverse run by a survey along the bank of a stream or lake is termed a *meander line*. The meander line is not generally a boundary in the usual sense, as ordinarily the bank itself marks the limits of the survey. All navigable bodies of water are meandered in the public land surveying practice, as well as many other important streams and lakes that have not been regarded as navigable in the broader sense.

All navigable rivers within the territory occupied by the public lands remain, and are deemed to be, public highways. Unless otherwise reserved for federal purposes, the beds of these waters were vested in the states at time of statehood. Under federal law, in all cases where the opposite banks of any stream not navigable belong to different persons, the stream and the bed thereof become common to both.

Grants by the United States of its public lands, including lands bounded by streams or other waters, are construed as to their effect according to federal law. This includes lands added to the grants by accretion.

The government conveyance of title to a fractional subdivision fronting upon a nonnavigable stream, unless specific reservations are indicated in the patent from the federal government, carries ownership to the middle of the stream.

Where surveys purport to meander a body of water where no such body exists, or where the meanders may be considered grossly erroneous, the United States may have a continuing public land interest in the lands within the segregated areas.

Where partition lines are to be run *across accretions*, the ordinary federal rule is to apportion the new frontage along the water boundary in the same ratio as the frontage along the line of the record meander courses. There are many variations to this rule where local conditions prevail and the added lands are not of great width or extent. The application of any rule, when surveying private lands, should be brought into harmony with the state law.

Where there is occasion to define the partition lines *within the beds* of nonnavigable streams, the usual rule is to begin at the property line at its intersection with the bank. From that point, run a line normal to the medial line of the stream that is located midway between the banks. Where the normals to the medial lines are deflecting rapidly, owing to abrupt changes in the course of the stream, suitable locations are selected above and below the doubtful positions, where acceptable normals may be placed. The several intervals along the medial line are then apportioned in the same ratio as the frontage along the bank.

The partition of the bed of nonnavigable lakes, whether water-covered or relicted, presents a more difficult problem because of the wide range of shapes of lake beds. In the simplest case of a circular bed, the partition lines can be run to the centroid, thus creating pie-shaped tracts fronting the individual holdings at the edge of the lake bed. Where odd-shaped beds are concerned, ingenuity will be required to divide the lake bed in such a manner that each shore proprietor will receive an equitable share of land in front of his or her holding. Any consideration of riparian rights inuring to private lands should be brought into line with appropriate state laws or decisions.

Part 12: The Legal Systems of the United States

156. ORGANIZATION

Each state in the United States is a sovereign state. Except for those powers delegated to the federal government, each state writes its own laws, including those pertaining to land titles. Each state has a constitution, a legislative body, and a court system.

Many state constitutions are patterned after the U.S. Constitution and contain a bill of rights. Many establish English common law as the basis of law.

In most states, the legislative body is known as the *legislature*, but in some states is known as the *general assembly* or the *general court*. The legislature in most states consists of two bodies: the senate and the house of representatives.

157. CRIMINAL AND CIVIL COURTS

Except for the lowest courts, most courts handle criminal cases but not civil cases or civil cases but not criminal cases. Cases that pertain to land titles are tried in civil courts. In general, civil courts are either trial courts (courts of the first instance) or courts of civil appeal (appellate courts).

158. TRIAL COURTS

Trial courts determine any questions of fact in dispute and then apply rules of law. Evidence is presented by both the plaintiff and the defendant.

If the parties do not ask for a jury trial, the court (meaning the judge) hears the proceedings and hands down conclusions and delivers the judgment (the official and authentic decision of the court of justice). If either party demands a jury, the judge charges the jury on the law of the case, and the jury decides the case.

159. COURTS OF CIVIL APPEAL

Courts of civil appeal (appellate courts) do not hear evidence from witnesses; they only decide questions of law from the record of the trial court. A suit in an appellate court is not a continuation of a suit to which it relates; it is a new suit to set aside judgment in a lower court. It is brought for supposed error in law apparent in the record.

The party who appeals a court decision is known as the appellant, whether he or she was the plaintiff or defendant in the lower court. The party who defends against the appeal is known as the *respondent*.

160. PETITION FOR WRIT OF ERROR

The appellant usually makes his or her application to an appeals court by a petition for writ of error, in which he or she contends that certain errors were made in the lower court.

The appellate court may take one of four actions on an application for writ of error:

(1) It may grant the request, which means it will hear the case.

(2) It may refuse the request, which means that the court will not hear the case. This indicates that the appellate court believes that the question of law was correctly decided in the lower court.

(3) It may refuse the request *no reversible error* (n.r.e.). This indicates that the court is not satisfied that the opinion of the lower court, in all respects, has correctly declared law, but it is of the opinion that the application presents no error that may require a reversal.

(4) It may dismiss the application because the parties have settled out of court or because the appellate court lacks jurisdiction.

161. OPINION OF THE COURT

The *opinion of the court* is an explanation of the court's decision. When the judges in a case reach a decision, one of them writes the opinion. A concurring opinion may be written by a judge in the majority who agrees with the decision but disagrees with the reasoning of the opinion. If the opinion is not unanimous, the judge who writes the opinion is selected from the judges in the majority. A dissenting opinion may be written by a judge in the minority. Another dissenting opinion may be written by a judge in the minority who agrees with the dissenting opinion but disagrees with the reasoning in that opinion. A *per curiam* opinion is an opinion of the whole court as opposed to an opinion written by a specific judge. The words "decision" and "judgment" are synonymous.

162. ELEMENTS OF A COURT DECISION

Reports of cases (i.e., *citations*) follow, in general, the same outline:

(1) the title (sometimes called the *style*) of the case

(2) the case (or docket) number

(3) the name of the court that hears the case

(4) the date of the decision

(5) a summary of points of law involved

(6) a synopsis of lower court findings

(7) the names of counsel for both parties

(8) the name of the judge writing the opinion

(9) the opinion of the court

(10) the decision: affirmed or reversed

Part 13: Subdivisions

163. DEFINITION

The act of subdivision is the division of any tract or parcel of land into two or more parts for the purpose of sale or building development.

164. REGULATION

Resurveys are made to determine what has taken place in the past. On the other hand, surveys made for the purpose of planning subdivisions are creative in nature, and the care and imagination used in planning affect the entire community for many years to come. There are many instances in the United States in the past where developers have created subdivisions that are a credit to their foresight and integrity without any regulatory laws. But, increasing population and decreasing availability of land for development have made it necessary for the states and the federal government to adopt laws regulating land divisions.

165. SUBDIVISION LAW AND PLATTING LAW

Subdivision law includes regulations for land use, types of streets and their dimensions, arrangement of lots and their sizes, land drainage, sewage disposal, protection of nature, and many other details. *Platting law* includes regulations for recording the subdivision plat, monumenting the parcels, establishing the accuracy of the survey, and means of identifying the parcels and their dimensions.

166. PURPOSE OF SUBDIVISION LAW

Creation of a subdivision involves more than furnishing a location for a home for a new member of a community and a profit for the developer. It requires planning for traffic, transportation, the location of schools, churches, and shopping centers, and the health and happiness of the citizens of the community. Poorly planned subdivisions have caused cities to spend excessive amounts for street widening and resurfacing, reconstruction of sewer lines, establishment of additional drainage facilities and increase in size of water mains. Poor planning in the past has caused an acute awareness among state, county, and municipal officials of the need for the regulation of subdivision development.

167. THE CITY AS THE REGULATORY AUTHORITY

The authority to enforce subdivision regulations lies for the most part with cities. Counties have legal authority to regulate subdivisions, but most of the regulation is enforced by cities. Cities have the authority to enforce regulations by means of an ordinance adopted by the city council.

168. CERTAINTY OF LAND LOCATION

Monuments set by the original surveyor and called for by the conveyance have no error of position. An interior monument in a subdivision, after it is used, is correct, but boundary monuments of a subdivision marking the line between an adjacent owner, if not original monuments, can be located in error. Establishing a subdivision does not take away the rights of the adjoiner. It is the surveyor's role to eliminate future boundary difficulties. An important part in eliminating future disputes is the establishment of a precise control traverse. It should be established before any detailed planning is made.

169. MONUMENTS

When the control traverse has been established and preliminary approval of street right-of-way widths and locations has been obtained from municipal authorities, monuments should be set on property corners and street right-of-way lines based on the control survey. Control monuments that are to be used to relocate lost corners should be permanent and indestructible.

170. BOUNDARY SURVEY

The first step in subdividing is the establishment of the control traverse, which is the base for determining the boundaries of the tract. Investigation as to any conflict with senior title holders should be made. After certainty of location is established, the corners should be monumented.

171. TOPOGRAPHIC MAP

The next step is to establish a system of bench marks for vertical control and to prepare a topographic map. The map is essential in planning the subdivision, especially in regard to drainage and sanitary sewer plans.

172. THE PLANNING COMMISSION

Most cities deal with developers of subdivisions through a planning commission. The next step for the developer is to contact the planning commission of the city (or any other designated approving agency) for consultation. Many planning commissions require, at the outset, a subdivider's data sheet that indicates the general features of the developer's plan and a location map that locates the proposed subdivision in relation to zoning regulation and to existing community facilities.

173. GENERAL DEVELOPMENT PLAN

After preliminary discussion, the developer should submit a penciled sketch showing contours, street locations, and lots. If the planning commission feels that the development plans fit into the city's overall plan and into the surrounding neighborhood, it will ask for a preliminary plat.

The *preliminary plat* is actually the detailed plan for the subdivision and includes names of the subdivider, engineer, or surveyor; a legal description of the tract, location and dimensions of all streets, lots, drainage structures, parks, and public areas, easements, lot and block numbers, contour lines, scale of map, north arrow, and date of preparation. After approval of the preliminary plat, the developer can proceed with stake-out and construction operations.

174. PRELIMINARY PLAT

The *preliminary plat* is actually the detailed plan for the subdivision and includes names of the subdivider, engineer, or surveyor; a legal description of the tract; location and dimensions of all streets, lots, drainage structures, parks, and public areas; easement, lot, and block numbers; contour lines; scale of map; north arrow; and date of preparation. After approval of the preliminary plat, the developer can proceed with stake-out and construction operations.

175. FINAL PLAT

The *final plat* conforms to the preliminary plat except for any changes imposed by the planning commission. It must be prepared and filed for record in accordance with platting laws of the state. It establishes a legal description of the streets, residential lots, and other sites in the subdivision.

Part 14: Planning the Residential Subdivision

176. STORM DRAINAGE

The first step in planning subdivision drainage is a careful study of the contour map. Storm sewers and sanitary sewers are designed for gravity flow. Streets act as drainage collectors for storm runoff. Therefore, a general plan for storm sewers and sanitary sewers should be formulated before the streets are finally located.

Lots should drain to the street or to some open drainage system in the area. Lots should never receive drainage from the street. The Federal Housing Administration has set up requirements for lot drainage for various types of topography where this agency guarantees loans.

These requirements call for the lot to slope away from the house for some distance in all directions. The ideal situation is for the lot to slope from the house to the street in the front of the lot and from the house to some drainage collector at the rear of the lot if one exists. Concrete alleyways with inverted crowns sometimes serve this purpose. Topography sometimes makes it necessary for the lot to slope from front to rear or from rear to front, but the house is always located so that the slope is away in all directions.

Storm water is carried along the gutter, which necessitates a curb. To avoid water collecting in pools, streets should have a minimum grade of 0.3% and preferably 0.5%. Water should not run across streets in valleys. Inlets should be planned for both sides of the streets at intersections.

The design of storm sewers depends on the amount of surface runoff to be carried in the storm sewers. The amount of rainfall that is absorbed by the surface soil depends on the perviousness of the soil, the intensity of the rainfall, the duration of the storm, and the slope of the surface. Water that is not absorbed is known as *surface runoff*. Prior to subdivision, farm or grazing land may absorb a large percentage of a given rainstorm. Streets, driveways, sidewalks, and rooftops, however, will cause most of the rain water to run off rather than be absorbed by the soil. The developer must be conscious of this fact in order to prevent damage to the development and to property downstream.

Factors included in the design of storm drains include area of drainage area in acres, shape of drainage area, slope of land in drainage area, use of land in drainage area (present and future), maximum intensity of rainfall, and frequency of maximum intensity.

177. SANITARY SEWERAGE

Septic tanks are undesirable in subdivisions and should not be used. The most desirable solution to sewage disposal is a collecting system connected to a municipal sewage disposal system. The second most desirable solution is a private collecting system and disposal plant.

Sanitary sewers flow by gravity, and their location depends on the topography of the land. They are often located under the streets, but they also may be located at the back of lots or in alleyways. The slope of the sewer should be such that the velocity will be between 2 ft/sec and 10 ft/sec. Manholes must be located at change of grade, junctions, and other points for inspection. Sewers must be laid in a straight line between manholes.

178. STREETS

Of major importance in the subdivision is the street system. Streets not only furnish an avenue for vehicle passage, they also furnish access to property for pedestrians, right-of-way for utility lines, channels for drainage, and access to fire plugs, garbage cans, and so on. The right street in the right place contributes to pleasant living. Insufficient street width creates traffic hazards, while excess width adds to the cost of construction and uses land that could be used for lots.

Local residential streets (also called *minor streets*) are designed to furnish access to private property. They are not designed to carry through traffic and should be designed to discourage it. Curved streets, loops, and cul-de-sacs all discourage speed. Long, curvilinear streets with block lengths up to 1800 ft have been found to be satisfactory. Where off-street parking is adequate, a roadway width of 26 ft (27 ft back-to-back of curbs) and a right-of-way of 50 ft are adequate for single-family residential neighborhoods. For multifamily neighborhoods, street widths should be 31 ft minimum (32 ft back-to-back of curbs) with 60 ft right-of-way. A street width of 26 ft will not provide parking on both sides of the street with two lanes for traffic, but many homeowners find weaving in and out between parked cars slows traffic.

Collector streets carry traffic between local streets and arterial streets. They also furnish access to private property along the street. A width of 36 ft (37 ft back-to-back of curbs) will furnish two parking lanes and two traffic lanes. Right-of-way width should be 60 ft.

Arterial streets (also called *major streets*) move heavy traffic at relatively high speeds. Intersections are usually controlled by traffic lights. Many arterial streets provide six moving lanes with no parking. A 100 ft right-of-way will provide two 33 ft (back-to-back of curbs) lanes and a 14 ft median.

Cul-de-sacs are dead-end streets with turnarounds at the end. They are popular for single-family residences because of the privacy and freedom from noise they provide. They should not be more than 1000 ft in length because of the long turnaround. The turnaround circle should be not less than 40 ft in radius.

Loop streets have the same advantages as cul-de-sacs, plus the advantage of better circulation for fire trucks, delivery trucks, and police cars. Loop streets and cul-de-sacs can be used in odd corners of subdivisions.

179. BLOCKS

Until recent years, most subdivisions were laid out on a gridiron pattern. Modern subdivisions use curved streets with blocks 1400 ft to 1800 ft in length. These have proven to be more economical, using less street area by eliminating many cross-streets, and safer because of fewer intersections and slower speed on curved streets. They also provide more lots per undeveloped acre.

180. LOTS

Lot size should depend on the type of development, the topography of the land, and the expected cost per housing unit. Lot dimensions vary in different parts of the United States, but a minimum width of 60 ft and a minimum depth of 100 ft (or about 6000 ft^2) is considered desirable. Large, ranch-style houses with multicar garages opening on the front require more than 75 ft in width.

181. COVENANTS

To protect the interests of future property owners, developers often include certain restrictive *covenants* as a part of the deed to the property. These include such things as type of construction, minimum size and cost, setback distance, restrictions against advertising signs, raising of animals, parking of mobile homes, conducting certain types of commercial enterprises, and any other restrictions that will insure that the neighborhood is used for the purpose for which it was designed.

182. SETBACK LINES

To prevent locating buildings in such a way as to mar the general view of the neighborhood, restrictive covenants often designate the minimum distance from the front property line to the front of the building. To prevent monotony, staggered setback lines are sometimes used.

183. DENSITY ZONING

Many cities that have formerly controlled crowding in new neighborhoods by controlling the dimensions of lots have now adopted *density control*, which controls the maximum number of dwellings per acre. This has allowed more imagination in planning and better use of natural features.

184. CLUSTER PLANNING

A recent development in subdivision planning is the *cluster pattern*, where residences are clustered together in small, private sections of the subdivision with common open space. Cul-de-sacs and loop streets are adaptable to the cluster plan.

PRACTICE PROBLEMS

1. English common law
   (A) was referred to as unwritten law
   (B) was evolutionary
   (C) evolved from court decisions over hundreds of years
   (D) all of the above

2. English common law
- (A) was enacted by Parliament
- (B) evolved from court decisions
- (C) was established by decree of the king of England
- (D) all of the above

3. Trial by jury, the grand jury, and freedom of religion had their beginning in
- (A) the Magna Carta
- (B) Roman law
- (C) English common law
- (D) none of the above

4. French and Spanish law had their beginning in
- (A) Roman law
- (B) Greek law
- (C) English common law
- (D) none of the above

5. The Statute of Frauds provides that
- (A) ownership of land by private individuals is fraudulent
- (B) a father may not pass title to his eldest son
- (C) transfer of title to land must be in writing
- (D) all of the above

6. Property is divided into the following classes:
- (A) farmland and ranchland
- (B) urban land and rural land
- (C) real and personal
- (D) none of the above

7. Most titles originate from
- (A) a sovereign after conquest
- (B) the United States Congress
- (C) the state legislatures
- (D) none of the above

8. A title abstract is
- (A) a conveyance that transfers title
- (B) a description of a piece of property
- (C) a history of transfers of title in chronological order
- (D) an insurance policy

9. Color of title is
- (A) any written instrument that appears to convey title but in fact does not
- (B) a claim on land that, if valid, would impair the title to the land but that can be proven invalid
- (C) an estate in real property
- (D) all of the above

10. Title, in the legal sense, is
- (A) the right to own real property
- (B) the evidence of the right to own real property
- (C) the legal instrument constituting the evidence to the right of ownership
- (D) all of the above

11. A conveyance is
- (A) a written instrument that transfers property
- (B) an estate in real property
- (C) a method of transferring property from the deceased to the heirs
- (D) a release of lien

12. An estate can be acquired by
- (A) purchase
- (B) inheritance or gift
- (C) power of the state
- (D) all of the above

13. An estate in fee simple absolute
- (A) implies a complete title
- (B) is the highest type of interest
- (C) gives the owner the right to dispose of the property in any way the owner sees fit
- (D) all of the above

14. A deed
- (A) is the most important document in the transfer of real property
- (B) is evidence in writing of the transfer of property
- (C) must be in writing
- (D) is all of the above

15. Conveyance of fee title to land is usually by
- (A) deed of trust
- (B) quitclaim deed
- (C) warranty deed
- (D) contract of sale

Law

16. Delivery of the property sold is considered to be

(A) delivery of the deed

(B) a statement by the seller that the seller has vacated the property

(C) a notice to the buyer from the county clerk that the transaction is complete

(D) the ceremony of livery of seisin

17. The grantee is

(A) the person to whom a grant is made

(B) the person who makes the grant

(C) a notary public

(D) the mortgagor

18. A patent is

(A) a conveyance of the sovereign's interest in a tract of land

(B) the state's warranty that title to a tract of land is good and clear

(C) a Spanish or Mexican grant

(D) the title to vacant land

19. An easement is

(A) title to surface rights only

(B) a lease or "estate for years"

(C) a fee simple title

(D) the right that the public or an individual has in the lands of another

20. A mortgage is a

(A) promissory note held by a lending agency

(B) conditional conveyance of an estate as a pledge of security of a debt

(C) request for payment of a note

(D) transfer of title

21. A deed of trust is a

(A) conditional conveyance of an estate as a pledge of security of a debt

(B) deed for excess acreage

(C) conveyance held in the hands of another for safekeeping

(D) conveyance made to a charitable organization

22. Dedication is

(A) a means of transferring title to property

(B) the acquisition of property without the owner's consent

(C) a ceremony marking the completion of streets

(D) a city ordinance relating to street specifications

23. Which of the following is not a requirement for obtaining possession by adverse possession?

(A) possession must be open and notorious

(B) possession must be hostile

(C) possession must be intermittent

(D) possession must be actual

24. Prescription is

(A) a method of acquiring right to use of property

(B) a tax lien

(C) an encumbrance

(D) a rendition of property

25. The power to exercise eminent domain must be authorized by the

(A) governor of the state

(B) U.S. Congress

(C) state legislature

(D) state department of highways

26. The Constitution of the United States provides that, before the power of eminent domain can be exercised,

(A) the owner must be guaranteed adequate compensation for the property

(B) the improvement must be for a public use

(C) the owner may not be deprived of property without due process of law

(D) all of the above

27. An encroachment is

(A) a cloud on title to property

(B) a second lien

(C) a written instrument that conveys title

(D) a gradual, stealthy, illegal acquisition of property

28. A quitclaim deed

(A) is a deed for excess acreage

(B) warrants a good and clear title

(C) passes any title the grantor may leave

(D) passes title to minerals only

29. Which of the following is not an essential of a deed?

(A) A deed must be signed.

(B) A deed must be acknowledged.

(C) A deed must contain a description of the property.

(D) A deed must be recovered.

30. A holographic will
(A) devises real property only
(B) is written entirely in the testator's own hand
(C) leaves all property to the eldest son
(D) cannot be overturned

31. Unwritten transfer of title to property implies
(A) transfer by a holographic will
(B) transfer of property from husband to wife by oral agreement
(C) acquisition of title by adverse possession
(D) none of the above

32. Recognition and acquiescence implies
(A) recognition of a junior title over a senior title
(B) recognition of and acquiescence in a boundary line by adjacent owners by agreement
(C) a court decision with regard to the boundary between adjacent owners
(D) all of the above

33. Title to beds of navigable streams within a state is generally held by
(A) the state
(B) the federal government
(C) the riparian owner
(D) the Department of the Interior

34. Land uncovered by reliction is owned by
(A) the state
(B) the federal government
(C) the littoral owner
(D) the Department of the Interior

35. The line between tidelands and private lands in most states is
(A) the line of mean sea level
(B) the line of highest tide in the winter
(C) the line of mean high tide
(D) the line of low tide

36. A riparian owner loses title to land by
(A) accretion
(B) reliction
(C) erosion
(D) avulsion

37. A riparian owner gains title to land by
(A) accretion
(B) reliction
(C) erosion
(D) avulsion

38. A riparian owner's right to, or in, a river or stream rests on
(A) his or her acquisition of a permit from the state
(B) his or her acquisition of a permit from the federal government
(C) the fact that his or her land abuts on the stream
(D) English common law

39. When avulsion occurs on a navigable stream and the old channel is abandoned by the river, title to the bed of the old channel goes to
(A) the state
(B) the federal government
(C) the riparian owner
(D) the Department of the Interior

40. When avulsion occurs on a navigable stream, title to the bed of the new channel lies in
(A) the state
(B) the federal government
(C) the riparian owner
(D) the Department of the Interior

41. When a navigable stream within a state cuts a new channel and continues to flow in the old channel also, the island formed belongs to
(A) the state
(B) the federal government
(C) the riparian owner
(D) the Department of the Interior

42. Riparian rights include
(A) use of water for domestic use
(B) use of water for cattle
(C) use of water for irrigation
(D) all of the above

43. Tidelands are owned by
(A) the state
(B) the federal government
(C) the littoral owner
(D) the Department of the Interior

44. Authority over navigation on navigable streams is in

(A) the state
(B) the federal government
(C) the riparian owner
(D) the Department of the Interior

45. The line BCDEFGHJ represents the boundary between the state-owned bed of a navigable stream and the private land of the riparian owner in 1940. The line BC'D'E'F'G'H'J represents the river line at the present time.

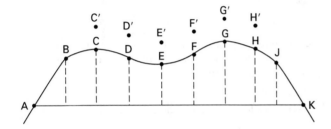

The legal boundary today would be which of the following?

(A) BCDEFGHJ
(B) BC'D'E'F'G'H'J

46. In evaluating calls in a deed, least weight is given to

(A) calls for artificial monuments
(B) quantity calls
(C) adjoiner calls
(D) course and distance calls

47. Of the four calls, least importance is given to a

(A) call for natural objects
(B) call for adjoiner
(C) call for course and distance
(D) call for artificial objects

48. Intent with respect to boundaries is to be ascertained from the

(A) face of the conveyance in the light of surrounding circumstances
(B) grantor
(C) grantee
(D) surveyor who made the survey

49. In surveying Bill Bell's 40 ac tract, which is east of senior Bob Allen 40 ac tract, the surveyor in charge finds all four original monuments for Bill Bell tract, but the west line of Bill Bell tract is found to be 10 ft east of the east line of Bob Allen tract (land not in U.S. System of Rectangular Surveys). The surveyor should

(A) establish west line of Bill Bell tract to be along east line of senior Bob Allen tract
(B) establish Bill Bell tract as defined by four original monuments as found

50. In surveying along Elm Street, the surveyor in charge finds that original monuments that mark right-of-way for Elm Street at each street intersection are actually 60 ft apart rather than 50 ft called on plat. The surveyor should

(A) add 5 ft to lots on each side of the street
(B) leave street widths and lots as they are
(C) contact the city engineer

51. In a resurvey of lots 1, 2, 3, 4, the actual distance between original monuments were found to be as shown on the sketch. Show corrected lot widths on the sketch.

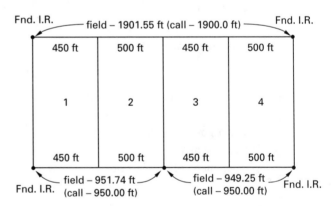

52. In 1824 the Mexican government granted a league of land in Austin's colony in Texas to Williams, to Allcorn, to Little, and to Jones. The Williams, Little, and Jones leagues were riparian to the Brazos River, a navigable river, and, at this location was not affected by the tide.

During the century following the grants, the Brazos River, by erosion cut into the Allcorn league, and by 1914 all of the leagues were riparian to the river. In 1914 the river, by avulsion, cut a new channel across a horseshoe bend and the abandoned bed became a freshwater lake, as can be seen in the figure.

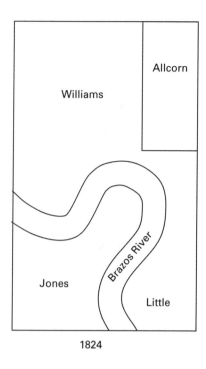

1824

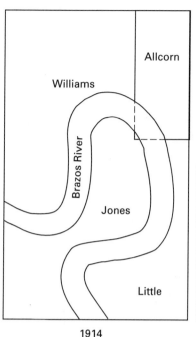

1914

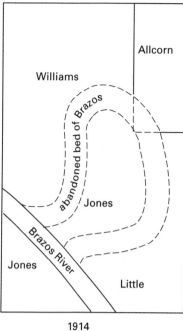

1914

Title to the abandoned bed went to

(A)   the heirs of Williams, Little, and Jones

(B)   the heirs of Williams, Little, Jones, and Allcorn

(C)   the State of Texas

(D)   the Department of the Interior

53. The plat of a subdivision shows five 100 ft lots fronting on a street with the east line of lot 5 along the east boundary of the subdivision. You have been retained to stake lot 5 and find the distance between original monuments to be 500.50 ft where called for 500.00 ft. You also find that the east line of the subdivision is actually 15 ft west of the true boundary of the subdivision.

Ash Lane
field – 500.50 ft (plat call – 500.0 ft)

100 ft	100 ft	100 ft	100 ft	100 ft
Fnd. I.R. 1	2	3	4	5 Fnd. I.R.
10	9	8	7	6

subdivision plat boundary
true boundary

15 ft

How many feet should you make the frontage of lot 5?

54. State whether single proportionate measure (S), single proportionate measure on a latitudinal (curved east-west) line (SC), or double proportionate measure (D) is used in restoring the following corners (assume none of the township boundaries are standard parallels).

(a)   northeast corner of section 10

(b)   northeast corner of section 12

(c)   south quarter-section corner of section 12

(d)   southeast corner of section 36

(e)   southwest corner of section 36

Problems 55–62 are based on Figures A–D shown on the next page.

55. Compute the field measurement for the east side of the southeast quarter of the fractional northeast quarter of section 3 shown in Fig. A.

56. Compute the field measurement for the west side of the southwest quarter of section 3 in Fig. A.

57. Compute the field measurement for the north side of the fractional northeast quarter of the northeast quarter of section 3 in Fig. A.

58. Compute the field coordinates to restore the lost south quarter-section corner of section 18 shown in Fig. B.

59. Compute the field coordinates to restore the lost northwest corner of section 16 shown in Fig. C.

60. Compute the coordinates to restore the lost north quarter-section corner of section 16 shown in Fig. C.

61. Compute the coordinates for the legal center of section 10 in Fig. D.

62. Assume coordinate values of 7,431.52′ N and 22,740.40′ E for the center of section 10 in Fig. D, compute the coordinates for the center of the northeast quarter of the section.

(Note: Distances shown in parentheses are from the GLO plat; "o" indicates proven GLO corner.)

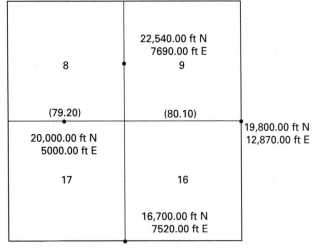

Figure C

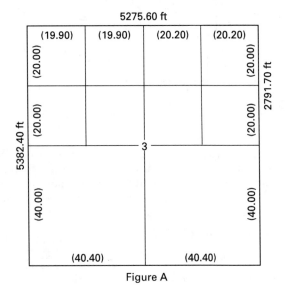

Figure A

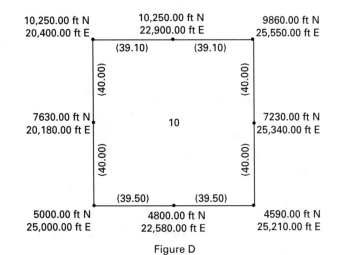

Figure D

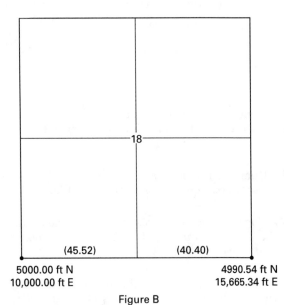

Figure B

SOLUTIONS

1. *The answer is D.*

2. *The answer is B.*

3. *The answer is C.*

4. *The answer is A.*

5. *The answer is C.*

6. *The answer is C.*

7. *The answer is A.*

8. *The answer is C.*

9. *The answer is A.*

10. *The answer is D.*

11. *The answer is A.*

12. *The answer is D.*

13. *The answer is D.*

14. *The answer is D.*

15. *The answer is C.*

16. *The answer is A.*

17. *The answer is A.*

18. *The answer is A.*

19. *The answer is D.*

20. *The answer is B.*

21. *The answer is A.*

22. *The answer is A.*

23. *The answer is C.*

24. *The answer is A.*

25. *The answer is C.*

26. *The answer is D.*

27. *The answer is D.*

28. *The answer is C.*

29. *The answer is D.*

30. *The answer is B.*

31. *The answer is C.*

32. *The answer is B.*

33. *The answer is A.*

34. *The answer is C.*

35. *The answer is C.*

36. *The answer is C.*

37. *The answer is C.*

38. *The answer is C.*

39. *The answer is C.*

40. *The answer is A.*

41. *The answer is C.*

42. *The answer is D.*

43. *The answer is A.*

44. *The answer is B.*

45. *The answer is B.*

46. *The answer is B.*

47. *The answer is C.*

48. *The answer is A.*

49. *The answer is B.*

Law

50. *The answer is B.*

51.

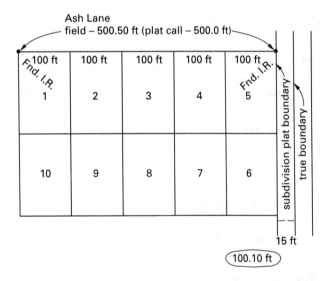

Fnd. I.R. — field – 1901.55 ft (call – 1900.0 ft) — Fnd. I.R.

450 ft	500 ft	450 ft	500 ft
(450.37)	(500.41)	(450.37)	(500.40)
1	2	3	4
(450.82)	(500.92)	(449.64)	(499.61)
450 ft	500 ft	450 ft	500 ft

field – 951.74 ft (call – 950.00 ft)   Fnd. I.R.   field – 949.25 ft (call – 950.00 ft)   Fnd. I.R.

Fnd. I.R.

52. *The answer is B.*

53.

Ash Lane
field – 500.50 ft (plat call – 500.0 ft)

100 ft	100 ft	100 ft	100 ft	100 ft
Fnd. I.R. 1	2	3	4	5 Fnd. I.R.
10	9	8	7	6

subdivision plat boundary
true boundary

15 ft

(100.10 ft)

54. (a) [D]

(b) [S]

(c) [S]

(d) [S]

(e) [SC]

55.
$$\frac{\text{field}}{2791.70 \text{ ft}} = \frac{20.00 \text{ ch}}{20.00 \text{ ch} + 22.60 \text{ ch}}$$

$$\text{field} = \left(\frac{20.00 \text{ ch}}{42.60 \text{ ch}}\right)(2791.70 \text{ ft})$$

$$= \boxed{1310.66 \text{ ft}}$$

56.
$$\frac{\text{field}}{5382.40 \text{ ft}} = \frac{40.00 \text{ ch}}{40.00 \text{ ch} + 20.00 \text{ ch} + 22.00 \text{ ch}}$$

$$\text{field} = \left(\frac{40.00 \text{ ch}}{82.00 \text{ ch}}\right)(5382.40 \text{ ft})$$

$$= \boxed{2625.56 \text{ ft}}$$

57.
$$\frac{\text{field}}{5275.60 \text{ ft}} = \frac{20.20 \text{ ch}}{19.90 \text{ ch} + 19.90 \text{ ch} + 20.20 \text{ ch} + 20.20 \text{ ch}}$$

$$\text{field} = \left(\frac{20.20 \text{ ch}}{80.20 \text{ ch}}\right)(5275.60 \text{ ft})$$

$$= \boxed{1328.77 \text{ ft}}$$

58. Field coordinates can be proportioned similar to field distances.

Proportioning left to right:

$$\frac{\Delta N}{4990.54 \text{ ft} - 5000.00 \text{ ft}} = \frac{45.52 \text{ ch}}{45.52 \text{ ch} + 40.00 \text{ ch}}$$

$$\Delta N = \left(\frac{45.52 \text{ ch}}{85.52 \text{ ch}}\right)(-9.46 \text{ ft})$$

$$= -5.04 \text{ ft}$$

$$\frac{\Delta E}{15,665.34 \text{ ft} - 10,000 \text{ ft}} = \frac{45.52 \text{ ch}}{45.52 \text{ ch} + 40.00 \text{ ch}}$$

$$\Delta E = \left(\frac{45.52 \text{ ch}}{85.52 \text{ ch}}\right)(5665.34 \text{ ft})$$

$$= 3015.51 \text{ ft}$$

$$N = 5000.00 \text{ ft} - 5.04 \text{ ft}$$

$$= \boxed{4994.96 \text{ ft}}$$

$$E = 10,000.00 \text{ ft} + 3015.51 \text{ ft}$$

$$= \boxed{13,015.51 \text{ ft}}$$

59. Using double proportionate measure, the northing of the restored corner will be the northing obtained by single proportionate measure in the north-south direction. Likewise, the easting will be the easting obtained by single proportionate measure in the east-west direction.

$$\frac{\Delta N}{22{,}540.00 \text{ ft} - 14{,}700.00 \text{ ft}} = \frac{80.00 \text{ ch}}{80.00 \text{ ch} + 40.00 \text{ ch}}$$

$$\Delta N = \left(\frac{80.00 \text{ ch}}{120.00 \text{ ch}}\right)(7840.00 \text{ ft})$$

$$= 5226.67 \text{ ft}$$

$$\frac{\Delta E}{12{,}870.00 \text{ ft} - 5000.00 \text{ ft}} = \frac{\dfrac{79.20 \text{ ch}}{2}}{\dfrac{79.20 \text{ ch}}{2} + 80.10 \text{ ch}}$$

$$\Delta E = \left(\frac{39.60 \text{ ch}}{119.70 \text{ ch}}\right)(7870.00 \text{ ft})$$

$$= 2603.61 \text{ ft}$$

$$N = 14{,}700.00 \text{ ft} + 5226.67 \text{ ft}$$

$$= \boxed{19{,}926.67 \text{ ft}}$$

$$E = 5000.00 \text{ ft} + 2603.61 \text{ ft}$$

$$= \boxed{7603.61 \text{ ft}}$$

60. A quarter corner is to be restored by single proportionate measure between adjacent section corners after they have been found or *restored*.

Using the restored section corner coordinates obtained in Prob. 60, and with the quarter corner being at a midpoint on a regular (normal) section,

$$N = \frac{19{,}926.67 \text{ ft} + 19{,}800.00 \text{ ft}}{2} = \boxed{19{,}863.33 \text{ ft}}$$

$$E = \frac{7603.61 \text{ ft} + 12{,}870.00 \text{ ft}}{2} = \boxed{10{,}236.80 \text{ ft}}$$

61. The center of the section is located at the intersection of the lines connecting the opposite quarter corners (bearing-bearing intersection).

The inverse from the west quarter corner to the east quarter corner yields an azimuth of $94°25'58''$.

The inverse from the south quarter corner to the north quarter corner yields an azimuth of $3°29'17''$.

Using bearing-bearing intersection software,

$$N = \boxed{7431.52 \text{ ft}}$$

$$E = \boxed{22{,}740.40 \text{ ft}}$$

62. The center of a quarter section is located at the intersection of lines connecting opposite quarter-quarter corners. For a regular (normal) section, quarter-quarter corners are to be located at midpoints between adjacent quarter corners.

For the southeast quarter section:

The north quarter-quarter corner is

$$N = \frac{10{,}050.00 \text{ ft} + 9860.00 \text{ ft}}{2} = 9955.00 \text{ ft}$$

$$E = \frac{22{,}900.00 \text{ ft} + 25{,}550.00 \text{ ft}}{2} = 24{,}225.00 \text{ ft}$$

The east quarter-quarter corner is

$$N = \frac{9860.00 \text{ ft} + 7230.00 \text{ ft}}{2} = 8545.00 \text{ ft}$$

$$E = \frac{25{,}550.00 \text{ ft} + 25{,}340.00 \text{ ft}}{2} = 25{,}445.00 \text{ ft}$$

The south quarter-quarter corner is

$$N = \frac{7431.52 \text{ ft} + 7230.00 \text{ ft}}{2} = 7330.76 \text{ ft}$$

$$E = \frac{22{,}740.40 \text{ ft} + 25{,}340.00 \text{ ft}}{2} = 24{,}040.20 \text{ ft}$$

The west quarter-quarter corner is

$$N = \frac{7431.52 \text{ ft} + 10{,}050.00 \text{ ft}}{2} = 8740.76 \text{ ft}$$

$$E = \frac{22{,}740.40 \text{ ft} + 22{,}900.00 \text{ ft}}{2} = 22{,}820.20 \text{ ft}$$

The coordinates of the intersection of lines connecting opposite midpoints on a four-sided figure will be at the average of the coordinates of these opposite midpoints.

Therefore, for east-west,

$$N = \frac{8545.00 \text{ ft} + 8740.76 \text{ ft}}{2} = 8642.88 \text{ ft}$$

$$E = \frac{25{,}445.00 \text{ ft} + 22{,}820.20 \text{ ft}}{2} = 24{,}132.60 \text{ ft}$$

Check:

For north-south,

$$N = \frac{9955.00 \text{ ft} + 7330.76 \text{ ft}}{2} = 8642.88 \text{ ft}$$

$$E = \frac{24{,}225.00 \text{ ft} + 24{,}040.20 \text{ ft}}{2} = 24{,}132.60 \text{ ft}$$

Law

22 Vertical Alignments

Nomenclature

e ordinate at PI in feet
g_1 gradient of first connecting tangent
g_2 gradient of second connecting tangent
L length of curve in stations
PC   beginning of curve
PI   point of intersection of two tangents
PT   end of curve
VC   vertical curve
x_1 horizontal distance in stations from PC
x_2 horizontal distance in stations from PT
y_1 ordinate of any station less than PI station
y_2 ordinate of any station greater than PI station

1. GRADE OR STEEPNESS

The *grade*, or *steepness*, of a road is the ratio of elevation to horizontal distance. If a highway rises 6 ft for every 100 ft of horizontal distance, the grade of the highway is 6 ft/100 ft = 0.06 ft/ft.

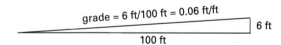

Figure 22.1 *Calculation of Grade*

2. SLOPE OF A LINE

The grade of a line representing the profile of a highway is also known as the *slope of the line*.

$$\text{slope} = \frac{\text{rise}}{\text{run}} \qquad 22.1$$

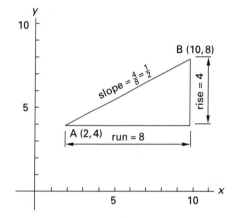

Figure 22.2 *Slope of a Line*

In Fig. 22.2, points A $(2, 4)$ and B $(10, 8)$ are connected by the straight line AB.

$$\text{slope of AB} = \frac{\text{ordinate of B} - \text{ordinate of A}}{\text{abscissa of B} - \text{abscissa of A}}$$
$$= \frac{8 - 4}{10 - 2} = \frac{4}{8} = \frac{1}{2} \qquad 22.2$$

The symbol m is used to denote the slope of the line between points 1 and 2.

$$m = \frac{y_2 - y_1}{x_2 - x_1} \qquad 22.3$$

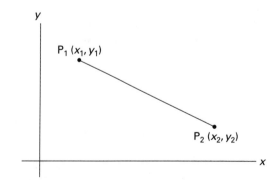

Figure 22.3 *Line Between Two Points*

In Fig. 22.4, the slope of line AB is

$$m = \frac{y_2 - y_1}{x_2 - x_1} = \frac{8 - 2}{6 - 3} = +2$$

Line BC has a slope of

$$m = \frac{y_2 - y_1}{x_2 - x_1} = \frac{2 - 8}{9 - 6} = -2$$

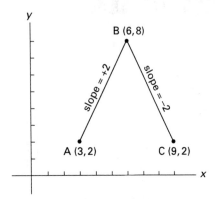

Figure 22.4 *Positive and Negative Slopes*

A line rising from left to right has a positive slope, and a line falling from left to right has a negative slope.

A horizontal line has a slope of zero, as shown in Fig. 22.5.

Figure 22.5 *Slope of Horizontal Line*

3. GRADE OR GRADIENT

In highway construction, the slope of the line that is the profile of the centerline is known as the *grade* or *gradient*.

The grade of a highway is computed in the same way as the slope of a line is computed. Horizontal distances are usually expressed in stations; vertical distances are expressed in feet.

Example 22.1

Determine the gradient of a highway that has a centerline elevation of 444.50 ft at sta 20+75.00 and a centerline elevation of 472.20 ft at sta 32+25.00.

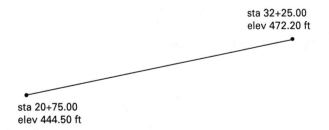

Solution

$$\text{gradient} = \frac{427.20 \text{ ft} - 444.50 \text{ ft}}{3225.00 \text{ ft} - 2075.00 \text{ ft}} = \frac{27.70 \text{ ft}}{1150.00 \text{ ft}}$$
$$= +0.02409 \text{ ft/ft}$$

If the numerator is expressed in feet and the denominator is expressed in stations, the decimal point in the gradient will move two places to the right. The gradient can then be expressed as a percent. In this form, gradient expresses change in elevation per station. For Ex. 22.1,

$$\text{gradient} = \frac{27.70 \text{ ft}}{11.50 \text{ sta}} = 0.02409 \text{ ft/sta} \quad (+2.409\%)$$

4. POINTS OF INTERSECTION

Vertical alignment for a highway is located similarly to horizontal alignment. Straight lines are located from point to point, and vertical curves are inserted. The points of intersecting gradients are known as *points of intersection*. These lines, after vertical curves have been inserted, are the centerline profile of the highway. Usually the profile is the finish elevation (pavement) profile, but it may be the subgrade (earthwork) profile.

5. TANGENT ELEVATIONS

After points of intersection have been located and connected by tangents (straight lines), elevations of each station on the tangent need to be determined before finding elevations on the vertical curve.

The gradient is the change in elevation per station. If the station number and elevation of each PI is known, the elevation at each station can be calculated. The gradient should be computed to three decimal places in percent. Finish elevations should be computed to two decimal places in feet (hundredths of a foot).

Example 22.2

From the information shown, compute the gradient of each tangent and the elevation at each full station on the tangents.

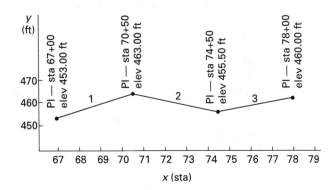

Solution

$$\text{gradient 1} = \frac{463.00 \text{ ft} - 453.00 \text{ ft}}{70.50 \text{ sta} - 67.00 \text{ sta}}$$

$$= 0.02857 \text{ ft/sta} \quad (+2.857\%)$$

$$\text{gradient 2} = \frac{463.00 \text{ ft} - 455.50 \text{ ft}}{74.50 \text{ sta} - 70.50 \text{ sta}}$$

$$= -0.01875 \text{ ft/sta} \quad (-1.875\%)$$

$$\text{gradient 3} = \frac{460.00 \text{ ft} - 455.50 \text{ ft}}{78.00 \text{ sta} - 74.50 \text{ sta}}$$

$$= 0.01286 \text{ ft/sta} \quad (+1.286\%)$$

Elevation Computations

point	station	computations	elevation (ft)
PI	67+00		= 453.00
	68+00	453.00 + (1)(2.857)	= 455.86
	69+00	453.00 + (2)(2.857)	= 458.71
	70+00	453.00 + (3)(2.857)	= 461.57
PI	70+50	453.00 + (3.5)(2.857)	= 463.00
	71+00	463.00 − (0.5)(1.875)	= 462.06
	72+00	463.00 − (1.5)(1.875)	= 460.19
	73+00	463.00 − (2.5)(1.875)	= 458.31
	74+00	463.00 − (3.5)(1.875)	= 456.44
PI	74+50	463.00 − (4.0)(1.875)	= 455.50
	75+00	455.50 + (0.5)(1.286)	= 456.14
	76+00	455.50 + (1.5)(1.286)	= 457.43
	77+00	455.50 + (2.5)(1.286)	= 458.72
PI	78+00	455.50 + (3.5)(1.286)	= 460.00

6. VERTICAL CURVES

Just as horizontal curves connect two tangents in horizontal alignment, vertical curves connect two tangents in vertical alignment. However, although the horizontal curve is usually an arc of a circle, the vertical curve is usually a parabola.

The simplest equation of a parabola is $y = ax^2$, where y is vertical distance and x is horizontal distance. This means that the vertical distances y from tangent to curve vary as the square of the horizontal distances x, measured from either the PC or the PT.

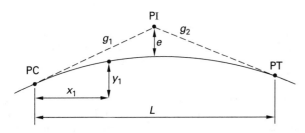

Figure 22.6 *Vertical Curve*

Equations for computing y-distances can be derived as follows.[1]

$$e = \frac{(g_1 - g_2)L}{8} \qquad 22.4$$

$$a = \frac{g_1 - g_2}{2L} \qquad 22.5$$

$$y = ax^2 \qquad 22.6$$

7. COMPUTATIONS FOR FINISH ELEVATIONS

Elevations along the tangent are first computed, and y-distances are then computed for stations on the vertical curve and added to or subtracted from the tangent elevations to determine the finish elevation. The y-distance is added on sag curves (Fig. 22.7) and subtracted on crest curves (Fig. 22.6).

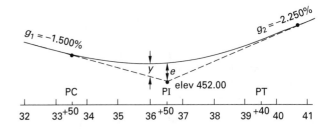

Figure 22.7 *Sag Curve*

Example 22.3

A −1.500% grade meets a +2.250% grade at sta 36+50, elev 452.00 ft. A vertical curve of length 600 ft (six stations) will be used. Compute the finish elevations for each full station and the PC and PT from sta 32+00 to sta 41+00. Refer to Fig. 22.7.

Solution

From Eqs. 22.4 and 22.5,

$$a = \frac{g_1 - g_2}{2L}$$

$$= \frac{-1.500\% - 2.250\%}{(2)(6 \text{ sta})}$$

$$= 0.312\%/\text{sta}$$

$$e = \frac{(g_1 - g_2)L}{8}$$

$$= \frac{(-1.500\% - 2.250\%)(6 \text{ sta})}{8}$$

$$= 2.81 \text{ ft}$$

---

[1]The symbol C is often used in place of a in Eq. 22.6. E may also be used instead of e.

Tangent Elevation Computations

point	station	computations	elevation
	32+00	452.00 + (4.5)(1.500)	= 458.75
	33+00	452.00 + (3.5)(1.500)	= 457.25
PC	33+50	452.00 + (3.0)(1.500)	= 456.50
	34+00	452.00 + (2.5)(1.500)	= 455.75
	35+00	452.00 + (1.5)(1.500)	= 454.25
	36+00	452.00 + (0.5)(1.500)	= 452.75
PI	36+50		= 452.00
	38+00	452.00 + (1.5)(2.250)	= 455.37
	39+00	452.00 + (2.5)(2.250)	= 457.62
PT	39+50	452.00 + (3.0)(2.250)	= 458.75
	40+00	452.00 + (3.5)(2.250)	= 459.88
	41+00	452.00 + (4.5)(2.250)	= 462.13

Computations for $y = 0.312x^2$

station	computations	y (ft)
33+50		= 0.00
34+00	= $(0.312)(0.15)^2$	= 0.08
35+00	= $(0.312)(1.5)^2$	= 0.70
36+00	= $(0.312)(2.5)^2$	= 1.95
36+50		= 2.81
37+00	= same as 36+00	= 1.95
38+00	= same as 35+00	= 0.70
39+00	= same as 34+00	= 0.08
39+50		= 0.00

The curve is symmetrical about the PI, so the distance x to sta 37+00 (from the PT) is the same as the distance x to sta 36+00 (from the PC). Therefore, the y-distances are the same. Likewise, the y-distances for sta 38+00 and 35+00 are the same, as well as for sta 39+00 and 34+00.

Finish Elevations

point	station	tangent elevation (ft)	y (ft)	finish elevation (ft)
	32+00	458.75		458.75
	33+00	457.25		457.25
PC	33+50	456.50	0.00	456.50
	34+00	455.75	+0.08	455.83
	35+00	454.25	+0.70	454.95
	36+00	452.75	+1.95	454.70
PI	36+50	452.00	+2.81	454.81
	37+00	453.12	+1.95	455.07
	38+00	455.37	+0.70	456.07
	39+00	457.62	+0.08	457.70
PT	39+50	458.75	0.00	458.75
	40+00	459.88		459.88
	41+00	462.13		462.13

8. PLAN-PROFILE SHEETS

On construction plans, a profile of the natural ground along the centerline of a highway project and the profile of the finish grade along the centerline are shown on *plan-profile sheets*. A plan view of the centerline with surrounding topography is shown on the top half of the sheet. The profiles are shown on the bottom half, together with PIs, PCs, PTs, gradients, and finish elevations. A plan-profile sheet is shown in Fig. 22.8.

Plan-profile sheets are also used in construction plans for streets, sanitary sewers, and storm sewers.

Example 22.4

Using the following information, compute finish elevations for each full station. Plot the finish elevation profile and show pertinent information needed for construction.

- gradient, sta 32+00 to 34+00 = −2.000%
- gradient, sta 39+50 to 43+00 = −2.125%
- PI, sta 34+00, elev 470.00, 300 ft (vertical curve)
- PI, sta 39+50, elev 482.00, 500 ft (vertical curve)

Solution

$$\text{gradient} = \frac{482.00 \text{ ft} - 470.00 \text{ ft}}{39.5 \text{ sta} - 34.00 \text{ sta}}$$

$$= 0.02182 \text{ ft/sta} \quad (+2.182\%)$$

$$a_1 = \frac{-2.000\% - 2.182\%}{(2)(3.00 \text{ sta})} = 0.697\%/\text{sta}$$

$$a_2 = \frac{2.182\% - (-2.125\%)}{(2)(5.00 \text{ sta})} = 0.431\%/\text{sta}$$

$$e_1 = \frac{(-2.000 - 2.182)(3.00)}{8} = 1.57 \text{ ft}$$

$$e_2 = \frac{\big(2.182\% - (-2.125\%)\big)(5.00 \text{ sta})}{8} = 2.69 \text{ ft}$$

Prior to performing finish elevation computations, the approximate finish elevation profile should be plotted on the plan-profile sheet. Vertical curves are symmetrical. To locate the PC and PT, measure one-half the length of the vertical curve in each direction from the PI. The midpoint of the vertical curve can be found by drawing a straight line from the PC to the PT and measuring one-half the distance from the PI to this line.

In determining whether y-distances should be added to or subtracted from tangent elevations, look at the plotted profile to see whether the curve is higher or lower than the tangent at a particular station.

Figure 22.8 would normally show the profile of the natural ground along the centerline, and the elevation at each station would be shown just under the finish elevation at the bottom of the sheet.

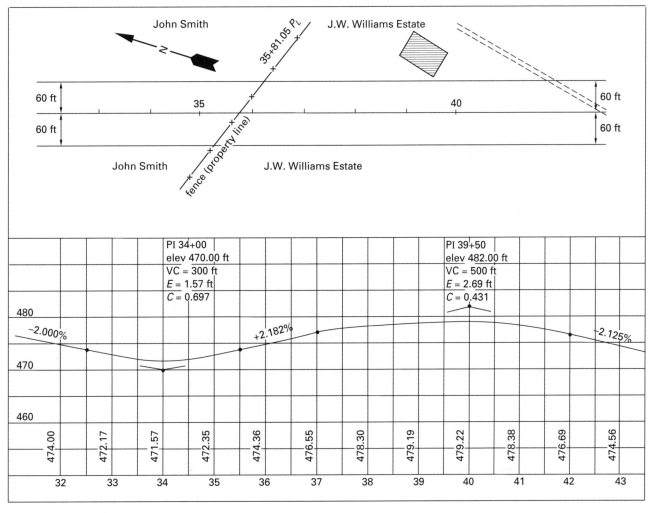

Figure 22.8 *Plan-Profile Sheet*

Computations for Finish Elevations

point	station	x (sta)	x^2 (sta^2)	tangent elevation (ft)	y (ft)	finish elevation (ft)
	32+00			474.00		474.00
PC	32+50					473.00
	33+00	0.5	0.25	472.00	+0.17	472.17
PI	34+00	1.5	2.25	470.00	+1.57	471.57
	35+00	0.5	0.25	472.18	+0.17	472.35
PT	35+50					
	36+00			474.36		474.36
PC	37+00			476.55		476.55
	38+00	1.0	1.00	478.73	−0.43	478.30
	39+00	2.0	4.00	480.91	−1.72	479.19
PI	39+50	2.5	6.25	482.00	−2.69	479.31
	40+00	2.0	4.00	480.94	−1.72	479.22
	41+00	1.0	1.00	478.81	−0.43	478.38
PC	42+00			476.69		476.69
	43+00			474.56		474.56

9. TURNING POINT ON SYMMETRICAL VERTICAL CURVE

The highest point of a crest curve (or the lowest point on a sag curve) is not usually vertically below (or above) the PI. This point is called the *turning point*. The distance x from the PC to the turning point can be found from Eq. 22.7.

$$x = \frac{g_1 L}{g_1 - g_2} \qquad 22.7$$

Example 22.5

A +1.500% grade meets a −2.500% grade at sta 12+50. Determine the distance from the PC to the turning point if a 600 ft vertical curve is used.

Solution

$$x = \frac{g_1 L}{g_1 - g_2} = \frac{(1.500\%)(6 \text{ sta})}{+1.500\% - (-2.500\%)}$$

$$= 2.25 \text{ sta} \quad (225 \text{ ft})$$

PRACTICE PROBLEMS

1. Determine the gradients between the points on the highway profiles in percent to three decimal places.

Example:

$$\text{PI} = 5+50$$
$$\text{elev} = 452.00 \text{ ft}$$
$$\text{PI} = 8+50$$
$$\text{elev} = 455.00 \text{ ft}$$
$$\text{PI} = 11+00$$
$$\text{elev} = 453.00 \text{ ft}$$

Solution:

$$g_1 = \frac{455.00 \text{ ft} - 452.00 \text{ ft}}{8.50 \text{ sta} - 5.50 \text{ sta}}$$

$$= \boxed{0.01 \text{ ft/sta} \quad (+1.000\%)}$$

$$g_2 = \frac{455.00 \text{ ft} - 453.00 \text{ ft}}{11.00 \text{ sta} - 8.50 \text{ sta}}$$

$$= \boxed{-0.008 \text{ ft/sta} \quad (-0.800\%)}$$

(a) PI = 20+70
elev = 504.00 ft

PI = 23+50
elev = 498.00 ft

PI = 26+60
elev = 503.00 ft

(b) PI = 40+00
elev = 461.00 ft

PI = 46+00
elev = 459.00 ft

PI = 52+00
elev = 465.00 ft

(c) PI = 55+00
elev = 474.00 ft

PI = 59+00
elev = 469.00 ft

PI = 67+00
elev = 477.50 ft

PI = 67+00
elev = 477.50 ft

(d) PI = 67+00
elev = 453.00 ft

PI = 70+50
elev = 463.00 ft

PI = 74+50
elev = 455.50 ft

PI = 79+00
elev = 461.70 ft

(e) PI = 29+25
elev = 445.00 ft

PI = 32+50
elev = 432.00 ft

PI = 37+75
elev = 432.00 ft

PI = 41+00
elev = 437.70 ft

2. Compute the gradient for each tangent of the highway profile and elevation of each full station on the tangents.

point	station	tangent elevation (ft)
PI	25+00	466.00
PI	31+00	458.00
PI	35+50	472.00
PI	39+00	465.00
PI	43+00	472.00

3. Compute a, e, and y for each station on the vertical curve.

(a) A $+2.234\%$ grade meets a -1.875% grade at sta 28+50, elev 436.00 ft. The vertical curve is 600 ft.

(b) A -3.467% grade meets a $+2.250\%$ grade at sta 45+00, elev 515.00 ft. The vertical curve is 800 ft.

4. From the information given, compute the finish elevation for each full station.

$$\text{PI} = 20+00$$
$$\text{elev} = 455.00 \text{ ft}$$
no vertical curve

$$\text{PI} = 23+50$$
$$\text{elev} = 448.50 \text{ ft}$$
400 ft vertical curve

$$\text{PI} = 33+00$$
$$\text{elev} = 469.00 \text{ ft}$$
1000 ft vertical curve

$$\text{PI} = 40+00$$
$$\text{elev} = 455.50 \text{ ft}$$
no vertical curve

5. Determine the distance x from the PC of the symmetrical curve to the high point of the crest curve or to the low point of the sag curve.

(a) A +4.000% grade meets a −3.000% grade at sta 35+00. The vertical curve is 600 ft in length.

(b) A +3.125% grade meets a −2.250% grade at sta 12+00. The vertical curve is 800 ft in length.

(c) A −2.750% grade meets a +3.500% grade at sta 22+00. The vertical curve is 500 ft in length.

(d) A −1.275% grade meets a +3.250% grade at sta 15+00. The vertical curve is 600 ft in length.

6. From the information given, complete the profile half of the plan-profile sheet.

station	finish elevation (ft)
32+00	476.00
33+00	474.00
34+00	472.18
35+00	471.65
36+00	472.58
37+00	474.80
38+00	476.82
39+00	478.09
40+00	478.61
41+00	478.37
42+00	477.37
43+00	475.63

Problem 6

gradient sta 32+00 to sta 35+00: −2.000% $e_1 = 1.65$ ft $a_1 = 0.733\%/\text{sta}$
gradient sta 35+00 to sta 40+00: +2.400% $e_2 = 3.39$ ft $a_2 = 0.377\%/\text{sta}$
gradient sta 40+00 to sta 43+00: −2.125% $VC_1 = 300$ ft $VC_2 = 600$ ft
PI = 35+00, elev 470.00 ft
PI = 40+00, elev 482.00 ft

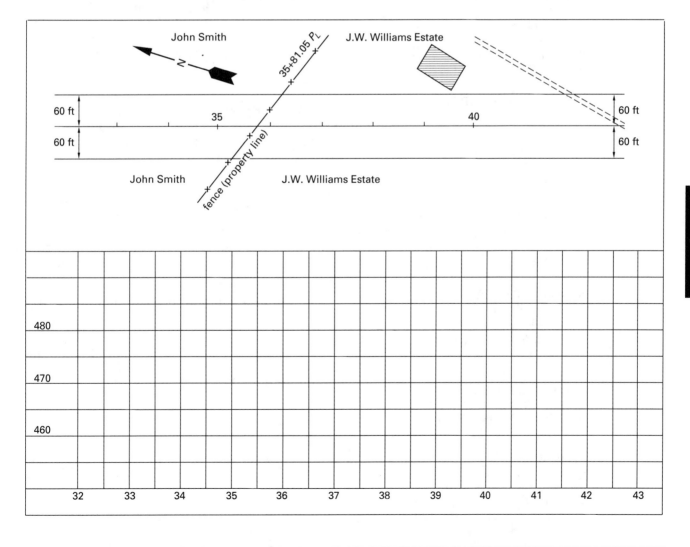

SOLUTIONS

1. (a) $g_1 = \dfrac{504.00 \text{ ft} - 498.00 \text{ ft}}{23.50 \text{ sta} - 20.70 \text{ sta}}$

$= \boxed{-0.02143 \text{ ft/sta} \quad (-2.143\%)}$

$g_2 = \dfrac{503.00 \text{ ft} - 498.00 \text{ ft}}{26.60 \text{ sta} - 23.50 \text{ sta}}$

$= \boxed{0.01613 \text{ ft/sta} \quad (+1.613\%)}$

(b) $g_1 = \dfrac{461.00 \text{ ft} - 459.00 \text{ ft}}{46.00 \text{ sta} - 40.00 \text{ sta}}$

$= \boxed{-0.00333 \text{ ft/sta} \quad (-0.333\%)}$

$g_2 = \dfrac{465.00 \text{ ft} - 459.00 \text{ ft}}{52.00 \text{ sta} - 46.00 \text{ sta}}$

$= \boxed{0.01000 \text{ ft/sta} \quad (+1.000\%)}$

(c) $g_1 = \dfrac{474.00 \text{ ft} - 469.00 \text{ ft}}{59.00 \text{ sta} - 55.00 \text{ sta}}$

$= \boxed{-0.01250 \text{ ft/sta} \quad (-1.250\%)}$

$g_2 = \dfrac{477.50 \text{ ft} - 469.00 \text{ ft}}{64.00 \text{ sta} - 59.00 \text{ sta}}$

$= \boxed{0.01700 \text{ ft/sta} \quad (+1.700\%)}$

$g_3 = \dfrac{477.50 \text{ ft} - 477.50 \text{ ft}}{67.00 \text{ sta} - 64.00 \text{ sta}}$

$= \boxed{0 \text{ ft/sta} \quad (0\%)}$

(d) $g_1 = \dfrac{463.00 \text{ ft} - 453.00 \text{ ft}}{70.50 \text{ sta} - 67.00 \text{ sta}}$

$= \boxed{0.02857 \text{ ft/sta} \quad (+2.857\%)}$

$g_2 = \dfrac{463.00 \text{ ft} - 455.50 \text{ ft}}{74.50 \text{ sta} - 70.50 \text{ sta}}$

$= \boxed{-0.01875 \text{ ft/sta} \quad (-1.875\%)}$

$g_3 = \dfrac{461.70 \text{ ft} - 455.50 \text{ ft}}{79.00 \text{ sta} - 74.50 \text{ sta}}$

$= \boxed{0.01378 \text{ ft/sta} \quad (+1.378\%)}$

(e) $g_1 = \dfrac{445.00 \text{ ft} - 432.00 \text{ ft}}{32.50 \text{ sta} - 29.25 \text{ sta}}$

$= \boxed{-0.04000 \text{ ft/sta} \quad (-4.000\%)}$

$g_2 = \dfrac{437.70 \text{ ft} - 432.00 \text{ ft}}{41.00 \text{ sta} - 37.75 \text{ sta}}$

$= \boxed{0.01754 \text{ ft/sta} \quad (+1.754\%)}$

$g_3 = \dfrac{432.00 \text{ ft} - 432.00 \text{ ft}}{37.75 \text{ sta} - 32.50 \text{ sta}}$

$= \boxed{0 \text{ ft/sta} \quad (0\%)}$

2.

point	station	tangent elevation (ft)
PI	25+00	466.00
	26+00	464.67
	27+00	463.33
$g_1 = -1.333\%$ 28+00		462.00
	29+00	460.67
	30+00	459.33
PI	31+00	458.00
	32+00	461.11
	33+00	464.22
$g_2 = +3.111\%$ 34+00		467.33
	35+00	470.44
PI	35+50	472.00
	36+00	471.00
$g_3 = -2.000\%$ 37+00		469.00
	38+00	467.00
PI	39+00	465.00
	40+00	466.75
$g_4 = +1.750\%$ 41+00		468.50
	42+00	470.25
PI	43+00	472.00

3. (a) $a = \dfrac{2.234\% - (-1.875\%)}{(2)(6 \text{ sta})} = \boxed{0.342\%/\text{sta}}$

$e = \dfrac{(2.234\% - (-1.875\%))(6 \text{ sta})}{8} = \boxed{3.08}$

y @ sta 26+00 $= \left(0.342 \dfrac{\%}{\text{sta}}\right)(0.5 \text{ sta})^2$

$= \boxed{0.09 \text{ ft}}$

y @ sta 27+00 $= \left(0.342 \dfrac{\%}{\text{sta}}\right)(1.5 \text{ sta})^2$

$= \boxed{0.77 \text{ ft}}$

y @ sta 28+00 $= \left(0.342 \dfrac{\%}{\text{sta}}\right)(2.5 \text{ sta})^2$

$= \boxed{2.14 \text{ ft}}$

$$y \text{ @ sta } 29+00 = \left(0.342 \ \frac{\%}{\text{sta}}\right)(2.5 \text{ sta})^2$$

$$= \boxed{2.14 \text{ ft}}$$

$$y \text{ @ sta } 30+00 = \left(0.342 \ \frac{\%}{\text{sta}}\right)(1.5 \text{ sta})^2$$

$$= \boxed{0.77 \text{ ft}}$$

$$y \text{ @ sta } 31+00 = \left(0.342 \ \frac{\%}{\text{sta}}\right)(0.5 \text{ sta})^2$$

$$= \boxed{0.09 \text{ ft}}$$

(b) $a = \dfrac{-3.467\% - 2.250\%}{(2)(8 \text{ sta})} = \boxed{0.357\%/\text{sta}}$

$e = \dfrac{(-3.467\% - 2.250\%)(8 \text{ sta})}{8} = \boxed{5.72}$

$$y \text{ @ sta } 42+00 = \left(0.357 \ \frac{\%}{\text{sta}}\right)(1 \text{ sta})^2$$

$$= \boxed{0.36 \text{ ft}}$$

$$y \text{ @ sta } 43+00 = \left(0.357 \ \frac{\%}{\text{sta}}\right)(2 \text{ sta})^2$$

$$= \boxed{1.43 \text{ ft}}$$

$$y \text{ @ sta } 44+00 = \left(0.357 \ \frac{\%}{\text{sta}}\right)(3 \text{ sta})^2$$

$$= \boxed{3.21 \text{ ft}}$$

$$y \text{ @ sta } 45+00 = \left(0.357 \ \frac{\%}{\text{sta}}\right)(4 \text{ sta})^2$$

$$= \boxed{5.72 \text{ ft}}$$

$$y \text{ @ sta } 46+00 = \left(0.357 \ \frac{\%}{\text{sta}}\right)(3 \text{ sta})^2$$

$$= \boxed{3.21 \text{ ft}}$$

$$y \text{ @ sta } 47+00 = \left(0.357 \ \frac{\%}{\text{sta}}\right)(2 \text{ sta})^2$$

$$= \boxed{1.43 \text{ ft}}$$

$$y \text{ @ sta } 48+00 = \left(0.357 \ \frac{\%}{\text{sta}}\right)(1 \text{ sta})^2$$

$$= \boxed{0.36 \text{ ft}}$$

4.

point	station	tangent elevation (ft)	y (ft)	finish elevation (ft)
PI	20+00	455.00		455.00
	21+00	453.14		453.14
PC	21+50	452.21		452.21
	22+00	451.29	0.13	451.42
	23+00	449.43	1.13	450.56
PI	23+50	448.50	2.01	450.51
	24+00	449.58	1.13	450.71
	25+00	451.74	0.13	451.87
PT	25+50	452.82		452.82
	26+00	453.90		453.90
	27+00	456.05		456.05
PC	28+00	458.21		458.21
	29+00	460.37	0.20	460.17
	30+00	462.53	0.82	461.71
	31+00	464.69	1.84	462.85
	32+00	466.84	3.26	463.58
PI	33+00	469.00	5.11	463.89
	34+00	467.07	3.26	463.81
	35+00	465.14	1.84	463.30
	36+00	463.21	0.82	462.39
	37+00	461.28	0.20	461.08
PT	38+00	459.36		459.36
	39+00	457.43		457.43
PI	40+00	455.50		455.50

5. (a) $x = \dfrac{(4.0\%)(6 \text{ sta})}{4\% + 3\%} = 343 \text{ ft}$

(b) $x = \dfrac{(3.125\%)(8 \text{ sta})}{3.125\% + 2.250\%} = 465 \text{ ft}$

(c) $x = \dfrac{(-2.750\%)(5 \text{ sta})}{-2.750\% - 1.500\%} = 220 \text{ ft}$

(d) $x = \dfrac{(1.275\%)(6 \text{ sta})}{-1.275\% - 3.250\%} = 169 \text{ ft}$

Vertical Alignments

6.

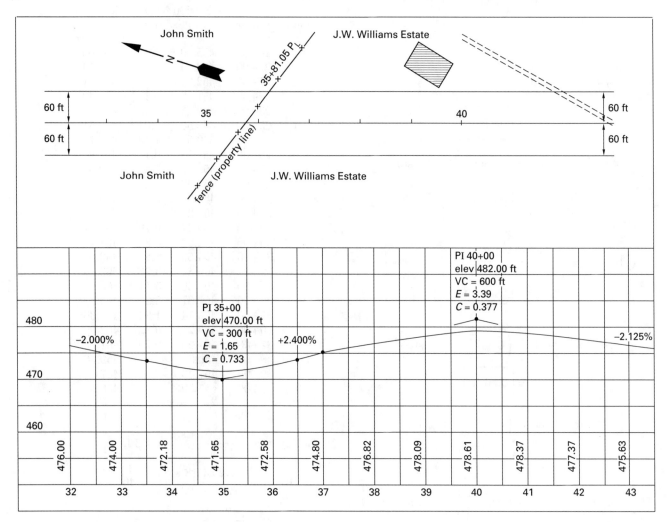

23 Construction Surveying

1. DEFINITION

Construction surveying involves locating and marking locations of structures that are to be built. It is often referred to as *giving line and grade*. A transit or theodolite is used in establishing line (horizontal alignment), and a level is used in establishing grade (elevation).[1]

2. CONVERSION BETWEEN INCHES AND DECIMALS OF A FOOT

Construction stakes are usually set to the nearest hundredth of a foot for concrete, asphalt, pipelines, and so on. For earthwork, stakes are set to the nearest tenth of a foot. Constructors use foot and inch rules. So, in deference to them, surveyors set the stakes for their convenience.

In converting measurements in feet, inches, and fractions to feet, it may be easier to convert the inches and fractions of an inch separately, and then to add the parts.

---

[1] The word *grade* is not used consistently, sometimes meaning slope and sometimes meaning elevation above a datum. In this text, *gradient* will be used for rate of slope, and *finish elevation* will be used for the elevation above a datum to which a part of the structure is to be built.

Example 23.1

Convert the measurements to feet and decimals of a foot.

(a) 1 ft, 4 in

(b) 11 ft, $9^1/_8$ in

(c) 7 ft, $5^3/_4$ in

(d) 2 ft, $8^7/_8$ in

(e) 5 ft, $11^1/_2$ in

Solution

(a) 1.33 ft

(b) 11.76 ft

(c) 7.48 ft

(d) 2.74 ft

(e) 5.96 ft

Example 23.2

Convert the following measurements to feet and inches.

(a) 3.79 ft

(b) 6.34 ft

(c) 5.65 ft

(d) 3.72 ft

Solution

(a) 3 ft, $9^1/_2$ in

(b) 6 ft, $4^1/_8$ in

(c) 5 ft, $7^3/_4$ in

(d) 3 ft, $8^5/_8$ in

3. STAKING OFFSET LINES FOR CIRCULAR CURVES

Stakes set for the construction of pavement or curbs must be set on an offset line so that they will not be destroyed by construction equipment. However, they must be close enough for short measurements to the actual line. The offset line may be 3 ft or 5 ft, or any convenient distance from the edge of pavement or back of curb. Stakes are set at 25 ft or 50 ft intervals, and tacks are set in the stakes to designate the offset line.

In setting stakes on a parallel circular arc, the central angle is the same for parallel arcs. The radius to the centerline of the street or road is usually the design radius. The PC of the design curve and the PC of a parallel offset curve, whether right or left, will fall on the same radial line. Likewise, the PTs of the parallel arcs will fall on the same radial line. In computing the stations for PCs and PTs, the design curve data (design radius) should be used. Then, the PC and PT stations for an offset line will be the same as for the design curve (centerline of road or street), even though the lengths of offset curves will not be the same as the length of the centerline curve.

The design curve data will be used to compute deflection angles. These angles will be the same for offset lines, since the central angle between any two radii is the same for the parallel arcs.

Because chord lengths are a function of the radius of an arc $\left(C = 2R\sin(\Delta/2)\right)$, the chord length between two stations on the design curve and two corresponding stations on the offset line will not be the same.

By using design curve data in computing PC and PT stations, deflection angles for curves can be recorded in the field book. Such angles will be the same whether a right offset line, a left offset line, or both right and left offset lines are used.

In performing field work, centerline PIs, PCs, and PTs are located on the ground before construction. Offset PCs and PTs are located at right angles to the centerline PCs and PTs. Offset PCs and PTs should be carefully referenced.

Example 23.3

Stakes are to be set on 4 ft offsets for each edge of pavement, which is 36 ft wide. The curve has a deflection angle Δ of 60° to the right, and a centerline radius of 300 ft, PI is at station 12+44.32, and stakes are to be set for each full-station, half-station, and at the PC and PT.

Compute PC and PT stations, deflection angles, and chord lengths. Set up field notes for the curve.

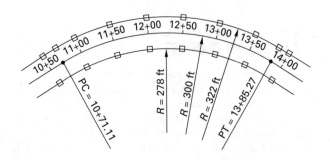

Solution

$\Delta = 60°00'$

$T = R\tan\dfrac{\Delta}{2} = (300\text{ ft})(\tan 30°) = 173.21\text{ ft}$

$L = \left(\dfrac{60°}{360°}\right)2\pi R = \left(\dfrac{60°}{360°}\right)2\pi(300\text{ ft}) = 314.16\text{ ft}$

$$PI = 12+44.32$$
$$T = \underline{-1+73.21}$$
$$PC = 10+71.11$$
$$L = \underline{3+14.16}$$
$$PT = 13+85.27$$

Deflection Angle Computations

sta 10+71.11: 0°00'

sta 11+00: $\left(\dfrac{11+00 - 10+71.11}{314.16\text{ ft}}\right)\left(\dfrac{60°}{2}\right)$

$\qquad = \left(\dfrac{28.89\text{ ft}}{314.16\text{ ft}}\right)(30°)$

$\qquad = 2.7588° \quad (2°45')$

sta 11+50: $\left(\dfrac{28.89 + 50\text{ ft}}{314.16\text{ ft}}\right)\left(\dfrac{60°}{2}\right)$

$\qquad = \left(\dfrac{78.89\text{ ft}}{314.16\text{ ft}}\right)(30°)$

$\qquad = 7.5334° \quad (7°32')$

sta 12+00: $\left(\dfrac{128.89\text{ ft}}{314.16\text{ ft}}\right)(30°) = 12.3081° \quad (12°18')$

sta 12+50: $\left(\dfrac{178.89\text{ ft}}{314.16\text{ ft}}\right)(30°) = 17.0827° \quad (17°05')$

sta 13+00: $\left(\dfrac{228.89\text{ ft}}{314.16\text{ ft}}\right)(30°) = 21.8573° \quad (21°51')$

sta 13+50: $\left(\dfrac{278.89\text{ ft}}{314.16\text{ ft}}\right)(30°) = 26.6320° \quad (26°38')$

sta 13+85.27: $\left(\dfrac{314.16\text{ ft}}{314.16\text{ ft}}\right)(30°) = 30.0000° \quad (30°00')$

Outside Chord Lengths

station to station	computation	length
10+71.11 to 11+00:	(2)(322 ft)(sin 2.7588°) =	31.00 ft
11+00 to 11+50:	(2)(322 ft)(sin 4.7746°) =	53.60 ft
13+50 to 13+85.27:	(2)(322 ft)(sin 3.3680°) =	37.83 ft

Inside Chord Lengths

station to station	computation	length
10+71.11 to 11+00:	(2)(278 ft)(sin 2.7588°) =	26.76 ft
11+00 to 11+50:	(2)(278 ft)(sin 4.7746°) =	46.28 ft
13+50 to 13+85.27:	(2)(278 ft)(sin 3.3680°) =	32.66 ft

Field Notes

point	station	angle	C-out (ft)	C-in (ft)	curve data
PT	13+85.27	30°00′			
			37.83	32.66	
	+50.00	26°38′			
			53.60	46.28	
	13+00.00	21°51′			$\Delta = 60°00′$
			53.60	46.28	
	+50.00	17°05′			$R = 300$ ft
			53.60	46.28	
	12+00.00	12°18′			$T = 173.21$ ft
			53.60	46.28	
	+50.00	7°32′			$L = 314.16$ ft
			53.60	46.28	
	11+00.00	2°45′			
			31.00	26.76	
PC	10+71.11	0°00′			

4. CURB RETURNS AT STREET INTERSECTIONS

Curb returns are the arcs made by the curbs at street intersections. The radius of the arc is selected by the designer with consideration given to the speed and volume of traffic. A radius of 30 ft to the back of curb is common. Streets that intersect at a right angle have curb returns of one-quarter circle. The arcs can be swung from a radius point (center of circle). The radius point can be located by finding the intersection of two lines, each of which is parallel to one of the centerlines of the streets and at a distance from the centerline equal to half the street width plus the radius, as shown in Fig. 23.1. One stake at the radius point is sufficient for a curb return.

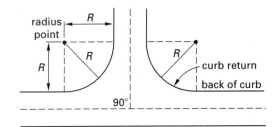

Figure 23.1 *Curb Returns*

5. STAKING OFFSET LINES AT STREET INTERSECTIONS

In setting stakes for curb and gutter for street construction on an offset line, stations for the PC and PT of a curb return at a street intersection are computed along the centerline of the street. In Ex. 23.4, stakes are to be set on a 5 ft offset line from the back of the left curb along Elm Street. In setting stakes at the intersection

of 24th Street, the PCs and PTs of the curb return are computed from the centerline stations. However, it must be remembered that in computing the long chord on the offset line, the radius R is not the design radius, but is the design radius minus the offset distance. Stakes are set at the PC, the PT, and the radius point of each curb return arc.

At street intersections not at 90°, there will be two deflection angles, one being the supplement of the other.

Example 23.4

Elm Street and 24th Street intersect as shown. Both streets are 26 ft wide, back to back of curb.

Compute the PC and PT stations, deflection angles from PC to PT, and long chord measured from the offset line for each return.

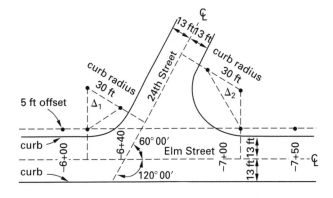

Solution

PC Stations Along Elm Street

$$\Delta_1 = 60°00′$$

$$T_1 = R_1 \tan \frac{\Delta_1}{2}$$

$$= (43 \text{ ft}) \tan \left(\frac{60°}{2} \right)$$

$$= 24.83 \text{ ft}$$

$$\text{PI} = 6+40.00$$

$$T_1 = \underline{-24.83}$$

$$\text{PC} = 6+15.17$$

$$\Delta_2 = 120°00′$$

$$T_2 = R_2 \tan \frac{\Delta_2}{2}$$

$$= (43 \text{ ft}) \tan \left(\frac{120°}{2} \right)$$

$$= 74.48 \text{ ft}$$

$$\text{PI} = 6+40.00$$

$$T_2 = \underline{+74.48}$$

$$\text{PC} = 7+14.48$$

PT Stations Along 24th Street

$$PI = 0+00.00$$
$$T_1 = \underline{+\ 24.83}$$
$$PT = 0+24.83$$
$$PI = 0+00.00$$
$$T_2 = \underline{+\ 74.48}$$
$$PT = 0+74.48$$

Deflection Angles and Long Chords

$$LC = 2R \sin \frac{\Delta_1}{2}$$
$$= (2)(25 \text{ ft}) \sin \left(\frac{60°}{2} \right)$$
$$= 25.00 \text{ ft}$$
$$\text{deflection angle} = \frac{60°}{2} = 30°$$
$$LC = 2R \sin \frac{\Delta_2}{2}$$
$$= (2)(25 \text{ ft}) \sin \left(\frac{120°}{2} \right)$$
$$= 43.30 \text{ ft}$$
$$\text{deflection angle} = \frac{120°}{2} = 60°$$

6. ESTABLISHING FINISH ELEVATIONS OR "GRADE"

Establishing the elevation above a datum to which a structure, or part of a structure, is to be built is usually accomplished by the following steps.

step 1: Set the top of a grade stake to the exact elevation (nearest one hundredth). Mark the top of the stake with blue keel.

step 2: Set the top of a grade stake at an exact distance above or below finish elevation. Mark the top of it with blue keel, and mark this exact distance above or below (called *cut* or *fill*) on another stake known as a *guard stake*, usually driven at an angle, beside the grade stake.

step 3: Use the line stake as a grade stake driven to a random elevation. Compute the difference in elevation between that elevation and the finish elevation, and mark this difference as cut or fill on a guard stake.

Marks on a wall, such as the wall of forms for a concrete structure, may be used instead of the tops of stakes. This is illustrated in Ex. 23.5.

7. GRADE ROD

A *grade rod* is the rod reading determined by finding the difference in elevation between the height of instrument (height of the level) and the finish elevation. In Fig. 23.2, the finish elevation is 441.23 ft, and the HI of the level is 445.55 ft. The grade rod is the difference in these two numbers, 4.32. A stake is driven so that when the level rod is placed on the top of it, the rod reading is 4.32.

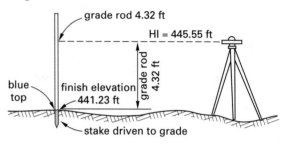

Figure 23.2 *Use of Grade Rod*

The procedure to set the stake is to place the rod on the ground where the stake is to be driven and determine the distance (in tenths) that the top of the stake should be above the ground. Then drive the stake until the top of the stake is at the finish elevation, stopping to check the rod reading so that the top of the stake will not be too low. When the grade rod reading is reached, the top of the stake is marked with blue keel, and thus the name *blue top* is given to this type of stake. A guard stake is driven beside the blue top in a slanting position. The guard stake is marked "G" to indicate the stake is driven to grade (finish elevation).

If the finish elevation is just below ground level, the blue top can be left above ground. The cut from the top of the stake to the finish elevation should be marked on a guard stake.

Where line stakes are also used as grade stakes driven to random elevations, a grade rod is not used. The elevation of the top of the stake is determined by leveling. The difference in elevation between the top of the stake and the finish elevation is determined and marked on the guard stake.

Example 23.5

The finish elevation is to be marked on the inside wall of the form for the concrete cap of a bridge. The finish elevation of the cap is 466.97 ft, and the HI is 468.72 ft.

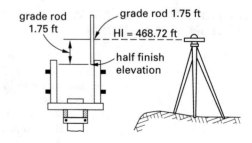

Solution

$$HI = 468.72 \text{ ft}$$
$$\text{finish elevation} = \underline{466.97} \text{ ft}$$
$$\text{grade rod} = 1.75 \text{ ft}$$

The rod is held against the side of the form and raised or lowered until the rod reading is 1.75 ft. A nail is driven at the bottom of the rod, and the rod is placed on the nail so that the rod reading can be checked to see that the nail is correctly placed. Another finish elevation nail is driven at the other end of the form and a string line is drawn between the two nails, then is chalked and snapped to mark the grade line on the form. A chamfer strip is nailed on the form along this line, and the strip is used to finish the concrete to grade (finish elevation).

8. SETTING STAKES FOR CURB AND GUTTER

Separate stakes are often set for line and grade. In Fig. 23.3, a hub stake is driven so that a tack is exactly 3 ft from the back of a curb. These line stakes are set on any convenient offset to avoid disturbance by construction equipment. A separate grade stake is driven so that the top of the stake is either at finish elevation or at an elevation that makes it an exact distance above or below finish elevation.

A guard stake is driven near the grade stake and marked to show this exact distance as cut or fill, and the top of

the grade stake is marked with blue keel. Grade stakes can be driven so that the cut or fill is in multiples of a half foot. If the grade stake is driven to finish elevation, the guard stake is marked "G" for grade. The cut or fill can be determined by considering the finish elevation and a ground rod reading at each station.

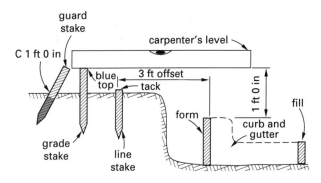

Figure 23.3 *Line and Grade Stakes for Curb and Gutter*

The top of the grade stake should be above ground. Then, the builder can lay a carpenter's level on top of the stake and measure from the established level line to establish the top of curb forms. In Ex. 23.6, the guard stake is marked for a cut as "C 1 ft, 0 in" so that the builder will measure 1 ft, 0 in down from the level line to the top of the forms. Horizontal alignment will be maintained by measuring 3 ft from each tack point to the back of curb line.

Table 23.1 *Field Notes for Curb and Gutter Grades*

(all measurements are in feet)

station	+	HI	−	rod	elevation	finish elevation	grade rod	ground	mark stake
BM no. 1	3.42	455.78			452.36	r.r. spike in 12″ oak-100′ lt. sta 0+00			
0+00				0.18	455.60	454.10	1.68	0.4	C 1 ft, 6 in
0+50				1.20	454.58	454.58	1.20	1.6	grade
1+00				2.72	453.06	455.06	0.72	3.2	F 2 ft, 0 in
1+50				2.75	453.03	455.53	0.25	3.3	F 2 ft, 6 in
2+00				2.27	453.51	456.01	−0.23	2.5	F 2 ft, 6 in
2+50				1.79	453.99	456.49	−0.71	1.8	F 2 ft, 6 in
3+00				0.31	455.47	455.97	−1.19	0.6	F 1 ft, 6 in
T.P.	8.21	460.24	3.75		452.03				
3+50				3.30	456.94	457.44	2.80	3.6	F 0 ft, 6 in
4+00				1.41	458.83	457.83	2.41	1.7	C 1 ft, 0 in
4+50				0.70	459.54	458.04	2.20	1.0	C 1 ft, 6 in
5+00				1.16	459.08	458.08	2.16	1.5	C 1 ft, 0 in
5+50				0.31	459.93	457.93	2.31	0.7	C 2 ft, 0 in
6+00				1.04	459.20	457.70	2.54	1.4	C 1 ft, 6 in
BM no. 2			2.06		458.18				
	11.63		5.81						
	11.63		458.18						
	5.81		452.36						
	5.82		5.82						

Example 23.6

Grade stakes for curb and gutter have been driven to grade or to a multiple of 6 in above or below grade. Part of the level notes recorded in setting the stakes is shown. Also shown are finish elevations that have been taken from construction plans. Computations for grade rod and for the cut or fill marks on guard stakes are to be made and recorded. Rod readings on grade stakes as driven are to be recorded in the column marked "rod." (Note: A more detailed explanation of this procedure can be found in Sec. 14.)

Curb and Gutter Grades

(all measurements are in feet)

station	+	−	elevation	finish elevation	ground rod
BM no. 1	3.42		452.36		
0+00				454.10	0.4
+50				454.58	1.6
1+00				455.06	3.2
+50				455.53	3.3
2+00				456.01	2.5
+50				456.49	1.8
3+00				456.97	0.6

Curb and Gutter Grades (continued)

(all measurements are in feet)

station	+	−	elevation	finish elevation	ground rod
TP	8.21	3.75			
+50				457.44	3.6
4+00				457.83	1.7
+50				458.04	1.0
5+00				458.08	1.5
+50				457.93	0.7
6+00				457.70	1.4
BM no. 2		2.06			

Solution

Finished field notes are shown in Table 23.1. A graphical solution is shown in Ex. 23.6, the finished grade line. Solutions for sta 0+00, 0+50, 1+00, and 2+50 are shown in Ex. 23.6, the sample staking.

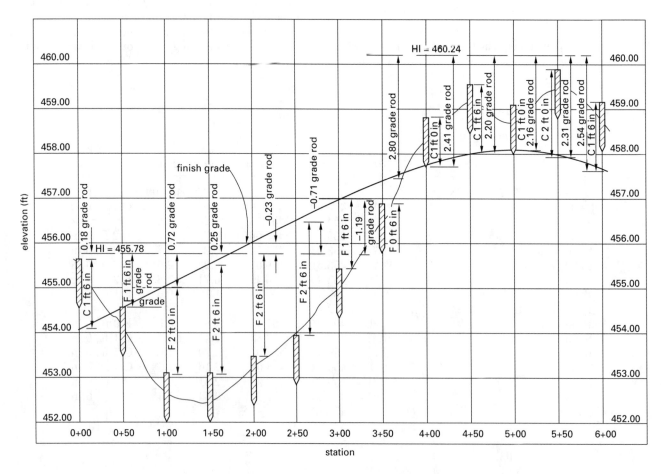

Finished Grade Line for Ex. 23.6

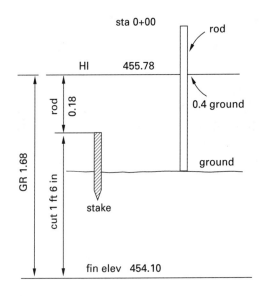

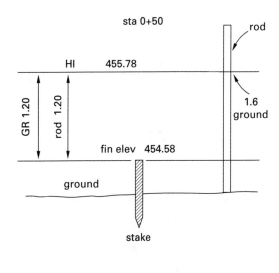

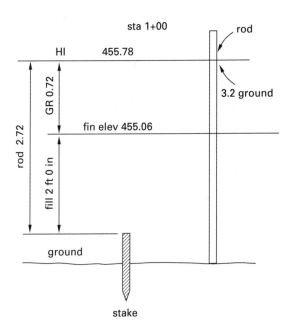

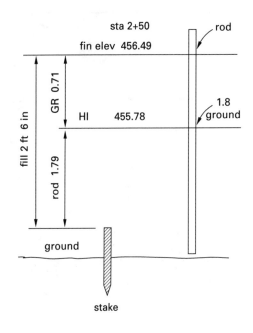

Sample Staking for Ex. 23.6

9. STAKING CONCRETE BOX CULVERTS ON HIGHWAYS

Tack points for concrete box culverts can be set on offsets from the outside corners of the culvert headwalls. For normal culverts (centerline of culvert at right angle to centerline of roadway), the distance from the centerline of the roadway to the outside of the headwall is equal to one-half the clear roadway width plus the width of the headwall. Tacks should also be set on the centerline of the roadway, offset from the outside of the culvert wall. The offset distance from the outside walls depends on the depth of cut. Stakes for wingwalls and aprons are not necessary, although stakes to establish the centerline of the culvert can be set if desired. Cuts to the flowline of the culvert can be marked on guard stakes at the tack points.

In staking skewed culverts, the distance from the centerline of the roadway to the outside of headwall (along the centerline of culvert) is equal to one half the clear roadway plus the headwall thickness divided by the cosine of the skew angle. The *skew angle* is the angle between the normal and the centerline of culvert.

Construction

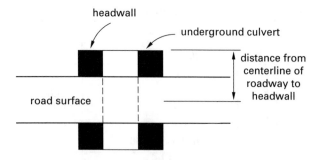

(a) normal culvert (headwall perpendicular to culvert)

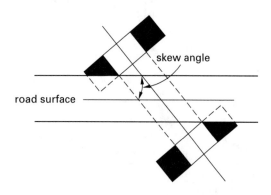

(b) skewed culvert (headwall perpendicular to culvert)

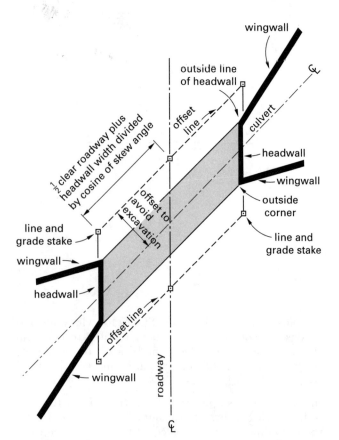

(c) skewed culvert (headwalls parallel to roadway)

Figure 23.4 *Staking Box Culverts (Plan View)*

10. SETTING SLOPE STAKES

Before earthwork construction is started, the extremities of a cut or fill must be located at numerous places for the benefit of machine operators engaged in the earthwork.

With the centerline as a reference, the edge (*toe*) of a fill must be established on the natural ground. This point is known as the *toe of slope*. Likewise, the top edge of a cut must be established on the natural ground.

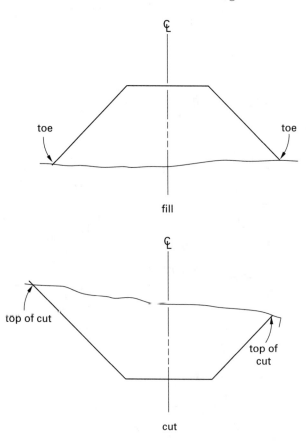

Figure 23.5 *Slope Staking*

Because the natural ground may slope from left to right or from right to left, the distance from the centerline to the left toe of slope of a fill is usually different from the distance from the centerline to the right toe of slope at any particular station. The same is true of the top of a cut. This fact, plus the fact that the height of fill or depth of cut varies along the centerline, makes toe and top lines irregular when seen in plan view, as is shown in Fig. 23.6.

The toe of a fill or the top of a cut is found by a measure-and-try method. The horizontal distance from centerline to toe or top is determined by horizontal tape measurements combined with vertical distance measurements derived by use of level and rod.

Dimensions of the top of a fill or bottom of a cut, and the slope of the sides of the fill or cut must be known. These are used in the measure and try method. They are shown on the "Typical Sections Sheet" of construction plans, as in Fig. 23.7.

The side slopes of a fill and the back slopes of a cut are expressed as a ratio of horizontal to vertical distance. Thus, a 4:1 slope means a rise or fall of 1 ft for each 4 ft of horizontal distance. Slopes of 1:1, 2:1, and 3:1 are illustrated in Fig. 23.8.

With the centerline finish elevation, width of top of fill or bottom of cut, and side slopes all known, the intersection of the side slopes and the natural ground is located at each station or intermediate point.

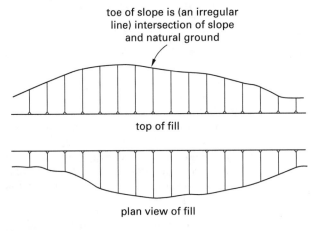

Figure 23.6 *Fill Views*

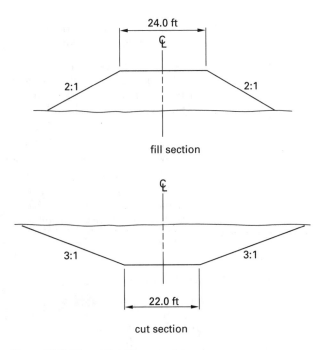

Figure 23.7 *Fill and Cut Dimensioning*

When the intersection is found, it is marked by a *slope stake*. The stake is driven so that it slopes away from the fill or cut and is marked with its horizontal distance (left or right) from the centerline and the vertical

distance from the ground at the stake to the finish elevation. A stake marked "C 3.2-48.2" means that the stake is 48.2 ft from the centerline, and the ground at the stake is 3.2 ft above the finish elevation. The station number is shown on the side of the stake facing the ground.

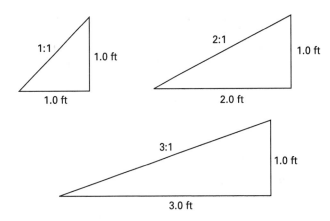

Figure 23.8 *Calculation of Slopes*

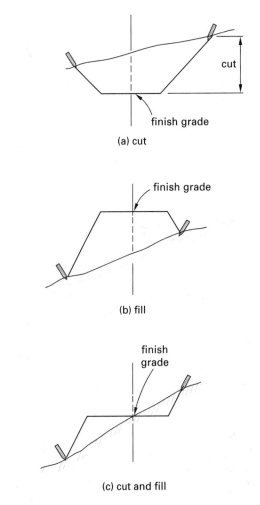

Figure 23.9 *Stake Orientations*

Construction

11. GRADE ROD

In setting slope stakes, as in setting finish elevation for pavement, sewer lines, and so on, the *grade rod* is used to determine the difference in elevation between the HI and the finish elevation. To determine the cut at a particular point, the rod is read on the ground, and the ground rod is subtracted from the grade rod at that point. To determine the fill at a particular point, the grade rod is subtracted from the ground rod if the HI is above the finish elevation. The grade rod is added to the ground rod if the HI is below the finish elevation. (See Fig. 23.10, Ex. 23.9, and Ex. 23.10.)

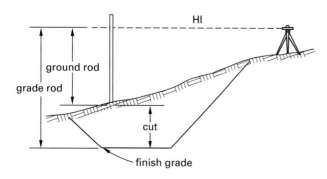

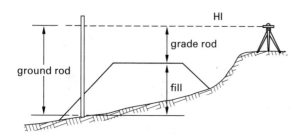

Figure 23.10 *Use of Grade Rod to Determine Cut and Fill*

12. SETTING SLOPE STAKES AT CUT SECTIONS

An explanation of setting slope stakes without the benefit of a demonstration in the field is difficult. In Ex. 23.7, a scale drawing is used at a cut section at which the HI and finish elevation are known and plotted on the drawing. The width of the ditch bottom and the side slopes (also referred to as *back slopes*) are also known. In this example, the level and rod are replaced by the plotted HI and an engineer's scale. The scale is used to measure vertical distance from HI to ground, just as the level and rod are used.

Example 23.7

The illustration on the following page shows the ground cross section at a station at which slope stakes are to be set for a ditch to be excavated. HI has been established and finish elevation, width of ditch bottom, and side slopes have been obtained from construction plans.

Centerline of ditch has also been established at this station. Two unsuccessful attempts have been made to locate the stakes, as shown. Known information is tabulated.

finish elevation of ditch bottom = 470.45 ft
bottom width = 12 ft
side slopes = 2:1
HI = 479.24 ft

Solution

step 1: Compute the grade rod (GR).

$$GR = HI - \text{finish elevation}$$
$$= 479.24 \text{ ft} - 470.46 \text{ ft} = 8.78 \text{ ft}$$

step 2: Read the rod on the ground (use scale) at the centerline. A rod reading on the ground is known as a *ground rod*. This centerline ground rod helps find the cut (vertical distance from ground to finish elevation) at the centerline and the horizontal distance from the centerline to the slope stake (on each side) if the ground were level. This distance will be used as a guide to find the actual distance to the slope stake where the ground is not level. The ground rod is 4.1 ft.

$$\text{cut at centerline} = GR - \text{ground rod}$$
$$= 8.78 \text{ ft} - 4.1 \text{ ft}$$
$$= 4.7 \text{ ft}$$

step 3: Find the cut and horizontal distance from centerline to slope stake on the left side.

(a) The horizontal distance from centerline to left stake is equal to one half the width of the ditch bottom plus the horizontal distance from the left edge of the ditch bottom to the stake.

(b) The slope is 2:1. Therefore, the side slope will rise (from ditch bottom) 1 ft vertically for each 2 ft horizontally. For level ground, the vertical rise is the cut at the centerline, which has been found to be 4.7 ft. Therefore, the horizontal distance for level ground is

$$(2)(4.7 \text{ ft}) = 9.4 \text{ ft}$$

The distance from the centerline is

$$6 \text{ ft} + (2)(4.7 \text{ ft}) = 15.4 \text{ ft}$$

(c) The ground is not level. The slope is down from left to right, and the left slope stake will be at a greater distance from the centerline than the right slope stake.

(d) Use the horizontal distance computed for level ground (15.4 ft) as a guide. Make a first try beyond it because of the slope of the ground.

(e) Try a distance of 19.0 ft (chosen arbitrarily) from the centerline and read the rod on the ground at this point. (Use scale for rod.) The ground rod is 2.7 ft. Then,

$$\text{cut} = 8.78 \text{ ft} - 2.7 \text{ ft}$$
$$= 6.1 \text{ ft}$$

$$\text{distance from centerline} = 6 \text{ ft} + (2)(6.1 \text{ ft})$$
$$= 18.2 \text{ ft}$$

This is not the correct location because the measured distance (19.0 ft) does not agree with the computed distance (18.2 ft).

(f) For the next try, move toward the centerline because 19.0 ft was too far. Try 17.0 ft where the ground rod is 3.2 ft. Then,

$$\text{cut} = 8.78 \text{ ft} - 3.2 \text{ ft}$$
$$= 5.6 \text{ ft}$$

$$\text{distance from centerline} = 6 \text{ ft} + (2)(5.6 \text{ ft})$$
$$= 17.2 \text{ ft}$$

(g) Try 17.2 ft, where the ground rod is 3.2 ft again.

$$\text{distance from centerline} = 6 \text{ ft} + (2)(5.6 \text{ ft})$$
$$= 17.2 \text{ ft}$$

The correct location for the slope stake has been found, so mark it "C 5.6 @ 17.2" on one side and the station number on the other. Drive the stake with the station number down and sloping away from the cut.

step 4: Find the cut and horizontal distance on the right side.

The slope stake on the right is set in the same manner. In arbitrarily selecting a horizontal distance for the first try, select a distance less than the 15.4 computed for level ground because the slope is down from left to right.

The correct cut and distance for the right slope stake is shown on the stake marking in Ex. 23.7.

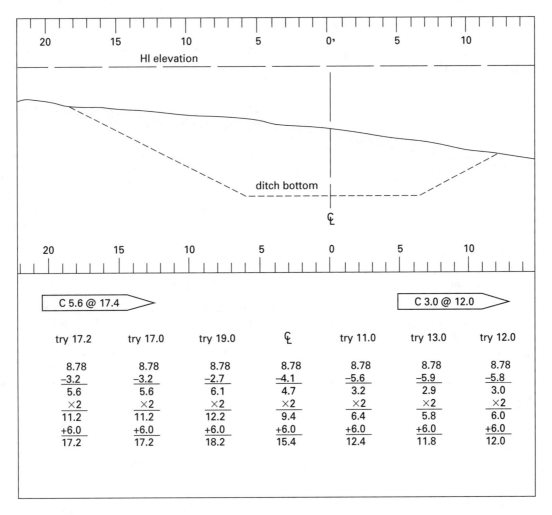

Cut Cross Section for Ex. 23.7

Example 23.8

Slope stakes are to be set at sta 3+00. The bottom of the cut is to be at elev 462.00 ft and is 10 ft wide. The side slopes are 2:1 (All measurements are in feet.).

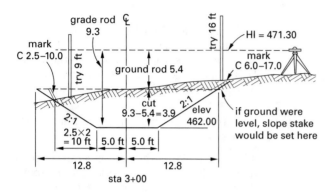

Solution

step 1: Establish the level near sta 3+00 and determine HI (471.30 ft in this example).

step 2: Compute the grade rod by subtracting the elevation at the bottom of the cut from HI.

$$471.30 \text{ ft} - 462.00 \text{ ft} = 9.30 \text{ ft}$$

step 3: Determine the ground rod by placing the rod on the ground at centerline. Read 5.4.

step 4: Compute the cut at the centerline by subtracting the ground rod from the grade rod.

$$9.3 \text{ ft} - 5.4 \text{ ft} = 3.9 \text{ ft}$$

step 5: Compute the distance to the left slope stake from the centerline as if the ground were level at this station.

$$5 \text{ ft} + (2)(3.9 \text{ ft}) = 12.8 \text{ ft}$$

step 6: Note that the ground on the left slopes down and the side of the cut slopes up, indicating that the distance to the stake will be less than that for level ground.

step 7: Try a distance less than 12.8 ft, say 9.0 ft, and read rod at this distance. The rod reading is 6.6 ft.

$$\text{grade rod} - \text{ground rod} = 9.3 \text{ ft} - 6.6 \text{ ft}$$
$$= 2.7 \text{ ft}$$

The distance computed from this rod reading is

$$5 \text{ ft} + (2)(2.7 \text{ ft}) = 10.4 \text{ ft}$$

Move toward 10.4 ft; try 10.0 ft. (Move less because slopes are opposite.)

step 8: The ground rod at 10.0 ft is 6.8 ft.

$$9.3 \text{ ft} - 6.8 \text{ ft} = 2.5 \text{ ft}$$

The computed distance is

$$5 \text{ ft} + (2)(2.5 \text{ ft}) = 10.0 \text{ ft}$$

The computed distance agrees with the measured distance.

step 9: Set the stake at 10.0 ft left of centerline and mark "C 2.5 @ 10.0" on top face of stake and "3+00" on bottom.

step 10: Move to the right side. Try a distance greater than that for level ground because the ground and sides both slope up.

step 11: Try 16.0; the ground rod is 3.4 ft.

$$9.3 \text{ ft} - 3.4 \text{ ft} = 5.9 \text{ ft}$$
$$5 \text{ ft} + (2)(5.9 \text{ ft}) = 16.8 \text{ ft}$$

step 12: Try 17.0 ft. Move beyond 16.8 ft because the slopes are in the same direction. The ground rod is 3.3 ft.

$$9.3 \text{ ft} - 3.3 \text{ ft} = 6.0 \text{ ft}$$
$$5 \text{ ft} + (2)(6.0 \text{ ft}) = 17.0 \text{ ft}$$

step 13: Set the stake at 17.0 ft. Mark "C 6.0 @ 17.0."

13. SETTING SLOPE STAKES AT FILL SECTION

In setting slope stakes for fills, two situations may arise: (a) the HI may be below the finish elevation as shown in Exs. 23.9 and 23.10, or (b) the HI may be above the finish elevation as shown in Fig. 23.10. If the HI is below the finish elevation, the fill is the sum of the grade rod and the ground rod. If the HI is above the finish elevation, the fill is the difference between the ground rod and the grade rod.

Example 23.9

The illustration on the following page shows the ground cross section at a station at which slope stakes are to be set for a fill. The HI has been established and the finish elevation, width of top of fill, and side slopes have been obtained from the construction plans. The centerline of fill has also been established. Known information is tabulated as follows.

finish elevation of top of fill = 452.36 ft
top of fill width = 4 ft
side slopes = 2:1
HI = 450.54 ft

Solution

The solution is shown on the following page. The correct cut and distance can be found on the marked stake.

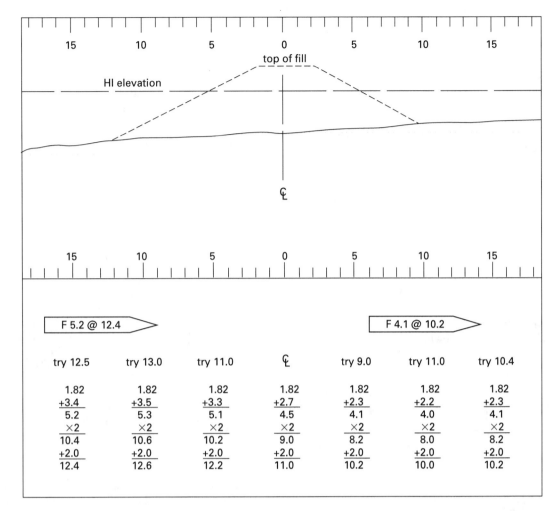

Fill Section for Ex. 23.9

Example 23.10

In the following illustration, slope stakes are to be set at sta 10+00. The top of fill is to be at elevation 468.00 ft and is 10 ft wide. Side slopes are 1¹/₂:1.

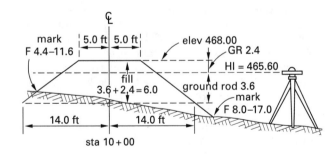

Solution

step 1: Establish the level near sta 10+00 and determine the HI (465.60 ft in this example).

step 2: Compute grade rod by subtracting the HI from the elevation of the top of the fill.

$$468.00 \text{ ft} - 465.60 \text{ ft} = 2.4 \text{ ft}$$

step 3: Determine ground rod by placing the rod on the ground at the centerline. Read 3.6 ft.

step 4: Compute the fill at the centerline by adding the grade rod and the ground rod.

$$3.6 \text{ ft} + 2.4 \text{ ft} = 6.0 \text{ ft}$$

step 5: Compute the distance to the left slope stake from the centerline as if the ground were level at this station.

$$5 \text{ ft} + (1.5)(6.0 \text{ ft}) = 14.0 \text{ ft}$$

step 6: Note that slopes are opposite, indicating that the distance will be less than that for level ground.

step 7: Try 11.0 ft. The rod reads 2.2 ft.

$$2.2 \text{ ft} + 2.4 \text{ ft} = 4.6 \text{ ft}$$

$$5 \text{ ft} + (1.5)(4.6 \text{ ft}) = 11.9 \text{ ft}$$

Move toward 11.9 ft, but less because slopes are opposite.

step 8: Try 11.5. The ground rod is 2.0 ft.

$$2.0 \text{ ft} + 2.4 \text{ ft} = 4.4 \text{ ft}$$

$$5 \text{ ft} + (1.5)(4.4 \text{ ft}) = 11.6 \text{ ft}$$

This is close enough.

step 9: Set the stake at 11.6 ft left of centerline and mark "F 4.4 @ 11.6." Mark "10+00" on the bottom face of the stake.

step 10: Move to the right side. Try a distance greater than that for level ground because ground and slope are in the same direction.

step 11: Try 15.0 ft. The ground rod is 5.3 ft.

$$5.3 \text{ ft} + 2.4 \text{ ft} = 7.7 \text{ ft}$$

$$5 \text{ ft} + (1.5)(7.7 \text{ ft}) = 16.6 \text{ ft}$$

Move toward 16.6 ft and beyond because the slopes are both down.

step 12: Try 17.0 ft. The ground rod is 5.6 ft.

$$5.6 \text{ ft} + 2.4 \text{ ft} = 8.0 \text{ ft}$$

$$5 \text{ ft} + (1.5)(8.0 \text{ ft}) = 17.0 \text{ ft}$$

step 13: Set the stake at 17.0 ft and mark "F 8.0 @ 17.0."

14. SETTING STAKES FOR UNDERGROUND PIPE

Stakes for line and grade for underground pipe, like stakes for roads and streets, are set on an offset line. One hub stake with tack can be used at each station for both line and grade, or separate stakes can be set for line and grade. If only one stake is to be used, the elevation of the top of that stake is determined. Then, the cut from the top of the stake to the flowline (*invert*) is computed and marked on a guard stake. This method is faster.

It is often desirable to set a grade stake close to the tacked line stake. This may be set so that the cut from the top of stake to the flowline is at some multiple of a half-foot.

In setting a cut stake for underground pipe, the surveyor first sets up the level and determines its HI. Using the flowline of the pipe at a particular station, the grade rod at that station is computed and recorded in the field book. A rod reading on the ground is taken at the point where the stake is to be driven. This is the ground rod. Using the grade rod and the ground rod, the rod reading on top of the stake that will give a half-foot cut from the top of the stake to the flowline is computed. A stake is driven to the rod reading that gives this cut. The stake is blued, and the cut is marked on the guard stake.

Example 23.11

A grade stake is to be set to show the cut to the flowline of a pipe. The HI is 472.36 ft, the flowline is 462.91 ft, and the ground rod at the point of stake is 5.1 ft. Determine the grade rod and rod reading that will give a half-foot cut to the flowline.

Solution

$$\text{HI} = 472.36 \text{ ft}$$
$$\text{flowline} = \underline{462.91 \text{ ft}}$$
$$\text{GR} = 9.45 \text{ ft}$$

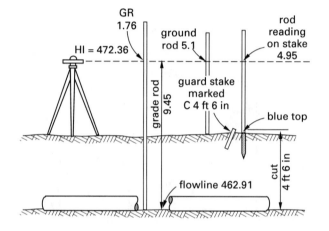

The rod reading on the grade stake to give a half-foot cut from the top of the stake to the flowline could be 8.95 ft, 8.45 ft, ..., 5.45 ft, 4.95 ft, and so on, and the corresponding cuts would be 0 ft 6 in, 1 ft 0 in, 4 ft 0 in, 4 ft 6 in, and so on. The ground rod is 5.1 ft; therefore, the cut is approximately 9.5 ft − 5.1 ft = 4.4 ft. Therefore, the rod reading for this cut will be either 5.45 ft (which will give a cut of 4 ft 0 in) or 4.95 ft (which will give a cut of 4 ft 6 in).

A rod reading of 5.45 ft cannot be used because the top of the stake would be 0.3 ft below the surface of the ground. A rod reading of 4.95 ft would place the top of the stake about 0.1 ft above ground, which is satisfactory. The stake is driven so that the rod reading is 4.95 ft. The top is marked with blue keel, and the guard stake is marked "C 4 ft, 6 in" since 9.45 ft − 4.95 ft = 4.50 ft (4 ft 6 in). It can be seen that the rod reading on the stake must be less than the ground rod in order that the top of the stake be above ground.

15. FLOWLINE AND INVERT

The bottom inside of a drainage pipe is known as the *flowline*.[2] It is also referred to as the *invert*. Invert is more commonly used to describe the bottom of the flow channel within a manhole.

---

[2] *Flowlines* are the lines used as finish elevation for pipes.

Vertical control is of prime importance in laying pipe for gravity flow, especially sanitary sewer pipe. In order to facilitate vertical alignment, excavation of the trench often extends a few inches below the bottom of the pipe so that a bedding material, such as sand, can be placed in the trench for the pipe to lie on. Because of various methods of using bedding material in laying pipe, stakes are always set for the flowline, or invert, of the pipes. Excavation depth allows for the amount of bedding specified.

16. MANHOLES

Sanitary sewers are not laid along horizontal or vertical curves. Horizontal and vertical alignments are straight lines. Where a change in horizontal alignment or a change in slope is necessary, a *manhole* is required at the point of change. Therefore, a vertical drop within the manhole is needed. In staking, two cuts are often recorded on guard stakes: one for the incoming sewer and one for the outgoing sewer.

Gravity lines, such as sanitary sewers, flow only partially full. The slope of the sewer determines the flow velocity, and the velocity and size of the pipe determine the quantity of flow. Manholes are used to provide a point of change in conditions. Sewers must be deep enough below the surface of the ground to prevent freezing of their contents and damage to the pipe by construction equipment.

PRACTICE PROBLEMS

1. Stakes are to be set on 4 ft offsets for each edge of pavement (which is 28 ft wide), for a curve that has a deflection angle of $55°00'$ and a centerline radius of 250 ft. The PI is at station 8+56.45. Stakes are to be set on full-stations and half-stations, and at the PC and PT. (a) Calculate T and L. (b) Determine the deflection angles used to stake the curve. (c) Calculate the outside and inside chord lengths.

2. Prepare a set of field notes to be used in staking a street curve on the quarter-stations from 3 ft offset lines on both sides of the street.

$$\text{PI} = 4+55.00$$
$$\Delta = 60°00' \text{ (angle to the left)}$$
$$R = 100 \text{ ft (centerline)}$$
$$\text{pavement width} = 28 \text{ ft}$$

3. The intersection of Ash Lane and 32nd Street is to be staked for paving from an offset line 4 ft left of the left edge of pavement. The pavement width is 28 ft, and the radius to the edge of the pavement is 30 ft. From this information and information shown on the following sketch, compute PC and PT stations and deflection angles along with chord lengths from PC to PT. The scale is $1/2$ in = 30 ft.

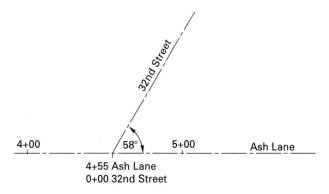

SOLUTIONS

1. (a) $T = (250 \text{ ft}) \tan\left(\dfrac{55°}{2}\right) = \boxed{130.14}$

$$L = \frac{(55°)2\pi(250 \text{ ft})}{360°} = \boxed{239.98 \text{ ft}}$$

$PI = 8+56.45$

$T = 1+30.14$

$PC = 7+26.31$

$L = 2+39.98$

$PT = 9+66.29$

(b)

point	station		deflection angle
PC	7+26.31		0° (0°00′)
	7+50	$\left(\dfrac{23.69 \text{ ft}}{239.98 \text{ ft}}\right)(27°30′)$	2.7147° (2°43′)
	8+00	$\left(\dfrac{73.69 \text{ ft}}{239.98 \text{ ft}}\right)(27°30′)$	8.4443° (8°27′)
	8+50	$\left(\dfrac{123.69 \text{ ft}}{239.98 \text{ ft}}\right)(27°30′)$	14.1740° (14°10′)
	9+00	$\left(\dfrac{173.69 \text{ ft}}{239.98 \text{ ft}}\right)(27°30′)$	19.9036° (19°54′)
	9+50	$\left(\dfrac{223.69 \text{ ft}}{239.98 \text{ ft}}\right)(27°30′)$	25.6333° (25°38′)
PT	9+66.29	$\left(\dfrac{239.98 \text{ ft}}{239.98 \text{ ft}}\right)(27°30′)$	27.5000° (27°30′)

(c) The outside chord lengths are

7+26.31 to 7+50: $(2)(268 \text{ ft})(\sin 2.7147°) = \boxed{25.39 \text{ ft}}$
7+50 to 8+00: $(2)(268 \text{ ft})(\sin 5.7296°) = \boxed{53.51 \text{ ft}}$
9+50 to 9+66.29: $(2)(268 \text{ ft})(\sin 1.8667°) = \boxed{17.46 \text{ ft}}$

The inside chord lengths are

7+26.31 to 7+50: $(2)(232 \text{ ft})(\sin 2.7147°) = \boxed{21.98 \text{ ft}}$
7+50 to 8+00: $(2)(232 \text{ ft})(\sin 5.7296°) = \boxed{46.32 \text{ ft}}$
9+50 to 9+66.29: $(2)(232 \text{ ft})(\sin 1.8667°) = \boxed{15.11 \text{ ft}}$

2.

point	station	deflection angle	C-in (ft)	C-out (ft)	curve data
					Elm Street
PT	5+01.98	30°00′			
			1.65	2.33	
	5+00	29°26′			$\Delta = 60°$ to the left
			20.70	29.17	$R = 100$ ft
	4+75	22°16′			$T = 57.72$ ft
			20.70	29.17	$L = 104.72$ ft
	4+50	15°07′			
			20.70	29.17	
	4+25	7°57′			
			20.70	29.17	
	4+00	0°47′			
			2.27	3.19	
PC	3+97.26	0°00′			

3. $\Delta_1 = 58°00′$ left

$T_1 = (44 \text{ ft}) \tan\left(\dfrac{58°}{2}\right) = 24.39 \text{ ft}$

$PI = 4+55.00$

$T_1 = 24.39 \text{ ft}$

$PC = \boxed{4+30.61}$

$PI = 0+00.00$

$T_1 = 24.39 \text{ ft}$

$PT = \boxed{0+24.39}$

$LC_1 = (2)(26 \text{ ft}) \sin\left(\dfrac{58°}{2}\right) = \boxed{25.21 \text{ ft}}$

deflection angle $= \boxed{29°00′}$

$\Delta_2 = 122°00′$

$T_2 = (44 \text{ ft}) \tan\left(\dfrac{122°}{2}\right) = 79.38 \text{ ft}$

$PI = 4+55.00$

$T_2 = 79.38 \text{ ft}$

$PC = \boxed{5+34.38}$

$PI = 0+00.00$

$T_2 = 79.38 \text{ ft}$

$PT = \boxed{0+79.38}$

$LC_2 = (2)(26 \text{ ft}) \sin\left(\dfrac{122°}{2}\right) = \boxed{45.48 \text{ ft}}$

deflection angle $= \boxed{61°00′}$

24 Earthwork

1. DEFINITION

Earthwork is the excavation, hauling, and placing of soil, rock, gravel, or other material found below the surface of the earth. This definition also includes the measurement of such material in the field, the computation in the office of the volume of such material, and the determination of the most economical method of performing such work.

2. UNIT OF MEASURE

The *cubic yard* (i.e., the "yard") is the unit of measure for earthwork. However, the volume and density of earth changes under natural conditions and during the operations of excavation, hauling, and placing.

3. SWELL AND SHRINKAGE

A cubic yard of earth measured in its natural position will be more than a cubic yard after it is excavated. If the earth is compacted after it is placed, the volume may be less than a cubic yard.

The volume of the earth in its natural state is known as *bank-measure*. The volume in the vehicle is known as *loose-measure*. The volume after compaction is known as *compacted-measure*.

The change in volume of earth from its natural to loose state is known as *swell*. Swell is expressed as a percent of the natural volume.

The change in volume of earth from its natural state to its compacted state is known as *shrinkage*. Shrinkage also is expressed as a percent decrease from the natural state.

As an example, 1 yd^3 in the ground may become 1.2 yd^3 loose-measure and 0.85 yd^3 after compaction. The swell would be 20%, and the shrinkage would be 15%. Swell and shrinkage vary with soil types.

4. CLASSIFICATION OF MATERIALS

Excavated material is usually classified as *common excavation* or *rock excavation*. Common excavation is soil.

In highway construction, common road excavation is soil found in the roadway. *Common borrow* is soil found outside the roadway and brought in to the roadway. Borrow is necessary where there is not enough material in the roadway excavation to provide for the embankment.

5. CUT AND FILL

Earthwork that is excavated, or is to be excavated, is known as *cut*. Excavation that is placed in embankment is known as *fill*.

Payment for earthwork is normally either for cut and not for fill, or for fill and not for cut. In highway work, payment is usually for cut; in dam work, payment is usually for fill. To pay for both would require measuring two different volumes and paying for moving the same earth twice.

Earthwork

6. FIELD MEASUREMENT

Cut and fill volumes can be computed from slope-stake notes, from plan cross sections, or by photogrammetric methods.

7. CROSS SECTIONS

Cross sections are profiles of the earth taken at right angles to the centerline of an engineering project (such as a highway, canal, dam, or railroad). A cross section for a highway is shown in Fig. 24.1.

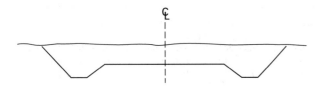

Figure 24.1 *Typical Highway Cross Section*

8. ORIGINAL AND FINAL CROSS SECTIONS

To obtain volume measurement, cross sections are taken before construction begins and after it is completed. By plotting the cross section at a particular station both before and after construction, a sectional view of the change in the profile of the earth along a certain line is obtained. The change along this line appears on the plan as an area. By using these areas at various intervals along the centerline, and by using distance between the areas, volume can be computed.

9. ESTIMATING EARTHWORK

Earthwork quantities for a highway, canal, or other project can be estimated by superimposing a template on the original plotted cross section, which is drawn to represent the final cross section. The template is obtained from the *typical section sheet* of the construction plans.

10. TYPICAL SECTIONS

Typical sections show the cross section view of the project as it will look on completion, including all dimensions. Highway projects usually show several typical sections including cut sections, fill sections, and sections showing both cut and fill. Interstate highway plans also show access-road sections and sections at ramps.

11. DISTANCE BETWEEN CROSS SECTIONS

Cross sections are usually taken at each full station and at breaks in the ground along the centerline. In taking cross sections, it must be assumed that the change in the earth's surface from one cross section to the next is uniform, and that a section halfway between the cross sections is an average of the two. If the ground breaks appreciably between any two full-stations, one or more cross sections between full-stations must be taken. This is referred to as *taking sections at pluses*. Figure 24.3 shows the stations at which cross sections should be taken.

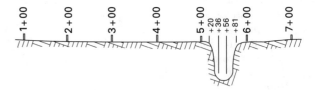

Figure 24.3 *Cross-Section Distances*

In rock excavation, or any other expensive operation, cross sections should be taken at intervals of 50 ft or less. Cross sections should always be taken at the PC and PT of a curve. Plans should also show a section on each end of a project (where no construction is to take place) so that changes caused by construction will not be abrupt.

Where a cut section of a highway is to change to a fill section, several additional cross sections are needed. Such sections are shown in Fig. 24.4.

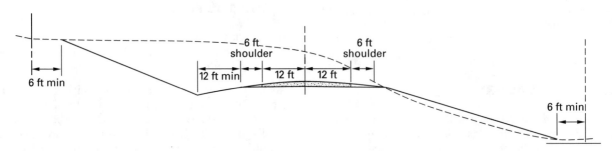

Figure 24.2 *Typical Completed Section*

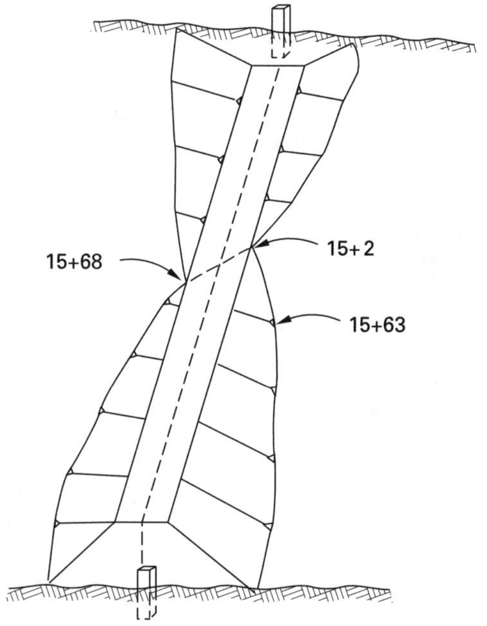

13. METHODS FOR COMPUTING VOLUME

The most common method for computing volume of earthwork is by the *average end area method*. (The *Prismoidal formula* furnishes more accurate results but is more complex.) The average end area method is accurate enough for most work.

14. AVERAGE END AREA METHOD

The average end area method is based on the assumption that the volume of earthwork between two vertical cross sections A_1 and A_2 is equal to the average of the two end areas multiplied by the horizontal distance L between them. Area is expressed in square feet, and distance is expressed in feet. So, the volume in cubic yards is

$$V = \frac{L(A_1 + A_2)}{(2)\left(27\,\frac{\text{ft}^3}{\text{yd}^3}\right)} = \frac{L}{54\,\frac{\text{ft}^3}{\text{yd}^3}}\,(A_1 + A_2) \qquad \textit{24.1}$$

12. GRADE POINT

The point where a fill section meets the natural ground (where a cut section begins) is known as a *grade point*.

15. FIELD NOTES

Figure 24.5 shows a sample sheet from cross-section field notes. The left half of the page is the same as for any

Figure 24.4 *Cut Changing to Fill*

station	+	HI	−	rod	elevation	₵L								
BM no. 12	3.30	468.21			464.91	r.r. spike in 12 in elm 125 rt sta 16+75								
11+00						$\frac{4.4}{50}$	$\frac{7.1}{15}$	$\frac{4.9}{12}$	$\frac{7.3}{5}$	7.9	$\frac{9.1}{20}$	$\frac{12.0}{25}$	$\frac{9.7}{30}$	$\frac{11.2}{50}$
TP$_4$	5.27	462.97	10.51		457.70									
12+00						$\frac{2.0}{50}$	$\frac{4.0}{20}$	$\frac{1.3}{15}$	$\frac{4.5}{10}$	5.0	$\frac{6.0}{15}$	$\frac{10.1}{20}$	$\frac{5.7}{27}$	$\frac{8.0}{50}$
13+00						$\frac{4.8}{50}$	$\frac{6.0}{25}$	$\frac{4.2}{20}$	$\frac{7.7}{15}$	9.9	$\frac{9.8}{8}$	$\frac{12.6}{15}$	$\frac{11.0}{24}$	$\frac{13.0}{50}$
TP$_5$	1.76	458.22	6.51		456.46									
13+50						$\frac{2.3}{50}$	$\frac{4.1}{30}$	$\frac{1.2}{25}$	$\frac{5.0}{20}$	6.7	$\frac{7.1}{3}$	$\frac{12.2}{10}$	$\frac{8.3}{19}$	$\frac{11.1}{50}$
14+00						$\frac{5.2}{50}$	$\frac{6.0}{42}$	$\frac{10.2}{35}$	$\frac{7.9}{20}$	10.1	$\frac{11.0}{20}$	$\frac{8.1}{50}$		
15+00						$\frac{5.0}{50}$	$\frac{5.8}{48}$	$\frac{9.6}{40}$	$\frac{7.3}{20}$	9.2	$\frac{10.1}{22}$	$\frac{7.5}{50}$		
BM no. 13			5.15		453.07									

Figure 24.5 *Typical Field Notes for Cross-Section Work*

Earthwork

set of level notes. The right half shows rod readings over horizontal distance measured from the centerline for each point on the ground that requires a reading. These readings should always include shots on the centerline, at each break in the ground, and at the right-of-way on each side.

16. PLOTTING CROSS SECTIONS

When drawn by hand, cross sections are plotted on specially printed cross-section paper. A scale of 1 in = 5 ft is usually used for both the horizontal and vertical. For wide sections, a scale of 1 in = 10 ft or 1 in = 20 ft can be used. The vertical scale can also be exaggerated if necessary.

A vertical line in the center of the sheet is drawn to represent the centerline of the project. Shots taken in the field are plotted to the proper elevation and distance from the centerline.

Each cross section is plotted as a separate section. Sufficient space is allowed between cross sections so that they do not overlap. The station number for each cross section is recorded just under the centerline shot, and the elevation at the centerline is recorded in a vertical direction just above the centerline shot.

The heavy lines on the paper are used to represent an elevation ending in 0 or 5 ft. With the elevation of the centerline recorded, these heavy lines can be identified as the elevation they represent.

The notes can be plotted by first reducing the level shots to elevation and then plotting by elevation. Alternatively, the rod shots can be plotted directly from a line on the paper representing the HI. As an example, if the HI is 447.6 ft (rounded off) and the rod shot is 5.4 ft, subtracting 5.4 ft from 7.6 ft gives an elevation of 442.2 ft.

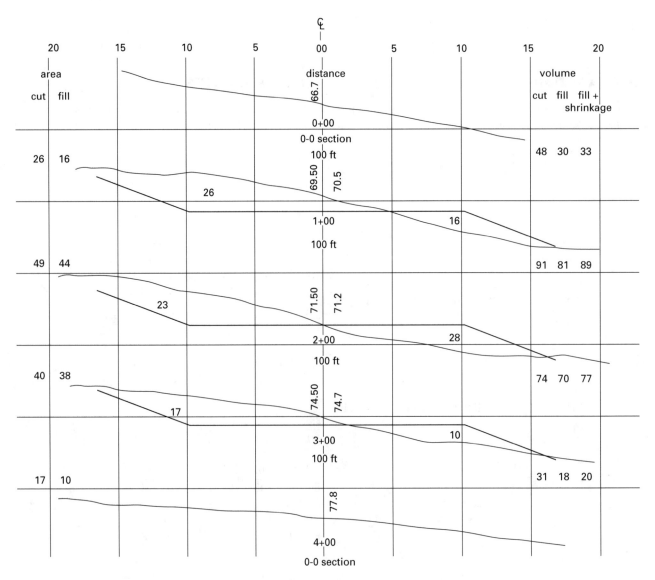

Figure 24.6 *Plotting Cross Sections*

17. DETERMINING END AREAS

End areas are commonly determined by planimetery or by dividing the area into triangles and trapezoids.

Cut areas and fill areas must be kept separate. After the areas have been determined, the sum of each two adjacent areas is placed in a column. The distance between two sections is recorded, and the volume for each sum is computed from Eq. 24.1.

After the volume has been computed, shrinkage must be added to fill quantities to balance with cut quantities. Shrinkage will vary from 30% for light cuts and fills to 10% for heavy cuts and fills.

Excavated rock will occupy larger volume when placed in a fill, and this swell will be subtracted from the fill quantity.

18. VOLUMES FROM PROFILES

For preliminary estimates of earthwork, volumes can be computed from the centerline profiles. After the ground profile and finish grade profile are plotted, the area of cut can be planimetered and the average determined by dividing by the length of cut. Using the average cut, a template can be drawn, and the end area can also be planimetered. This area times the length of the cut will give the volume.

19. BORROW PIT

As mentioned previously, it is often necessary to borrow earth from an adjacent area to construct embankments.

Normally, the *borrow pit* area is laid out in a rectangular grid with 10 ft, 50 ft, or even 100 ft squares. Elevations are determined at the corners of each square by leveling before and after excavation so that the cut at each corner can be computed.

Points outside the cut area are established on the grid lines so that the lines can be reestablished after excavation is completed.

As an example, volumes for two of the prisms shown in Fig. 24.7 are computed by multiplying the average cut by the area of the figure. The volume of the prism A0-B0-B1-A1 is

$$V = \left(\dfrac{(50 \text{ ft})(50 \text{ ft})}{27 \; \dfrac{\text{ft}^3}{\text{yd}^3}} \right)$$
$$\times \left(\dfrac{3.2 \text{ ft} + 3.4 \text{ ft} + 3.0 \text{ ft} + 2.6 \text{ ft}}{4} \right)$$
$$= 282 \text{ yd}^3$$

The volume of the triangle E2-F2-E3 is

$$V = \left(\dfrac{(50 \text{ ft})(15 \text{ ft})}{(2) \left(27 \; \dfrac{\text{ft}^3}{\text{yd}^3} \right)} \right) \left(\dfrac{2.3 \text{ ft} + 2.4 \text{ ft} + 2.4 \text{ ft}}{3} \right)$$
$$= 33 \text{ yd}^3$$

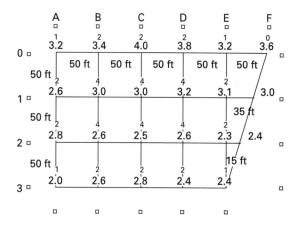

Figure 24.7 *Borrow Pit Areas*

Instead of computing volumes of prisms represented by squares separately, all square-based prisms can be computed collectively by multiplying the area of one square by the sum of the cut at each corner times the number of times that cut appears in any square, divided by 4. For instance, on the second line from the top in Fig. 24.7, which is line 1, 2.6 appears in two squares, 3.0 appears in four squares, 3.0 appears in four squares, 3.2 appears in four squares, and 3.1 appears in two squares. In the figure, the small number above the cut indicates the number of times the cut is used in averaging the cuts for the prisms.

Example 24.1

Calculate the volume of earth excavated from the borrow pit shown in Fig. 24.7.

Solution

The volume of the squares is

$$V = \left(\dfrac{50 \text{ ft}}{\left(27 \; \dfrac{\text{ft}^3}{\text{yd}^3} \right)(4)} \right)(50 \text{ ft})$$
$$\times \left(\begin{array}{l} 3.2 \text{ ft} + (2)(3.4 \text{ ft}) + (2)(4.0 \text{ ft}) \\ \quad + (2)(3.8 \text{ ft}) + 3.2 \text{ ft} + (2)(2.6 \text{ ft}) \\ \quad + (4)(3.0 \text{ ft}) + (4)(3.0 \text{ ft}) + (4)(3.2 \text{ ft}) \\ \quad + (2)(3.1 \text{ ft}) + (2)(2.8 \text{ ft}) + (4)(2.6 \text{ ft}) \\ \quad + (4)(2.5 \text{ ft}) + (4)(2.6 \text{ ft}) + (2)(2.3 \text{ ft}) \\ \quad + 2.0 + (2)(2.6 \text{ ft}) + (2)(2.8 \text{ ft}) \\ \quad + (2)(2.4 \text{ ft}) + 2.4 \text{ ft} \end{array} \right)$$
$$= 3194 \text{ yd}^3$$

The volume of the trapezoids is

$$V = \left(\frac{50 \text{ ft} + 35 \text{ ft}}{(2) \left(27 \, \frac{\text{ft}^3}{\text{yd}^3} \right)} \right) (50 \text{ ft})$$
$$\times \left(\frac{3.2 \text{ ft} + 3.6 \text{ ft} + 3.1 \text{ ft} + 3.0 \text{ ft}}{4} \right)$$
$$= 254 \text{ yd}^3$$

$$V = \left(\frac{35 \text{ ft} + 15 \text{ ft}}{(2) \left(27 \, \frac{\text{ft}^3}{\text{yd}^3} \right)} \right) (50 \text{ ft})$$
$$\times \left(\frac{3.1 \text{ ft} + 3.0 \text{ ft} + 2.3 \text{ ft} + 2.4 \text{ ft}}{4} \right)$$
$$= 125 \text{ yd}^3$$

The volume of the triangle is

$$V = \left(\frac{(15 \text{ ft})}{(2) \left(27 \, \frac{\text{ft}^3}{\text{yd}^3} \right)} \right) (50 \text{ ft})$$
$$\times \left(\frac{2.3 \text{ ft} + 2.4 \text{ ft} + 2.4 \text{ ft}}{3} \right)$$
$$= 33 \text{ yd}^3$$

The total volume of earth excavated is

$$3194 \text{ yd}^3 + 254 \text{ yd}^3 + 125 \text{ yd}^3 + 33 \text{ yd}^3 = 3606 \text{ yd}^3$$

20. HAUL

In some contracts for highways and railroads, the contractor is paid per cubic yard for excavation (which includes the cost of excavation, hauling, placing in embankment, and compaction of embankment). However, the cost of hauling 1 yd^3 of earth over a long distance can easily become greater than the cost of excavation, so that it is often practical to pay a contractor for excavating and hauling earth.

21. FREE HAUL

It is common not to pay for hauling if the material is hauled less than a certain distance, usually 500 ft to 1000 ft. An additional price is paid for hauling the earth beyond the prescribed limit. The haul distance for which no pay is received is known as *free haul*.

22. OVERHAUL

The hauling of material beyond the free haul limit is known as *overhaul*. The unit of overhaul measure is yard-stations or yard-quarters. A *yard-quarter* is the hauling of 1 yd^3 of earth a distance of $^1/_4$ mi. For example, if 6 yd of earth were hauled 4 mi, the overhaul would be 24 yard-quarters.

Thus, the word "haul" may have two meanings. It may mean linear distance or volume times distance.

It should be mentioned that the distance is measured along the centerline. Distance from the extremity of the right-of-way to the centerline is not considered.

23. BALANCE POINTS

It is important in planning and construction to know the points along the centerline a particular section of cut that will balance a particular section of fill. For example, assume that a cut section extends from sta 12+25 to sta 18+65, and a fill section extends from sta 18+65 to sta 26+80. Also, assume that the excavated material will exactly provide the material needed to make the embankment. Then, cut balances fill, and sta 12+25 and 26+80 are *balance points*.

24. MASS DIAGRAMS

A method of determining economical handling of material, quantities of overhaul, and location of balance points is the mass-diagram method.

The *mass diagram* is a graph that has distance in stations as the abscissa and the cumulative earthwork (i.e., the algebraic sums of cut and fill) as ordinate. The x-axis parallels the centerline, and the cut and fill (plus shrinkage) quantities are taken from the cross-section sheets. Often, the mass diagram is plotted below the centerline profile so that like stations are vertically in line.

To add cut and fill algebraically, cut is given a plus sign, and fill is given a minus sign.

25. PLOTTING THE MASS DIAGRAM

After volumes of cut and fill between stations have been computed, they are tabulated as shown in Table 24.1. The cuts and fills are then added, and the cumulative yardage at each station is recorded in the table. It is this cumulative yardage that is plotted as an ordinate. In Fig. 24.8, the baseline serves as the x-axis and cumulative yardage that has a plus sign is plotted above the baseline. Cumulative yardage that has a minus sign is plotted below the baseline.

The scale is not important. In Fig. 24.8, the horizontal scale is 1 in = 5 stations, and the vertical scale is 1 in = 5000 yd^3. A larger scale would be more practical in actual computations.

In Fig. 24.8, the mass diagram is plotted on the lower half of the sheet, and the centerline profile of the project is plotted on the upper half.

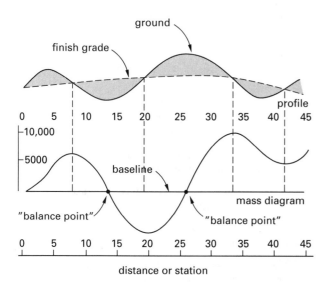

Figure 24.8 *Baseline and Centerline Profile Mass Diagram*

26. BALANCE LINE

Any horizontal line cutting off a loop of the mass curve intersects the curve at two points, between which the cut is equal to the fill. Figure 24.9 shows a portion of the mass diagram of Fig. 24.8 with an enlarged scale.

27. SUB-BASES

Sub-bases are horizontal balance lines that divide an area of the mass diagram between two balance points into trapezoids for the purpose of more accurately computing overhaul. In Fig. 24.9, the top sub-base is less than 600 ft in length. Therefore, the volume of earth represented by the area above this line will be hauled a distance less than the free-haul distance, and no payment will be made for overhaul. All the volume represented by the area below this top sub-base will receive payment for overhaul.

The area between two sub-bases is very nearly trapezoidal. The average length of the bases of a trapezoid can be measured in feet, and the altitude of the trapezoid can be measured in cubic yards of earth. The product of these two quantities can be expressed in yard-quarters. If free haul is subtracted from the length, the quantity can be expressed as overhaul.

Sub-bases are drawn at distinct breaks in the mass curve. Distinct breaks in Fig. 24.9 can be seen at sta 1+00(184), 2+00(806), and 5+00(4340). After sub-bases are drawn, a horizontal line is drawn midway between the sub-bases. This line represents the average haul for the volume of earth between the two sub-bases.

If a horizontal scale of 1 in = 100 ft is used, the length of the average haul can be determined by scaling. This line, shown as a dashed line in Fig. 24.9, scales 875 ft for the area between the top sub-bases. The free haul is subtracted from this in Table 24.1.

Table 24.1 *Typical Cut and Fill Calculations*

(all volumes are in cubic yards)

sta	cut +	fill −	cum sum	sta	cut +	fill −	cum sum
0			0	23			−4710
	184				1377		
1			+184	24			−3034
	622				1676		
2			+806	25			−1358
	1035				1860		
3			+1841	26			+502
	1268				1917		
4			+3109	27			+2419
	1231				1839		
5			+4340	28			+4258
	919				1611		
6			+5259	29			+5869
	503				1338		
7			+5762	30			+7207
	164	21			1029		
8			+5905	31			+8236
	12	190			652		
9			+5727	32			+8888
		616			357		
10			+5111	33			+9245
		942			150	39	
11			+4169	34			+9356
		1150			52	236	
12			+3019	35			+9172
		1500				465	
13			+1519	36			+8707
		1773				712	
14			−254	37			+7995
		1755				904	
15			−2009	38			+7091
		1540				904	
16			−3549	39			+6187
		1262				757	
17			−4811	40			+5430
		932				516	
18			−5743	41			+4914
		546				280	
19			−6289	42			+4634
		203				127	
20			−6461	43			+4507
		101				98	
21			−6283	44			+4409
		18				20	
22			−5715	45			+4389

The volume of earth between the two sub-bases is found by subtracting the ordinate of the lower sub-base from the ordinate of the upper sub-base. These ordinates are found in Table 24.1.

Multiplying average haul minus free haul in feet by volume of earth in cubic yards gives overhaul in yard-feet. Dividing by 1320 ft gives yard-quarters.

28. LOCATING BALANCE POINTS

A balance point occurs where the mass curve crosses the baseline. It can be seen that a balance point falls between sta 13 and 14. The ordinate of 13 is +1519; the ordinate of 14 is −254. Therefore, the curve fell $1519 + 254 = 1773$ yd^3 in 100 ft or 17.73 yd^3/ft. The curve crosses the baseline at a distance of $1519/17.73 = 86$ ft from sta 13 (13+86).

29. CHARACTERISTICS OF THE MASS DIAGRAM

Important characteristics that should be considered in using the mass diagram in planning the economical hauling of earth are as follows.

- A horizontal line connecting two points on the mass curve cuts off a loop in which the cut equals the fill.

- A loop that rises and then falls from left to right indicates that the haul from cut to fill will be from left to right.

- A loop that falls and then rises from left to right indicates the haul will be from right to left.

- A high point on a mass curve indicates a change from cut to fill.

- A low point on a mass curve indicates a change from fill to cut.

- High and low points on the mass curve occur at or near grade points on the profile.

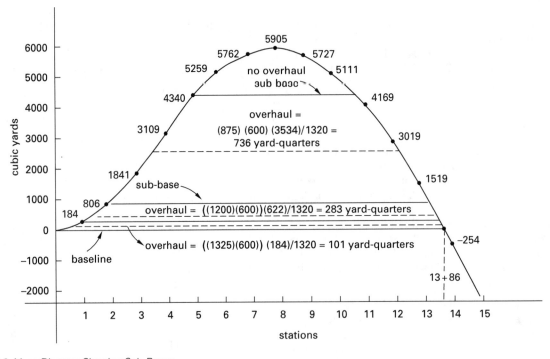

Figure 24.9 *Mass Diagram Showing Sub-Bases*

25 Numbering Systems

1. POSITIONAL NUMBERING SYSTEMS

A *base-b number*, N_b, is made up of individual *digits*. In a *positional numbering system*, the position of a digit in the number determines that digit's contribution to the total value of the number. Specifically, the position of the digit determines the power to which the *base* (also known as the *radix*), b, is raised. For *decimal numbers*, the radix is 10, hence the description *base-10 numbers*.

$$(a_n a_{n-1} \cdots a_2 a_1 a_0)_b = a_n b^n + a_{n-1} b^{n-1} + \cdots$$
$$+ a_2 b^2 + a_1 b + a_0 \quad \textbf{25.1}$$

The leftmost digit, a_n, contributes the greatest to the number's magnitude and is known as the *most significant digit*, MSD. The rightmost digit, a_0, contributes the least and is known as the *least significant digit*, LSD.

2. CONVERTING BASE-b NUMBERS TO BASE-10

Equation 25.1 converts base-b numbers to base-10 numbers. The calculation of the right-hand side of Eq. 25.1 is performed in the base-10 arithmetic and is known as the *expansion method*.[1]

Converting base-b fractions to base-10 is similar to converting whole numbers and is accomplished by Eq. 25.2.

$$(0.a_1 a_2 \cdots a_m)_b = a_1 b^{-1} + a_2 b^{-2} + \cdots + a_m b^{-m} \quad \textbf{25.2}$$

---

[1]Equation 25.1 works with any base number. The *double-dabble (double and add) method* is a specialized method of converting from base-2 to base-10 numbers.

3. CONVERTING BASE-10 NUMBERS TO BASE-b

The *remainder method* is used to convert base-10 numbers to base-b numbers. This method consists of successive divisions by the base, b, until the quotient is zero. The base-b number is found by taking the remainders in the reverse order from which they were found. This method is illustrated by Exs. 25.1 and 25.3.

Converting a base-10 fraction to base-b requires multiplication of the base-10 fraction and subsequent fractional parts by the base. The base-b fraction is formed from the integer parts of the products taken in the same order in which they were determined.

4. BINARY NUMBER SYSTEM

There are only two *binary digits (bits)* in the *binary number system*: *zero* and *one*.[2] Thus, all binary numbers consist of strings of bits (i.e., zeros and ones). The leftmost bit is known as the *most significant bit (MSB)*, and rightmost bit is the *least significant bit (LSB)*.

As with digits from other numbering systems, bits can be added, subtracted, multiplied, and divided, although only digits 0 and 1 are allowed in the results. The rules of bit addition are

$$0 + 0 = 0$$
$$0 + 1 = 1$$
$$1 + 0 = 1$$
$$1 + 1 = 0 \text{ carry } 1$$

Example 25.1

(a) Convert $(1011)_2$ to base-10.

(b) Convert $(75)_{10}$ to base-2.

Solution

(a) Using Eq. 25.1 with $b = 2$,

$$(1)(2)^3 + (0)(2)^2 + (1)(2)^1 + 1 = 11$$

---

[2]Alternatively, the binary states may be called *true* and *false*, *on* and *off*, *high* and *low*, or *positive* and *negative*.

(b) Use the remainder method. (See Sec. 3.)

$$75 \div 2 = 37 \text{ remainder } 1$$
$$37 \div 2 = 18 \text{ remainder } 1$$
$$18 \div 2 = 9 \text{ remainder } 0$$
$$9 \div 2 = 4 \text{ remainder } 1$$
$$4 \div 2 = 2 \text{ remainder } 0$$
$$2 \div 2 = 1 \text{ remainder } 0$$
$$1 \div 2 = 0 \text{ remainder } 1$$

The binary representation of $(75)_{10}$ is $(1001011)_2$.

5. OCTAL NUMBER SYSTEM

The *octal (base-8) system* is one of the alternatives to working with long binary numbers. Only the digits 0 through 7 are used. The rules for addition in the octal system are the same as for the decimal system except that the digits 8 and 9 do not exist. For example,

$$7 + 1 = 6 + 2 = 5 + 3 = (10)_8$$
$$7 + 2 = 6 + 3 = 5 + 4 = (11)_8$$
$$7 + 3 = 6 + 4 = 5 + 5 = (12)_8$$

Example 25.2

Perform the following operations:

(a) $(2)_8 + (5)_8$

(b) $(7)_8 + (6)_8$

(c) Convert $(75)_{10}$ to base-8.

(d) Convert $(0.14)_{10}$ to base-8.

(e) Convert $(13)_8$ to base-10.

(f) Convert $(27.52)_8$ to base-10.

Solution

(a) The sum of 2 and 5 in base-10 is 7, which is less then 8 and, therefore, a valid number in the octal system. The answer is $(7)_8$.

(b) The sum of 7 and 6 in base-10 is 13, which is greater than 8 (and, therefore, needs to be converted). Using the remainder method (Sec. 3),

$$13 \div 8 = 1 \text{ remainder } 5$$
$$1 \div 8 = 0 \text{ remainder } 1$$

The answer is $(15)_8$.

(c) Use the remainder method (Sec. 3),

$$75 \div 8 = 9 \text{ remainder } 3$$
$$9 \div 8 = 1 \text{ remainder } 1$$
$$1 \div 8 = 0 \text{ remainder } 1$$

The answer is $(113)_8$.

(d) Refer to Sec. 3.

$$0.14 \times 8 = 1.12$$
$$0.12 \times 8 = 0.96$$
$$0.96 \times 8 = 7.68$$
$$0.68 \times 8 = 5.44$$
$$0.44 \times 8 = \text{etc.}$$

The answer, $(0.1075\cdots)_8$, is constructed from the integer parts of the products.

(e) Use Eq. 25.1.

$$(1)(8) + 3 = (11)_{10}$$

(f) Use Eqs. 25.1 and 25.2.

$$(2)(8)^1 + (7)(8)^0 + (5)(8)^{-1} + (2)(8)^{-2}$$
$$= 16 + 7 + 5/8 + 2/64$$
$$= (23.656)_{10}$$

6. HEXADECIMAL NUMBER SYSTEM

The *hexadecimal (base-16) system* is a shorthand method of representing the value of four binary digits at a time.[3] Since 16 distinctly different characters are needed, the capital letters A through F are used to represent the decimal numbers 10 through 15. The progression of hexadecimal numbers is illustrated in Table 25.1.

Example 25.3

(a) Convert $(4D3)_{16}$ to base-10.

(b) Convert $(1475)_{10}$ to base-16.

(c) Convert $(0.8)_{10}$ to base-16.

Solution

(a) The hexadecimal number D is 13 in base-10. Using Eq. 25.1,

$$(4)(16)^2 + (13)(16)^1 + 3 = (1235)_{10}$$

(b) Use the remainder method. (See Sec. 3.)

$$1475 \div 16 = 92 \text{ remainder } 3$$
$$92 \div 16 = 5 \text{ remainder } 12$$
$$5 \div 16 = 0 \text{ remainder } 5$$

Since $(12)_{10}$ is $(C)_{16}$, (or hex C), the answer is $(5C3)_{16}$.

---

[3]The term *hex number* is often heard.

(c) Refer to Sec. 3.

$$0.8 \times 16 = 12.8$$
$$0.8 \times 16 = 12.8$$
$$0.8 \times 16 = \text{etc.}$$

Since $(12)_{10} = (C)_{16}$, the answer is $(0.CCCCC\cdots)_{16}$.

7. CONVERSIONS AMONG BINARY, OCTAL, AND HEXADECIMAL NUMBERS

The octal system is closely related to the binary system since $(2)^3 = 8$. Conversion from a binary to an octal number is accomplished directly by starting at the LSB (right-hand bit) and grouping the bits in threes. Each group of three bits corresponds to an octal digit. Similarly, each digit in an octal number generates three bits in the equivalent binary number.

Conversion from a binary to a hexadecimal number starts by grouping the bits (starting at the LSB) into fours. Each group of four bits corresponds to a hexadecimal digit. Similarly, each digit in a hexadecimal number generates four bits in the equivalent binary number.

Table 25.1 Binary, Octal, Decimal, and Hexadecimal Equivalents

binary	octal	decimal	hexadecimal
0	0	0	0
1	1	1	1
10	2	2	2
11	3	3	3
100	4	4	4
101	5	5	5
110	6	6	6
111	7	7	7
1000	10	8	8
1001	11	9	9
1010	12	10	A
1011	13	11	B
1100	14	12	C
1101	15	13	D
1110	16	14	E
1111	17	15	F
10000	20	16	10

Conversion between octal and hexadecimal numbers is easiest when the number is first converted to a binary number.

Example 25.4

(a) Convert $(5431)_8$ to base-2.

(b) Convert $(1001011)_2$ to base-8.

(c) Convert $(1011111101111001)_2$ to base-16.

Solution

(a) Convert each octal digit to binary digits.

$$(5)_8 = (101)_2$$
$$(4)_8 = (100)_2$$
$$(3)_8 = (011)_2$$
$$(1)_8 = (001)_2$$

The answer is $(101100011001)_2$.

(b) Group the bits into threes starting at the LSB.

$$1 \quad 001 \quad 011$$

Convert these groups into their octal equivalents.

$$(1)_2 = (1)_8$$
$$(001)_2 = (1)_8$$
$$(011)_2 = (3)_8$$

The answer is $(113)_8$.

(c) Group the bits into fours starting at the LSB.

$$1011 \quad 1111 \quad 0111 \quad 1001$$

Convert these groups into their hexadecimal equivalents.

$$(1011)_2 = (B)_{16}$$
$$(1111)_2 = (F)_{16}$$
$$(0111)_2 = (7)_{16}$$
$$(1001)_2 = (9)_{16}$$

The answer is $(BF79)_{16}$.

8. COMPLEMENT OF A NUMBER

The *complement*, N^*, of a number, N, depends on the machine (i.e., computer, calculator, etc.) being used. Assuming that the machine has a maximum number, n, of digits per integer number stored, the b's and $(b-1)$'s complements are

$$N_b^* = (b)^n - N \qquad 25.3$$
$$N_{b-1}^* = N_b^* - 1 \qquad 25.4$$

For a machine that works in base-10 arithmetic, the *tens* and *nines complements* are

$$N_{10}^* = 10^n - N \qquad 25.5$$
$$N_9^* = N_{10}^* - 1 \qquad 25.6$$

For a machine that works in base-2 arithmetic, the *twos* and *ones complements* are

$$N_2^* = 2^n - N \qquad 25.7$$
$$N_1^* = N_2^* - 1 \qquad 25.8$$

The ones complement is easily found by switching all of the ones and zeros to zeros and ones, respectively. It can be combined with a technique known as *end-around carry* to perform subtraction. End-around carry is the addition of the *overflow bit* to the sum of N and its ones complement.

9. APPLICATION OF COMPLEMENTS TO COMPUTER ARITHMETIC

Equations 25.9 and 25.10 are the practical applications of complements to computer arithmetic.

$$(N^*)^* = N \qquad 25.9$$
$$M - N = M + N^* \qquad 25.10$$

Example 25.5

(a) Simulate the operation of a base-10 machine with a capacity of four digits per number and calculate the difference $(18)_{10} - (6)_{10}$ with tens complements.

(b) Simulate the operation of a base-2 machine with a capacity of five digits per number and calculate the difference $(01101)_2 - (01010)_2$ with twos complements.

(c) Solve part (b) with a ones complement and end-around carry.

Solution

(a) The tens complement of 6 is

$$(6)^*_{10} = (10)^4 - 6 = 10,000 - 6 = 9994$$

Using Eq. 25.10,

$$18 - 6 = 18 + (6)^*_{10} = 18 + 9994 = 10,012$$

However, the machine has a maximum capacity of four digits. Therefore, the leading 1 is dropped, leaving 0012 as the answer.

(b) The twos complement of $(01010)_2$ is

$$N^*_2 = (2)^5 - N = (32)_{10} - N$$
$$= (100000)_2 - (01010)_2$$
$$= (10110)_2$$

From Eq. 25.10,

$$(01101)_2 - (01010)_2 = (01101)_2 + (10110)_2$$
$$= (100011)_2$$

Since the machine has a capacity of only five bits, the leftmost bit is dropped, leaving $(00011)_2$ as the difference.

(c) The ones complement is found by reversing all the digits.

$$(01010)^*_1 = (10101)_2$$

Adding the ones complement,

$$(01101)_2 + (10101)_2 = (100010)_2$$

The leading bit is the overflow bit, which is removed and added to give the difference.

$$(00010)_2 + (1)_2 = (00011)_2$$

10. COMPUTER REPRESENTATION OF NEGATIVE NUMBERS

On paper, we write a minus sign to indicate a negative number. This representation is not possible in a machine. Hence, one of the n digits, usually the MSB, is reserved for sign representation. (This reduces the machine's capacity to represent numbers to $n-1$ bits per number.) It is arbitrary whether the sign bit is 1 or 0 for negative numbers as long as the MSB is different for positive and negative numbers.

The ones complement is ideal for forming a negative number since it automatically reverses the MSB. For example, $(00011)_2$ is a five-bit representation of decimal 3. The ones complement is $(11100)_2$, which is recognized as a negative number because the MSB is 1.

Example 25.6

Simulate the operation of a six-digit binary machine that uses ones complements for negative numbers.

(a) What is the machine representation of $(-27)_{10}$?

(b) What is the decimal equivalent of the twos complement of $(-27)_{10}$?

(c) What is the decimal equivalent of the ones complement of $(-27)_{10}$?

Solution

(a) $(27)_{10} = (011011)_2$. The negative of this number is the same as the ones complement: $(100100)_2$.

(b) The twos complement is one more than the ones complement. (See Eqs. 25.7 and 25.8.) Therefore, the twos complement is

$$(100100)_2 + 1 = (100101)_2$$

This represents $(-26)_{10}$.

(c) From Eq. 25.9, the complement of a complement of a number is the original number. Therefore, the decimal equivalent is -27.

Computers

PRACTICE PROBLEMS

1. Perform the following binary operations. Check your work by converting to decimal.

(a) $101 + 011$ (d) $0100 - 1100$ (g) 111×11

(b) $101 + 110$ (e) $1110 - 1000$ (h) 100×11

(c) $101 + 100$ (f) $010 - 101$ (i) 1011×1101

2. Perform the following octal operations. Check your work by converting to decimal.

(a) $466 + 457$ (d) $71 - 27$ (g) 77×66

(b) $1007 + 6661$ (e) $1143 - 367$ (h) 325×36

(c) $321 + 465$ (f) $646 - 677$ (i) 3251×161

3. Perform the following hexadecimal operations. Check your work by converting to decimal.

(a) $BA + C$ (d) $FF - E$ (g) $4A \times 3E$

(b) $BB + A$ (e) $74 - 4A$ (h) $FE \times EF$

(c) $BE + 10 + 1A$ (f) $FB - BF$ (i) $17 \times 7A$

4. Convert the following numbers to decimal numbers.

(a) $(674)_8$ (c) $(734.262)_8$

(b) $(101101)_2$ (d) $(1011.11)_2$

5. Convert the following numbers to octal numbers.

(a) $(75)_{10}$ (c) $(121.875)_{10}$

(b) $(0.375)_{10}$ (d) $(1011100.01110)_2$

6. Convert the following numbers to binary numbers.

(a) $(83)_{10}$ (c) $(0.97)_{10}$

(b) $(100.3)_{10}$ (d) $(321.422)_8$

SOLUTIONS

1. (a)
$$\begin{array}{r} 101 \\ +\ 011 \\ \hline \boxed{1000} \end{array}$$

(b)
$$\begin{array}{r} 101 \\ +\ 110 \\ \hline \boxed{1011} \end{array}$$

(c)
$$\begin{array}{r} 101 \\ +\ 100 \\ \hline \boxed{1001} \end{array}$$

(d)
$$-\left(\begin{array}{r} 1100 \\ -\ 0100 \\ \hline 1000 \end{array}\right) = \boxed{-1000}$$

(e)
$$\begin{array}{r} 1110 \\ -\ 1000 \\ \hline \boxed{0110} \end{array}$$

(f)
$$-\left(\begin{array}{r} 101 \\ -\ 010 \\ \hline 011 \end{array}\right) = \boxed{-011}$$

(g)
$$\begin{array}{r} 111 \\ \times\ 11 \\ \hline 111 \\ 111 \\ \hline \boxed{10101} \end{array}$$

(h)
$$\begin{array}{r} 100 \\ \times\ 11 \\ \hline 100 \\ 100 \\ \hline \boxed{1100} \end{array}$$

(i)

$$
\begin{array}{r}
1011 \\
\times \quad 1101 \\
\hline
1011 \\
1011 \\
1011 \\
\hline
\end{array}
$$

$$\boxed{10001111}$$

2. (a)

$$
\begin{array}{r}
466 \\
+ \quad 457 \\
\hline
\end{array}
$$

$$\boxed{1145}$$

(b)

$$
\begin{array}{r}
1007 \\
+ \quad 6661 \\
\hline
\end{array}
$$

$$\boxed{7670}$$

(c)

$$
\begin{array}{r}
321 \\
+ \quad 465 \\
\hline
\end{array}
$$

$$\boxed{1006}$$

(d)

$$
\begin{array}{r}
71 \\
- \quad 27 \\
\hline
\end{array}
\; = \;
\begin{array}{r}
6 \;\; 11 \\
- \; 2 \;\; 7 \\
\hline
\end{array}
$$

$$\boxed{4 \quad 2}$$

(e)

$$
\begin{array}{r}
1143 \\
- \; 367 \\
\hline
\end{array}
\; = \;
\begin{array}{r}
113 \;\; 13 \\
- \; 36 \;\; 7 \\
\hline
\end{array}
\; = \;
\begin{array}{r}
10 \;\; 13 \;\; 13 \\
- \; 3 \;\; 6 \;\; 7 \\
\hline
\end{array}
$$

$$\boxed{5 \quad 5 \quad 4}$$

(f)

$$
- \left(
\begin{array}{r}
677 \\
- \; 646 \\
\hline
31
\end{array}
\right)
= \boxed{-31}
$$

(g)

$$
\begin{array}{r}
77 \\
\times \quad 66 \\
\hline
572 \\
572 \\
\hline
\end{array}
$$

$$\boxed{6512}$$

(h)

$$
\begin{array}{r}
325 \\
\times \quad 36 \\
\hline
2376 \\
1177 \\
\hline
\end{array}
$$

$$\boxed{14366}$$

(i)

$$
\begin{array}{r}
3251 \\
\times \quad 161 \\
\hline
3251 \\
23766 \\
3251 \\
\hline
\end{array}
$$

$$\boxed{570231}$$

3. (a)

$$
\begin{array}{r}
BA \\
+ \quad C \\
\hline
\end{array}
$$

$$\boxed{C6}$$

(b)

$$
\begin{array}{r}
BB \\
+ \quad A \\
\hline
\end{array}
$$

$$\boxed{C5}$$

(c)

$$
\begin{array}{r}
BE \\
10 \\
+ \; 1A \\
\hline
\end{array}
$$

$$\boxed{E8}$$

(d)

$$
\begin{array}{r}
FF \\
- \quad E \\
\hline
\end{array}
$$

$$\boxed{F1}$$

(e)

$$
\begin{array}{r}
74 \\
- \; 4A \\
\hline
\end{array}
\; = \;
\begin{array}{r}
6 \;\; 14 \\
- \; 4 \;\; A \\
\hline
\end{array}
$$

$$\boxed{2 \quad A}$$

(f)
$$\begin{array}{r} FB \\ - \ BF \end{array} = \begin{array}{r} E \ 1B \\ - \ B \ \ F \end{array}$$

$$\boxed{3 \quad C}$$

(g)
$$\begin{array}{r} 4A \\ \times \quad 3E \\ \hline 40C \\ DE \quad \\ \hline \end{array}$$

$$\boxed{11EC}$$

(h)
$$\begin{array}{r} FE \\ \times \quad EF \\ \hline EE2 \\ DE4 \quad \\ \hline \end{array}$$

$$\boxed{ED22}$$

(i)
$$\begin{array}{r} 17 \\ \times \quad 7A \\ \hline E6 \\ A1 \quad \\ \hline \end{array}$$

$$\boxed{AF6}$$

4. (a) $(674)_8 = (6)(8)^2 + (7)(8)^1 + (4)(8)^0$

$$= \boxed{444}$$

(b) $(101101)_2 = (1)(2)^5 + (0)(2)^4 + (1)(2)^3$
$$+ \ (1)(2)^2 + (0)(2)^1 + (1)(2)^0$$

$$= \boxed{45}$$

(c) $(734.262)_8 = (7)(8)^2 + (3)(8)^1 + (4)(8)^0$
$$+ \ (2)(8)^{-1} + (6)(8)^{-2} + (2)(8)^{-3}$$

$$= \boxed{476.348}$$

(d) $(1011.11)_2 = (1)(2)^3 + (0)(2)^2 + (1)(2)^1$
$$+ \ (1)(2)^0 + (1)(2)^{-1} + (1)(2)^{-2}$$

$$= \boxed{11.75}$$

5. (a)
$$75 \div 8 = 9 \quad \text{remainder } 3$$
$$9 \div 8 = 1 \quad \text{remainder } 1$$
$$1 \div 8 = 0 \quad \text{remainder } 1$$
$$(75)_{10} = \boxed{(113)_8}$$

(b)
$$0.375 \times 8 = 3 \quad \text{remainder } 0$$
$$(0.375)_{10} = \boxed{(0.3)_8}$$

(c)
$$121 \div 8 = 15 \quad \text{remainder } 1$$
$$15 \div 8 = 1 \quad \text{remainder } 7$$
$$1 \div 8 = 0 \quad \text{remainder } 1$$
$$0.875 \times 8 = 7 \quad \text{remainder } 0$$
$$(121.875)_{10} = \boxed{(171.7)_8}$$

(d) Since $(2)^3 = 8$, group the bits into groups of three starting at the decimal point and working outward in both directions.

$$001011100.011100 = 001 \ 011 \ 100.011 \ 100$$

Convert each of the groups into its octal equivalent.

$$001 \ 011 \ 100.011 \ 100 = 1 \ 3 \ 4.3 \ 4$$

$$\boxed{(134.34)_8}$$

6. (a)
$$83 \div 2 = 41 \quad \text{remainder } 1$$
$$41 \div 2 = 20 \quad \text{remainder } 1$$
$$20 \div 2 = 10 \quad \text{remainder } 0$$
$$10 \div 2 = 5 \quad \text{remainder } 0$$
$$5 \div 2 = 2 \quad \text{remainder } 1$$
$$2 \div 2 = 1 \quad \text{remainder } 0$$
$$1 \div 2 = 0 \quad \text{remainder } 1$$
$$(83)_{10} = \boxed{(1010011)_2}$$

(b)
$$100 \div 2 = 50 \quad \text{remainder } 0$$
$$50 \div 2 = 25 \quad \text{remainder } 0$$
$$25 \div 2 = 12 \quad \text{remainder } 1$$

$$12 \div 2 = 6 \quad \text{remainder } 0$$
$$6 \div 2 = 3 \quad \text{remainder } 0$$
$$3 \div 2 = 1 \quad \text{remainder } 1$$
$$1 \div 2 = 0 \quad \text{remainder } 1$$
$$0.3 \times 2 = 0 \quad \text{remainder } 0.6$$
$$0.6 \times 2 = 1 \quad \text{remainder } 0.2$$
$$0.2 \times 2 = 0 \quad \text{remainder } 0.4$$
$$0.4 \times 2 = 0 \quad \text{remainder } 0.8$$
$$0.8 \times 2 = 1 \quad \text{remainder } 0.6$$
$$0.6 \times 2 = 1 \quad \text{remainder } 0.2$$
$$\vdots$$

$$(100.3)_{10} = \boxed{(1100100.010011\cdots)_2}$$

(c)
$$0.97 \times 2 = 1 \quad \text{remainder } 0.94$$
$$0.94 \times 2 = 1 \quad \text{remainder } 0.88$$
$$0.88 \times 2 = 1 \quad \text{remainder } 0.76$$
$$0.76 \times 2 = 1 \quad \text{remainder } 0.52$$
$$0.52 \times 2 = 1 \quad \text{remainder } 0.04$$
$$0.04 \times 2 = 0 \quad \text{remainder } 0.08$$
$$\vdots$$

$$(0.97)_{10} = \boxed{(0.111110\cdots)_2}$$

(d) Since $8 = (2)^3$, convert each octal digit into its binary equivalent.

$$321.422: \quad 3 = 011$$
$$2 = 010$$
$$1 = 001$$
$$4 = 100$$
$$2 = 010$$
$$2 = 010$$

$$\boxed{(321.422)_8 = 011010001.100010010}$$

26 Computer Hardware

1. EVOLUTION OF COMPUTER HARDWARE

The term *hardware* encompasses the equipment and devices that perform data preparation, input, computation, control, primary and secondary storage, and output functions, but it does not include the programs, routines, and applications (i.e., computer *software*) that control the computer.[1]

Digital computers are generally acknowledged to have gone through five major evolutionary stages.[2,3]

- *first generation*: electromechanical calculators
- *second generation*: vacuum tube computers
- *third generation*: transistor computers
- *fourth generation*: integrated circuit computers
- *fifth generation*: VLSI (very large-scale integration) computers

The term *fifth generation* also is used to refer to the efforts (largely on the part of Japanese researchers) to produce computers that are easier to program and use. These efforts have generally been unsuccessful and have been replaced by research into the *sixth generation* of computers, devices that rely on parallel processing in order to appear more human-like in their programming and processing.

2. COMPUTER SIZE

Computers used for data processing (excluding process control devices) are classified into four categories depending on size and cost.

- *Microcomputers* (*personal computers*, PC's) are small, generally single-user computers without extensive peripherals or storage.
- *Minicomputers* are larger computers, usually dedicated to business data processing at a single site. They have the ability to support multiple terminals and a wide range of peripherals.
- *Mainframe computers* are general-purpose computers used in large, centralized data processing complexes and departments in which many programs are running simultaneously.
- *Supercomputers* are extremely powerful computers that usually have specific functions (e.g., engineering design, number crunching, analysis of strategies).

The distinction between these categories is rapidly becoming indistinct. Some current microcomputers have the same capabilities as minicomputers and mainframes of five years ago. The categorization must be made on the basis of physical size and cost, not on memory size or processing speed as was done in the past.

3. COMPUTER ARCHITECTURE

All digital computers, from giant mainframes to the smallest microcomputers, contain three main components—a central processing unit (CPU), main memory, and external (peripheral) devices. Figure 26.1 illustrates a typical integration of these functions.

---

[1]The term "software program" is redundant.
[2]The categorization depends on who is counting and what characteristics are considered evolutionary. Some writers omit electromechanical calculators from the evolution.
[3]Before the days of calculators and computers, some companies were large enough to have employees who did nothing but crank through calculations for engineers and designers. These people were called the company's "computers".

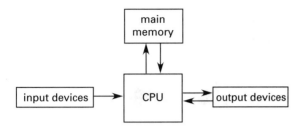

Figure 26.1 *Simplified Computer Architecture*

Microprocessors can be designed to operate on 4-bit, 8-bit, 16-bit, 32-bit, and 64-bit words, although microprocessors with 4- and 8-bit words are now used primarily only in process control applications. Some microprocessors can be combined, and the resulting larger unit is known as a *bit-slice microprocessor*. (For example, four 4-bit microprocessors might be combined into a 16-bit-slice microprocessor.)

4. MICROPROCESSORS

A *microprocessor* is a central processing unit (CPU) on a single chip. With *large-scale integration* (LSI), most microprocessors are contained on one chip, although other chips in the set can be used for memory and input/output control. The most popular microprocessor families are produced by Intel (Pentium, Pentium Pro, Pentium II, and Pentium III families) and Motorola (680X0 family), although work-alike clones of these chips may be produced by other companies (e.g., AMD's K6 and Cyrix's GX and M2 processors).

Microprocessor CPUs consist of an arithmetic and logic unit, several accumulators, one or more registers, stacks, and a control unit. The *control unit* fetches and decodes the incoming instructions and generates the signals necessary for the arithmetic and logic unit to perform the intended function. The *arithmetic and logic unit* (ALU) executes commands and manipulates data.

Accumulators hold data and instructions for further manipulation in the ALU. Registers are used for temporary storage of instructions or data. The *program counter* (PC) is a special register that always points to (contains) the address of the next instruction to be executed. Another special register is the *instruction register* (IR), which holds the current instruction during its execution. *Stacks* provide temporary data storage in sequential order—usually on a last-in, first-out (LIFO) basis. Because their operation is analogous to spring-loaded tray holders in cafeterias, the name *pushdown stack* is also used.

Microprocessors communicate with support chips and peripherals through connections in a *bus* or *channel*, which is logically subdivided into three different functions.[5] The *address bus* directs memory and input/output device transfers. The *data bus* carries the actual data and is the busiest bus. The *control bus* communicates control and status information. The number of lines in the address bus determines the amount of random access memory that can be directly addressed. When there are n address lines in the bus, 2^n words of memory can be addressed.

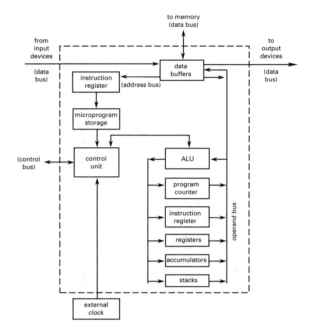

Figure 26.2 *Microprocessor Architecture*

All microprocessors use a crystal-controlled *clock* to control instruction and data movements. The *clock rate* is specified in microprocessor cycles per second (e.g., 200 MHz). Ideally, the clock rate is the number of instructions the microprocessor can execute per second. However, one or more cycles may be required for complex instructions (*macrocommands*). For example, executing a complex instruction may require one or more cycles each to fetch the instruction from memory, decode the instruction to see what to do, execute the instruction, and store (write) the result.[6] Since operations on floating point numbers are macrocommands, the speed of a microprocessor can also be specified in *flops*, the number of floating point operations it can perform per second. Similar units of processing speed are *mips* (millions of instructions per second).

Most microprocessors are rich in complex executable commands (i.e., the *command set*) and are known as *complex instruction-set computing* (CISC) microprocessors. In order to increase the operating speed, however, *reduced instruction-set computing* (RISC) microprocessors are limited to performing simple, standardized format instructions but are otherwise fully featured. Unlike CISC units, RISC microprocessors

---

[5]The term *bus* refers to the physical path (i.e., wires or circuit board traces) along which the signal travels. The term *channel* refers to the logical path.

[6]In some cases, other chips (e.g., *memory management units*) can perform some of these tasks.

require four separate instructions for the common fetch-decode-execute-write sequence.

Some microprocessors can emulate the operation of other microprocessors. For example, an 80486 chip could operate in virtual 8086 mode. *Emulation mode* is also referred to as *virtual mode*.

5. CONTROL OF COMPUTER OPERATION

The user interface and basic operation of a computer are controlled by the *operating system* (OS), also known as the *monitor program*. The operating system is a program that controls the computer at its most basic level and provides the environment for application programs. The operating system manages the memory, schedules processing operations, accesses peripheral devices, communicates with the user/operator, and (in multi-tasking environments) resolves conflicting requirements for resources. Since one of the functions of the operating system is to coordinate use of the peripheral devices (disk drives, keyboards, etc.), the term *basic input/output system* (BIOS) is also used.

All or part of the operating system can be stored in read-only memory (ROM). In early computers, a small part of executable code that was used to initiate data transfers and logical operations when the computer was first started was known as a *bootstrap loader*. Although modern start-up operations are more sophisticated, the phrase "booting the computer" is still used today.

During program operation, peripherals and other parts of the computer signal the operating system through interrupts. An *interrupt* is a signal that stops the execution of the current instruction (or, in some cases, the current program) and transfers control to another memory location, subroutine, or program. Other interrupts signal error conditions such as division by zero, overflow and underflow, and syntax errors. The operating system intercepts, decodes, and acts on these interrupts.

6. COMPUTER MEMORY

Computer memory consists of many equally sized storage locations, each of which has an associated address. The contents of a storage location may change, but the address does not.

The total number of storage locations in a computer can be measured in various ways. A *bit* (binary digit) is the smallest changeable data unit. Bits can only have values of 1 or 0. Bits are combined into *nibbles* (4 bits), *bytes* (8 bits, the smallest number of bits that can represent one alphanumeric character), *half-words* (8 and 16 bits), *words* (8, 16, and 32 bits) and *doublewords* (16, 32, and 64 bits).[7]

The number of memory storage locations is always a multiple of two. The abbreviations K, M, and G are used to designate the quantities 2^{10} (1024), 2^{20} (1,048,576), and 2^{30} (1,073,741,824), respectively. For example, a 6M memory would contain 6×2^{20} bytes. (K and M do not mean one thousand and one million exactly.)

Most of the memory locations are used for user programs and data. However, portions of the memory may be used for video memory, I/O cache memory, the BIOS, and other purposes. *Video memory* (known as VRAM) contains the text displayed on the screen of a terminal. Since the screen is refreshed many times per second, the screen information must be repeatedly read from video memory. *Cache memory* holds the most recently read and frequently read data in memory, making subsequent retrieval of that data much faster than reading from a tape or disk drive, or even from main memory.[8] *OS memory* contains the BIOS that is read in when the computer is first started. *Scratchpad memory* is high-speed memory used to store a small amount of data temporarily so that the data can be retrieved quickly.

Modern memory hardware is semiconductor based.[9] Memory is designated as RAM (random access memory), ROM (read-only memory), PROM (programmable read-only memory), and EPROM (erasable programmable read-only memory). While data in RAM is easily changed, data in ROM cannot be altered. PROMs are initially blank but once filled, they cannot be changed. EPROMs are initially blank but can be filled, erased, and refilled repeatedly.[10] The term *firmware* is used to describe programs stored in ROMs and EPROMs.

The contents of a *volatile memory* are lost when the power is turned off. RAM is usually volatile, while ROM, PROM, and EPROM are *non-volatile*. With *static memory*, data does not need to be refreshed and remains as long as the power stays on. With *dynamic memory*, data must be continually refreshed. Static and dynamic RAM (i.e., DRAM) are both volatile.

Virtual memory (storage) (VS) is a technique by which programs and data larger than main memory can be accessed by the computer. (Virtual memory is not synonymous with *virtual machine*, described in Sec. 12.) In virtual memory systems, some of the disk space is used as an extension of the semiconductor memory. A large application or program is divided into modules of equal size called *pages*. Each page is switched into (and out of) RAM from (and back to) disk storage as

---

[7]The distinction between doublewords, words, and half-words depends on the computer. Sixteen bits would be a word in a 16-bit computer but would be a half-word in a 32-bit computer. Furthermore, *double-precision (double-length) words* double the number of bytes normally used. The abbreviations KB (*kilobytes*) and KW (*kilowords*) used by some manufacturers do not help much to clarify the ambiguity.

[8]A high-speed mainframe computer may require 200–500 nanoseconds to access main memory but only 20–50 nanoseconds to access cache memory.

[9]The term core, derived from the ferrite cores used in early computers, is seldom used today.

[10]Most EPROMs can be erased by exposing them to ultraviolet light.

needed, a process known as *paging*. This interchange is largely transparent to the user. Of course, access to data stored on a disk drive is much slower than semiconductor memory access. *Thrashing* is a deadlock situation that occurs when a program references a different page for almost every instruction, and there is not even enough real memory to hold most of the virtual memory.

Most memory locations are filled and managed by the CPU. However, *direct memory access* (DMA) is a powerful I/O technique that allows peripherals (e.g., tape and disk drives) to transfer data directly into and out of memory without affecting the CPU. Although special DMA hardware is required, DMA does not require explicit program instructions, making data transfer faster.

7. PARITY

Parity is a technique used to ensure that the bits within a memory byte are correct. For every eight data bits, there is a ninth bit—the parity bit—that serves as a *check bit*. The nine bits together constitute a *frame*. In *odd-parity recording*, the parity bit will be set so there is an odd number of one-bits. In *even-parity recording*, the parity bit will be set so that there is an even number of one-bits. When the data are read, the nine bits are checked to ensure valid data.[11]

8. INPUT/OUTPUT DEVICES

Devices that feed data to, or receive data from, the computer are known as *input/output* (I/O) *devices*. Terminals, light pens, digitizers, printers and plotters, and tape and disk drives are common peripherals.[12] Point-of-sale (POS) devices, bar code readers, and magnetic ink character recognition (MICR) and optical character recognition (OCR) readers are less common devices.

Peripherals are connected to their computer through multi-line cables. With a *parallel interface* (used in a *parallel device*), there are as many separate lines in the cable as there are bits (typically seven, eight, or nine) in the code representing a character. An additional line is used as the *strobe signal* to carry a timing signal. With a *serial interface* (used in a *serial device*), all bits pass one at a time along a single line in the cable. The *transmission speed (baud rate)* in bps is the number of bits that pass through the data line per second.[13]

Peripherals such as terminals and printers typically do not have large memories. They only need memories large enough to store the information before the data

are displayed or printed. The small memories are known as *buffers*. The peripheral can send the status (i.e., full, empty, off-line, etc.) of its buffer to the computer in several different ways. This is known as *flow control* or *handshaking*.

If the computer and peripheral are configured so that each can send and receive data, the peripheral can send a single character (e.g., the XOFF *character* for transmission off) to the computer when its buffer is full. Similarly, a different character (e.g., the XON *character* for transmission on) can be sent when the peripheral is ready for more data. The computer must monitor the incoming data line for these characters. This is known as *software flow control* or *software handshaking*.

If there are enough separate lines between the computer and the peripheral, one or more of them can be used for *hardware handshaking*.[14] In this method of flow control, the peripheral keeps the voltage on one of the lines high (or low) when it is able to accept more data. The computer monitors the voltage on this line.

Most peripheral devices are connected to the computer by a dedicated channel (cable). However, a pair of *multiplexers* (*statistical multiplexers* or *concentrators*) can be used to carry data for several peripherals along a single cable known as the *composite link*. There are two methods of achieving multiplexed transmission: *frequency division multiplexing* (FDM) and *time division multiplexing* (TDM). With FDM, the available transmission band is divided into narrower bands, each used for a separate channel. In TDM, the connecting channel is operated at a much higher clock rate (proportional to the capacity of the multiplexer), and each peripheral shares equally in the available cycles.

The Electronics Industries Association (EIA) RS-232 standard was developed in an attempt to standardize the connectors and pin uses in serial device cables.

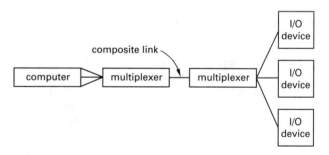

Figure 26.3 *Multiplexed Peripherals*

9. RANDOM SECONDARY STORAGE DEVICES

Random access (direct access) storage devices, also known as *mass storage devices*, include magnetic and

---

[11]This does not detect two of the bits in the frame being incorrect, however.
[12]CRT, the abbreviation for *cathode ray tube*, is often used to mean a terminal.
[13]The name *baud rate* is derived from the use of the *Baudot code* (see Ch. 27). One *baud* is one modulation per second. If there is a one-to-one correspondence between modulations and bits, one baud unit is the same as one bit per second (bps). In general, the unit bps should always be used.

[14]Terminals and printers usually require only two or three lines—data in, data out, and ground. Most computer cables contain more lines than this, and one of the extra lines can be used for DTR (*data terminal ready*) or CTS (*clear to send*) handshaking.

optical disk drive units. They are random access because individual records can be accessed without having to read through the entire file.

Magnetic disk drives (*hard drives*) are composed of several *platters*, each with one or more read/write heads. The platters typically turn at 4500–7200 rpm. Data on a surface are organized into tracks, sectors, and cylinders. *Tracks* are the concentric storage areas. *Sectors* are pie-shaped subdivisions of each track. A *cylinder* consists of the same numbered track on all drive platters. Some platters and *disk packs* are removable, but most hard drives are fixed (i.e., non-removable).

Depending on the media, *optical disk drives* can be *read only* (R/O) or *read/write* (R/W) in nature. WORM drives (write once, read many) can be written by the user, while others such as CD-ROM (compact disk read-only memory) can only be read.[15]

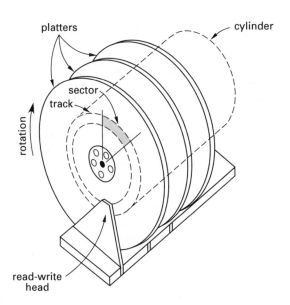

Figure 26.4 *Tracks, Sectors, and Cylinders*

In addition to *storage capacity* (usually specified in megabytes—MB) there are several parameters that describe the performance of a disk drive. *Areal density* is a measure of the number of data bits stored per square inch of disk surface. It is calculated by multiplying the number of bits per track by the number of tracks per (radial) inch. The *average seek time* is the average time it takes to move a head from one location to a new location. The *track-to-track seek time* is the time required to move a head from one track to an adjacent track. *Latency* or *rotational delay* is the time it takes for a disk to spin a particular sector under the head for reading. On the average, latency is one-half of the time to spin a full revolution. The *average access time* is the time to move to a new sector and read the data. Access time is the sum of average latency and average seek time.

Floppy disks (diskettes) are suitable for low-capacity random-access storage. Although their capacities are comparatively low (typically less than one million characters for magnetic media), they are used to transfer programs and data between computers (primarily microcomputers). The capacity of a diskette depends on its size, the recording density, number of tracks, and number of sides.

10. SEQUENTIAL SECONDARY STORAGE DEVICES

Tape units are *sequential access devices* because a computer cannot access information stored at the end of a tape without first reading or passing by the information stored at the front of the tape. Some tapes use *indexed sequential formats* in which a directory of files on the tape is placed at the start of the tape. The tape can be rapidly wound to (near) the start of the target file without having to read everything in between.

There are a number of tape formats in use. One of the few standardized commercial formats is the *nine-track* format.[16] The tape is divided into nine tracks running the length of the tape. The width of the tape is divided into frames (characters). Eight tracks are used to record the data in either ASCII or EBCDIC format. (See Ch. 27.) The remaining track is used to record the parity bit. 1600 bpi (bits or frames per inch) is still a common *recording density* for sharing data, although densities of 9600 bpi and above are in use.

Frames are grouped into fixed-length *blocks* separated by *interblock gaps* (IBG). Reflective spots for photoelectric detection are used to indicate the beginnings and ends of magnetic tapes. These spots are called *load point* and *end-of-file* (EOF) *markers*, respectively.

Magnetic tape is often used to back up hard disks. A *streaming tape* operates in a continuously running—or streaming—mode, with data being written or read while the tape is running.

11. REAL-TIME AND BATCH PROCESSING

Programs run on a computer in one of two main ways: batch mode and real-time mode. In some data processing environments, programs are held (either by the operator or by the operating system) and eventually grouped into efficient categories requiring the same peripherals and resources. This is called *batch mode processing* since all jobs of a particular type are batched together for subsequent processing. There is usually no interaction between the user and the computer once batch processing begins. In *real-time (interactive) processing*, a program runs when it is submitted, often with user interaction during processing.

---

[15]The WORM acronym is also interpreted as write once—read mostly.

[16]Other formats that have received some acceptance include *quarter inch cartridge* (QIC) and *digital audio tape* (DAT).

12. MULTI-TASKING AND TIME-SHARING

If a computer's main memory is large enough and the CPU is fast enough, it may be possible to allocate the main memory among several users running applications simultaneously. This is the concept of a *virtual machine* (VM)—each user appears to have his own computer. This is also known as *multi-tasking* and *multi-programming* since multiple tasks can be performed simultaneously. A *multi-user system* is similar to a multi-tasking system in that several users can use the computer simultaneously, although that term also means that all users are using the same program.

Time-sharing (*swapping*) is a technique where each user takes turns (under the control of the operating system) using the entire computer main memory for a certain length of time (usually less than a second). At the end of that time period, all of the active memory is written to a private area, and the memory for the next user is loaded. The swapping occurs so frequently that all users are able to accomplish useful work on a real-time basis.

13. BACKGROUND AND FOREGROUND PROCESSING

A program running in real time is an example of *foreground processing*. There are times, however, when it is convenient to start a long program running while the same computer is used for a second program. The first program continues to run unseen in the background. *Background processing* can be accomplished by segmenting the main computer memory (i.e., establishing virtual machines as defined in Sec. 12) or by time-sharing (i.e., allowing the background application to have all cycles not used by the foreground application).

14. TELEPROCESSING

Teleprocessing is the access of a computer from a remote station, usually over a telephone line (although fiber optic, coaxial, and microwave links can also be used). Since these media transmit analog signals, a *modem* (*modulator-demodulator*) is used to convert to/from the digital signals required by a computer.

With *simplex communication*, transmission is only in one direction. With *half-duplex communication*, data can be transmitted in both directions, but only in one direction at a time. With *duplex* or *full duplex communication*, data can be transmitted in both directions simultaneously. 56,000 bps is the maximum practical transmission speed over voice-grade lines without data compression. Much higher rates, however, are possible over dedicated data lines and wide-band lines.

In *asynchronous* or *start-stop transmission*, each character is preceded and followed by special signals (i.e., *start* and *stop bits*). Thus, every 8-bit character is actually transmitted as 10 bits, and the character transmission rate is one-tenth of the transmission speed in bits per second.[17] With asynchronous transmission, it is possible to distinguish the beginning and end of each character from the bit stream itself.

Synchronous equipment transmits a block of data continuously without pause and requires a built-in clock to maintain synchronization. Synchronous transmission is preceded and interwoven with special clock-synchronizing characters, and the separation of a bit stream into individual characters is done by counting bits from the start of the previous character. Since start and stop bits are not used, synchronous communication is approximately 20 percent faster than asynchronous communication.

There are three classes of communication lines—narrowband, voice-grade, and wide-band—depending on the bandwidth (i.e., range of frequencies) available for signaling.

Narrow-band may only support a single channel of communication, as the bandwidth is too narrow for modulation. *Wide-band channels* support the highest transfer rates, since the bandwidth can be divided into individual channels. *Voice-grade lines*, supporting frequencies between 300 and 3000–3300 Hz, are midrange in bandwidth.

Errors in transmission can easily occur over voice grade lines at the rate of 1 in 10,000. In general, methods of ensuring the accuracy of transmitted and received data are known as *communications protocols* and *transmission standards*. A simple way of checking the transmission is to have the receiver send each block of data back to the sender. This process is known as *loop checking* or *echo checking*. If the characters in a block do not match, they are re-sent. While accurate, this method requires sending each block of data twice.

Another method of checking the accuracy of transmitted data is for both the receiver and sender to calculate a *check digit* or *block check character* derived from each block of *characters* sent. (A common transmission block size is 128 characters.) With CRC (*cycling redundancy checking*), the block check character is the remainder after dividing all the serialized bits in a transmission block by a predetermined binary number. Then, the block check character is sent and compared after each block of data.

15. DISTRIBUTED SYSTEMS AND LOCAL-AREA NETWORKS

Distributed data processing systems assume many configurations. In the traditional situation, a centrally located main computer interacts with, and is fed by, smaller computers in other locations. In a second configuration, many identical computers are linked together in order to share storage and printing resources. This

---

[17]For historical reasons, a second stop bit is used when data are sent at ten characters per second. This is referred to as 110 bps.

latter case is known as a *local-area network* (LAN). Local-area networks typically communicate at speeds between 200 kbps and 50 Mbps.

PRACTICE PROBLEMS

1. A disk has 1000 tracks with 10,000 bits per track. The disk diameter is 6 inches. The average seek time is 20 ms, and the rotational speed is 2000 rpm. (a) What is the areal density of the disk? (b) Estimate the average access time.

2. What length of nine-track tape with density 1600 bpi is required to store 1 Mb of data?

3. The entire contents of a full 20-Mb disk are sent in an asynchronous transmission at a 2400 baud rate. What is the time required for transmission?

SOLUTIONS

1. (a) areal density

$$= \text{(no. bits per track)}$$
$$\times \text{(no. tracks per radial inch)}$$

$$= \left(10{,}000 \ \frac{\text{bits}}{\text{track}}\right) \left(\frac{1000 \text{ tracks}}{\frac{D}{2}}\right)$$

$$= \left(10{,}000 \ \frac{\text{bits}}{\text{track}}\right) \left(\frac{2000 \text{ tracks}}{6 \text{ in}}\right)$$

$$= \boxed{3.33 \times 10^6 \text{ bits/radial in}}$$

(b) access time = average latency + average seek time

The average latency can be approximated as one-half time for the full revolution.

$$\left(\frac{1}{2} \text{ rev}\right) \left(\frac{1}{2000} \frac{\text{min}}{\text{rev}}\right) \left(60 \ \frac{\text{sec}}{\text{min}}\right) \left(1000 \ \frac{\text{ms}}{\text{sec}}\right) = 15 \text{ ms}$$

$$\text{average access time} = 15 \text{ ms} + 20 \text{ ms}$$

$$= \boxed{35 \text{ ms}}$$

2. Since eight tracks are used for data recording, 1 byte of data is stored across the width of the tape. Therefore, 1600 bytes of data are stored per inch of length. The length required for 1 megabyte is

$$\frac{1{,}048{,}576 \text{ bytes}}{\left(1600 \ \frac{\text{bytes}}{\text{in}}\right) \left(12 \ \frac{\text{in}}{\text{ft}}\right)} = \boxed{54.6 \text{ ft}}$$

This neglects any interblock gaps.

3. The time required is

$$t = \frac{(20 \text{ Mb}) \left(1{,}048{,}576 \ \frac{\text{bytes}}{\text{Mb}}\right) \left(10 \ \frac{\text{bits}}{\text{byte}}\right)}{\left(2400 \ \frac{\text{bits}}{\text{sec}}\right) \left(60 \ \frac{\text{sec}}{\text{min}}\right) \left(60 \ \frac{\text{min}}{\text{hr}}\right)}$$

$$= \boxed{24.3 \text{ hrs}}$$

Computers

27 Data Structures and Program Design

1. CHARACTER CODING

Alphanumeric data refers to characters that can be displayed or printed, including numerals and symbols ($, %, &, etc.) but excluding *control characters* (tab, carriage return, form feed, etc.). Since computers can handle binary numbers only, all symbolic data must be represented by binary codes. *Coding* refers to the manner in which alphanumeric data and control characters are represented by sequences of bits.

The standard method for coding data on 80-column, 12-row cards is the *Hollerith code*.[1]

The *Baudot code* is a five-bit code that has long been used in Telex and teletypewriter (TWX and TTY) communications. By shifting to an alternate character set (numerals versus letters), it has a maximum of 64 (2×2^5) characters. Primarily due to its slow transmission speed, but also due to its limited character set, the Baudot code is no longer favored by new equipment manufacturers.

In some early computers, characters were represented by six-bit combinations known as *Binary Coded Decimal* (BCD).[2] The 64 (2^6) different combinations, however, proved insufficient to represent all necessary characters.

The *American Standard Code for Information Interchange* (ASCII) is a seven-bit code permitting 128 (2^7) different combinations. It is commonly used in microcomputers, although use of the high order (eighth) bit is not standardized. ASCII-coded magnetic tape and disk files are used to transfer data and documents between computers of all sizes that would otherwise be unable to share data structures.

The *Extended Binary Coded Decimal Interchange Code* (EBCDIC) is in widespread use in mainframe computers.[3] It uses eight bits (a byte) for each character, allowing a maximum of 256 (2^8) different characters.

Since strings of bits are difficult to read, the *packed decimal* format is used to simplify working with EBCDIC data. Each byte is converted into two strings of four bits each. The two strings are then converted to hexadecimal format. Since $(1111)_2 = (15)_{10} = (F)_{16}$, the largest possible EBCDIC character is coded FF in packed decimal.

Example 27.1

The number $(7)_{10}$ is represented as 11110111 in EBCDIC. What is this in packed decimal?

Solution

The first four bits are 1111, which is $(15)_{10}$ or $(F)_{16}$. The last four bits are 0111, which is $(7)_{10}$ or $(7)_{16}$. The packed decimal representation is F7.

2. PROGRAM DESIGN

A *program* is a sequence of computer instructions that performs some function. The program is designed to implement an algorithm, which is a procedure consisting of a finite set of well-defined steps. Each step in the algorithm usually is implemented by one or more instructions (e.g., READ, GOTO, OPEN, etc.) entered by the programmer. These original "human-readable" instructions are known as source code statements.

---

[1] Punch cards are now seldom used.

[2] BCD was reintroduced when IBM started using 96-column cards.
[3] EBCDIC is pronounced eb'-sih-dik.

Except in rare cases, a computer will not understand source code statements. Therefore, the source code is translated into machine-readable object code and absolute memory locations. Eventually, an executable program is produced.

If the executable program is kept on disk or tape, it is normally referred to as *software*. If the program is placed in ROM or EPROM, it is referred to as *firmware*. The computer mechanism itself is known as the *hardware*.

3. FLOWCHARTING SYMBOLS

A *flowchart* is a step-by-step drawing representing a specific procedure or algorithm. Figure 27.1 illustrates the most common flowcharting symbols. The terminal symbol begins and ends a flowchart. The input/output symbol defines an I/O operation, including those to and from keyboard, printer, memory, and permanent data storage. The processing symbol and predefined process symbol refer to calculations or data manipulation. The decision symbol indicates a point where a decision must be made or two items are compared. The connector symbol indicates that the flowchart continues elsewhere. The off-page symbol indicates that the flowchart continues on the following page. Comments can be added in an annotation symbol.

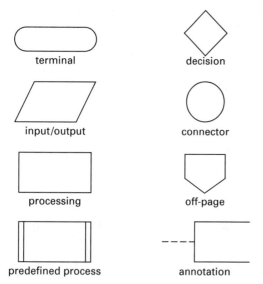

Figure 27.1 *Flowcharting Symbols*

4. LOW-LEVEL LANGUAGES

Programs are written in specific languages, of which there are two general types: low-level and high-level. Low-level languages include machine language and assembly language.

Machine language instructions are intrinsically compatible with and understood by the computer's CPU. They are the CPU's native language. An instruction normally consists of two parts: the operation to be performed (*op-code*) and the operand expressed as a storage location. Each instruction must ultimately be expressed as a series of bits, a form known as *intrinsic machine code*. However, octal and hexadecimal coding are more convenient. In either case, coding a machine language program is tedious and seldom done by hand.

Table 27.1 *Comparison of Typical ADD Commands*

language	instruction
intrinsic machine code	1111 0001
machine language	1A
assembly language	AR
FORTRAN	+

Assembly language is more sophisticated (i.e., is more symbolic) than machine language. Mnemonic codes are used to specify the operations. The operands are referred to by variable names rather than the addresses. Blocks of code that are to be repeated verbatim at multiple locations in the program are known as *macros* (*macro instructions*). Macros are written only once and are referred to by a symbolic name in the source code.

Assembly language code is translated into machine language by an *assembler* (*macro-assembler* if macros are supported). After assembly, portions of other programs or function libraries may be combined by a *linker*. In order to run, the program must be placed in the computer's memory by a *loader*. Assembly language programs are preferred for highly efficient programs. However, the coding inconvenience outweighs this advantage for most applications.

5. HIGH-LEVEL LANGUAGES

High-level languages are easier to use than low-level languages because the instructions resemble English. High-level statements are translated into machine language by either an interpreter or a compiler. A *compiler* performs the checking and conversion functions on all instructions only when the compiler is invoked. A true stand-alone executable program is created. An *interpreter*, however, checks the instructions and converts them line by line into machine code during execution but produces no stand-alone program capable of being used without the interpreter.[4,5]

There are many high-level languages, although only a few are in widespread use. Some of these languages are listed in Table 27.2.

---

[4]Some interpreters check syntax as each statement is entered by the programmer.

[5]Some languages and implementations of other languages blur the distinction between interpreters and compilers. Terms such as *pseudo-compiler* and *incremental compiler* are used in these cases.

Table 27.2 *High-Level Languages*

ADA: a rigidly standardized language similar to Pascal

ALGOL: acronym for ALGOrithmic Language

APL: acronym for A Programming Language; a language with a special character set

BASIC: Beginner's All-purpose Symbolic Instruction Code; English-like instructions; similar to FORTRAN

C: a structured high-level, function-oriented language capable of low-level machine control

COBOL: acronym for COmmon Business Oriented Languages; used in business programs; very English-like

FORTH: a language originally designed for telescope process control; programs are called words

FORTRAN: acronym for FORmula TRANslation; rich in scientific functions.

Modula-2: a structured language derived from Pascal

Pascal: a highly structured, portable language

PL/I: acronym for Programming Language One; programs are called functions

RPG: acronym for RePort Generator

6. SPECIAL PURPOSE LANGUAGES

There are many special purpose languages, some of which are listed in Table 27.3.

Special applications include AI (artificial intelligence—see Sec. 17), CAD (computer-aided design), CAM (computer-aided manufacturing), CAI (computer-aided instruction), DB (database), DBMS (database management system), EIS (executive information system), JCL (job control language), and MIS (management information system).

Table 27.3 *Special Purpose Languages*

name	use or strength
COMIT	string processing
GPSS	simulation
LISP	list processing; artificial intelligence
LOGO	childrens' introductory language
PROLOG	artificial intelligence
SMALLTALK	artificial intelligence
SNOBOL	string processing
SPSS	statistical analysis

7. RELATIVE COMPUTATIONAL SPEED

Certain languages are more efficient (i.e., execute faster) than others.[6] While it is impossible to be specific, and exceptions abound, assembly language programs are fastest, followed in order of decreasing speed by compiled, pseudo-compiled, and interpreted programs.

Similarly, certain program structures are more efficient than others. For example, when performing a repetitive operation, the most efficient structure will be a single equation, followed in order of decreasing speed by a stand-alone loop and a loop within a subroutine. Incrementing the loop variables and managing the exit and entry points is known as *overhead* and takes time during execution.

8. STRUCTURE, DATA TYPING, AND PORTABILITY

A language is said to be *structured* if subroutines and other procedures each have one specific entry point and one specific return point.[7] A language has *strong data types* if integer and real numbers cannot be combined in arithmetic statements.

A *portable language* can be implemented on different machines. Most portable languages are either sufficiently rigidly defined (as in the cases of ADA and C) to eliminate variants and extensions, or (as in the case of Pascal) are compiled into an intermediate, machine-independent form. This so-called *pseudo-code (p-code)* is neither source nor object code. The language is said to have been "ported to a new machine" when an interpreter is written that converts p-code to the appropriate machine code and supplies specific drivers for input, output, printers, and disk use.[8]

9. STRUCTURED PROGRAMMING

Structured programming (also known as top-down programming, procedure-oriented programming, and GOTO-less programming) divides a procedure or algorithm into parts known as subprograms, subroutines, modules, blocks, or procedures.[9] Internal subprograms are written by the programmer; external subprograms are supplied in a library by another source. Ideally, the mainline program will consist entirely of a series of calls (references) to these subprograms. Liberal use is made of FOR/NEXT, DO/WHILE, and DO/UNTIL

[6]Efficiency can also, but seldom does, refer to the size of the program.
[7]Contrast this with BASIC, which permits (1) a GOSUB to a specific subroutine with a return from anywhere within the subroutine and (2) unlimited GOTO statements to anywhere in the main program.
[8]Some companies have produced Pascal engines that run p-code directly.
[9]The format and readability of the source code—improved by indenting nested structures, for example—do not define structured programming.

commands. Labels and GOTO commands are avoided as much as possible.

Very efficient programs can be constructed in languages that support *recursive calls* (i.e., permit a subprogram to call itself). Some languages (e.g., Pascal and PL/I) permit recursion; others do not.

Variables whose values are accessible strictly within the subprogram are *local variables*. *Global variables* can be referred to by the main program and all other subprograms.

10. FIELDS, RECORDS, AND FILE TYPES

A collection of *fields* is known as a *record*. For example, name, age, and address might be fields in a personnel record. Groups of records are stored in a *file*.

A *sequential file* structure (e.g., typical of data on magnetic tape) contains consecutive records and must be read starting at the beginning. An *indexed sequential file* is one for which a separate index file (see Sec. 11) is maintained to help locate records.

With a *random (direct access) file structure*, any record can be accessed without starting at the beginning of the file.

11. FILE INDEXING

It is usually inefficient to place the records of an entire file in order. (A good example is a mailing list with thousands of names. It is more efficient to keep the names in the order of entry than to sort the list each time names are added or deleted.) Indexing is a means of specifying the order of the records without actually changing the order of those records.

An index (key or keyword) file is analogous to the index at the end of this book. It is an ordered list of items with references to the complete record. One field in the data record is selected as the key field (record index).[10] The sorted keys are usually kept in a file separate from the data file. One of the standard search techniques is used to find a specific key.

12. SORTING

Sorting routines place data in ascending or descending numerical or alphabetical order.

With the method of *successive minima*, a list is searched sequentially until the smallest element is found and brought to the top of the list. That element is then ignored, and the remaining elements are searched for

the smallest element, which, when found, is placed after previous minimum, and so on. A total of $n(n-1)/2$ comparisons will be required.[11]

key file

key	ref.
ADAMS	3
JONES	2
SMITH	1
THOMAS	4

data file

record	last name	first name	age
1	SMITH	JOHN	27
2	JONES	WANDA	39
3	ADAMS	HENRY	58
4	THOMAS	SUSAN	18

Figure 27.2 *Key and Data Files*

In a *bubble sort*, each element in the list is compared with the element immediately following it. If the first element is larger, the positions of the two elements are reversed (swapped). In effect, the smaller element "bubbles" to the top of the list. The comparisons continue to be made until the bottom of the list is reached. If no swaps are made in a pass, the list is sorted. A total of approximately $n^2/2$ comparisons are needed, on the average, to sort a list in this manner.[12]

In an *insertion sort*, the elements are ordered by rewriting them in the proper sequence. After the proper position of an element is found, all elements below that position are bumped down one place in the sequence. The resulting vacancy is filled by the inserted element. At worst, approximately $n^2/2$ comparisons will be required. On the average, there will be approximately $n^2/4$ comparisons.

Disregarding the number of swaps, the number of comparisons required by the successive minima, bubble, and insertion sorts is on the order of n^2. When n is large, these methods are two slow. The *quicksort* is more complex but reduces the average number of comparisons (with random data) to approximately $n \times \log(n)/\log(2)$, generally considered as being on the order of $n \times \log(n)$.[13] The maximum number of comparisons for a heap sort is $n \times \log(n)/\log(2)$, but it is likely that even fewer comparisons will be needed.

---

[10]More than one field can be indexed. However, each field will require its own index file.
[11]When n is large, $n^2/2$ is sometimes given as the number of comparisons.

[12]This is the same as for the successive minima approach. However, swapping occurs more frequently in the bubble sort, slowing it down.
[13]However, the quicksort falters (in speed) when the elements are in near-perfect order.

13. SEARCHING[14]

If a group of records (i.e., a list) is randomly organized, a particular element in the list can be found only by a linear search (sequential search). At best, only one comparison and, at worst, n comparisons will be required to find something (an event known as a *hit*) in a list of n elements. The average is $n/2$ comparisons, described as being on the order of n.

If the records are in ascending or descending order, a binary search will be superior.[15] The search begins by looking at the middle element in the list. If the middle element is the sought-for element, the search is over. If not, half of the list can be disregarded in further searching since elements in that portion will be either too large or too small. The middle element in the remaining part of the list is investigated, and the procedure continues until a hit occurs or the list is exhausted. The number of required comparisons in a list of n elements will be $\log(n)/\log(2)$ (i.e., on the order of $\log(n)$).

14. HASHING

An index file is not needed if the record number (i.e., the storage location for a read or write operation) can be calculated directly from the key, a technique known as *hashing*.[16] The procedure by which a numeric or nonnumeric key (e.g., a last name) is converted into a record number is called the hashing function or hashing algorithm. Most hashing algorithms use a remaindering modulus—the remainder after dividing the key by the number of records, n, in the list. Excellent results are obtained if n is a prime number; poor results occur if n is a power of 2.

Not all hashed record numbers will be correct. A *collision* occurs when an attempt is made to use a record number that is already in use. Chaining, linear probing, and double hashing are techniques used to resolve such collisions.

15. DATABASE STRUCTURES

Databases can be implemented as indexed files, linked lists, and tree structures; in all three cases, the records are written and remain in the order of entry.

An indexed file such as that shown in Fig. 27.2 keeps the data in one file and maintains separate index files (usually in sorted order) for each key field. The index file must be recreated each time records are added to the file. A *flat file* has only one key field by which records can be located. Searching techniques (see Sec. 13) are used to locate a particular record. In a *linked list*

(*threaded list*), each record has an associated *pointer* (usually a record number or memory address) to the next record in key sequence. Only two pointers are changed when a record is added or deleted. Generally, a linear search following the links through the records is used.

Pointers are also used in *tree structures*. Each record has one or more pointers to other records that meet certain criteria. In a binary tree structure, each record has two pointers—usually to records that are lower and higher, respectively, in key sequence. In general, records in a tree structure are referred to as *nodes*. The first record in the file is called the *root node*. A particular node will have one node above it (the *parent* or *ancestor*) and one or more nodes below it (the *daughters* or *offspring*). Records are found in a tree by starting at the root node and moving sequentially according to the tree structure. The number of comparisons required to find a particular element is $1 + (\log(n)/\log(2))$, which is on the order of $\log(n)$.

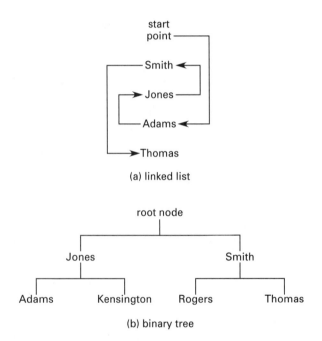

(a) linked list

(b) binary tree

Figure 27.3 *Database Structures*

16. HIERARCHICAL AND RELATIONAL DATA STRUCTURES

A *hierarchical database* contains records in an organized, structured format. Records are organized according to one or more of the indexing schemes. However, each field within a record is not normally accessible.

A *relational database* stores all information in the equivalent of a matrix. Nothing else (no index files, pointers, etc.) is needed to find, read, edit, or select information. Any piece of information can be accessed directly by referring to the field name and field value.

---

[14]The term *probing* is synonymous with searching.

[15]A binary search is unrelated to a binary tree. A binary tree structure (see Sec. 15) greatly reduces search time but does not use a sorted list.

[16]Of course, finding a record in this manner requires it to have been written in a location determined by the same hashing routine.

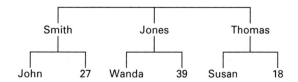

Figure 27.4 *A Hierarchical Personnel File*

rec. no.	last	first	age
1	Smith	John	27
2	Jones	Wanda	39
3	Thomas	Susan	18

Figure 27.5 *A Relational Personnel File*

17. ARTIFICIAL INTELLIGENCE

Artificial intelligence (AI) in a machine implies that the machine is capable of absorbing and organizing new data, learning new concepts, reasoning logically, and responding to inquiries. AI is implemented in a category of programs known as *expert systems* that "learn" rules from sets of events that are entered whenever they occur. (The manner in which the entry is made depends on the particular system.) Once the rules are learned, an expert system can participate in a dialogue to give advice, make predictions and diagnoses, or draw conclusions.

18. HIERARCHY OF OPERATIONS

Operations in an arithmetic statement are performed in the order of exponentiation first, multiplication and division second, and addition and subtraction third. In the event there are two consecutive operations with the same hierarchy (e.g., a multiplication followed by a division), the operations are performed in the order encountered, normally left to right (except for exponentiation, which is right to left).[17] Parentheses can modify this order; operations within parentheses are always evaluated before operations outside. If nested parentheses are present in an expression, the expression is evaluated outward starting from the innermost pair.

Example 27.2

Evaluate J in the following expression. Mixed-mode arithmetic is permitted, and expressions are scanned left to right.

$$J = (6.0 + 3.0)*3.0/6.0 + 5.0 - 6.0**2.0$$

---

[17]In most implementations, a statement will be scanned from left to right. Once a left-to-right scan is complete, some implementations then scan from right to left; others return to the equals sign and start a second left-to-right scan. Parentheses should be used to define the intended order of operations.

Solution

The expression within the parentheses is evaluated first.

$$9.0*3.0/6.0 + 5.0 - 6.0**2.0$$

The exponentiation is performed next.

$$9.0*3.0/6.0 + 5.0 - 36.0$$

The multiplication and division are performed next.

$$4.5 + 5.0 - 36.0$$

The addition and subtraction are performed last.

$$-26.5$$

However, J is an integer variable, and the assignment of $J = -26.5$ results in truncating the fractional part. Ultimately, J has the value of -26.

19. LOGIC GATES

A *gate* performs a logical operation on one or more inputs. The inputs, labeled A, B, C, etc., and output are limited to the values of 0 or 1. A listing of the output value for all possible input values is known as a *truth table*. Table 27.4 combines the symbols, names, and truth tables for the most common gates.

Table 27.4 *Logic Gates*

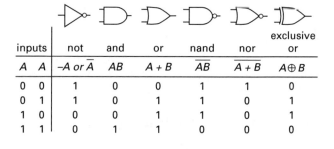

inputs		not	and	or	nand	nor	exclusive or
A	A	$-A$ or $\overline{A}$	AB	$A+B$	$\overline{AB}$	$\overline{A+B}$	$A \oplus B$
0	0	1	0	0	1	1	0
0	1	1	0	1	1	0	1
1	0	0	0	1	1	0	1
1	1	0	1	1	0	0	0

20. BOOLEAN ALGEBRA

The rules of *Boolean Algebra* are used to write and simplify expressions of binary variables (i.e., variables constrained to two values). The basic laws governing Boolean variables are listed here.

commutative:	$A + B = B + A$
	$A \cdot B = B \cdot A$
associative:	$A + (B + C) = (A + B) + C$
	$A \cdot (B \cdot C) = (A \cdot B) \cdot C$
distributive:	$A \cdot (B + C) = (A \cdot B) + (A \cdot C)$
	$A + (B \cdot C) = (A + B) \cdot (A + C)$
absorptive:	$A + (A \cdot B) = A$
	$A \cdot (A + B) = A$

De Morgan's theorems are

$$\overline{(A + B)} = \overline{A} \cdot \overline{B}$$
$$\overline{A \cdot B} = \overline{A} + \overline{B}$$

The following basic identities are used to simplify Boolean expressions.

$$0 + 0 = 0$$
$$0 + 1 = 1$$
$$1 + 0 = 1$$
$$1 + 1 = 1$$

$$0 \cdot 0 = 0$$
$$0 \cdot 1 = 1$$
$$1 \cdot 0 = 0$$
$$1 \cdot 1 = 1$$

$$A + 0 = A$$
$$A + 1 = 1$$
$$A + A = A$$
$$A + \overline{A} = 1$$

$$A \cdot 0 = 0$$
$$A \cdot 1 = A$$
$$A \cdot A = A$$
$$A \cdot \overline{A} = 0$$

$$-0 = 1$$
$$-1 = 0$$
$$-\overline{A} = A$$

Example 27.3

Simplify and write the truth tables for the following network of logic gates.

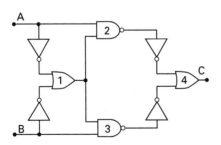

Solution

Determine the inputs and output of each gate in turn.

gate	inputs	output
1	$\overline{A}$, $\overline{B}$	$(\overline{A} + \overline{B})$
2	A, $(\overline{A} + \overline{B})$	$\overline{A \cdot (\overline{A} + \overline{B})}$
3	B, $(\overline{A} + \overline{B})$	$\overline{B \cdot (\overline{A} + \overline{B})}$
4	$A \cdot (\overline{A} + \overline{B})$, $B \cdot (\overline{A} + \overline{B})$	$A \cdot (\overline{A} + \overline{B}) + B \cdot (\overline{A} + \overline{B})$

Simplify the output of gate 4.

$A \cdot (\overline{A} + \overline{B}) + B \cdot (\overline{A} + \overline{B})$	[original]
$A \cdot \overline{A} + A \cdot \overline{B} + B \cdot \overline{A} + B \cdot \overline{B}$	[distributive]
$A \cdot \overline{B} + B \cdot \overline{A}$	[since $A \cdot \overline{A} = 0$]
$A \oplus B$	[definition]

The truth table is

A	B	C
0	0	0
0	1	1
1	0	1
1	1	0

PRACTICE PROBLEMS

1. Flowchart the following procedure: If A is greater than B and A is greater than C, then add B to C and go to PLACE1. If B is greater than A and B is greater than C, then add A and C and go to PLACE2. If B is greater than A but less than C, then subtract A and B and go to PLACE1.

2. Flowchart the following procedure: If A is greater than 10 and if A is less than 14, subtract X from A. If A is less than 10 or if A is greater than 14, exit the program.

3. Draw a flowchart representing an algorithm that finds the real roots of a second-degree equation with real coefficients. The coefficients are to be read from an external peripheral, and the results are to be printed.

4. Draw a flowchart representing the bubble sort algorithm. (Sort into ascending order.)

5. For the binary search procedure, derive the number of required comparisons in a list of n elements.

SOLUTIONS

1.

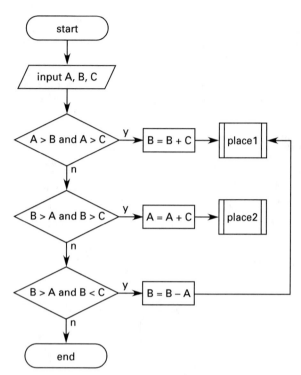

2.

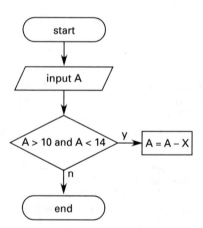

3.

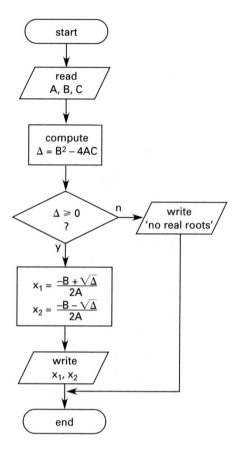

4.

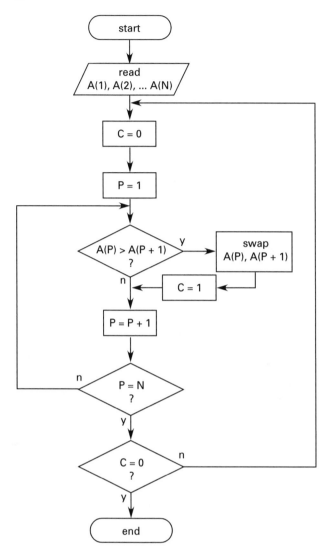

5. To simplify, assume that the number of elements in the list is some integer power of 2.

$$n = 2^p$$

Assume that the sought-for element is the first element in the list. This will produce a worst-case search. Let $a(n)$ be the number of comparisons needed to find the element in the list. Since the binary search procedure discards (disregards) half of the list after each comparison,

$$a(n) = 1 + a\left(\frac{n}{2}\right) \qquad \text{[I]}$$

But, $n = 2^p$.

$$a(n) = a\left(2^p\right) = 1 + a\left(\frac{2^p}{2}\right) = 1 + a\left(2^{p-1}\right) \quad \text{[II]}$$

Equation I can be applied to the second term of Eq. II.

$$a\left(2^{p-1}\right) = 1 + a\left(\frac{2^{p-1}}{2}\right) = 1 + a\left(2^{p-2}\right) \quad \text{[III]}$$

Combining Eqs. II and III,

$$a(n) = a\left(2^p\right) = 1 + 1 + a\left(2^{p-2}\right)$$

Equation I can continue to be applied to the last term until the list contains only two elements, after which the search becomes trivial. This results in a string of p 1's.

$$a(n) = 1 + 1 + 1 + \cdots + 1 = p$$

(An argument for $a(n) = p + 1$ can be made depending on the how the search logic handles a miss with 2 elements in the list. If it merely selects the remaining element without comparing it, then $a(n) = p$. If a singular element is investigated, then $a(n) = p + 1$.)

Now,

$$n = 2^p$$

$$\log_2(n) = \log_2\left(2^p\right) = p$$

But $p = a(n) = \log_2(n)$.

Using logarithm identities,

$$p = \log_2(n) = \frac{\log_{10}(n)}{\log_{10}(2)}$$

APPENDIX A
Tangents and Externals for Horizontal Curves

$$\Delta = \text{interior angle} \qquad T^* = 5729.578 \tan \frac{\Delta}{2} \qquad E^* = 5729.578 \left(\sec \frac{\Delta}{2} - 1 \right)$$

For arc definition, calculate T and E for any degree of curve D by dividing both table values by D.

$$T = \frac{T^*}{D} \qquad E = \frac{E^*}{D}$$

minutes	$\Delta = 0°$ T^*	E^*	$\Delta = 1°$ T^*	E^*	$\Delta = 2°$ T^*	E^*	$\Delta = 3°$ T^*	E^*	$\Delta = 4°$ T^*	E^*	$\Delta = 5°$ T^*	E^*	$\Delta = 6°$ T^*	E^*	minutes
0	0.00	0.00	50.00	0.22	100.01	0.87	150.03	1.96	200.08	3.49	250.16	5.46	300.27	7.86	0
1	0.83	0.00	50.83	0.23	100.84	0.89	150.87	1.99	200.92	3.52	250.99	5.50	301.11	7.91	1
2	1.67	0.00	51.67	0.23	101.68	0.90	151.70	2.01	201.75	3.55	251.83	5.53	301.95	7.95	2
3	2.50	0.00	52.50	0.24	102.51	0.92	152.54	2.03	202.58	3.58	252.66	5.57	302.78	7.99	3
4	3.33	0.00	53.33	0.25	103.34	0.93	153.37	2.05	203.42	3.61	253.50	5.60	303.62	8.04	4
5	4.17	0.00	54.17	0.26	104.18	0.95	154.20	2.08	204.25	3.64	254.33	5.64	304.45	8.08	5
6	5.00	0.00	55.00	0.26	105.01	0.96	155.04	2.10	205.09	3.67	255.17	5.68	305.29	8.13	6
7	5.83	0.00	55.84	0.27	105.85	0.98	155.87	2.12	205.92	3.70	256.00	5.72	306.12	8.17	7
8	6.67	0.00	56.67	0.28	106.68	0.99	156.71	2.14	206.76	3.73	256.84	5.75	306.96	8.22	8
9	7.50	0.00	57.50	0.29	107.51	1.01	157.54	2.17	207.59	3.76	257.67	5.79	307.80	8.26	9
10	8.33	0.01	58.34	0.30	108.35	1.02	158.37	2.19	208.43	3.79	258.51	5.83	308.63	8.31	10
11	9.17	0.01	59.17	0.31	109.18	1.04	159.21	2.21	209.26	3.82	259.34	5.87	309.47	8.35	11
12	10.00	0.01	60.00	0.31	110.01	1.06	160.04	2.23	210.09	3.85	260.18	5.90	310.30	8.40	12
13	10.83	0.01	60.84	0.32	110.85	1.07	160.88	2.26	210.93	3.88	261.01	5.94	311.14	8.44	13
14	11.67	0.01	61.67	0.33	111.68	1.09	161.71	2.28	211.76	3.91	261.85	5.98	311.97	8.49	14
15	12.50	0.01	62.50	0.34	112.51	1.10	162.54	2.30	212.60	3.94	262.68	6.02	312.81	8.53	15
16	13.33	0.02	63.34	0.35	113.35	1.12	163.38	2.33	213.43	3.97	263.52	6.06	313.65	8.58	16
17	14.17	0.02	64.17	0.36	114.18	1.14	164.21	2.35	214.27	4.00	264.35	6.09	314.48	8.62	17
18	15.00	0.02	65.00	0.37	115.02	1.15	165.05	2.38	215.10	4.04	265.19	6.13	315.32	8.67	18
19	15.83	0.02	65.84	0.38	115.85	1.17	165.88	2.40	215.94	4.07	266.02	6.17	316.15	8.72	19
20	16.67	0.02	66.67	0.39	116.68	1.19	166.71	2.42	216.77	4.10	266.86	6.21	316.99	8.76	20
21	17.50	0.03	67.50	0.40	117.52	1.20	167.55	2.45	217.60	4.13	267.69	6.25	317.83	8.81	21
22	18.33	0.03	68.34	0.41	118.35	1.22	168.38	2.47	218.44	4.16	268.53	6.29	318.66	8.85	22
23	19.17	0.03	69.17	0.42	119.18	1.24	169.22	2.50	219.27	4.19	269.36	6.33	319.50	8.90	23
24	20.00	0.03	70.00	0.43	120.02	1.26	170.05	2.52	220.11	4.23	270.20	6.37	320.33	8.95	24
25	20.83	0.04	70.84	0.44	120.85	1.27	170.88	2.55	220.94	4.26	271.04	6.41	321.17	8.99	25
26	21.67	0.04	71.67	0.45	121.68	1.29	171.72	2.57	221.78	4.29	271.87	6.45	322.01	9.04	26
27	22.50	0.04	72.50	0.46	122.52	1.31	172.55	2.60	222.61	4.32	272.71	6.49	322.84	9.09	27
28	23.33	0.05	73.34	0.47	123.35	1.33	173.39	2.62	223.45	4.36	273.54	6.53	323.68	9.14	28
29	24.17	0.05	74.17	0.48	124.19	1.34	174.22	2.65	224.28	4.39	274.38	6.57	324.51	9.18	29
30	25.00	0.05	75.00	0.49	125.02	1.36	175.05	2.67	225.12	4.42	275.21	6.61	325.35	9.23	30
31	25.83	0.06	75.84	0.50	125.85	1.38	175.89	2.70	225.95	4.45	276.05	6.65	326.19	9.28	31
32	26.67	0.06	76.67	0.51	126.69	1.40	176.72	2.72	226.78	4.49	276.88	6.69	327.02	9.32	32
33	27.50	0.07	77.50	0.52	127.52	1.42	177.56	2.75	227.62	4.52	277.72	6.73	327.86	9.37	33
34	28.33	0.07	78.34	0.54	128.35	1.44	178.39	2.78	228.45	4.55	278.55	6.77	328.69	9.42	34
35	29.17	0.07	79.17	0.55	129.19	1.46	179.23	2.80	229.29	4.59	279.39	6.81	329.53	9.47	35
36	30.00	0.08	80.01	0.56	130.02	1.48	180.06	2.83	230.12	4.62	280.22	6.85	330.37	9.52	36
37	30.83	0.08	80.84	0.57	130.86	1.49	180.89	2.86	230.96	4.65	281.06	6.89	331.20	9.56	37
38	31.67	0.09	81.67	0.58	131.69	1.51	181.73	2.88	231.79	4.69	281.89	6.93	332.04	9.61	38
39	32.50	0.09	82.51	0.59	132.52	1.53	182.56	2.91	232.63	4.72	282.73	6.97	332.87	9.66	39
40	33.33	0.10	83.34	0.61	133.36	1.55	183.40	2.93	233.46	4.75	283.56	7.01	333.71	9.71	40
41	34.17	0.10	84.17	0.62	134.19	1.57	184.23	2.96	234.30	4.79	284.40	7.05	334.55	9.76	41
42	35.00	0.11	85.01	0.63	135.02	1.59	185.06	2.99	235.13	4.82	285.24	7.09	335.38	9.81	42
43	35.83	0.11	85.84	0.64	135.86	1.61	185.90	3.01	235.97	4.86	286.07	7.14	336.22	9.86	43
44	36.67	0.12	86.67	0.66	136.69	1.63	186.73	3.04	236.80	4.89	286.91	7.18	337.05	9.91	44
45	37.50	0.12	87.51	0.67	137.53	1.65	187.57	3.07	237.64	4.93	287.74	7.22	337.89	9.95	45
46	38.33	0.13	88.34	0.68	138.36	1.67	188.40	3.10	238.47	4.96	288.58	7.26	338.73	10.00	46
47	39.17	0.13	89.17	0.69	139.19	1.69	189.24	3.12	239.31	4.99	289.41	7.30	339.56	10.05	47
48	40.00	0.14	90.01	0.71	140.03	1.71	190.07	3.15	240.14	5.03	290.25	7.35	340.40	10.10	48
49	40.83	0.15	90.84	0.72	140.86	1.73	190.90	3.18	240.98	5.07	291.08	7.39	341.24	10.15	49
50	41.67	0.15	91.67	0.73	141.70	1.75	191.74	3.21	241.81	5.10	291.92	7.43	342.07	10.20	50
51	42.50	0.16	92.51	0.75	142.53	1.77	192.57	3.23	242.64	5.13	292.75	7.47	342.91	10.25	51
52	43.33	0.16	93.34	0.76	143.36	1.79	193.41	3.26	243.48	5.17	293.59	7.52	343.74	10.30	52
53	44.17	0.17	94.18	0.77	144.20	1.81	194.24	3.29	244.31	5.21	294.43	7.56	344.58	10.35	53
54	45.00	0.18	95.01	0.79	145.03	1.84	195.08	3.32	245.15	5.24	295.26	7.60	345.42	10.40	54
55	45.83	0.18	95.84	0.80	145.86	1.86	195.91	3.35	245.98	5.28	296.10	7.65	346.25	10.45	55
56	46.67	0.19	96.68	0.82	146.70	1.88	196.74	3.38	246.82	5.31	296.93	7.69	347.09	10.50	56
57	47.50	0.20	97.51	0.83	147.53	1.90	197.58	3.41	247.65	5.35	297.77	7.73	347.93	10.55	57
58	48.33	0.20	98.34	0.84	148.37	1.92	198.41	3.43	248.49	5.39	298.60	7.78	348.76	10.61	58
59	49.17	0.21	99.18	0.86	149.20	1.94	199.25	3.46	249.32	5.42	299.44	7.82	349.60	10.66	59

(continued)

APPENDIX A *(continued)*
Tangents and Externals for Horizontal Curves

$$\Delta = \text{interior angle} \qquad T^* = 5729.578 \tan\frac{\Delta}{2} \qquad E^* = 5729.578\left(\sec\frac{\Delta}{2} - 1\right)$$

For arc definition, calculate T and E for any degree of curve D by dividing both table values by D.

$$T = \frac{T^*}{D} \qquad E = \frac{E^*}{D}$$

minutes	$\Delta = 7°$ T^*	E^*	$\Delta = 8°$ T^*	E^*	$\Delta = 9°$ T^*	E^*	$\Delta = 10°$ T^*	E^*	$\Delta = 11°$ T^*	E^*	$\Delta = 12°$ T^*	E^*	$\Delta = 13°$ T^*	E^*	minutes
0	350.44	10.71	400.65	13.99	450.93	17.72	501.27	21.89	551.70	26.50	602.20	31.56	652.80	37.07	0
1	351.27	10.76	401.49	14.05	451.77	17.78	502.11	21.96	552.54	26.58	603.05	31.65	653.65	37.16	1
2	352.11	10.81	402.33	14.11	452.60	17.85	502.95	22.03	553.38	26.66	603.89	31.74	654.49	37.26	2
3	352.95	10.86	403.16	14.17	453.44	17.91	503.79	22.11	554.22	26.74	604.73	31.82	655.34	37.36	3
4	353.78	10.91	404.00	14.23	454.28	17.98	504.63	22.18	555.06	26.82	605.57	31.91	656.18	37.45	4
5	354.62	10.96	404.84	14.28	455.12	18.05	505.47	22.25	555.90	26.90	606.42	32.00	657.02	37.55	5
6	355.45	11.02	405.68	14.34	455.96	18.11	506.31	22.33	556.74	26.99	607.26	32.09	657.87	37.64	6
7	356.29	11.07	406.51	14.40	456.80	18.18	507.15	22.40	557.58	27.07	608.10	32.18	658.71	37.74	7
8	357.13	11.12	407.35	14.46	457.64	18.25	507.99	22.48	558.42	27.15	608.94	32.27	659.56	37.84	8
9	357.96	11.17	408.19	14.52	458.47	18.31	508.83	22.55	559.27	27.23	609.79	32.36	660.40	37.93	9
10	358.80	11.22	409.03	14.58	459.31	18.38	509.67	22.62	560.11	27.31	610.63	32.45	661.25	38.03	10
11	359.64	11.28	409.86	14.64	460.15	18.45	510.51	22.70	560.95	27.39	611.47	32.54	662.09	38.13	11
12	360.47	11.33	410.70	14.70	460.99	18.51	511.35	22.77	561.79	27.48	612.32	32.63	662.93	38.22	12
13	361.31	11.38	411.54	14.76	461.83	18.58	512.19	22.85	562.63	27.56	613.16	32.72	663.78	38.32	13
14	362.15	11.43	412.38	14.82	462.67	18.65	513.03	22.92	563.47	27.64	614.00	32.80	664.62	38.42	14
15	362.98	11.49	413.21	14.88	463.51	18.72	513.87	23.00	564.31	27.72	614.84	32.89	665.47	38.52	15
16	363.82	11.54	414.05	14.94	464.35	18.79	514.71	23.07	565.16	27.81	615.69	32.98	666.31	38.61	16
17	364.66	11.59	414.89	15.00	465.18	18.85	515.55	23.15	566.00	27.89	616.53	33.08	667.16	38.71	17
18	365.49	11.65	415.73	15.06	466.02	18.92	516.39	23.22	566.84	27.97	617.37	33.17	668.00	38.81	18
19	366.33	11.70	416.56	15.12	466.86	18.99	517.23	23.30	567.68	28.05	618.22	33.26	668.85	38.91	19
20	367.17	11.75	417.40	15.18	467.70	19.06	518.07	23.37	568.52	28.14	619.06	33.35	669.69	39.00	20
21	368.00	11.81	418.24	15.24	468.54	19.13	518.91	23.45	569.36	28.22	619.90	33.44	670.54	39.10	21
22	368.84	11.86	419.08	15.31	469.38	19.19	519.75	23.53	570.20	28.30	620.75	33.53	671.38	39.20	22
23	369.68	11.91	419.92	15.37	470.22	19.26	520.59	23.60	571.05	28.39	621.59	33.62	672.23	39.30	23
24	370.52	11.97	420.75	15.43	471.06	19.33	521.43	23.68	571.89	28.47	622.43	33.71	673.07	39.40	24
25	371.35	12.02	421.59	15.49	471.90	19.40	522.27	23.75	572.73	28.55	623.27	33.80	673.92	39.50	25
26	372.19	12.08	422.43	15.55	472.73	19.47	523.11	23.83	573.57	28.64	624.12	33.89	674.76	39.60	26
27	373.03	12.13	423.27	15.61	473.57	19.54	523.95	23.91	574.41	28.72	624.96	33.98	675.61	39.69	27
28	373.86	12.18	424.11	15.67	474.41	19.61	524.79	23.98	575.25	28.81	625.80	34.08	676.45	39.79	28
29	374.70	12.24	424.94	15.74	475.25	19.68	525.63	24.06	576.10	28.89	626.65	34.17	677.30	39.89	29
30	375.54	12.29	425.78	15.80	476.09	19.75	526.47	24.14	576.94	28.97	627.49	34.26	678.14	39.99	30
31	376.37	12.35	426.62	15.86	476.93	19.82	527.31	24.21	577.78	29.06	628.33	34.35	678.99	40.09	31
32	377.21	12.40	427.46	15.92	477.77	19.89	528.16	24.29	578.62	29.14	629.18	34.44	679.83	40.19	32
33	378.05	12.46	428.30	15.99	478.61	19.96	529.00	24.37	579.46	29.23	630.02	34.52	680.68	40.29	33
34	378.88	12.51	429.13	16.05	479.45	20.02	529.84	24.45	580.31	29.31	630.86	34.63	681.52	40.39	34
35	379.72	12.57	429.97	16.11	480.29	20.10	530.68	24.52	581.15	29.40	631.71	34.72	682.37	40.49	35
36	380.56	12.62	430.81	16.17	481.13	20.17	531.52	24.60	581.99	29.48	632.55	34.81	683.21	40.59	36
37	381.40	12.68	431.65	16.24	481.97	20.24	532.36	24.68	582.83	29.57	633.39	34.90	684.06	40.69	37
38	382.23	12.74	432.49	16.30	482.80	20.31	533.20	24.76	583.67	29.65	634.24	35.00	684.90	40.79	38
39	383.07	12.79	433.32	16.36	483.64	20.38	534.04	24.83	584.52	29.74	635.08	35.09	685.75	40.89	39
40	383.91	12.85	434.16	16.43	484.48	20.45	534.88	24.91	585.36	29.82	635.93	35.18	686.59	40.99	40
41	384.74	12.90	435.00	16.49	485.32	20.52	535.72	24.99	586.20	29.91	636.77	35.28	687.44	41.09	41
42	385.58	12.96	435.84	16.55	486.16	20.59	536.56	25.07	587.04	29.99	637.61	35.37	688.28	41.19	42
43	386.42	13.02	436.68	16.62	487.00	20.66	537.40	25.15	587.88	30.08	638.46	35.46	689.13	41.29	43
44	387.25	13.07	437.51	16.68	487.84	20.73	538.24	25.23	588.73	30.17	639.30	35.56	689.97	41.39	44
45	388.09	13.13	438.35	16.74	488.68	20.80	539.08	25.30	589.57	30.25	640.14	35.65	690.82	41.50	45
46	388.93	13.18	439.19	16.81	489.52	20.87	539.92	25.38	590.41	30.34	640.99	35.74	691.66	41.60	46
47	389.77	13.24	440.03	16.87	490.36	20.94	540.76	25.46	591.25	30.43	641.83	35.84	692.51	41.70	47
48	390.60	13.30	440.87	16.94	491.20	21.02	541.60	25.54	592.09	30.51	642.68	35.93	693.36	41.80	48
49	391.44	13.36	441.71	17.00	492.04	21.09	542.45	25.62	592.94	30.60	643.52	36.03	694.20	41.90	49
50	392.28	13.41	442.54	17.06	492.88	21.16	543.29	25.70	593.78	30.69	644.36	36.12	695.05	42.00	50
51	393.12	13.47	443.38	17.13	493.72	21.23	544.13	25.78	594.62	30.77	645.21	36.21	695.89	42.11	51
52	393.95	13.53	444.22	17.19	494.56	21.30	544.97	25.86	595.46	30.86	646.05	36.31	696.74	42.21	52
53	394.79	13.59	445.06	17.26	495.40	21.38	545.81	25.94	596.31	30.95	646.89	36.40	697.58	42.31	53
54	395.63	13.64	445.90	17.32	496.24	21.45	546.65	26.02	597.15	31.03	647.74	36.50	698.43	42.41	54
55	396.46	13.70	446.74	17.39	497.07	21.52	547.49	26.10	597.99	31.12	648.58	36.59	699.27	42.51	55
56	397.30	13.76	447.57	17.46	497.91	21.59	548.33	26.18	598.83	31.21	649.43	36.69	700.12	42.62	56
57	398.14	13.82	448.41	17.52	498.75	21.67	549.17	26.26	599.68	31.30	650.27	36.78	700.97	42.72	57
58	398.98	13.87	449.25	17.59	499.59	21.74	550.01	26.34	600.52	31.38	651.11	36.88	701.81	42.82	58
59	399.81	13.93	450.09	17.65	500.43	21.81	550.85	26.42	601.36	31.47	651.96	36.97	702.66	42.92	59

(continued)

APPENDIX A *(continued)*
Tangents and Externals for Horizontal Curves

$$\Delta = \text{interior angle} \qquad T^* = 5729.578 \tan \frac{\Delta}{2} \qquad E^* = 5729.578 \left(\sec \frac{\Delta}{2} - 1 \right)$$

For arc definition, calculate T and E for any degree of curve D by dividing both table values by D.

$$T = \frac{T^*}{D} \qquad E = \frac{E^*}{D}$$

minutes	$\Delta = 14°$ T^*	E^*	$\Delta = 15°$ T^*	E^*	$\Delta = 16°$ T^*	E^*	$\Delta = 17°$ T^*	E^*	$\Delta = 18°$ T^*	E^*	$\Delta = 19°$ T^*	E^*	$\Delta = 20°$ T^*	E^*	minutes
0	703.50	43.03	754.31	49.44	805.24	56.31	856.29	63.63	907.48	71.42	958.80	79.67	1010.28	88.39	0
1	704.35	43.13	755.16	49.55	806.09	56.43	857.14	63.76	908.33	71.55	959.66	79.81	1011.14	88.54	1
2	705.20	43.23	756.01	49.66	806.94	56.55	858.00	63.88	909.18	71.69	960.52	79.95	1012.00	88.69	2
3	706.04	43.34	756.86	49.77	807.79	56.66	858.85	64.01	910.04	71.82	961.37	80.09	1012.86	88.84	3
4	706.89	43.44	757.70	49.88	808.64	56.78	859.70	64.14	910.89	71.96	962.23	80.24	1013.72	88.99	4
5	707.73	43.55	758.55	49.99	809.49	56.90	860.55	64.26	911.75	72.09	963.09	80.38	1014.58	89.14	5
6	708.58	43.65	759.40	50.11	810.34	57.02	861.40	64.39	912.60	72.22	963.94	80.52	1015.44	89.29	6
7	709.43	43.75	760.25	50.22	811.19	57.14	862.26	64.52	913.46	72.36	964.80	80.66	1016.29	89.44	7
8	710.27	43.86	761.10	50.33	812.04	57.26	863.11	64.64	914.31	72.49	965.66	80.81	1017.15	89.59	8
9	711.12	43.96	761.94	50.44	812.89	57.38	863.96	64.77	915.17	72.63	966.51	80.95	1018.01	89.74	9
10	711.96	44.07	762.79	50.55	813.74	57.50	864.81	64.90	916.02	72.76	967.37	81.09	1018.87	89.89	10
11	712.81	44.17	763.64	50.67	814.59	57.62	865.66	65.03	916.88	72.90	968.23	81.23	1019.73	90.04	11
12	713.66	44.27	764.49	50.78	815.44	57.74	866.52	65.15	917.73	73.03	969.09	81.38	1020.59	90.19	12
13	714.50	44.38	765.34	50.89	816.29	57.86	867.37	65.28	918.58	73.17	969.94	81.52	1021.45	90.34	13
14	715.35	44.48	766.19	51.00	817.14	57.98	868.22	65.41	919.44	73.30	970.80	81.66	1022.31	90.49	14
15	716.20	44.59	767.03	51.11	817.99	58.10	869.07	65.54	920.29	73.44	971.66	81.81	1023.17	90.64	15
16	717.04	44.69	767.88	51.23	818.84	58.22	869.93	65.66	921.15	73.57	972.51	81.95	1024.03	90.79	16
17	717.89	44.80	768.73	51.34	819.69	58.34	870.78	65.79	922.00	73.71	973.37	82.09	1024.89	90.94	17
18	718.73	44.90	769.58	51.45	820.54	58.46	871.63	65.92	922.86	73.85	974.23	82.24	1025.75	91.09	18
19	719.58	45.01	770.43	51.57	821.39	58.58	872.48	66.05	923.71	73.98	975.09	82.38	1026.61	91.25	19
20	720.43	45.11	771.28	51.68	822.24	58.70	873.34	66.18	924.57	74.12	975.94	82.52	1027.47	91.40	20
21	721.27	45.22	772.12	51.79	823.09	58.82	874.19	66.31	925.42	74.25	976.80	82.67	1028.33	91.55	21
22	722.12	45.33	772.97	51.91	823.94	58.94	875.04	66.43	926.28	74.39	977.66	82.81	1029.19	91.70	22
23	722.97	45.43	773.82	52.02	824.79	59.06	875.90	66.56	927.13	74.53	978.52	82.96	1030.05	91.85	23
24	723.81	45.54	774.67	52.13	825.64	59.18	876.75	66.69	927.99	74.66	979.37	83.10	1030.91	92.01	24
25	724.66	45.64	775.52	52.25	826.50	59.30	877.60	66.82	928.84	74.80	980.23	83.25	1031.77	92.16	25
26	725.51	45.75	776.37	52.36	827.35	59.43	878.45	66.95	929.70	74.94	981.09	83.39	1032.63	92.31	26
27	726.35	45.86	777.22	52.47	828.20	59.55	879.31	67.08	930.56	75.08	981.95	83.53	1033.49	92.46	27
28	727.20	45.96	778.06	52.59	829.05	59.67	880.16	67.21	931.41	75.21	982.81	83.68	1034.36	92.62	28
29	728.05	46.07	778.91	52.70	829.90	59.79	881.01	67.34	932.27	75.35	983.66	83.83	1035.22	92.77	29
30	728.89	46.18	779.76	52.82	830.75	59.91	881.87	67.47	933.12	75.49	984.52	83.97	1036.08	92.92	30
31	729.74	46.28	780.61	52.93	831.60	60.03	882.72	67.60	933.98	75.62	985.38	84.12	1036.94	93.08	31
32	730.59	46.39	781.46	53.05	832.45	60.16	883.57	67.73	934.83	75.76	986.24	84.26	1037.80	93.23	32
33	731.44	46.50	782.31	53.16	833.30	60.28	884.43	67.86	935.69	75.90	987.10	84.41	1038.66	93.38	33
34	732.28	46.61	783.16	53.28	834.15	60.40	885.28	67.99	936.54	76.04	987.95	84.55	1039.52	93.54	34
35	733.13	46.71	784.01	53.39	835.00	60.53	886.13	68.12	937.40	76.18	988.81	84.70	1040.38	93.69	35
36	733.98	46.82	784.85	53.51	835.86	60.65	886.98	68.25	938.25	76.31	989.67	84.84	1041.24	93.84	36
37	734.82	46.93	785.70	53.62	836.71	60.77	887.84	68.38	939.11	76.45	990.53	84.99	1042.10	94.00	37
38	735.67	47.04	786.55	53.74	837.56	60.89	888.69	68.51	939.97	76.59	991.39	85.14	1042.96	94.15	38
39	736.52	47.14	787.40	53.85	838.41	61.02	889.55	68.64	940.82	76.73	992.24	85.28	1043.82	94.31	39
40	737.36	47.25	788.25	53.97	839.26	61.14	890.40	68.77	941.68	76.87	993.10	85.43	1044.68	94.46	40
41	738.21	47.36	789.10	54.08	840.11	61.26	891.25	68.90	942.53	77.01	993.96	85.58	1045.55	94.62	41
42	739.06	47.47	789.95	54.20	840.96	61.39	892.11	69.03	943.39	77.15	994.82	85.72	1046.41	94.77	42
43	739.91	47.58	790.80	54.32	841.81	61.51	892.96	69.17	944.25	77.28	995.68	85.87	1047.27	94.93	43
44	740.75	47.69	791.65	54.43	842.66	61.63	893.81	69.30	945.10	77.42	996.54	86.02	1048.13	95.08	44
45	741.60	47.79	792.50	54.55	843.52	61.76	894.67	69.43	945.96	77.56	997.40	86.16	1048.99	95.23	45
46	742.45	47.90	793.35	54.66	844.37	61.88	895.52	69.56	946.81	77.70	998.25	86.31	1049.85	95.39	46
47	743.29	48.01	794.20	54.78	845.22	62.01	896.37	69.69	947.67	77.84	999.11	86.46	1050.71	95.54	47
48	744.14	48.12	795.04	54.90	846.07	62.13	897.23	69.83	948.53	77.98	999.97	86.61	1051.57	95.70	48
49	744.99	48.23	795.89	55.01	846.92	62.26	898.08	69.96	949.38	78.12	1000.83	86.75	1052.44	95.86	49
50	745.84	48.34	796.74	55.13	847.77	62.38	898.94	70.09	950.24	78.26	1001.69	86.90	1053.30	96.01	50
51	746.68	48.45	797.59	55.25	848.63	62.51	899.79	70.22	951.09	78.40	1002.55	87.05	1054.16	96.17	51
52	747.53	48.56	798.44	55.37	849.48	62.63	900.64	70.35	951.95	78.54	1003.41	87.20	1055.02	96.32	52
53	748.38	48.67	799.29	55.48	850.33	62.76	901.50	70.49	952.81	78.68	1004.27	87.35	1055.88	96.48	53
54	749.23	48.78	800.14	55.60	851.18	62.88	902.35	70.62	953.66	78.82	1005.12	87.49	1056.74	96.64	54
55	750.07	48.89	800.99	55.72	852.03	63.01	903.21	70.75	954.52	78.97	1005.98	87.64	1057.61	96.79	55
56	750.92	49.00	801.84	55.84	852.88	63.13	904.06	70.89	955.38	79.11	1006.84	87.79	1058.47	96.95	56
57	751.77	49.11	802.69	55.95	853.74	63.26	904.91	71.02	956.23	79.25	1007.70	87.94	1059.33	97.11	57
58	752.62	49.22	803.54	56.07	854.59	63.38	905.77	71.15	957.09	79.39	1008.56	88.09	1060.19	97.26	58
59	753.47	49.33	804.39	56.19	855.44	63.51	906.62	71.29	957.95	79.53	1009.42	88.24	1061.05	97.42	59

(continued)

Appendices

APPENDIX A *(continued)*
Tangents and Externals for Horizontal Curves

$$\Delta = \text{interior angle} \qquad T^* = 5729.578 \tan \frac{\Delta}{2} \qquad E^* = 5729.578 \left(\sec \frac{\Delta}{2} - 1 \right)$$

For arc definition, calculate T and E for any degree of curve D by dividing both table values by D.

$$T = \frac{T^*}{D} \qquad E = \frac{E^*}{D}$$

minutes	$\Delta = 21°$ T^*	E^*	$\Delta = 22°$ T^*	E^*	$\Delta = 23°$ T^*	E^*	$\Delta = 24°$ T^*	E^*	$\Delta = 25°$ T^*	E^*	$\Delta = 26°$ T^*	E^*	$\Delta = 27°$ T^*	E^*	minutes
0	1061.91	97.58	1113.72	107.24	1165.70	117.38	1217.86	128.00	1270.22	139.11	1322.78	150.71	1375.55	162.81	0
1	1062.78	97.73	1114.58	107.40	1166.56	117.55	1218.73	128.18	1271.09	139.30	1323.66	150.91	1376.43	163.01	1
2	1063.64	97.89	1115.45	107.57	1167.43	117.73	1219.60	128.36	1271.97	139.49	1324.53	151.11	1377.31	163.22	2
3	1064.50	98.05	1116.31	107.73	1168.30	117.90	1220.47	128.55	1272.84	139.68	1325.41	151.30	1378.19	163.42	3
4	1065.36	98.21	1117.18	107.90	1169.17	118.07	1221.34	128.73	1273.71	139.87	1326.29	151.50	1379.08	163.63	4
5	1066.22	98.36	1118.04	108.07	1170.04	118.25	1222.22	128.91	1274.59	140.06	1327.17	151.70	1379.96	163.84	5
6	1067.09	98.52	1118.91	108.23	1170.90	118.42	1223.09	129.09	1275.46	140.25	1328.04	151.90	1380.84	164.04	6
7	1067.95	98.68	1119.77	108.40	1171.77	118.59	1223.96	129.27	1276.34	140.44	1328.92	152.10	1381.72	164.25	7
8	1068.81	98.84	1120.64	108.56	1172.64	118.77	1224.83	129.45	1277.21	140.63	1329.80	152.30	1382.60	164.46	8
9	1069.67	99.00	1121.50	108.73	1173.51	118.94	1225.70	129.64	1278.09	140.82	1330.68	152.49	1383.48	164.66	9
10	1070.54	99.15	1122.37	108.90	1174.38	119.12	1226.57	129.82	1278.96	141.01	1331.56	152.69	1384.37	164.87	10
11	1071.40	99.31	1123.23	109.06	1175.25	119.29	1227.44	130.00	1279.84	141.20	1332.44	152.89	1385.25	165.08	11
12	1072.26	99.47	1124.10	109.23	1176.11	119.46	1228.32	130.18	1280.71	141.39	1333.31	153.09	1386.13	165.29	12
13	1073.12	99.63	1124.96	109.40	1176.98	119.64	1229.19	130.37	1281.59	141.58	1334.19	153.29	1387.01	165.49	13
14	1073.99	99.79	1125.83	109.56	1177.85	119.81	1230.06	130.55	1282.46	141.77	1335.07	153.49	1387.90	165.70	14
15	1074.85	99.95	1126.69	109.73	1178.72	119.99	1230.93	130.73	1283.34	141.96	1335.95	153.69	1388.78	165.91	15
16	1075.71	100.11	1127.56	109.90	1179.59	120.16	1231.80	130.92	1284.21	142.16	1336.83	153.89	1389.66	166.12	16
17	1076.57	100.27	1128.43	110.06	1180.46	120.34	1232.67	131.10	1285.09	142.35	1337.71	154.09	1390.54	166.33	17
18	1077.44	100.42	1129.29	110.23	1181.33	120.52	1233.55	131.28	1285.96	142.54	1338.59	154.29	1391.42	166.53	18
19	1078.30	100.58	1130.16	110.40	1182.19	120.69	1234.42	131.47	1286.84	142.73	1339.47	154.49	1392.31	166.74	19
20	1079.16	100.74	1131.02	110.57	1183.06	120.87	1235.29	131.65	1287.71	142.92	1340.34	154.69	1393.19	166.95	20
21	1080.03	100.90	1131.89	110.73	1183.93	121.04	1236.16	131.83	1288.59	143.12	1341.22	154.89	1394.07	167.16	21
22	1080.89	101.06	1132.76	110.90	1184.80	121.22	1237.03	132.02	1289.47	143.31	1342.10	155.09	1394.96	167.37	22
23	1081.75	101.22	1133.62	111.07	1185.67	121.39	1237.91	132.20	1290.34	143.50	1342.98	155.29	1395.84	167.58	23
24	1082.61	101.38	1134.49	111.24	1186.54	121.57	1238.78	132.39	1291.22	143.69	1343.86	155.49	1396.72	167.79	24
25	1083.48	101.54	1135.35	111.41	1187.41	121.75	1239.65	132.57	1292.09	143.88	1344.74	155.69	1397.60	167.99	25
26	1084.34	101.70	1136.22	111.57	1188.28	121.92	1240.52	132.76	1292.97	144.08	1345.62	155.89	1398.49	168.20	26
27	1085.20	101.87	1137.09	111.74	1189.15	122.10	1241.40	132.94	1293.84	144.27	1346.50	156.09	1399.37	168.41	27
28	1086.07	102.03	1137.95	111.91	1190.02	122.28	1242.27	133.13	1294.72	144.46	1347.38	156.29	1400.25	168.62	28
29	1086.93	102.19	1138.82	112.08	1190.88	122.45	1243.14	133.31	1295.60	144.66	1348.26	156.50	1401.14	168.83	29
30	1087.79	102.35	1139.68	112.25	1191.75	122.63	1244.01	133.50	1296.47	144.85	1349.14	156.70	1402.02	169.04	30
31	1088.66	102.51	1140.55	112.42	1192.62	122.81	1244.89	133.68	1297.35	145.04	1350.02	156.90	1402.90	169.25	31
32	1089.52	102.67	1141.42	112.59	1193.49	122.98	1245.76	133.87	1298.22	145.24	1350.90	157.10	1403.79	169.46	32
33	1090.38	102.83	1142.28	112.76	1194.36	123.16	1246.63	134.05	1299.10	145.43	1351.78	157.30	1404.67	169.67	33
34	1091.25	102.99	1143.15	112.93	1195.23	123.34	1247.50	134.24	1299.98	145.62	1352.66	157.50	1405.55	169.88	34
35	1092.11	103.15	1144.02	113.10	1196.10	123.52	1248.38	134.42	1300.85	145.82	1353.53	157.71	1406.44	170.09	35
36	1092.98	103.32	1144.88	113.27	1196.97	123.69	1249.25	134.61	1301.73	146.01	1354.41	157.91	1407.32	170.30	36
37	1093.84	103.48	1145.75	113.44	1197.84	123.87	1250.12	134.80	1302.60	146.21	1355.29	158.11	1408.20	170.51	37
38	1094.70	103.64	1146.62	113.61	1198.71	124.05	1251.00	134.98	1303.48	146.40	1356.17	158.31	1409.09	170.73	38
39	1095.57	103.80	1147.48	113.78	1199.58	124.23	1251.87	135.17	1304.36	146.59	1357.05	158.52	1409.97	170.94	39
40	1096.43	103.97	1148.35	113.95	1200.45	124.41	1252.74	135.35	1305.23	146.79	1357.93	158.72	1410.85	171.15	40
41	1097.29	104.13	1149.22	114.12	1201.32	124.59	1253.62	135.54	1306.11	146.98	1358.81	158.92	1411.74	171.36	41
42	1098.16	104.29	1150.08	114.29	1202.19	124.76	1254.49	135.73	1306.99	147.18	1359.70	159.13	1412.62	171.57	42
43	1099.02	104.45	1150.95	114.46	1203.06	124.94	1255.36	135.91	1307.86	147.37	1360.58	159.33	1413.51	171.78	43
44	1099.89	104.62	1151.82	114.63	1203.93	125.12	1256.24	136.10	1308.74	147.57	1361.46	159.53	1414.39	171.99	44
45	1100.75	104.78	1152.68	114.80	1204.80	125.30	1257.11	136.29	1309.62	147.76	1362.34	159.74	1415.27	172.21	45
46	1101.61	104.94	1153.55	114.97	1205.67	125.48	1257.98	136.48	1310.49	147.96	1363.22	159.94	1416.16	172.42	46
47	1102.48	105.10	1154.42	115.14	1206.54	125.66	1258.86	136.66	1311.37	148.16	1364.10	160.14	1417.04	172.63	47
48	1103.34	105.27	1155.29	115.31	1207.41	125.84	1259.73	136.85	1312.25	148.35	1364.98	160.35	1417.93	172.84	48
49	1104.21	105.43	1156.15	115.48	1208.28	126.02	1260.60	137.04	1313.13	148.55	1365.86	160.55	1418.81	173.06	49
50	1105.07	105.60	1157.02	115.66	1209.15	126.20	1261.48	137.23	1314.00	148.74	1366.74	160.76	1419.70	173.27	50
51	1105.94	105.76	1157.89	115.83	1210.02	126.38	1262.35	137.41	1314.88	148.94	1367.62	160.96	1420.58	173.48	51
52	1106.80	105.92	1158.76	116.00	1210.89	126.56	1263.22	137.60	1315.76	149.14	1368.50	161.17	1421.47	173.69	52
53	1107.66	106.09	1159.62	116.17	1211.76	126.74	1264.10	137.79	1316.63	149.33	1369.38	161.37	1422.35	173.91	53
54	1108.53	106.25	1160.49	116.34	1212.63	126.92	1264.97	137.98	1317.51	149.53	1370.26	161.57	1423.23	174.12	54
55	1109.39	106.41	1161.36	116.52	1213.51	127.10	1265.85	138.17	1318.39	149.73	1371.14	161.78	1424.12	174.33	55
56	1110.26	106.58	1162.22	116.69	1214.38	127.28	1266.72	138.36	1319.27	149.92	1372.02	161.99	1425.00	174.55	56
57	1111.12	106.74	1163.09	116.86	1215.25	127.46	1267.59	138.54	1320.14	150.12	1372.91	162.19	1425.89	174.76	57
58	1111.99	106.91	1163.96	117.03	1216.12	127.64	1268.47	138.73	1321.02	150.32	1373.79	162.40	1426.77	174.98	58
59	1112.85	107.07	1164.83	117.21	1216.99	127.82	1269.34	138.92	1321.90	150.51	1374.67	162.60	1427.66	175.19	59

(continued)

APPENDIX A *(continued)*
Tangents and Externals for Horizontal Curves

$$\Delta = \text{interior angle} \qquad T^* = 5729.578 \tan \frac{\Delta}{2} \qquad E^* = 5729.578 \left(\sec \frac{\Delta}{2} - 1 \right)$$

For arc definition, calculate T and E for any degree of curve D by dividing both table values by D.

$$T = \frac{T^*}{D} \qquad E = \frac{E^*}{D}$$

minutes	$\Delta = 28°$ T^*	E^*	$\Delta = 29°$ T^*	E^*	$\Delta = 30°$ T^*	E^*	$\Delta = 31°$ T^*	E^*	$\Delta = 32°$ T^*	E^*	$\Delta = 33°$ T^*	E^*	$\Delta = 34°$ T^*	E^*	minutes
0	1428.54	175.40	1481.77	188.51	1535.24	202.12	1588.95	216.25	1642.93	230.90	1697.18	246.08	1751.71	261.79	0
1	1429.43	175.62	1482.66	188.73	1536.13	202.35	1589.85	216.49	1643.83	231.15	1698.08	246.34	1752.62	262.06	1
2	1430.31	175.83	1483.55	188.95	1537.02	202.58	1590.75	216.73	1644.73	231.40	1698.99	246.59	1753.53	262.33	2
3	1431.20	176.05	1484.44	189.17	1537.92	202.81	1591.65	216.97	1645.64	231.64	1699.90	246.85	1754.44	262.59	3
4	1432.09	176.26	1485.33	189.40	1538.81	203.04	1592.54	217.21	1646.54	231.89	1700.80	247.11	1755.35	262.86	4
5	1432.97	176.48	1486.22	189.62	1539.77	203.27	1593.44	217.45	1647.44	232.14	1701.71	247.37	1756.26	263.13	5
6	1433.86	176.69	1487.11	189.84	1540.60	203.51	1594.34	217.69	1648.34	232.39	1702.62	247.63	1757.18	263.40	6
7	1434.74	176.91	1487.99	190.07	1541.49	203.74	1595.24	217.93	1649.24	232.64	1703.53	247.88	1758.09	263.66	7
8	1435.63	177.12	1488.88	190.29	1542.38	203.97	1596.13	218.17	1650.15	232.89	1704.43	248.14	1759.00	263.93	8
9	1436.51	177.34	1489.77	190.51	1543.28	204.20	1597.03	218.41	1651.05	233.14	1705.34	248.40	1759.91	264.20	9
10	1437.40	177.55	1490.66	190.74	1544.17	204.44	1597.93	218.65	1651.95	233.39	1706.25	248.66	1760.82	264.47	10
11	1438.28	177.77	1491.55	190.96	1545.06	204.67	1598.83	218.89	1652.85	233.64	1707.15	248.92	1761.74	264.73	11
12	1439.17	177.98	1492.44	191.19	1545.96	204.90	1599.73	219.14	1653.76	233.89	1708.06	249.18	1762.65	265.00	12
13	1440.06	178.20	1493.33	191.41	1546.85	205.13	1600.63	219.38	1654.66	234.14	1708.97	249.44	1763.56	265.27	13
14	1440.94	178.41	1494.22	191.64	1547.75	205.37	1601.52	219.62	1655.56	234.39	1709.88	249.70	1764.47	265.54	14
15	1441.83	178.63	1495.11	191.86	1548.64	205.60	1602.42	219.86	1656.47	234.64	1710.78	249.96	1765.39	265.81	15
16	1442.71	178.85	1496.00	192.08	1549.54	205.83	1603.32	220.10	1657.37	234.89	1711.69	250.22	1766.30	266.08	16
17	1443.60	179.06	1496.89	192.31	1550.43	206.07	1604.22	220.35	1658.27	235.15	1712.60	250.48	1767.21	266.35	17
18	1444.49	179.28	1497.78	192.53	1551.32	206.30	1605.12	220.59	1659.18	235.40	1713.51	250.74	1768.12	266.61	18
19	1445.37	179.50	1498.67	192.76	1552.22	206.54	1606.02	220.83	1660.08	235.65	1714.42	251.00	1769.04	266.88	19
20	1446.26	179.71	1499.56	192.98	1553.11	206.77	1606.92	221.07	1660.98	235.90	1715.32	251.26	1769.95	267.15	20
21	1447.15	179.93	1500.45	193.21	1554.01	207.00	1607.81	221.32	1661.89	236.15	1716.23	251.52	1770.86	267.42	21
22	1448.03	180.15	1501.35	193.44	1554.90	207.24	1608.71	221.56	1662.79	236.40	1717.14	251.78	1771.77	267.69	22
23	1448.92	180.36	1502.24	193.66	1555.80	207.47	1609.61	221.80	1663.69	236.66	1718.05	252.04	1772.69	267.96	23
24	1449.81	180.58	1503.13	193.89	1556.69	207.71	1610.51	222.04	1664.60	236.91	1718.96	252.30	1773.60	268.23	24
25	1450.69	180.80	1504.02	194.11	1557.59	207.94	1611.41	222.29	1665.50	237.16	1719.86	252.56	1774.51	268.50	25
26	1451.58	181.02	1504.91	194.34	1558.48	208.18	1612.31	222.53	1666.40	237.41	1720.77	252.82	1775.43	268.77	26
27	1452.47	181.24	1505.80	194.57	1559.38	208.41	1613.21	222.77	1667.31	237.66	1721.68	253.08	1776.34	269.04	27
28	1453.35	181.45	1506.69	194.79	1560.27	208.65	1614.11	223.02	1668.21	237.92	1722.59	253.35	1777.25	269.31	28
29	1454.24	181.67	1507.58	195.02	1561.17	208.88	1615.01	223.26	1669.12	238.17	1723.50	253.61	1778.17	269.58	29
30	1455.13	181.89	1508.47	195.25	1562.06	209.12	1615.91	223.51	1670.02	238.42	1724.41	253.87	1779.08	269.85	30
31	1456.01	182.11	1509.36	195.47	1562.96	209.35	1616.81	223.75	1670.92	238.68	1725.32	254.13	1779.99	270.13	31
32	1456.90	182.33	1510.25	195.70	1563.85	209.59	1617.71	224.00	1671.83	238.93	1726.22	254.39	1780.91	270.40	32
33	1457.79	182.55	1511.15	195.93	1564.75	209.82	1618.61	224.24	1672.73	239.18	1727.13	254.66	1781.82	270.67	33
34	1458.68	182.76	1512.04	196.16	1565.64	210.06	1619.51	224.49	1673.64	239.44	1728.04	254.92	1782.74	270.94	34
35	1459.56	182.98	1512.93	196.38	1566.54	210.30	1620.41	224.73	1674.54	239.69	1728.95	255.18	1783.65	271.21	35
36	1460.45	183.20	1513.82	196.61	1567.44	210.53	1621.31	224.97	1675.45	239.94	1729.86	255.44	1784.56	271.48	36
37	1461.34	183.42	1514.71	196.84	1568.33	210.77	1622.21	225.22	1676.35	240.20	1730.77	255.71	1785.48	271.76	37
38	1462.23	183.64	1515.60	197.07	1569.23	211.01	1623.11	255.47	1677.26	240.45	1731.68	255.97	1786.39	272.03	38
39	1463.11	183.86	1516.49	197.29	1570.12	211.24	1624.01	225.71	1678.16	240.71	1732.59	256.23	1787.31	272.30	39
40	1464.00	184.08	1517.39	197.52	1571.02	211.48	1624.91	225.96	1679.06	240.96	1733.50	256.50	1788.22	272.57	40
41	1464.89	184.30	1518.28	197.75	1571.91	211.72	1625.81	226.20	1679.97	241.21	1734.41	256.76	1789.14	272.84	41
42	1465.78	184.52	1519.17	197.98	1572.81	211.95	1626.71	226.45	1680.88	241.47	1735.32	257.02	1790.05	273.12	42
43	1466.67	184.74	1520.06	198.21	1573.71	212.19	1627.61	226.69	1681.78	241.72	1736.23	257.29	1790.97	273.39	43
44	1467.55	184.96	1520.95	198.44	1574.60	212.43	1628.51	226.94	1682.69	241.98	1737.14	257.55	1791.88	273.66	44
45	1468.44	185.18	1521.85	198.67	1575.50	212.67	1629.41	227.19	1683.59	242.23	1738.05	257.82	1792.80	273.94	45
46	1469.33	185.40	1522.74	198.90	1576.40	212.90	1630.31	227.43	1684.50	242.49	1738.96	258.08	1793.71	274.21	46
47	1470.22	185.62	1523.63	199.12	1577.29	213.14	1631.21	227.68	1685.40	242.74	1739.87	258.34	1794.63	274.48	47
48	1471.11	185.84	1524.52	199.35	1578.19	213.38	1632.11	227.93	1686.31	243.00	1740.78	258.61	1795.54	274.76	48
49	1471.99	186.06	1525.42	199.58	1579.09	213.62	1633.01	228.17	1687.21	243.26	1741.69	258.87	1796.46	275.03	49
50	1472.88	186.29	1526.31	199.81	1579.98	213.86	1633.92	228.42	1688.12	243.51	1742.60	259.14	1797.37	275.30	50
51	1473.77	186.51	1527.20	200.04	1580.88	214.09	1634.82	228.67	1689.02	243.77	1743.51	259.40	1798.29	275.58	51
52	1474.66	186.73	1528.09	200.27	1581.78	214.33	1635.72	228.92	1689.93	244.02	1744.42	259.67	1799.20	275.85	52
53	1475.55	186.95	1528.99	200.50	1582.67	214.57	1636.62	229.16	1690.84	244.28	1745.33	259.93	1800.12	276.13	53
54	1476.44	187.17	1529.88	200.73	1583.57	214.81	1637.52	229.41	1691.74	244.54	1746.24	260.20	1801.03	276.40	54
55	1477.32	187.39	1530.77	200.96	1584.47	215.05	1638.42	229.66	1692.65	244.79	1747.15	260.46	1801.95	276.68	55
56	1478.21	187.62	1531.66	201.19	1585.36	215.29	1639.32	229.91	1693.55	245.05	1748.06	260.73	1802.87	276.95	56
57	1479.10	187.84	1532.56	201.42	1586.26	215.53	1640.22	230.15	1694.46	245.31	1748.97	261.00	1803.78	277.23	57
58	1479.99	188.06	1533.45	201.66	1587.16	215.77	1641.13	230.40	1695.37	245.56	1749.89	261.26	1804.70	277.50	58
59	1480.88	188.28	1534.34	201.89	1588.06	216.01	1642.03	230.65	1696.27	245.82	1750.80	261.53	1805.61	277.78	59

(continued)

APPENDIX A *(continued)*
Tangents and Externals for Horizontal Curves

$$\Delta = \text{interior angle} \qquad T^* = 5729.578 \tan \frac{\Delta}{2} \qquad E^* = 5729.578 \left(\sec \frac{\Delta}{2} - 1 \right)$$

For arc definition, calculate T and E for any degree of curve D by dividing both table values by D.

$$T = \frac{T^*}{D} \qquad E = \frac{E^*}{D}$$

minutes	$\Delta = 35°$ T^*	E^*	$\Delta = 36°$ T^*	E^*	$\Delta = 37°$ T^*	E^*	$\Delta = 38°$ T^*	E^*	$\Delta = 39°$ T^*	E^*	$\Delta = 40°$ T^*	E^*	$\Delta = 41°$ T^*	E^*	minutes
0	1806.53	278.05	1861.65	294.86	1917.09	312.22	1972.85	330.14	2028.96	348.64	2085.40	367.71	2142.20	387.37	0
1	1807.45	278.33	1862.57	295.14	1918.02	312.51	1973.78	330.45	2029.89	348.95	2086.34	368.03	2143.15	387.71	1
2	1808.36	278.60	1863.50	295.43	1918.94	312.81	1974.72	330.75	2030.83	349.26	2087.28	368.36	2144.10	388.04	2
3	1809.28	278.88	1864.42	295.71	1919.87	313.10	1975.65	331.05	2031.76	349.58	2088.23	368.68	2145.05	388.37	3
4	1810.19	279.15	1865.34	296.00	1920.80	313.40	1976.58	331.36	2032.70	349.89	2089.17	369.00	2146.00	388.70	4
5	1811.11	279.43	1866.26	296.28	1921.72	313.69	1977.51	331.66	2033.64	350.20	2090.12	369.33	2146.95	389.04	5
6	1812.03	279.71	1867.18	296.57	1922.65	313.98	1978.45	331.97	2034.58	350.52	2091.06	369.65	2147.90	389.37	6
7	1812.94	279.98	1868.10	296.85	1923.58	314.28	1979.38	332.27	2035.52	350.83	2092.00	369.97	2148.85	389.71	7
8	1813.86	280.26	1869.03	297.14	1924.51	314.58	1980.31	332.57	2036.46	351.15	2092.95	370.30	2149.80	390.04	8
9	1814.78	280.54	1869.95	297.43	1925.43	314.87	1981.24	332.88	2037.39	351.46	2093.89	370.62	2150.75	390.37	9
10	1815.70	280.81	1870.87	297.71	1926.36	315.17	1982.18	333.18	2038.33	351.78	2094.84	370.95	2151.70	390.71	10
11	1816.61	281.09	1871.79	298.00	1927.29	315.46	1983.11	333.49	2039.27	352.09	2095.78	371.27	2152.66	391.04	11
12	1817.53	281.37	1872.71	298.28	1928.22	315.76	1984.04	333.80	2040.21	352.41	2096.73	371.60	2153.61	391.38	12
13	1818.45	281.65	1873.64	298.57	1929.14	316.05	1984.98	334.10	2041.15	352.72	2097.67	371.92	2154.56	391.71	13
14	1819.36	281.92	1874.56	298.86	1930.07	316.35	1985.91	334.41	2042.09	353.04	2098.62	372.25	2155.51	392.05	14
15	1820.28	282.20	1875.48	299.14	1931.00	316.65	1986.84	334.71	2043.03	353.35	2099.56	372.57	2156.46	392.38	15
16	1821.20	282.48	1876.41	299.43	1931.93	316.94	1987.78	335.02	2043.97	353.67	2100.51	372.90	2157.41	392.72	16
17	1822.12	282.76	1877.33	299.72	1932.86	317.24	1988.71	335.32	2044.91	353.98	2101.45	373.22	2158.36	393.05	17
18	1823.03	283.03	1878.25	300.01	1933.78	317.53	1989.65	335.63	2045.85	354.30	2102.40	373.55	2159.31	393.39	18
19	1823.95	283.31	1879.17	300.29	1934.71	317.83	1990.58	335.94	2046.79	354.61	2103.34	373.87	2160.27	393.72	19
20	1824.87	283.59	1880.10	300.58	1935.64	318.13	1991.51	336.24	2047.73	354.93	2104.29	374.20	2161.22	394.06	20
21	1825.79	283.87	1881.02	300.87	1936.57	318.43	1992.45	336.55	2048.67	355.25	2105.24	374.53	2162.17	394.39	21
22	1826.71	284.15	1881.94	301.16	1937.50	318.72	1993.38	336.86	2049.61	355.56	2106.18	374.85	2163.12	394.73	22
23	1827.62	284.43	1882.87	301.45	1938.43	319.02	1994.32	337.16	2050.55	355.88	2107.13	375.18	2164.07	395.07	23
24	1828.54	284.71	1883.79	301.73	1939.36	319.32	1995.25	337.47	2051.49	356.20	2108.07	375.50	2165.03	395.40	24
25	1829.46	284.99	1884.71	302.02	1940.28	319.62	1996.18	337.78	2052.43	356.51	2109.02	375.83	2165.98	395.74	25
26	1830.38	285.27	1885.64	302.31	1941.21	319.92	1997.12	338.09	2053.37	356.83	2109.97	376.16	2166.93	396.08	26
27	1831.30	285.54	1886.56	302.60	1942.14	320.21	1998.05	338.39	2054.31	357.15	2110.91	376.49	2167.88	396.41	27
28	1832.22	285.83	1887.48	302.89	1943.07	320.51	1998.99	338.70	2055.25	357.47	2111.86	376.81	2168.84	396.75	28
29	1833.13	286.10	1888.41	303.18	1944.00	320.81	1999.92	339.01	2056.19	357.78	2112.81	377.14	2169.79	397.09	29
30	1834.05	286.39	1889.33	303.47	1944.93	321.11	2000.86	339.32	2057.13	358.10	2113.75	377.47	2170.74	397.43	30
31	1834.97	286.67	1890.26	303.76	1945.86	321.41	2001.79	339.63	2058.07	358.42	2114.70	377.80	2171.70	397.76	31
32	1835.89	286.95	1891.18	304.05	1946.79	321.71	2002.73	339.93	2059.01	358.74	2115.65	378.12	2172.65	398.10	32
33	1836.81	287.23	1892.10	304.34	1947.72	322.01	2003.66	340.24	2059.95	359.06	2116.59	378.45	2173.60	398.44	33
34	1837.73	287.51	1893.03	304.63	1948.65	322.31	2004.60	340.55	2060.89	359.37	2117.54	378.78	2174.56	398.78	34
35	1838.65	287.79	1893.95	304.92	1949.58	322.60	2005.53	340.86	2061.83	359.69	2118.49	379.11	2175.51	399.12	35
36	1839.57	288.07	1894.88	305.21	1950.51	322.90	2006.47	341.17	2062.78	360.01	2119.44	379.44	2176.46	399.46	36
37	1840.49	288.35	1895.80	305.50	1951.44	323.20	2007.41	341.48	2063.72	360.33	2120.38	379.77	2177.42	399.79	37
38	1841.41	288.63	1896.73	305.79	1952.37	323.50	2008.34	341.79	2064.66	360.65	2121.33	380.10	2178.37	400.13	38
39	1842.32	288.91	1897.65	306.08	1953.30	323.80	2009.28	342.10	2065.60	360.97	2122.28	380.42	2179.32	400.47	39
40	1843.24	289.19	1898.58	306.37	1954.23	324.10	2010.21	342.41	2066.54	361.29	2123.23	380.75	2180.28	400.81	40
41	1844.16	289.48	1899.50	306.66	1955.16	324.40	2011.15	342.72	2067.48	361.61	2124.17	381.08	2181.23	401.15	41
42	1845.08	289.76	1900.43	306.95	1956.09	324.71	2012.08	343.03	2068.42	361.93	2125.12	381.41	2182.19	401.49	42
43	1846.00	290.04	1901.35	307.24	1957.02	325.01	2013.02	343.34	2069.37	362.25	2126.07	381.74	2183.14	401.83	43
44	1846.92	290.32	1902.28	307.53	1957.95	325.31	2013.96	343.65	2070.31	362.57	2127.02	382.07	2184.09	402.17	44
45	1847.84	290.60	1903.20	307.83	1958.88	325.61	2014.89	343.96	2071.25	362.89	2127.97	382.40	2185.05	402.51	45
46	1848.76	290.89	1904.13	308.12	1959.81	325.91	2015.83	344.27	2072.19	363.21	2128.91	382.73	2186.00	402.85	46
47	1849.68	291.17	1905.05	308.41	1960.74	326.21	2016.77	344.58	3072.14	363.53	2129.86	383.06	2186.96	403.19	47
48	1850.60	291.45	1905.98	308.70	1961.67	326.51	2017.70	344.89	2074.08	363.85	2130.81	383.39	2187.91	403.53	48
49	1851.52	291.73	1906.90	308.99	1962.60	326.81	2018.64	345.20	2075.02	364.17	2131.76	383.72	2188.87	403.87	49
50	1852.44	292.02	1907.83	309.29	1963.54	327.11	2019.58	345.51	2075.96	364.49	2132.71	384.05	2189.82	404.21	50
51	1853.36	292.30	1908.75	309.58	1964.47	327.42	2020.51	345.83	2076.91	364.81	2133.66	384.39	2190.78	404.55	51
52	1854.29	292.58	1909.68	309.87	1965.40	327.72	2021.45	346.14	2077.85	365.13	2134.61	384.72	2191.73	404.89	52
53	1855.21	292.87	1910.61	310.16	1966.33	328.02	2022.39	346.45	2078.79	365.46	2135.56	385.05	2192.69	405.24	53
54	1856.13	293.15	1911.53	310.46	1967.26	328.32	2023.32	346.76	2079.74	365.78	2136.50	385.38	2193.64	405.58	54
55	1857.05	293.44	1912.46	310.75	1968.19	328.63	2024.26	347.07	2080.68	366.10	2137.45	385.71	2194.60	405.92	55
56	1857.97	293.72	1913.38	311.04	1969.12	328.93	2025.20	347.39	2081.62	366.42	2138.40	386.04	2195.56	406.26	56
57	1858.89	294.00	1914.31	311.34	1970.06	329.23	2026.14	347.70	2082.57	366.74	2139.35	386.38	2196.51	406.60	57
58	1859.81	294.29	1915.24	311.63	1970.99	329.53	2027.07	348.01	2083.51	367.07	2140.30	386.71	2197.47	406.95	58
59	1860.73	294.57	1916.16	311.92	1971.92	329.84	2028.01	348.32	2084.45	367.39	2141.25	387.04	2198.42	407.29	59

(continued)

APPENDIX A *(continued)*
Tangents and Externals for Horizontal Curves

$$\Delta = \text{interior angle} \qquad T^* = 5729.578 \tan \frac{\Delta}{2} \qquad E^* = 5729.578 \left(\sec \frac{\Delta}{2} - 1 \right)$$

For arc definition, calculate T and E for any degree of curve D by dividing both table values by D.

$$T = \frac{T^*}{D} \qquad E = \frac{E^*}{D}$$

minutes	$\Delta = 42°$ T^*	E^*	$\Delta = 43°$ T^*	E^*	$\Delta = 44°$ T^*	E^*	$\Delta = 45°$ T^*	E^*	minutes
0	2199.38	407.63	2256.94	428.49	2314.90	449.97	2373.27	472.07	0
1	2200.34	407.97	2257.90	428.85	2315.87	450.33	2374.25	472.45	1
2	2201.29	408.32	2258.87	429.20	2316.84	450.70	2375.22	472.82	2
3	2202.25	408.66	2259.83	429.55	2317.81	451.06	2376.20	473.19	3
4	2203.20	409.00	2260.79	429.91	2318.78	451.42	2377.18	473.57	4
5	2204.16	409.35	2261.76	430.26	2319.75	451.79	2378.15	473.94	5
6	2205.12	409.69	2262.72	430.61	2320.72	452.15	2379.13	474.32	6
7	2206.07	410.03	2263.68	430.97	2321.69	452.52	2380.11	474.69	7
8	2207.03	410.38	2264.65	431.32	2322.66	452.88	2381.08	475.07	8
9	2207.99	410.72	2265.61	431.68	2323.63	453.25	2382.06	475.44	9
10	2208.95	411.06	2266.57	432.03	2324.60	453.61	2383.04	475.82	10
11	2209.90	411.41	2267.54	432.38	2325.57	453.98	2384.02	476.19	11
12	2210.86	411.75	2268.50	432.74	2326.54	454.34	2384.99	476.57	12
13	2211.82	412.10	2269.46	433.09	2327.51	454.71	2385.97	476.94	13
14	2212.78	412.44	2270.43	433.45	2328.48	455.07	2386.95	477.32	14
15	2213.73	412.79	2271.39	433.81	2329.45	455.44	2387.93	477.70	15
16	2214.69	413.13	2272.36	434.16	2330.42	455.80	2388.90	478.07	16
17	2215.65	413.48	2273.32	434.52	2331.40	456.17	2389.88	478.45	17
18	2216.61	413.83	2274.29	434.87	2332.37	456.54	2390.86	478.83	18
19	2217.56	414.17	2275.25	435.23	2333.34	456.90	2391.84	479.20	19
20	2218.52	414.52	2276.22	435.58	2334.31	457.27	2392.82	479.58	20
21	2219.48	414.86	2277.18	435.94	2335.28	457.63	2393.80	479.96	21
22	2220.44	415.21	2278.15	436.30	2336.25	458.00	2394.78	480.34	22
23	2221.40	415.56	2279.11	436.65	2337.23	458.37	2395.76	480.71	23
24	2222.36	415.90	2280.08	437.01	2338.20	458.74	2396.73	481.09	24
25	2223.32	416.25	2281.04	437.37	2339.17	459.10	2397.71	481.47	25
26	2224.27	416.60	2282.01	437.72	2340.14	459.47	2398.69	481.85	26
27	2225.23	416.94	2282.97	438.08	2341.11	459.84	2399.67	482.22	27
28	2226.19	417.29	2283.94	438.44	2342.09	460.21	2400.65	482.60	28
29	2227.15	417.64	2284.90	438.80	2343.06	460.57	2401.63	482.98	29
30	2228.11	417.99	2285.87	439.15	2344.03	460.94	2402.61	483.36	30
31	2229.07	418.33	2286.84	439.51	2345.01	461.31	2403.59	483.74	31
32	2230.03	418.68	2287.80	439.87	2345.98	461.68	2404.57	484.12	32
33	2230.99	419.03	2288.77	440.23	2346.95	462.05	2405.55	484.50	33
34	2231.95	419.38	2289.73	440.59	2347.92	462.42	2406.53	484.88	34
35	2232.91	419.73	2290.70	440.95	2348.90	462.79	2407.51	485.26	35
36	2233.87	420.08	2291.67	441.31	2349.87	463.16	2408.49	485.64	36
37	2234.83	420.42	2292.63	441.66	2350.84	463.53	2409.47	486.02	37
38	2235.79	420.77	2293.60	442.02	2351.82	463.89	2410.45	486.40	38
39	2236.75	421.12	2294.57	442.38	2352.79	464.27	2411.44	486.78	39
40	2237.71	421.47	2295.54	442.74	2353.77	464.64	2412.42	487.16	40
41	2238.67	421.82	2296.50	443.10	2354.74	465.00	2413.40	487.54	41
42	2239.63	422.17	2297.47	443.46	2355.71	465.38	2414.38	487.92	42
43	2240.59	422.52	2298.44	443.82	2356.69	465.75	2415.36	488.30	43
44	2241.55	422.87	2299.40	444.18	2357.66	466.12	2416.34	488.68	44
45	2242.51	423.22	2300.37	444.54	2358.64	466.49	2417.32	489.06	45
46	2243.47	423.57	2301.34	444.90	2359.61	466.86	2418.30	489.45	46
47	2244.44	423.92	2302.31	445.26	2360.59	467.23	2419.29	489.83	47
48	2245.40	424.27	2303.28	445.63	2361.56	467.60	2420.27	490.21	48
49	2246.36	424.62	2304.24	445.99	2362.54	467.97	2421.25	490.59	49
50	2247.32	424.97	2305.21	446.35	2363.51	468.34	2422.23	490.97	50
51	2248.28	425.32	2306.18	446.71	2364.49	468.72	2423.22	491.36	51
52	2249.24	425.68	2307.15	447.07	2365.46	469.09	2424.20	491.74	52
53	2250.21	426.03	2308.12	447.43	2366.44	469.46	2425.18	492.12	53
54	2251.17	426.38	2309.09	447.79	2367.41	469.83	2426.16	492.51	54
55	2252.13	426.73	2310.05	448.16	2368.39	470.21	2427.15	492.89	55
56	2253.09	427.08	2311.02	448.52	2369.36	470.58	2428.13	493.27	56
57	2254.05	427.44	2311.99	448.88	2370.34	470.95	2429.11	493.66	57
58	2255.02	427.79	2312.96	449.24	2371.32	471.32	2430.09	494.04	58
59	2255.98	428.14	2313.93	449.61	2372.29	471.70	2431.08	494.42	59

APPENDIX B
Radius When Degree of Curve Is Known

$$R = \frac{5729.578}{D}$$

D	R (ft)	D	R (ft)	D	R (ft)	D	R (ft)	D	R (ft)
0°00′		9°00′	636.62	18°00′	318.31	27°00′	212.21	36°00′	159.15
15′	22918.31	15′	619.41	15′	313.95	15′	210.26	15′	158.06
30′	11459.16	30′	603.11	30′	309.71	30′	208.35	30′	156.97
45′	7639.44	45′	587.65	45′	305.58	45′	206.47	45′	155.91
1°00′	5729.58	10°00′	572.96	19°00′	301.56	28°00′	204.63	37°00′	154.85
15′	4583.66	15′	558.98	15′	297.64	15′	202.82	15′	153.81
30′	3819.72	30′	545.67	30′	293.82	30′	201.04	30′	152.79
45′	3274.04	45′	532.98	45′	290.11	45′	199.29	45′	151.78
2°00′	2864.79	11°00′	520.87	20°00′	286.48	29°00′	197.57	38°00′	150.78
15′	2546.48	15′	509.30	15′	282.94	15′	195.88	15′	149.79
30′	2291.83	30′	498.22	30′	279.49	30′	194.22	30′	148.82
45′	2083.48	45′	487.62	45′	276.12	45′	192.59	45′	147.86
3°00′	1909.86	12°00′	477.46	21°00′	272.84	30°00′	190.99	39°00′	146.91
15′	1762.95	15′	467.72	15′	269.63	15′	189.41	15′	145.98
30′	1637.02	30′	458.37	13′	266.49	30′	187.86	30′	145.05
45′	1527.89	45′	449.38	45′	263.43	45′	186.33	45′	144.14
4°00′	1432.39	13°00′	440.74	22°00′	260.44	31°00′	184.83	40°00′	143.24
15′	1348.14	15′	432.42	15′	257.51	15′	183.35	15′	142.35
30′	1273.24	30′	424.41	30′	254.65	30′	181.89	30′	141.47
45′	1206.23	45′	416.70	45′	251.85	45′	180.46	45′	140.60
5°00′	1145.92	14°00′	409.26	23°00′	249.11	32°00′	179.05	41°00′	139.75
15′	1091.35	15′	402.08	15′	246.43	15′	177.66	15′	138.90
30′	1041.74	30′	395.14	30′	243.81	30′	176.29	30′	138.06
45′	996.45	45′	388.45	45′	241.25	45′	174.95	45′	137.24
6°00′	954.93	15°00′	381.97	24°00′	238.73	33°00′	173.62	42°00′	136.42
15′	916.73	15′	375.71	15′	236.27	15′	172.32	15′	135.61
30′	881.47	30′	369.65	30′	233.86	30′	171.03	30′	134.81
45′	848.83	45′	363.78	45′	231.50	45′	169.77	45′	134.03
7°00′	818.51	16°00′	358.10	25°00′	229.18	34°00′	168.52	43°00′	133.25
15′	790.29	15′	352.59	15′	226.91	15′	167.29	15′	132.48
30′	763.94	30′	347.25	30′	224.69	30′	166.07	30′	131.71
45′	739.30	45′	342.06	45′	222.51	45′	164.88	45′	130.96
8°00′	716.20	17°00′	337.03	26°00′	220.37	35°00′	163.70	44°00′	130.22
15′	694.49	15′	332.15	15′	218.27	15′	162.54	15′	129.48
30′	674.07	30′	327.40	30′	216.21	30′	161.40	30′	128.75
45′	654.81	45′	322.79	45′	214.19	45′	160.27	45′	128.04
								45°00′	127.32
								15′	126.62
								30′	125.92
								45′	125.24

(continued)

APPENDIX B *(continued)*
Radius When Degree of Curve Is Known

$$R = \frac{5729.578}{D}$$

D	R (ft)	D	R (ft)	D	R (ft)	D	R (ft)	D	R (ft)
46°00′	124.56	55°00′	104.17	64°00′	89.52	73°00′	78.49	82°00′	69.87
15′	123.88	15′	103.70	15′	89.18	15′	78.22	15′	69.66
30′	123.22	30′	103.24	30′	88.83	30′	77.95	30′	69.45
45′	122.56	45′	102.77	45′	88.49	45′	77.69	45′	69.24
47°00′	121.91	56°00′	102.31	65°00′	88.15	74°00′	77.43	83°00′	69.03
15′	121.26	15′	101.86	15′	87.81	15′	77.17	15′	68.82
30′	120.62	30′	101.41	30′	87.47	30′	76.91	30′	68.62
45′	119.99	45′	100.96	45′	87.14	45′	76.65	45′	68.41
48°00′	119.37	57°00′	100.52	66°00′	86.81	75°00′	76.39	84°00′	68.21
15′	118.75	15′	100.08	15′	86.48	15′	76.14	15′	68.01
30′	118.14	30′	99.64	30′	86.16	30′	75.89	30′	67.81
45′	117.53	45′	99.21	45′	85.84	45′	75.64	45′	67.61
49°00′	116.93	58°00′	98.79	67°00′	85.52	76°00′	75.39	85°00′	67.41
15′	116.34	15′	98.36	15′	85.20	15′	75.14	15′	67.21
30′	115.75	30′	97.94	30′	84.88	30′	74.90	30′	67.01
45′	115.17	45′	97.52	45′	84.57	45′	74.65	45′	66.82
50°00′	114.59	59°00	97.11	68°00′	84.26	77°00′	74.41	86°00′	66.62
15′	114.02	15′	96.70	15′	83.95	15′	74.17	15′	66.43
30′	113.46	30′	96.30	30′	83.64	30′	73.93	30′	66.24
45′	112.90	45′	95.89	45′	83.34	45′	73.69	45′	66.05
51°00′	112.34	60°00′	95.49	69°00′	83.04	78°00′	73.46	87°00′	65.86
15′	111.80	15′	95.10	15′	82.74	15′	73.22	15′	65.67
30′	111.25	30′	94.70	30′	82.44	30′	72.99	30′	65.48
45′	110.72	45′	94.31	45′	82.14	45′	72.76	45′	65.29
52°00′	110.18	61°00′	93.93	70°00′	81.85	79°00′	72.53	88°00′	65.11
15′	109.66	15′	93.54	15′	81.56	15′	72.30	15′	64.92
30′	109.13	30′	93.16	30′	81.27	30′	72.07	30′	64.74
45′	108.62	45′	92.79	45′	80.98	45′	71.84	45′	64.56
53°00′	108.11	62°00′	92.41	71°00′	80.70	80°00′	71.62	89°00	64.38
15′	107.60	15′	92.04	15′	80.42	15′	71.40	15′	64.20
30′	107.09	30′	91.67	30′	80.13	30′	71.17	30′	64.02
45′	106.60	45′	91.31	45′	79.85	45′	70.95	45′	63.84
54°00′	106.10	63°00′	90.95	72°00′	79.58	81°00′	70.74	90°00′	63.66
15′	105.61	15′	90.59	15′	79.30	15′	70.52	15′	63.49
30′	105.13	30′	90.23	30′	79.03	30′	70.30	30′	63.31
45′	104.65	45′	89.88	45′	78.76	45′	70.09	45′	63.14

APPENDIX C
Chord Lengths of Circular Arcs (Arc Definition)

$$C = 2R\sin\left(\frac{D}{2}\right) = \frac{(2)(360°)(100\text{ ft})}{2\pi D}\sin\left(\frac{D}{2}\right) = \frac{5729.578}{D}\sin\left(\frac{D}{2}\right)$$

(all distances in feet)

degree of curve	for arc of								degree of curve
	100	95	90	85	80	75	70	60	
1°	100	95	90	85	80	75	70	60	1°
2°	100	95	90	85	80	75	70	60	2°
3°	99.99	94.99	89.99	85	80	75	70	60	3°
4°	99.98	94.98	89.98	84.99	79.99	74.99	70	60	4°
5°	99.97	94.97	89.98	84.98	79.98	74.99	69.99	59.99	5°
6°	99.95	94.96	89.97	84.97	79.98	74.98	69.98	59.99	6°
7°	99.94	94.95	89.96	84.96	79.96	74.97	69.98	59.99	7°
8°	99.92	94.93	89.94	84.95	79.96	74.97	69.97	59.98	8°
9°	99.90	94.91	89.93	84.94	79.95	74.96	69.96	59.98	9°
10°	99.87	94.89	89.91	84.92	79.94	74.95	69.96	59.97	10°

degree of curve	for arc of								degree of curve
	100	60	55	50	45	40	35	30	
11°	99.85	59.97	54.97	49.98	44.99	39.99	34.99	29.99	11°
12°	99.82	59.96	54.97	49.98	44.98	39.99	34.99	29.99	12°
13°	99.79	59.95	54.96	49.97	44.98	39.99	34.99	29.99	13°
14°	99.75	59.95	54.96	49.97	44.98	39.98	34.99	29.99	14°
15°	99.71	59.94	54.95	49.96	44.97	39.98	34.99	29.99	15°
16°	99.68	59.93	54.95	49.96	44.97	39.98	34.99	29.99	16°
17°	99.63	59.92	54.94	49.95	44.97	39.98	34.98	29.99	17°
18°	99.59	59.91	54.93	49.95	44.96	39.97	34.98	29.99	18°
19°	99.54	59.90	54.92	49.94	44.96	39.97	34.98	29.99	19°
20°	99.49	59.89	54.92	49.94	44.95	39.97	34.98	29.99	20°
21°	99.44	59.88	54.91	49.93	44.95	39.96	34.98	29.98	21°
22°	99.39	59.87	54.90	49.92	44.94	39.96	34.97	29.98	22°
23°	99.33	59.85	54.89	49.92	44.94	39.96	34.97	29.98	23°
24°	99.27	59.84	54.88	49.91	44.93	39.95	34.97	29.98	24°
25°	99.21	59.83	54.87	49.90	44.93	39.95	34.97	29.98	25°

APPENDIX D
Conversion Factors

multiply	by	to obtain
acres	0.4047	hectares
	43,560.0	square feet
	1.5625×10^{-3}	square miles
ampere-hours	3600.0	coulombs
angstrom units	3.937×10^{-9}	inches
	1×10^{-4}	microns
astronomical units	1.496×10^{8}	kilometers
atmospheres	76.0	centimeters of mercury
atomic mass unit	9.316×10^{8}	electron-volts
	1.492×10^{-10}	joules
	1.66×10^{-27}	kilograms
BeV (also GeV)	10^{9}	electron-volts
Btu	3.93×10^{-4}	horsepower-hours
	778.3	foot-pounds
	2.93×10^{-4}	kilowatt hours
	1.0×10^{-5}	therms
Btu/hr	0.2161	foot-pounds/sec
	3.929×10^{-4}	horsepower
	0.293	watts
bushels	2150.4	cubic inches
calories, gram (mean)	3.9683×10^{-3}	Btu (mean)
centares	1.0	square meters
centimeters	1×10^{-5}	kilometers
	1×10^{-2}	meters
	10.0	millimeters
	3.281×10^{-2}	feet
	0.3937	inches
chains	792.0	inches
coulombs	1.036×10^{-5}	faradays
cubic centimeters	0.06102	cubic inches
	2.113×10^{-3}	pints (U.S. liquid)
cubic feet	0.02832	cubic meters
	7.4805	gallons
cubic feet/min	62.43	pounds H_2O/min
cubic feet/sec	448.831	gallons/min
	0.64632	millions of gallons per day
cubits	18.0	inches
days	86,400.0	seconds
degrees (angle)	1.745×10^{-2}	radians
degrees/sec	0.1667	revolutions/min
dynes	1×10^{-5}	newtons
electron-volts	1.074×10^{-9}	atomic mass units
	10^{-9}	BeV (also GeV)
	1.602×10^{-19}	joules
	1.783×10^{-36}	kilograms
	10^{-6}	MeV
faradays/sec	96,500	amperes (absolute)
fathoms	6.0	feet
feet	30.48	centimeters
	0.3048	meters
	1.645×10^{-4}	miles (nautical)
	1.894×10^{-4}	miles (statute)
feet/min	0.5080	centimeters/sec
feet/sec	0.592	knots
	0.6818	miles/hr

multiply	by	to obtain
foot-pounds	1.285×10^{-3}	Btu
	5.051×10^{-7}	horsepower-hours
	3.766×10^{-7}	kilowatt-hours
foot-pound/sec	4.6272	Btu/hr
	1.818×10^{-3}	horsepower
	1.356×10^{-3}	kilowatts
furlongs	660.0	feet
	0.125	miles (statute)
gallons	0.1337	cubic feet
	3.785	liters
gallons H_2O	8.3453	pounds H_2O
gallons/min	8.0208	cubic feet/hr
	0.002228	cubic feet/sec
GeV (also BeV)	10^{9}	electron-volts
grams	10^{-3}	kilograms
	3.527×10^{-2}	ounces (avoirdupois)
	3.215×10^{-2}	ounces (troy)
	2.205×10^{-3}	pounds
hectares	2.471	acres
	1.076×10^{5}	square feet
horsepower	2545.0	Btu/hr
	42.44	Btu/min
	550	foot-pounds/sec
	0.7457	kilowatts
	745.7	watts
horsepower-hours	2545.0	Btu
	1.976×10^{-6}	foot-pounds
	0.7457	kilowatt-hours
hours	4.167×10^{-2}	days
	5.952×10^{-3}	weeks
inches	2.540	centimeters
	1.578×10^{-5}	miles
inches, H_2O	5.199	pounds force/ft^2
	0.0361	psi
	0.0735	inches, mercury
inches, mercury	70.7	pounds force/ft^2
	0.491	pounds force/in^2
	13.60	inches, H_2O
joules	6.705×10^{9}	atomic mass units
	9.480×10^{-4}	Btu
	1×10^{7}	ergs
	6.242×10^{18}	electron-volts
	1.113×10^{-17}	kilograms
kilograms	6.025×10^{26}	atomic mass units
	5.610×10^{35}	electron-volts
	8.987×10^{16}	joules
	2.205	pounds
kilometers	3281.0	feet
	1000.0	meters
	0.6214	miles
kilometers/hr	0.5396	knots
kilowatts	3412.9	Btu/hr
	737.6	foot-pounds/sec
	1.341	horsepower
kilowatt-hours	3413.0	Btu

(continued)

APPENDIX D *(continued)*
Conversion Factors

multiply	by	to obtain
knots	6076.0	feet/hr
	1.0	nautical miles/hr
	1.151	statute miles/hr
light years	5.9×10^{12}	miles
links (surveyor)	7.92	inches
liters	1000.0	cubic centimeters
	61.02	cubic inches
	0.2642	gallons (U.S. liquid)
	1000.0	milliliters
	2.113	pints
MeV	10^6	electron-volts
meters	100.0	centimeters
	3.281	feet
	1×10^{-3}	kilometers
	5.396×10^{-4}	miles (nautical)
	6.214×10^{-4}	miles (statute)
	1000.0	millimeters
microns	1×10^{-6}	meters
miles (nautical)	6076	feet
	1.853	kilometers
	1.1516	miles (statute)
miles (statute)	5280.0	feet
	1.609	kilometers
	0.8684	miles (nautical)
miles/hr	88.0	feet/min
milligrams/liter	1.0	parts/million
milliliters	1×10^{-3}	liters
millimeters	3.937×10^{-2}	inches
newtons	1×10^5	dynes
ohms (international)	1.0005	ohms (absolute)
ounces	28.349527	grams
	6.25×10^{-2}	pounds
ounces (troy)	1.09714	ounces (avoirdupois)
parsecs	3.086×10^{13}	kilometers
	1.9×10^{13}	miles
pascal-sec	1000	centipoise
	10	poise
	0.02089	pound force-sec/ft^2
	0.6720	pound mass/ft-sec
	0.02089	slug/ft-sec
pints (liquid)	473.2	cubic centimeters
	28.87	cubic inches
	0.125	gallons
	0.5	quarts (liquid)

multiply	by	to obtain
poise	0.002089	pound-sec/ft^2
pounds	0.4536	kilograms
	16.0	ounces
	14.5833	ounces (troy)
	1.21528	pounds (troy)
pounds/ft^2	0.006944	pounds/in^2
pounds/in^2	2.308	feet, H_2O
	27.7	inches, H_2O
	2.037	inches, mercury
	144	pounds/ft^2
quarts (dry)	67.20	cubic inches
quarts (liquid)	57.75	cubic inches
	0.25	gallons
	0.9463	liters
radians	57.30	degrees
	3438.0	minutes
revolutions	360.0	degrees
revolutions/min	6.0	degrees/sec
rods	16.5	feet
	5.029	meters
rods (surveyor)	5.5	yards
seconds	1.667×10^{-2}	minutes
square meters/sec	10^6	centistokes
	10.76	square feet/sec
	10^4	stokes
slugs	32.174	pounds mass
stokes	0.0010764	square feet/sec
tons (long)	1016.0	kilograms
	2240.0	pounds
	1.120	tons (short)
tons (short)	907.1848	kilograms
	2000.0	pounds
	0.89287	tons (long)
volts (absolute)	3.336×10^{-3}	statvolts
watts	3.4129	Btu/hr
	1.341×10^{-3}	horsepower
yards	0.9144	meters
	4.934×10^{-4}	miles (nautical)
	5.682×10^{-4}	miles (statute)

APPENDIX E
Surveying Conversion Factors

multiply	by	to obtain
ac	43,560	ft^2
	10	chains^2
	4046.87	m^2
ac-ft	43,560	ft^3
	1233.49	m^3
chain	66	ft
	22	yd
	4	rods
day (mean solar)	86,400	sec
day (sidereal)	86,164.09	sec
deg (angle)	0.0174533	rad
	17.77778	mils
engineer's link	1	ft
ft (U.S. Survey)	0.3048006	m
grads	0.9	degrees (angle)
	0.01570797	rad
hectare	2.47104	ac
	10,000	m^2
in	25.4	mm
labors	177.14	ac
leagues	4428.40	ac
link (see engineer's link and surveyor's link)		
mils	0.05625	degrees (angle)
	3,037,500	min
mi (statute)	5280	ft
	80	chains (surveyor's)
	320	rods
	0.86839	mi (nautical)
mi^2	640	ac
	27,878,400	ft^2
min (angle)	0.29630	mils
	0.000290888	rad
min (mean solar)	60	sec
min (sidereal)	59.83617	sec
outs	330	ft
	10	33 ft chains
rad	57.2957795	degrees (angle)
	57°17′44.806″	degrees (angle)
rods	16.5	ft
	1	perches
	1	poles
sec (angle)	4.848137×10^{-6}	rad
sec (sidereal)	0.9972696	sec (mean solar)
surveyor's link	0.66	ft
	7.92	in
VARA (California)	33	in
VARA (Texas)	$33\frac{1}{3}$	in
yd (U.S.)	0.914402	m

Appendices

APPENDIX F
Miscellaneous Constants and Conversions

0.0000001 per °F	= coefficient of expansion for invar tape
0.00000645 per °F	= coefficient of expansion for steel tape
0.6745	= coefficient for 50% standard deviation
1.15 mi	= 1 minute of latitude
1.6449	= coefficient for 90% standard deviation
6 mi	= length and width of township
10 square chains	= 1 ac
15° longitude	= width of one time zone
23°26.5′	= maximum declination of the sun at solstice
24 hr	= 360° of longitude
36	= number of sections in a township
69.1 mi	= 1° latitude
100	= usual stadia ratio
101 ft	= 1 second of latitude
400 grads	= 360°
480 chains	= width and length of township
640 ac	= 1 normal section
4046.9 mi^2	= 1 ac
6400 mils	= 360°
43,560 ft^2	= 1 ac
20,906,000 ft	= mean radius of earth

APPENDIX G
Glossary

Not all words are covered in this glossary.
Refer to the index for additional words that are discussed in the text.

Abstract – A summary of facts.

Abstract of title – A condensed history of the title to land.

Accessory to corner – A physical object that is adjacent to a corner. An accessory is usually considered part of the monument.

Acclivity – An upward slope of ground.

Accretion – The gradual accumulation of land by natural causes.

Acknowledgment – A declaration by a person before an official (usually a notary public) that he or she executed a legal document.

Acquiescence – Implied consent to a transaction, to the accrual of a right, or to any act, by one's silence (or without express assent).

Adjudication – The giving or pronouncing of a judgment or decree.

Adverse possession – A method of acquiring property by holding it for a period of time under certain conditions.

Affidavit – A written declaration under oath before an authorized official (usually a notary public).

Alienation – The transfer of property and/or possessions from one person to another.

Aliquot – A portion contained in something else a whole number of times.

Alluvium – Sand or soil deposited by streams.

Appellant – The party that takes an appeal from one court or jurisdiction to another.

Appurtenance – A right, privilege, or improvement belonging to and passing with a piece of property when it is conveyed.

Assigns – Those to whom property is transferred.

Avulsion – A sudden and perceptible change of shoreline by the violent action of water.

Bayou – An outlet from a swamp or lagoon to the sea.

Bed of stream – The depression between the banks of a water course worn by the regular and usual flow of the water.

Bequest – A gift by will of personal property.

Bounty lands – Portions of the public domain given or donated as a bounty for services rendered.

Chain of title – A chronological list of documents that comprise the record history of title of real property.

Civil law – That part of the law pertaining to civil rights, as distinguished from criminal law. Civil law and Roman law have the same meaning. In contradistinction to English common law, civil law is enacted by legislative bodies.

Clear title – Good title. Title free from encumbrances.

Cloud on title – A claim or encumbrance on a title to land that may or may not be valid.

Color of title – Any written instrument that appears to convey title, even though it does not.

Common law – Principles and rules of action determined by court decisions that have been accepted by generation after generation, and that are distinguished from laws enacted by legislative bodies.

Consideration – Something of value given to make an agreement binding.

Conveyance – Any instrument in writing by which an interest in real property is transferred.

Covenant – When used in deeds, restrictions imposed on the grantee as to the use of land conveyed.

Crown – The sovereign power in a monarchy.

Cut bank – The watershed and relatively permanent elevation or acclivity that separates the bed of a river from its adjacent upland.

Decree – A judgment by the court in a legal proceeding.

Dedication – An appropriation of land to some public use made by the owner, and accepted for such use by or on behalf of the public.

Deed – Evidence in writing of the transfer of real property.

Deed of trust – An instrument taking the place of a mortgage, by which the legal title to real property is placed in one or more trustees to secure repayment of a sum of money.

Demurrer – In legal pleading, the formal mode of disputing the sufficiency of the pleading of the other side.

Devise – A gift of real property by the last will and testament of the donor.

Easement – The right that the public, an individual, or individuals have in the lands of another.

Egress – The right or permission to go out from a place; right of exit.

Eminent domain – The right or power of government or certain other agencies to take private property for public use on payment of just compensation to the owner.

Appendices

APPENDIX G *(continued)*
Glossary

Encroachment – An obstruction that intrudes upon the land of another. The gradual, stealthy, illegal acquisition of property.

Encumbrance – Any burden or claim on property, such as a mortgage or delinquent taxes.

Equity – The excess of the market value over any indebtedness.

Erosion – The process by which the surface of the earth is worn away by the action of waters, glaciers, wind, or waves.

Escheat – Reversion of property to the state where there is no competent or available person to inherit it.

Escrow – Something placed in the keeping of a third person for delivery to a given party upon fulfillment of some condition.

Estate – An interest in property, real or personal.

Estoppel – A bar or impediment that precludes allegation or denial of a certain fact or state of facts in consequence of a final adjudication.

Et al. – An abbreviation for "and others."

Et Mode Ad Hune Diem – An abbreviation for "and now at this day."

Et ux – An abbreviation for "and wife."

Evidence aliunde – Evidence from outside or from another source.

Extrinsic evidence – Evidence NOT contained in the deed, but offered to clear up an ambiguity found to exist when applying the description to the ground.

Grant – A transfer of property.

Grantee – The person to whom a grant is made.

Grantor – The person by whom a grant is made.

Good faith – An honest intention to abstain from taking advantage of another.

Gradient – An inclined surface. The change in elevation per unit of horizontal distance.

Hereditament – Something capable of being inherited, be it real or personal property.

Hiatus – An area between two surveys of record described as having one or more common boundary lines with no omission.

Holograph – A will written entirely by the testator in his or her own handwriting.

Incumbrance – A right, interest in, or legal liability upon real property that does not prohibit passing title to the land but that diminishes its value.

Ingress – The right or permission to go upon a place; right of entrance.

Intent – The true meaning (from the written words of an instrument).

Intestate – Without making a will.

Judgment – The official and authentic decision of a court of justice.

Leasehold – An estate in realty held under a lease; an estate for a fixed term of years.

Lessee – The person to whom a lease is made.

Lessor – The person who grants a lease.

Lien – A claim or charge on property for payment of some debt, obligation, or duty.

Lis pendens – A pending suit. A notice of lis pendens is filed for the purpose of warning all persons that a suit is pending.

Litigation – Contest in a court of justice for the purpose of enforcing a right.

Littoral – Belonging to the shore, as of seas and lakes.

Logical relevancy – A relationship in logic between the fact for which evidence is offered and a fact in issue such that the existence of the former renders probable or improbable the existence of the latter.

Mean – Intermediate; the middle between two extremes.

Memorial – That which contains the particulars of a deed, and so on. In practice, a memorial is a short note, abstract, memorandum, or rough draft of the orders of the court, from which the records thereof may at any time be fully made up.

Mortgage – A conditional conveyance of an estate as a pledge for the security of a debt.

Muniment – Documentary evidence of title.

Option – The right as granted in a contract or by an initial payment of acquiring something in the future.

Parcel – A part of a piece of land that cannot be identified by a lot or tract number.

Parol evidence – Evidence that is given verbally.

Patent – A government grant of land. The instrument by which a government conveys title to land.

Plat – A scaled diagram showing boundaries of a tract of land or subdivisions. May constitute a legal description of the land and be used in lieu of a written description.

Power of attorney – A written document given by one person to another authorizing the latter to act for the former.

Prescription – Creation of an easement under claim of right by use of land that has been open, continuous, and exclusive for a period of time prescribed by law.

APPENDIX G *(continued)*
Glossary

Prima facie evidence – Facts presumed to be true unless disproved by evidence to the contrary.

Privity – The relationship that exists between parties to a contract. Mutual or successive relationship to the same rights of property, such as the relationship of heir with ancestor or donee with donor.

Privy – A person who is in privity with another.

Probate – The act or process of validating a will.

Quiet title – Action of law to remove an adverse claim or cloud on title.

Quitclaim deed – A conveyance that passes any title, interest, or claim that the grantor may have.

Reliction – A gradual and imperceptible recession of water, resulting in increased shoreline, beach, or property.

Relinquishment – The forsaking, abandonment, renouncement, or gift of a right.

Remand – To send a cause back to the same court out of which it came for the purpose of having some action taken upon it there.

Riparian – Belonging or relating to the bank of a river.

Royalty – A share of the profit from sale of minerals paid to the owner of the property by the lessee.

Said – Refers to one previously mentioned.

Scrivener – A person whose occupation is to draw up contracts, write deeds and mortgages, and prepare other written instruments.

Shore – The place lying between the line of ordinary high tide and the line of lowest tide.

Sovereign – A person, body, or state in which independent and supreme authority is vested.

Squatter – One who settles on another's land without legal authority.

Statute – A particular law established by the legislative branch of government.

Statutory – Relating to a statute.

Submerged land – In tidal areas, land that extends seaward from the shore and is continuously covered during the ebb and flow of the tide.

Substantive evidence – Evidence used to prove a fact (as opposed to evidence given for the purpose of discrediting a claim).

Tenancy by the entirety – Husband and wife each possesses the entire estate in order that, upon the death of either spouse, the survivor is entitled to the estate in its entirety.

Tenancy, Joint – The holding of property by two or more persons, each of whom has an undivided interest. After the death of one of the joint tenants, the surviving tenant(s) receive the descendent's share.

Tenant – One who has the temporary use and occupation of real property owned by another person (the landlord).

Tenements – Property held by a tenant. Everything of a permanent nature. In a more restrictive sense, a house or dwelling.

Testament – A will of personal property.

Testator – One who makes a testament or will. One who dies leaving a will.

Thalweg – The deepest part of a channel.

Thence – From that place; the following course is continuous from the one before it.

Title policy – Insurance against loss or damage resulting from defects or failure of title to a particular parcel of land.

To wit – That is to say; namely.

Upland – Land above mean high water and subject to private ownership (as distinguished from tidelands, which are in the state). Also used as meaning nonriparian.

Watercourse – A running stream of water fed from permanent or natural sources running in a particular direction and having a channel formed by a well-defined bed and banks (though it need not flow continuously).

Warranty deed – A deed in which the grantor proclaims that he or she is the lawful owner of real property and will forever defend the grantee against any claim on the property.

Will – The legal declaration of a person's wishes as to the disposition of his or her property after his or her death.

Witness mark – A mark placed at a known location to aid in recovery and identification of a monument or corner.

Writ – A mandatory order issued from a court of justice.

Writ of coram nobis – A common law writ, the purpose of which is to correct an error in a judgment in the same court in which it was rendered.

Appendices

APPENDIX H
Areas Under the Standard Normal Curve
(0 to z)

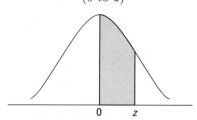

z	0	1	2	3	4	5	6	7	8	9
0.0	0.0000	0.0040	0.0080	0.0120	0.0160	0.0199	0.0239	0.0279	0.0319	0.0359
0.1	0.0398	0.0438	0.0478	0.0517	0.0557	0.0596	0.0636	0.0675	0.0714	0.0754
0.2	0.0793	0.0832	0.0871	0.0910	0.0948	0.0987	0.1026	0.1064	0.1103	0.1141
0.3	0.1179	0.1217	0.1255	0.1293	0.1331	0.1368	0.1406	0.1443	0.1480	0.1517
0.4	0.1554	0.1591	0.1628	0.1664	0.1700	0.1736	0.1772	0.1808	0.1844	0.1879
0.5	0.1915	0.1950	0.1985	0.2019	0.2054	0.2088	0.2123	0.2157	0.2190	0.2224
0.6	0.2258	0.2291	0.2324	0.2357	0.2389	0.2422	0.2454	0.2486	0.2518	0.2549
0.7	0.2580	0.2612	0.2642	0.2673	0.2704	0.2734	0.2764	0.2794	0.2823	0.2852
0.8	0.2881	0.2910	0.2939	0.2967	0.2996	0.3023	0.3051	0.3078	0.3106	0.3133
0.9	0.3159	0.3186	0.3212	0.3238	0.3264	0.3289	0.3315	0.3340	0.3365	0.3389
1.0	0.3413	0.3438	0.3461	0.3485	0.3508	0.3531	0.3554	0.3577	0.3599	0.3621
1.1	0.3643	0.3665	0.3686	0.3708	0.3729	0.3749	0.3770	0.3790	0.3810	0.3830
1.2	0.3849	0.3869	0.3888	0.3907	0.3925	0.3944	0.3962	0.3980	0.3997	0.4015
1.3	0.4032	0.4049	0.4066	0.4082	0.4099	0.4115	0.4131	0.4147	0.4162	0.4177
1.4	0.4192	0.4207	0.4222	0.4236	0.4251	0.4265	0.4279	0.4292	0.4306	0.4319
1.5	0.4332	0.4345	0.4357	0.4370	0.4382	0.4394	0.4406	0.4418	0.4429	0.4441
1.6	0.4452	0.4463	0.4474	0.4484	0.4495	0.4505	0.4515	0.4525	0.4535	0.4545
1.7	0.4554	0.4564	0.4573	0.4582	0.4591	0.4599	0.4608	0.4616	0.4625	0.4633
1.8	0.4641	0.4649	0.4656	0.4664	0.4671	0.4678	0.4686	0.4693	0.4699	0.4706
1.9	0.4713	0.4719	0.4726	0.4732	0.4738	0.4744	0.4750	0.4756	0.4761	0.4767
2.0	0.4772	0.4778	0.4783	0.4788	0.4793	0.4798	0.4803	0.4808	0.4812	0.4817
2.1	0.4821	0.4826	0.4830	0.4834	0.4838	0.4842	0.4846	0.4850	0.4854	0.4857
2.2	0.4861	0.4864	0.4868	0.4871	0.4875	0.4878	0.4881	0.4884	0.4887	0.4890
2.3	0.4893	0.4896	0.4898	0.4901	0.4904	0.4906	0.4909	0.4911	0.4913	0.4916
2.4	0.4918	0.4920	0.4922	0.4925	0.4927	0.4929	0.4931	0.4932	0.4934	0.4936
2.5	0.4938	0.4940	0.4941	0.4943	0.4945	0.4946	0.4948	0.4949	0.4951	0.4952
2.6	0.4953	0.4955	0.4956	0.4957	0.4959	0.4960	0.4961	0.4962	0.4963	0.4964
2.7	0.4965	0.4966	0.4967	0.4968	0.4969	0.4970	0.4971	0.4972	0.4973	0.4974
2.8	0.4974	0.4975	0.4976	0.4977	0.4977	0.4978	0.4979	0.4979	0.4980	0.4981
2.9	0.4981	0.4982	0.4982	0.4983	0.4984	0.4984	0.4985	0.4985	0.4986	0.4986
3.0	0.4987	0.4987	0.4987	0.4988	0.4988	0.4989	0.4989	0.4989	0.4990	0.4990
3.1	0.4990	0.4991	0.4991	0.4991	0.4992	0.4992	0.4992	0.4992	0.4993	0.4993
3.2	0.4993	0.4993	0.4994	0.4994	0.4994	0.4994	0.4994	0.4995	0.4995	0.4995
3.3	0.4995	0.4995	0.4996	0.4996	0.4996	0.4996	0.4996	0.4996	0.4996	0.4997
3.4	0.4997	0.4997	0.4997	0.4997	0.4997	0.4997	0.4997	0.4997	0.4997	0.4998
3.5	0.4998	0.4998	0.4998	0.4998	0.4998	0.4998	0.4998	0.4998	0.4998	0.4998
3.6	0.4998	0.4998	0.4999	0.4999	0.4999	0.4999	0.4999	0.4999	0.4999	0.4999
3.7	0.4999	0.4999	0.4999	0.4999	0.4999	0.4999	0.4999	0.4999	0.4999	0.4999
3.8	0.4999	0.4999	0.4999	0.4999	0.4999	0.4999	0.4999	0.4999	0.4999	0.4999
3.9	0.5000	0.5000	0.5000	0.5000	0.5000	0.5000	0.5000	0.5000	0.5000	0.5000

APPENDIX I
*Representative Plane Coordinate Projection Tables for Texas

Constants for Texas zones

Constant	Zone	
	North	North Central
C	2,000,000.00	2,000,000.00
Central Meridian	101° 30' 00".000	97° 30' 00".000
R_D	29,972,959.94	32,691,654.54
y_0	516,052.65	503,844.96
ℓ	0.57953 58654	0.54539 44146
$\dfrac{1}{2\rho_0^2 \sin 1"}$	2.360×10^{-10}	2.362×10^{-10}
$\log \dfrac{1}{2\rho_0^2 \sin 1"}$	0.373 0025 − 10	0.373 2288 − 10
$\log \ell$	9.76308 03181 − 10	9.73671 06860 − 10
$\log K$	7.63475 78652	7.65172 89823

Constant	Zone
	Central
C	2,000,000.00
Central Meridian	100° 20' 00".000
R_D	35,337,121.23
y_0	485,417.77
ℓ	0.51505 88857
$\dfrac{1}{2\rho_0^2 \sin 1"}$	2.363×10^{-10}
$\log \dfrac{1}{2\rho_0^2 \sin 1"}$	0.373 4182 − 10
$\log \ell$	9.71185 68921 − 10
$\log K$	7.66885 39642

Lambert Projection for Texas (North Central)
Table I (Cont'd)

Lat.	R (feet)	Y' y value on central meridian (feet)	Tabular difference for 1 sec. of lat. (feet)	Scale in units of 7th place of logs	Scale expressed as a ratio
33° 26'	32048935.45	642719.09	101.06550	−457.1	0.9998947
27	32042871.52	648783.02	101.06617	−448.6	0.9998967
28	32036807.55	654846.99	101.06650	−439.6	0.9998988
29	32030743.56	660910.98	101.06700	−430.3	0.9999009
30	32024679.54	666975.00	101.06750	−420.6	0.9999032
33° 31'	32018615.49	673039.05	101.06817	−410.6	0.9999055
32	32012551.40	679103.14	101.06850	−400.2	0.9999079
33	32006487.29	685167.25	101.06917	−389.4	0.9999103
34	32000423.14	691231.40	101.06967	−378.2	0.9999129
35	31994358.96	697295.58	101.07017	−366.7	0.9999156
33° 36'	31988294.75	703359.79	101.07067	−354.8	0.9999183
37	31982230.51	709424.03	101.07133	−342.6	0.9999211
38	31976166.23	715488.31	101.07183	−329.9	0.9999240
39	31970101.92	721552.62	101.07250	−316.9	0.9999270
40	31965037.57	727616.97	101.07317	−303.6	0.9999301
33° 41'	31957973.18	733681.36	101.07350	−289.8	0.9999333
42	31951908.77	739745.77	101.07433	−275.7	0.9999365
43	31945844.31	745810.23	101.07483	−261.3	0.9999398
44	31939779.82	751874.72	101.07550	−246.4	0.9999433
45	31933715.29	757939.25	101.07617	−231.2	0.9999468
33° 46'	31927650.72	764003.82	101.07683	−215.7	0.9999503
47	31921586.11	770068.43	101.07750	−199.7	0.9999540
48	31915521.46	776133.08	101.07800	−183.4	0.9999578
49	31909456.78	782197.76	101.07883	−166.7	0.9999616
50	31903392.05	788262.49	101.07950	−149.7	0.9999655
33° 51'	31897327.28	794327.26	101.08017	−132.3	0.9999695
52	31891262.47	800392.07	101.08083	−114.5	0.9999736
53	31885197.62	806456.92	101.08150	−96.3	0.9999778
54	31879132.73	812521.81	101.08217	−77.8	0.9999821
55	31873067.80	818586.74	101.08300	−58.9	0.9999864
33° 56'	31867002.82	824651.72	101.08367	−39.6	0.9999909
57	31860937.80	830716.74	101.08450	−20.0	0.9999954
58	31854872.73	836781.81	101.08517	0	1.0000000
59	31848807.62	842846.92	101.08600	+20.4	1.0000047
34° 00'	31842742.46	848912.08	101.08667	+41.1	1.0000095

Lambert Projection for Texas (North Central)
Table II (Cont'd)
1" of Long. = 0".54539441 of Θ

Long.		Θ	Long.		Θ	Long.		Θ
98° 31'	−0°	33' 16".1436	99° 06'	−0°	52' 21".4718	99° 41'	−1°	11' 26".8001
32	−0	33 48.8672	07	−0	52 54.1955	42	−1	11 59.5238
33	−0	34 21.5909	08	−0	53 26.9192	43	−1	12 32.2474
34	−0	34 54.3146	09	−0	53 59.6428	44	−1	13 04.9711
35	−0	35 27.0382	10	−0	54 32.3665	45	−1	13 37.6948
98° 36'	−0	35 59.7619	99° 11'	−0	55 05.0902	99° 46'	−1	14 10.4184
37	−0	36 32.4855	12	−0	55 37.8138	47	−1	14 43.1421
38	−0	37 05.2092	13	−0	56 10.5375	48	−1	15 15.8658
39	−0	37 37.9329	14	−0	56 43.2611	49	−1	15 48.5894
40	−0	38 10.6565	15	−0	57 15.9848	50	−1	16 21.3131
98° 41'	−0	38 43.3802	99° 16'	−0	57 48.7085	99° 51'	−1	16 54.0367
42	−0	39 16.1039	17	−0	58 21.4321	52	−1	17 26.7604
43	−0	39 48.8275	18	−0	58 54.1558	53	−1	17 59.4841
44	−0	40 21.5512	19	−0	59 26.8795	54	−1	18 32.2077
45	−0	40 54.2749	20	−0	59 59.6031	55	−1	19 04.9314
98° 46'	−0	41 26.9985	99° 21'	−1	00 32.3268	99° 56'	−1	19 37.6551
47	−0	41 59.7222	22	−1	01 05.0505	57	−1	20 10.3787
48	−0	42 32.4459	23	−1	01 37.7741	58	−1	20 43.1024
49	−0	43 05.1695	24	−1	02 10.4978	59	−1	21 15.8261
50	−0	43 37.8932	25	−1	02 43.2215	100° 00'	−1	21 48.5497
98° 51'	−0	44 10.6169	99° 26'	−1	03 15.9451	100° 01'	−1	22 21.2734
52	−0	44 43.3405	27	−1	03 48.6688	02	−1	22 53.9971
53	−0	45 16.0642	28	−1	04 21.3925	03	−1	23 26.7207
54	−0	45 48.7878	29	−1	04 54.1161	04	−1	23 59.4444
55	−0	46 21.5115	30	−1	05 26.8398	05	−1	24 32.1681
98° 56'	−0	46 54.2352	99° 31'	−1	05 59.5635	100° 06'	−1	25 04.8917
57	−0	47 26.9588	32	−1	06 32.2871	07	−1	25 37.6154
58	−0	47 59.6825	33	−1	07 05.0108	08	−1	26 10.3391
59	−0	48 32.4062	34	−1	07 37.7344	09	−1	26 43.0627
99° 00'	−0	49 05.1298	35	−1	08 10.4581	10	−1	27 15.7864
99° 01'	−0	49 37.8535	99° 36'	−1	08 43.1818	100° 11'	−1	27 48.5100
02	−0	50 10.5772	37	−1	09 15.9054	12	−1	28 21.2337
03	−0	50 43.3008	38	−1	09 48.6291	13	−1	28 53.9574
04	−0	51 16.0245	39	−1	10 21.3528	14	−1	29 26.6810
05	−0	51 48.7482	40	−1	10 54.0764	15	−1	29 59.4047

Lambert Projection for Texas (Central)
Table I (Cont'd)

Lat.	R (feet)	Y' y value on central meridian (feet)	Tabular difference for 1 sec. of lat. (feet)	Scale in units of 7th place of logs	Scale expressed as a ratio
31° 26'	34694604.31	642516.92	101.03467	−390.6	0.9999101
27	34688542.23	648579.00	101.03500	−380.9	0.9999123
28	34682480.13	654641.10	101.03567	−370.9	0.9999146
29	34676417.99	660703.24	101.03617	−360.4	0.9999170
30	34670355.82	666765.41	101.03667	−349.7	0.9999195
31° 31'	34664293.62	672827.61	101.03733	−338.5	0.9999221
32	34658231.38	678889.85	101.03767	−327.0	0.9999247
33	34652169.12	684952.11	101.03833	−315.1	0.9999274
34	34646106.52	691014.41	101.03883	−302.5	0.9999303
35	34640044.49	697076.74	101.03950	−290.2	0.9999332
31° 36'	34633982.12	703139.11	101.04000	−277.2	0.9999362
37	34627919.72	709201.51	101.04050	−263.9	0.9999392
38	34621857.29	715263.94	101.04117	−250.1	0.9999424
39	34615794.82	721326.41	101.04183	−236.0	0.9999457
40	34609732.31	727388.92	101.04233	−221.6	0.9999490
31° 41'	34603669.77	733451.46	101.04300	−206.7	0.9999524
42	34597607.19	739514.04	101.04367	−191.5	0.9999559
43	34591544.57	745576.66	101.04417	−176.0	0.9999595
44	34585481.92	751639.31	101.04483	−160.0	0.9999632
45	34579419.23	757702.00	101.04567	−143.7	0.9999669
31° 46'	34573356.49	763764.74	101.04617	−127.1	0.9999707
47	34567293.72	769827.51	101.04683	−110.0	0.9999747
48	37561230.91	775890.32	101.04750	−92.6	0.9999787
49	34555168.06	781953.17	101.04833	−74.8	0.9999828
50	34549105.16	788016.07	101.04883	−56.7	0.9999869
31° 51'	34543042.23	794079.00	101.04967	−38.1	0.9999912
52	34536979.25	800141.98	101.05033	−19.3	0.9999956
53	34530916.23	806205.00	101.05100	0.0	1.0000000
54	34524853.17	812268.06	101.05183	+19.6	1.0000045
55	34518790.06	818331.17	101.05250	+39.6	1.0000091
31° 56'	34512726.91	824394.32	101.05333	+60.0	1.0000138
57	34506663.71	830457.52	101.05400	+80.7	1.0000186
58	34500600.47	836520.76	101.05433	+101.8	1.0000234
59	34494537.18	842584.05	101.05550	+123.3	1.0000284
32° 00'	34488473.85	848647.38	101.05650	+145.2	1.0000334

APPENDIX I *(continued)*
*Representative Plane Coordinate Projection Tables for Texas

Lambert Projection for Texas (Central)
Table II (Cont'd)
1" of Long. = 0".51505889

Long.	θ	Long.	θ	Long.	θ
96° 51'	+1° 47' 38".8384	97° 26'	+1° 29' 37".2148	98° 01'	+1° 11' 35".5911
52	+1 47 07.9349	27	+1 29 06.3112	02	+1 11 04.6876
53	+1 46 37.0314	28	+1 28 35.4077	03	+1 10 33.7840
54	+1 46 06.1278	29	+1 28 04.5042	04	+1 10 02.8805
55	+1 45 35.2243	30	+1 27 33.6006	05	+1 09 31.9770
96° 56'	+1 45 04.3208	97° 31'	+1 27 02.6971	98° 06'	+1 09 01.0734
57	+1 44 33.4172	32	+1 26 31.7936	07	+1 08 30.1699
58	+1 44 02.5137	33	+1 26 00.8900	08	+1 07 59.2664
59	+1 43 31.6102	34	+1 25 29.9865	09	+1 07 28.3628
97° 00	+1 43 00.7066	35	+1 24 59.0830	10	+1 06 57.4593
97° 01'	+1 42 29.8031	97° 36'	+1 24 28.1794	98° 11'	+1 06 26.5558
02	+1 41 58.8996	37	+1 23 57.2759	12	+1 05 55.6522
03	+1 41 27.9960	38	+1 23 26.3724	13	+1 05 24.7487
04	+1 40 57.0925	39	+1 22 55.4688	14	+1 04 53.8452
05	+1 40 26.1890	40	+1 22 24.5653	15	+1 04 22.9416
97° 06'	+1 39 55.2854	97° 41'	+1 21 53.6618	98° 16'	+1 03 52.0381
07	+1 39 24.3819	42	+1 21 22.7582	17	+1 03 21.1346
08	+1 38 53.4784	43	+1 20 51.8547	18	+1 02 50.2310
09	+1 38 22.5748	44	+1 20 20.9512	19	+1 02 19.3275
10	+1 37 51.6713	45	+1 19 50.0476	20	+1 01 48.4240
97° 11'	+1 37 20.7678	97° 46'	+1 19 19.1441	98° 21'	+1 01 17.5204
12	+1 36 49.8642	47	+1 18 48.2406	22	+1 00 46.6169
13	+1 36 18.9607	48	+1 18 17.3370	23	+1 00 15.7134
14	+1 35 48.0572	49	+1 17 46.4335	24	+0 59 44.8098
15	+1 35 17.1536	50	+1 17 15.5300	25	+0 59 13.9063
97° 16'	+1 34 46.2501	97° 51'	+1 16 44.6264	98° 26'	+0 58 43.0028
17	+1 34 15.3466	52	+1 16 13.7229	27	+0 58 12.0992
18	+1 33 44.4430	53	+1 15 42.8194	28	+0 57 41.1957
19	+1 33 13.5395	54	+1 15 11.9158	29	+0 57 10.2922
20	+1 32 42.6360	55	+1 14 41.0123	30	+0 56 39.3886
97° 21'	+1 32 11.7324	97° 56'	+1 14 10.1088	98° 31'	+0 56 08.4851
22	+1 31 40.8289	57	+1 13 39.2052	32	+0 55 37.5816
23	+1 31 09.9254	58	+1 13 08.3017	33	+0 55 06.6780
24	+1 30 39.0218	59	+1 12 37.3982	34	+0 54 35.7745
25	+1 30 08.1183	98° 00	+1 12 06.4946	35	+0 54 04.8710

Lambert Projection for Texas (Central)
Table II (Cont'd)
1" of Long. = 0".51505889

Long.	θ	Long.	θ	Long.	θ
102° 06'	−0° 54' 35".7745	102° 41'	−1° 12' 37".3982	103° 16'	−1° 30' 39".0218
07	−0 55 06.6780	42	−1 13 08.3017	17	−1 31 09.9254
08	−0 55 37.5816	43	−1 13 39.2052	18	−1 31 40.8289
09	−0 56 08.4851	44	−1 14 10.1088	19	−1 32 11.7324
10	−0 56 39.3886	45	−1 14 41.0123	20	−1 32 42.6360
102° 11'	−0 57 10.2922	102° 46'	−1 15 11.9158	103° 21'	−1 33 13.5395
12	−0 57 41.1957	47	−1 15 42.8194	22	−1 33 44.4430
13	−0 58 12.0992	48	−1 16 13.7229	23	−1 34 15.3466
14	−0 58 43.0028	49	−1 16 44.6264	24	−1 34 46.2501
15	−0 59 13.9063	50	−1 17 15.5300	25	−1 35 17.1536
102° 16'	−0 59 44.8098	102° 51'	−1 17 46.4335	103° 26'	−1 35 48.0572
17	−1 00 15.7134	52	−1 18 17.3370	27	−1 36 18.9607
18	−1 00 46.6169	53	−1 18 48.2406	28	−1 36 49.8642
19	−1 01 17.5204	54	−1 19 19.1441	29	−1 37 20.7678
20	−1 01 48.4240	55	−1 19 50.0476	30	−1 37 51.6713
102° 21'	−1 02 19.3275	102° 56'	−1 20 20.9512	103° 31'	−1 38 22.5748
22	−1 02 50.2310	57	−1 20 51.8547	32	−1 38 53.4784
23	−1 03 21.1346	58	−1 21 22.7582	33	−1 39 24.3819
24	−1 03 52.0381	59	−1 21 53.6618	34	−1 39 55.2854
25	−1 04 22.9416	103° 00'	−1 22 24.5653	35	−1 40 26.1890
102° 26'	−1 04 53.8452	103° 01'	−1 22 55.4688	103° 36'	−1 40 57.0925
27	−1 05 24.7487	02	−1 23 26.3724	37	−1 41 27.9960
28	−1 05 55.6522	03	−1 23 57.2759	38	−1 41 58.8996
29	−1 06 26.5558	04	−1 24 28.1794	39	−1 42 29.8031
30	−1 06 57.4593	05	−1 24 59.0830	40	−1 43 00.7066
102° 31'	−1 07 28.3628	103° 06'	−1 25 29.9865	103° 41'	−1 43 31.6102
32	−1 07 59.2664	07	−1 26 00.8900	42	−1 44 02.5137
33	−1 08 30.1669	08	−1 26 31.7936	43	−1 44 33.4172
34	−1 09 01.0734	09	−1 27 02.6971	44	−1 45 04.3208
35	−1 09 31.9770	10	−1 27 33.6006	45	−1 45 35.2243
102° 36'	−1 10 02.8805	103° 11'	−1 28 04.5042	103° 46'	−1 46 06.1278
37	−1 10 33.7840	12	−1 28 35.4077	47	−1 46 37.0314
38	−1 11 04.6876	13	−1 29 06.3112	48	−1 47 07.9349
39	−1 11 35.5911	14	−1 29 37.2148	49	−1 47 38.8384
40	−1 12 06.4946	15	−1 30 08.1183	50	−1 48 09.7420

*Not for actual use.

APPENDIX J
Mensuration of Two-Dimensional Areas

Nomenclature

A	total surface area
b	base
c	chord length
d	distance
h	height
L	length
p	perimeter
r	radius
s	side (edge) length, arc length
θ	vertex angle, in radians
ϕ	central angle, in radians

Circular Sector

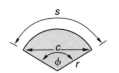

$$A = \tfrac{1}{2}\phi r^2 = \tfrac{1}{2}sr$$

$$\phi = \frac{s}{r}$$

$$s = r\phi$$

$$c = 2r\sin\left(\frac{\phi}{2}\right)$$

Triangle

equilateral   right   oblique

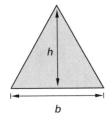 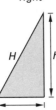 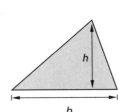

$$A = \tfrac{1}{2}bh = \frac{\sqrt{3}}{4}b^2 \qquad A = \tfrac{1}{2}bh \qquad A = \tfrac{1}{2}bh$$

$$h = \frac{\sqrt{3}}{2}b \qquad\qquad H^2 = b^2 + h^2$$

Parabola

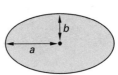

$$A = \tfrac{2}{3}bh$$

$$A = \tfrac{1}{3}bh$$

Circle

$$p = 2\pi r$$

$$A = \pi r^2 = \frac{p^2}{4\pi}$$

Ellipse

$$A = \pi ab$$

$$p = 2\pi\sqrt{\tfrac{1}{2}(a^2 + b^2)}$$

Circular Segment

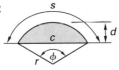

$$A = \tfrac{1}{2}r^2(\phi - \sin\phi)$$

$$\phi = \frac{s}{r} = 2\left(\arccos\frac{r-d}{r}\right)$$

$$c = 2r\sin\left(\frac{\phi}{2}\right)$$

(continued)

APPENDIX J *(continued)*
Mensuration of Two-Dimensional Areas

Trapezoid

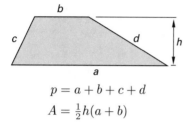

$$p = a + b + c + d$$
$$A = \tfrac{1}{2}h(a + b)$$

The trapezoid is isosceles if $c = d$.

Parallelogram

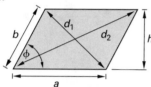

$$p = 2(a + b)$$
$$d_1 = \sqrt{a^2 + b^2 - 2ab\cos\phi}$$
$$d_2 = \sqrt{a^2 + b^2 + 2ab\cos\phi}$$
$$d_1^2 + d_2^2 = 2(a^2 + b^2)$$
$$A = ah = ab\sin\phi$$

If $a = b$, the parallelogram is a rhombus.

Regular Polygon (*n* equal sides)

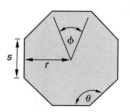

$$\phi = \frac{2\pi}{n}$$
$$\theta = \frac{\pi(n - 2)}{n} = \pi - \phi$$
$$p = ns$$
$$s = 2r\tan\left(\frac{\theta}{2}\right)$$
$$A = \tfrac{1}{2}nsr$$

sides	name	area (A) when diameter of inscribed circle = 1	area (A) when side = 1	radius (r) of circumscribed circle when side = 1	length (L) of side when radius (r) of circumscribed circle = 1	length (L) of side when perpendicular to center = 1	perpendicular (p) to center when side = 1
3	triangle	1.299	0.433	0.577	1.732	3.464	0.289
4	square	1.000	1.000	0.707	1.414	2.000	0.500
5	pentagon	0.908	1.720	0.851	1.176	1.453	0.688
6	hexagon	0.866	2.598	1.000	1.000	1.155	0.866
7	heptagon	0.843	3.634	1.152	0.868	0.963	1.038
8	octagon	0.828	4.828	1.307	0.765	0.828	1.207
9	nonagon	0.819	6.182	1.462	0.684	0.728	1.374
10	decagon	0.812	7.694	1.618	0.618	0.650	1.539
11	undecagon	0.807	9.366	1.775	0.563	0.587	1.703
12	dodecagon	0.804	11.196	1.932	0.518	0.536	1.866

APPENDIX K
Mensuration of Three-Dimensional Volumes

Nomenclature
A area
b base
h height
r radius
R radius
s side (edge) length
V volume

Sphere

$$V = \frac{4\pi r^3}{3}$$
$$A = 4\pi r^2$$

Right Circular Cone

$$V = \frac{\pi r^2 h}{3}$$
$$A = \pi r \sqrt{r^2 + h^2}$$

(does not include base area)

Right Circular Cylinder

$$V = \pi r^2 h$$
$$A = 2\pi r h$$

(does not include end area)

Spherical Segment

Surface area of a spherical segment of radius r cut out by an angle θ_0 rotated from the center about a radius, r, is

$$A = 2\pi r^2 \left(1 - \cos\theta_0\right)$$
$$\phi = \frac{A}{r^2} = 2\pi \left(1 - \cos\theta_0\right)$$

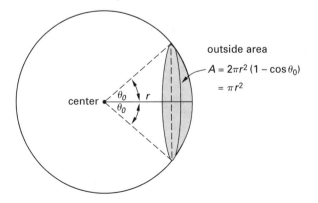

center

θ_0
θ_0

r

outside area
$A = 2\pi r^2\,(1 - \cos\theta_0)$
$= \pi r^2$

Paraboloid of Revolution

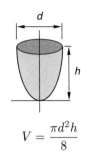

d

h

$$V = \frac{\pi d^2 h}{8}$$

Torus

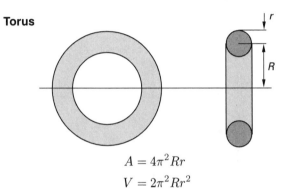

r

R

$$A = 4\pi^2 R r$$
$$V = 2\pi^2 R r^2$$

Regular Polyhedra (identical faces)

name	number of faces	form of faces	total surface area	volume
tetrahedron	4	equilateral triangle	$1.7321s^2$	$0.1179s^3$
cube	6	square	$6.0000s^2$	$1.0000s^3$
octahedron	8	equilateral triangle	$3.4641s^2$	$0.4714s^3$
dodecahedron	12	regular pentagon	$20.6457s^2$	$7.6631s^3$
isosahedron	20	equilateral triangle	$8.6603s^2$	$2.1817s^3$

The radius of a sphere inscribed within a regular polyhedron is

$$r = \frac{3V_{\text{polyhedron}}}{A_{\text{polyhedron}}}$$

APPENDIX L
Manning Equation Nomograph

$$v = \left(\frac{1.486}{n}\right)(r_h)^{\frac{2}{3}}\sqrt{S}$$

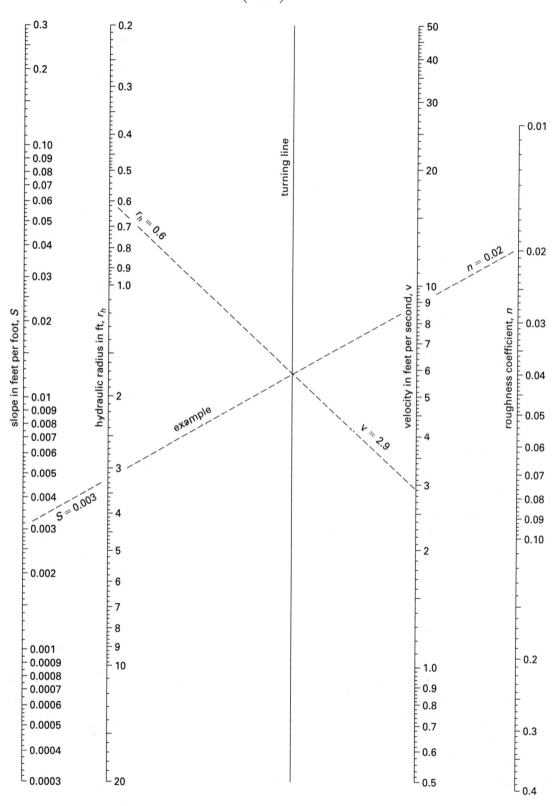

Index

Index

Index